Essentials of
Polymer Science and Engineering

Essentials of
Polymer Science
and Engineering

PAUL C. PAINTER
Professor, The Pennsylvania State University

MICHAEL M. COLEMAN
Professor Emeritus, The Pennsylvania State University

DEStech Publications, Inc.

Essentials of Polymer Science and Engineering

DEStech Publications, Inc.
439 North Duke Street
Lancaster, Pennsylvania 17602 U.S.A.

Printed in the United States of America
10 9 8 7 6 5 4 3 2 1

Main entry under title:
    Essentials of Polymer Science and Engineering

A DEStech Publications book
Bibliography: p.
Includes index p. 517

Library of Congress Catalog Card No. 2008925701
ISBN: 978-1-932078-75-6

Table of Contents

Preface

We have written books before, but this one may be our farewell performance, our swan song. According to legend, the Mute Swan (Cygnus olor) is mute during its lifetime, but sings a single, heartbreakingly beautiful song just before it dies. Of course, your authors have been compared to many things, but the list does not include swans. Also, a polymer science textbook is anything but heartbreakingly beautiful, although we've attempted to tart this one up by making it in living color. The problem is that we're aging fast and perhaps there's not another book in us.

> **Old age is when you resent the swimsuit issue of Sports Illustrated**
> **because there are fewer articles to read.**
>
> *George Burns*

We are finding that holding a collection of facts and ideas in one's mind in order to organize them in some logical, thematic fashion gets harder.

> **There are three things an aging golfer loses.**
> **His nerve, his memory and I can't remember the third thing.**
>
> *Lee Trevino*

Seemingly minor tasks like proof reading become more difficult.

> **Always proof-read carefully to see if you any words out.**
>
> *Author Unknown*

And without some rules we learned long ago, we would forget how to spell important, everyday words.

> **Remember: 'i' before 'e', except in Budweiser.**
>
> *Author Unknown*

But there are compensations. We can look back over a subject and thanks to those years of experience and effort find a depth of understanding that eluded us as young men.

> **I had a feeling once about Mathematics, that I saw it all—**
> **Depth beyond depth was revealed to me—the Byss and the Abyss.**
> **I saw, as one might see the transit of Venus—**
> **a quantity passing through infinity and changing its sign from plus to minus.**
> **I saw exactly how it happened and why the tergiversation was inevitable:**
> **and how the one step involved all the others.**
> **But it was after dinner and I let it go.**
>
> *Winston Churchill*

This, too, is perhaps an illusion and things just don't add up to us anymore.

Two thirds of Americans can't do fractions. The other half, just don't care.
Author Unknown

But our recreation is golf and gardening, so what are we to do in the depths of a Pennsylvania Winter? Might as well write another book. So here it is. It covers much of the same ground as our previous textbook, including the usual topics of polymer organic synthesis;

Organic chemistry is the study of organs;
inorganic chemistry is the study of the insides of organs.
Max Shulman

and some basic polymer physics and physical chemistry.

Physical Chemistry is research on everything for which
the negative logarithm is linear with 1/T.
D.L. Bunker

We also included two new topics, on natural polymers;

Rubber looks good but it doesn't do anything for me.
No, let's have the real thing-give me some leather gear.
Cat Deeley

and on polymer processing.

You know you've made it when you've been molded in miniature plastic.
But you know what children do with Barbie dolls—it's a bit scary, actually.'
Cate Blanchett

But, dear reader, this one got hard, particularly towards the end and without the loving support and encouragement of our wives we may never have completed it.

Why don't you write books people can read?
Nora Joyce to her husband James (1882–1941)

We just hope that you feel the same way as Francis Bacon when it comes to things that are getting old.

Age [and retirement] appear to be best in four things — old wood best to burn,
old wine to drink, old friends to trust, and old authors to read.
Francis Bacon

PAUL PAINTER AND MIKE COLEMAN

Acknowledgements

Most of the superior figures in this book would not have been produced without the support of the National Science Foundation, through grants from the Advanced Technological Education Program, the Course, Curriculum and Laboratory Improvement Program and the Polymers Program. Any figure that displays any semblance of artistic talent or technical skill was almost certainly produced by Mike Fleck or Alex Bierly. Many of the ideas for the original figures in the chapter on processing originated with our colleagues at Penn College, Tim Weston and Kirk Cantor. Kirk also wrote the original version of the chapter on processing, which we subsequently mangled into its present form. Thanks, guys, you will get your reward in heaven, because we're too cheap to give you a share of the royalties!

1

Introduction

PRELIMINARY RANT

This is a textbook about polymer science, a field your authors have studied and labored in for most of their professional lives. We recall that it was not that long ago, when we were just getting started in this subject, that polymers were regarded as an unalloyed blessing and a boon to mankind. (OK, on second thought, maybe it was that long ago.) Then, sometime in the 1960s, there was a turning point in popular culture, first reflected in the disdain expressed by the writer, Norman Mailer (see quotes). For our money, though, the real tipping point came with the one-word advice, *"Plastics!"* given to Dustin Hoffman in the movie *The Graduate*. Hearing this, shudders of disgust and nausea would pass like a wave through the audience. Plastics were no longer cool.

We beg to disagree. First, we think that the subject of polymer science and engineering is so broad and diverse that scientists and engineers of every stripe can find topics that have an engaging richness and depth. This is reflected in the history of the subject, which includes not only the usual list of Nobel Prize–winning scientists of the highest caliber, but also many interesting characters and even some outright scoundrels. We also believe that a feel for the history of this subject and some of the controversies surrounding the use of certain polymer materials can help bring the subject alive to many students. Accordingly, in writing

"I am inclined to think that the development of polymerization is, perhaps, the biggest thing that chemistry has done, where it has the biggest effect on everyday life."

Lord Todd
President, Royal Society of London

"I sometimes think there is a malign force loose in the universe that is the social equivalent of cancer, and it's plastic."

Norman Mailer

"We divorced ourselves from the materials of the earth, the rock, the wood, the iron ore. We looked to new materials, which were cooked in vats, long complex derivatives of urine, which we call plastic."

Norman Mailer

"From packaging materials, through fibers, foams and surface coatings, to continuous extrusions and large-scale moldings, plastics have transformed almost every aspect of life. Without them, much of modern medicine would be impossible and the consumer electronics and computer industries would disappear. Plastic sewage and water pipes alone have made an immeasurable contribution to public health worldwide."

Dr. N. C. Billingham

1

this textbook we have sought to do things a little differently than in many conventional texts (including our previous efforts). We will discuss the usual topics, polymer synthesis, structure and morphology, etc., but here and there you will also find what we will call *Polymer Milestones*, brief summaries of the contributions of some great scientists and interesting characters. In addition, books such as this, which are largely concerned with general principles, often do not give students a feel for the nature and use of specific polymers, so we will also sprinkle in some brief reviews of *Fascinating Polymers*. These reviews are not intended to be comprehensive, but are aimed at just giving you a feel for the depth of this field and its rich history. (Those of you who like this stuff and want to explore it in more detail, may want to check out our CD, *The Incredible World of Polymers: Tales of Innovation, Serendipity and Perseverance,* also published by DEStech.) In order to set this up, the rest of this chapter will briefly review the early history of polymer materials, so as to provide a context for our *Polymer Milestones*, together

with our view of some of the controversies surrounding the role of polymers in modern society, just to get you worked up and into the spirit of the thing when we get to certain topics in our *Fascinating Polymers* sections. But first, for those of you who have wandered into polymers from other disciplines, we'll provide some basic definitions.

What Is a Polymer?

If you are new to this field, the first thing you are probably wondering after reading our preliminary rant is "what's the difference between a plastic and a polymer, or are they the same thing?" A plastic is a type of material, whereas a polymer is a type of molecule, a very, very large molecule (Figure 1-1)! The word plastic comes from the Greek, *plastikos*, meaning shapeable. The word polymer also comes from the Greek, meaning many parts. Plastics are polymers, but so are many other things, as we will mention below under "classification." For now we will just note that all disciplines develop their own ways of abusing the English language. Some, psychology and sociology come to mind, seem to take pleasure in describing the obvious in complex terms that appear to be simply designed to impress and confuse those unfamiliar with their subject. But, we digress, and before our friends in the sociology department start to get too upset, let's come to the point and consider some plain, common or garden, polymer nomenclature. Some of the most common terms are listed in the boxes opposite and we believe these definitions are essentially self-evident.

How Big Are Polymers?

It would be nice if we could assume that anyone with a high school diploma would be familiar with the molecular structure of say, water or benzene (Figure 1-2). But, to our jaundiced eye, the current staple of many high school science curriculums seems to be "Saving the Rainforest" and molecular science appears to be largely an afterthought. But that's a different rant. The point we wish

---

POLY.........MER
Many   Units

-M-M-M-M-M-M-M-

or

$-(M)_n-$

SOME BASIC NOMENCLATURE

POLYMER—a large molecule made up of smaller building blocks

MONOMER—the building blocks

HOMOPOLYMER—when all the monomers are the same

COPOLYMER—a polymer composed of different monomers

BLENDS—a mixture of polymers

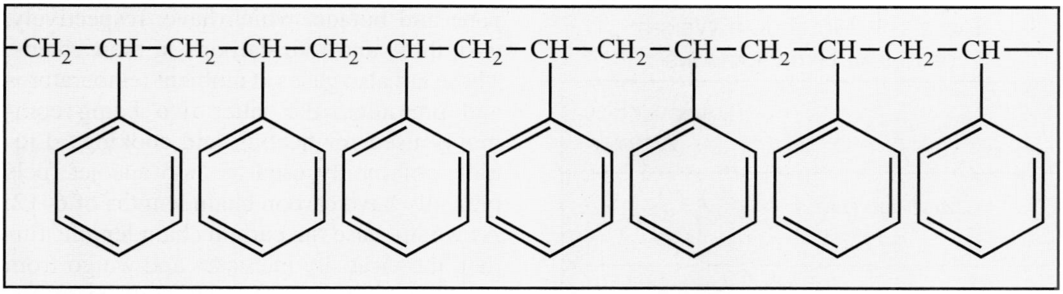

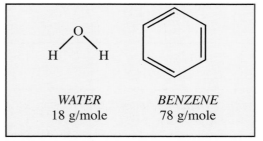

FIGURE 1-1 Part of a polystyrene chain (Typical molecular weight: 10,000 to 5,000,000 g/mole).

to make is that molecules like water, benzene and the like are generally called "low molecular weight" or "low molar mass" materials by polymer scientists. As a rule of thumb, molecules having molecular weights of, say, less than 500 g/mole are considered low molecular weight materials. High molecular weight polymers, on the other hand, are covalently bound, chain-like molecules that generally have molecular weights that exceed 10,000 g/mole and can be as high as 10^7 g/mole. Between these extremes of low and high molecular weights, there is a poorly defined region of moderately high molecular weight materials and such molecules are often referred to as *oligomers*.

To give you a feel for the difference between a low molecular weight material and a high molecular weight polymer, let's next put them on a scale that we can all relate to. If we assume that a single methylene group, CH_2, may be represented by a single bead in Figure 1-3, then the gas, ethylene, is simply two beads joined together. The chain of beads shown in the figure would represent not a polymer (it's too short), but an oligomer made up of about 100 CH_2 units. Let's further assume that the length of the ethylene molecule is 1 cm. Now, if we consider a polyethylene molecule that has a molecular weight of 700,000 g/mole, it would be made up of 700,000/14 or roughly 50,000 methylene groups. On our scale this would be equivalent to a chain roughly a quarter of a kilometer long. These are very big molecules indeed!

Many of the physical properties of polymers are simply a consequence of their large

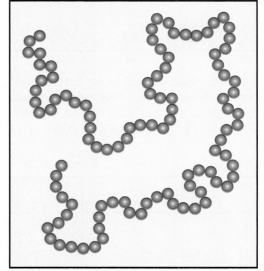

FIGURE 1-2 Chemical structures of water and benzene.

WATER 18 g/mole *BENZENE* 78 g/mole

FIGURE 1-3 Depiction of an oligomer composed of 100 beads.

size. To also get a feel for this, let's consider building a simple hydrocarbon chain one carbon atom at a time, filling up all the unsatisfied valences with hydrogen atoms (Table 1-1). If we have just one carbon atom (and hence four hydrogens), we have the gas

TABLE 1-1 MOLECULAR WEIGHT OF LINEAR HYDROCARBONS

	MOLECULAR WEIGHT
Methane (Gas) CH_4	16
Ethane (Gas) CH_3-CH_3	30
Propane (Gas) $CH_3-CH_2-CH_3$	44
Butane (Gas) $CH_3-CH_2-CH_2-CH_3$	58
Octane (Liquid) $CH_3-(CH_2)_6-CH_3$	114
Oligomer (Semi-solid) $CH_3-(CH_2)_{30}-CH_3$	450
Polyethylene (Solid) $CH_3-(CH_2)_{3000}-CH_3$	420,030

pane and butane, which have, respectively, two, three and four carbons in their chains. These are also gases at ambient temperatures and pressures, the latter two being commonly used for heating and cooking. Liquids, commonly used as auto and jet fuels typically have carbon chain lengths of 6–12. As we increase the carbon chain lengths further, the viscosity increases and we go from liquid materials used for baby oils, to "semi-solid" materials used as soft and hard candle waxes. At even higher carbon chain lengths, typically exceeding 30,000, we encounter hard, solid polyethylenes.

Classification

It is important to recognize that all polymers, be they natural or synthetic, are simply giant long chain molecules or macromolecules. It's all chemistry, and there really is no basic difference between a natural or synthetic polymer, in that both obey the same physical laws. Cellulose, for example, the natural polymer that is found in cotton and numerous other plants (and the most abundant polymer on the face of the planet), is a long chain macromolecule composed of carbon, hydrogen and oxygen. So is the synthetic polyester fiber, poly(ethylene terephthalate) (PET). It's just that the arrangement (architecture) of the atoms in the two

methane, often referred to as "natural gas," perhaps because it not only emanates from the ground, but also the rear end of cows. The next three in the series are ethane, pro-

TABLE 1-2 EXAMPLES OF NATURAL AND SYNTHETIC POLYMERS

USE	NATURAL	SYNTHETIC
Fibers	Wool, Silk, Cellulose	Nylon, PET, Lycra®
Elastomers	Natural Rubber, Elastin	SBR, Silicones, Polybutadiene
Plastics	Gutta Percha, DNA, Polypeptides	Polyethylene, Polypropylene, Polystyrene
Composites	Wood, Bone, Teeth	Polyester/Glass Epoxy/Carbon Fibers, Formica
Adhesives	Barnacles	Elmer's® Glue, Super® Glue
Paints	Shellac	Acrylics

polymer chains is different. However, it is common practice to distinguish between natural and synthetic polymers and also classify them in terms of plastics, fibers, elastomers, composites, paints, adhesives, etc., as shown in Table 1-2. Not only are plastics made out of polymers, so are all the other things listed in this table.

THE EARLY HISTORY OF POLYMERS

Synthetic polymer materials are so ubiquitous in modern life that we now take them for granted. But, the first commercially significant, completely synthetic plastic was only introduced at the beginning of the 20th century. This was Bakelite, invented by Leo Baekeland and a short account of his contributions will form the subject of one of our *Polymer Milestones* in the next chapter. The introduction of this new material was preceded by roughly 40 years of the development of what can be called semi-synthetics based on chemically modified forms of cellulose.

If you look at the miscellany of Victorian objects shown in Figure 1-4, molded from a material called Parkesine in the early 1860s, you may not be impressed. But, think about it for a while. What if the ordinary things of everyday life, buttons, combs, toothbrushes, hand mirrors, and so on, had to be made of bone or ivory (for those with money), or fashioned (by a skilled craftsman) out of wood? Could somebody making minimum wage afford them? This is where things stood at the middle of the 19th century. While scientists were struggling with the nature of atoms, molecules and bonding, a bunch of chemical entrepreneurs were starting to make useful stuff without worrying too much about chemical structure. One such entrepreneur was Charles Goodyear, who discovered how to "cure" natural rubber and make it useful. But that is a story that we will tell later. Here we will consider the events that led to the development of the first true synthetic plastic, Bakelite, and the recognition of the existence of very large molecules or polymers.

FIGURE 1-4 Articles made of Parkesine (Courtesy: London Science Museum).

Schönbein and Nitrated Cellulose

Natural polymeric materials, like cotton, silk, wool, flax, wood, leather, bone, ivory, and feathers, have been used for clothing, utensils and shelter since humankind appeared on the face of the earth. These were the types of organic materials available in the 19th century. In England (the birthplace of your two authors), this is a period that to us appears to be dominated by Dickensian images of the industrial revolution; those "dark, satanic mills"; depressing row houses encased in shrouds of smoke; coal mines and the exploitation of children. It was a society built on coal, iron, steel, and the power of steam. But this also bought enormous social change. Wealth was no longer largely based on owning land, but on producing goods. An emerging middle class was acquiring disposable income and wanted some of the finer things in life. This resulted in a strong demand for increasingly rare natural materials, like silk and ivory, which, in turn, trans-

POLYMER MILESTONES—CHRISTIAN SCHÖNBEIN

Christian Schönbein (Courtesy: Deutsches Museum)

Cellulose is, to all intent and purposes, totally insoluble. It is also infusible, decomposing well before its melting point and cannot be processed in the melt or solution. It can be modified chemically, however, as the Swiss chemist, Christian Schönbein, discovered. The story has it (it may be apocryphal) that he spilled a mixture of concentrated nitric and sulfuric acids, powerful mineral acids that dissolve copper and other metals and readily destroy organic tissues, such as skin! Because the mixture is so corrosive, he grabbed the nearest item, his wife's cotton apron, to wipe up the mess. After washing it out, he put the apron in front of the fire to dry. It flared brightly, almost explosively, and disappeared. He had made something that would be called guncotton. Up until this discovery, the most powerful explosive available was gunpowder, but this produced so much smoke that generals could not see the people they were slaughtering and were often unable to tell if they were winning or losing a battle. Guncotton was not only three times more powerful than gunpowder, it produced far less smoke. So what had Schönbein done to his wife's cotton apron? Schönbein had nitrated the –OH groups of cellulose. To give you a feel for what this means, consider glycerol, $HO-CH_2-CH(OH)-CH_2OH$ (sometimes called glycerin or gylcerine). This is an innocuous material that can be found in many hand and body lotions, where it acts to soften skin. However, if you replace the hydrogen atoms on the –OH groups with nitro, NO_2, groups, you make nitroglycerin. You may have seen this stuff in old western movies. John Wayne or Clint Eastwood are carefully driving some rickety old wagon, chock full of bottles of this stuff, along some really terrible, rutted track. Sometimes there's a precipice on one side, just for good measure. They are sweating profusely, because this stuff is shock-sensitive. Just shaking the bottle can cause it to explode. It decomposes into nitrogen, carbon dioxide and water (plus a little oxygen) and releases a whole lot of energy during the process. And so it is with guncotton. Each hydrogen on the –OH groups of cellulose can be replaced with an NO_2 group. This stuff is a lot more stable than nitroglycerin, but once ignited it will also release a lot of energy. But it's an ill wind that blows nobody any good. In addition to making cellulose explosive, nitrating the –OH groups also made it soluble (because the crystal structure is disrupted). Its explosive character still made it practically impossible to melt process (think about it!), but solutions could now be cast to form films. One of the first uses of a solution of the partially nitrated cellulose (in ethanol/ether solvents) was on the battlefield. The solution, called collodion, was applied to open wounds, forming a protective film of polymer (not unlike a synthetic skin) after the solvents evaporated. A collodionlike solution containing salicylic acid (common aspirin) is still used nowadays to remove warts and can be purchased over-the-counter from most pharmacies. Collodion films were also used as a photographic film base.

lated into a strong demand for substitutes when these became difficult or expensive to obtain. Cellulose was to feature prominently in the start of a new materials revolution that grew to satisfy these needs.

Cellulose is a polysaccharide, a natural polymer made up of glucose units strung together in the form of a linear chain. The chemical structure of a glucose unit (in a cellulose chain) is shown in Figure 1-5. The important part is the three –OH groups, which can react with other groups fairly readily. The number of repeating units, n, in a typical cellulose molecule varies with the source, but is usually in the range of 1000–7000. Cellulose is the major constituent of all plant cell walls and for industrial purposes, cotton fiber and wood, which contain between 80–90% and about 50% cellulose, respectively, are the most important sources. The cellulose chain is "linear" (i.e., there are no branches sticking off the side) and very regular. This, together with the strong intermolecular hydrogen bonding that occurs between chains (defined later!), results in a tightly packed, highly crystalline polymeric material. This makes it insoluble in nearly all solvents and gives it such a high melting point that it degrades (turns black and burns) before softening or melting. This, in turn, means it cannot be processed in the melt or solution. You can't mold a bucket or extrude a film from naturally occurring cellulose. However, it can be modified chemically, as the Swiss chemist, Christian Schönbein, discovered (see opposite in our first *Polymer Milestone*).

Parkes, Hyatt and the Elephants

In the 1870s the demand for ivory, which was "harvested" from the trunks of a rapidly dwindling supply of hapless elephants, was high and increasing. Ivory was a favorite of the rich and famous, who used it for items such as jewelry, piano keys, billiard balls, etc., and there was a desperate need to find new materials for some of these applications. Alexander Parkes, a prolific English inventor, and John Wesley Hyatt, a great American

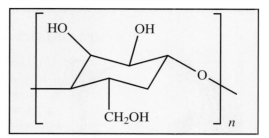

FIGURE 1-5 Chemical structure of cellulose.

entrepreneur, built upon Schönbein's discovery to make the next major advances. This culminated in the introduction of two new revolutionary materials that could be considered reasonable substitutes for ivory called *Parkesine* and *Celluloid* (see Figure 1-6). Both these gentlemen had fascinating lives and both deserve, at the very least, to have individual *Polymer Milestones* devoted to them, as we present on the next two pages.

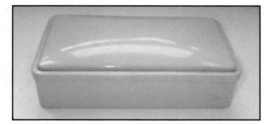

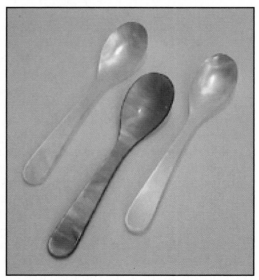

FIGURE 1-6 Celluloid box and spoons (Courtesy: www.sintetica.de).

POLYMER MILESTONES—ALEXANDER PARKES

Alexander Parks (Source: www. me.umist.ac.uk).

Alexander Parkes, a prolific English inventor, made a major advance in 1862. He was way ahead of his time, a true material scientist and a "generalist," respected for his accomplishments by metallurgists, ceramists and organic chemists (a rare feat, even today!). Parkes had no formal education in chemistry. However, he caused a stir when he displayed solid articles molded out of a mixture of cellulose nitrate, various oils and solvents, at the Great International Exhibition in Crystal Palace (London). He modestly called this material, Parkesine (see Figure 1-4). The introduction of Parkesine is generally recognized as the birth of the plastics industry. Parkes first had to make cellulose nitrate on a reasonably large scale, but not the essentially fully nitrated, highly inflammable guncotton material that Schönbein had produced. In fact, Parkes experimented with nitric acid/sulfuric acid mixtures and succeeded in making a partially nitrated cellulose that was still soluble in a variety of solvents, but far less flammable. This material was called pyroxyline. After removal of the acid and washing, the pyroxyline was mixed with a variety of solvents to make a viscous mass that could be molded into different shapes. A typical formulation for Parkesine was:

Pyroxyline (moistened with naphtha)	100 parts
Nitrobenzole (or aniline or camphor)	10–50 parts
Vegetable oil	150–200 parts

In another leap of insight, Parks also patented the use of metal chlorides to reduce the flammability of Parkesine. After the volatile solvents had evaporated, a solid Parkesine article was formed. Camphor, which has a special role in this story, was evidently considered by Parkes to be just another non-volatile solvent—this became a bone of contention later. Following his success at the Crystal Palace Exhibition, Parkes continued to experiment and by 1865 had reduced the price of Parkesine to about 1/- per pound (that's 1 shilling—a now defunct currency that was perversely 1/20 of an English £—a lot of money in those days). He decided that it was time to form a company and produce Parkesine on a commercial scale. Thus the Parkesine Company was formed in 1866 with a capital investment of £10,000. But, like many a fledgling entrepreneur, Parkes encountered scale-up and quality control problems. In attempting to keep the cost of Parkesine down, he evidently used inferior starting materials, sacrificed quality and was unable to consistently reproduce the material that he had previously made. The company folded two years later. Daniel Spill, who had first encountered Parkesine at the Crystal Palace Exhibition, became works manager for the Parkesine Company at its inception (Parkes himself was managing director). Upon dissolution of the Parkesine Company, Spill continued operations and raised money to form the Xylonite Company. Unfortunately, although this new company did produce a reasonable amount of acceptable material, it suffered the same fate and dissolved in 1874.

POLYMER MILESTONES—JOHN WESLEY HYATT

In the 1870s, ivory was a favorite of the rich and famous, who used it for items of jewelry, piano keys, billiard balls, etc.; and there was a desperate need to find new materials for some of these applications. In fact, a substantial prize ($10,000) was offered for the first person who could find a substitute for ivory billiard balls. John Wesley Hyatt (now there's a name that conjures up an image of an aging preacher or country and western singer— well at least to your authors!), a bona fide American entrepreneur, was motivated, but did not win this prize. Nonetheless he is credited with the invention of Celluloid; the first true thermoplastic. Hyatt and his brother were awarded a patent for the preparation and molding of a plastic material that was based upon a mixture of cellulose nitrate and camphor. In modern terms, Hyatt was the first to recognize the essential plasticizing effect that

John W. Hyatt (Source: The Plastics Institute, London).

camphor had on cellulose nitrate, which facilitated processing from the melt. By incorporating the plasticizer (camphor) to facilitate the molding of cellulose nitrate, the Hyatt brothers had removed volatile solvents from the mixture, which essentially eliminated the shrinking and warping of articles as these solvents evaporated. Hyatt formed the Celluloid Manufacturing Company (after several others) and Celluloid rapidly became a huge commercial success. Its applications included everyday items like combs, dentures, knife handles, toys and spectacle frames. In later years, film for the fledgling movie industry was made of Celluloid. All of these articles are now more likely to be made of other synthetic plastics, but your authors are old enough to remember the Celluloid ping-pong ball. Before we leave the subject of Celluloid, it is perhaps interesting to note that the paths of Hyatt, Parkes and Spill were to cross in the American courts. Hyatt's Celluloid Manufacturing Company was going great guns and then the boom dropped. Daniel Spill sued the Hyatts, claiming that, among other issues, the use of camphor and alcohol as solvents for pyroxyline infringed upon his patents. Alexander Parkes testified on behalf of the Hyatts and maintained that Spill's patents were based upon his earlier work (in legal parlance, prior art) and were thus invalid. (Needless to say Parkes and Still were not exactly bosom buddies at this time!) Nonetheless, after some three years, the judge ruled in favor of Spill. This was a crushing blow to Hyatt. Attempting to circumvent Spill's patents, the Hyatts substituted methanol (wood alcohol) for ethanol (sprits of wine) and continued manufacturing. Spill sued again, but this time the Hyatts won, albeit not decisively. A long drawn-out fight ensued and it wasn't until 1884 that the same judge reversed his original decision and ruled that Parkes was indeed the true inventor of the camphor/alcohol process. Now anyone could use the technology!

Readers interested in more details of the history of Celluloid are referred to the superb monograph by M. Kaufman, *The First Century of Plastics*, published by the Plastics Institute, London, 1963.

Regenerated Cellulose

The next inventor to make a mark was Louis Marie Hilaire Bernigaud, the Comte de Chardonnet. He was searching for a way to make a synthetic silk (as we describe in the *Polymer Milestone* below). Hopefully,

FIGURE 1-7 Silkworms (By Shoshanah: www.flickr.com/photos/shoshanah/19025983).

you have now picked up on the main theme. Cellulose, although a wonderful naturally occurring polymer, cannot be processed into useful objects. Cotton, itself a comfortable fiber, is coarse relative to silk which is produced from silk worms (Figure 1-7). But once cellulose is chemically modified, it can be spun into much finer fibers or processed into films or molded objects. With cellulose nitrate, however, there is still the problem of flammability. But what if you could do what Chardonnet did, chemically modify cellulose to make it processable, but then regenerate cellulose by removing whatever group you had attached? This was accomplished in the 1890s by Cross, Bevan and Beadle (sounds like a 60s rock band!). They found that cellulose could be rendered soluble by sodium hydroxide and carbon disulfide. In very simple terms, this reaction forms a basic solution of cellulose xanthate (a rather evil-smelling concoction), which upon acidification regen-

---

POLYMER MILESTONES—CHARDONNET'S SILK

Louis Marie Hilaire Bernigaud, the Comte de Chardonnet (Source: Science Museum/Science & Society Picture Library).

Remember, pure cellulose is insoluble, but cellulose nitrate dissolves in an ether/ethanol mixture (collodion). In 1884, Louis Marie Hilaire Bernigaud, the Comte de Chardonnet, was searching for a way to make a synthetic silk. Silk, in the latter part of the 19th century, was in great demand from both the upper and the emerging middle classes. Chardonnet, who had been studying silkworms, noticed that they secreted a liquid through a narrow orifice that hardened upon exposure to air and transformed to silk fiber. He came up with the idea of passing a collodion solution through a small metal orifice or die to produce, once the solvents had evaporated, a new fiber. And it worked! He obtained a patent in 1884. The fibers had great "feel," but Chardonnet's workers, displaying a wonderfully mordant sense of humor, started calling his material "mother-in-law silk" (which evidently originated from something like: "Buy your mother-in-law this nightgown and stand her in front of the fire!"). These materials were highly flammable. Chardonnet overcame this problem by "denitrating" the fibers by passing them through a solution of aqueous ammonium hydrogen sulfide and, in effect, regenerating cellulose. Chardonnet's silk was manufactured (in steadily declining quantities) until the 1940s.

erates cellulose. The spinning of fibers from this solution lends itself to the large-scale production of rayon fibers of varied and precise deniers (sizes). And, not only that, continuous transparent films, called cellophane, can be readily produced (Figure 1-8).

It is hard to overstate the impact of rayon and cellophane. Could you imagine a modern songwriter including a favorable reference to a polymeric material in one of their songs? But that is exactly what Cole Porter did in his song, *"You're the Top"* (see box below).

"You're the top!
You're Mahatma Gandhi.
You're the top!
You're Napoleon Brandy.
You're the purple light
Of a summer night in Spain,
You're the National Gallery
You're Garbo's salary,
You're cellophane.
You're sublime,
You're turkey dinner,
You're the time, the time of a
Derby winner
I'm a toy balloon that's fated soon
to pop
But if, baby, I'm the bottom,
You're the top!"

Cole Porter

The Growth of Polymers and the Birth of Polymer Science

At this time, around the turn of the 19th century, Bakelite was also introduced and (as mentioned above) we will spotlight this story in one of our *Polymer Milestones* in Chapter 2. A lot was also happening in the rubber industry and we will also come back to those stories later. What might seem astonishing today is that all these materials were developed in an "Edisonian" fashion, by trial and error, and in the absence of any fundamental understanding of the macromolecular nature of polymers. In fact, the prevailing view at that time was that materials like silk, cellu-

FIGURE 1-8 Commercial cellophane films (Source: www.s-walter.com)

lose and natural rubber were colloidal aggregates of small molecules. Not only that, the chemistry of what were then called "high molecular compounds," which could not be purified by crystallization or distillation, was regarded with disdain by academic chemists and referred to as "grease chemistry." It was therefore an act of considerable courage for Hermann Staudinger to decide to devote the bulk of his efforts to these "high molecular compounds." He asserted in lectures and papers (published in the period 1917 to 1920) that these materials were covalently bonded long chain molecules. This is another great story and we touch on the main points in the *Polymer Milestone* on the next page.

Following the work of Staudinger and Wallace Carothers, the inventor of (among other things) nylon, whose great but tragic story will be featured in Chapter 3, there was an explosion in the discovery and production of new synthetic polymers. The graph in Figure 1-9 shows the approximate US production of synthetic polymers over the last 100 years (in 1998 it was fast approaching 80 billion pounds!). In addition, the approximate dates of the introduction of various commercial

POLYMER MILESTONES—HERMANN STAUDINGER

Hermann Staudinger (Courtesy: Deutches Museum).

Hermann Staudinger was born in Worms, Germany in 1881. He had a very distinguished career, holding academic positions at prestigious universities in Karlsruhe, Zurich and Freiburg. But his greatest achievement was that he was finally able to convince a skeptical scientific community that macromolecules did, indeed, exist. And that wasn't easy. Staudinger recollects in his book published in his 80th year, that he was advised in the 1920s by a famous organic chemist of the day, H. Wieland: "My dear chap, give up your ideas on big molecules. There are no organic molecules with a molecular weight of more than 5000. Just clean up your products and they will crystallize and reveal themselves as low-molecular-weight compounds." Now there's vision! Basically, Staudinger did not accept the "association of small molecules" idea to explain the physical properties of rubber, proteins, cellulose and the like. He maintained that these materials were composed simply of molecules that were fundamentally the same as any other organic molecule. The atoms were still connected together by covalent chemical bonds, as they are in any other "crystallizable" low molecular weight compound (e.g., water, benzene, octane etc.). It's just the size (molar mass) of the molecule that is different. He was, of course, correct, and he was finally recognized for his achievements in 1953 when he was awarded the Nobel Prize in Chemistry. At first Staudinger had to dispel the "aggregation of small molecules" idea. He showed that, unlike inorganic colloids, changing the solvent does not affect the properties of "colloidal" organic materials. Moreover, and more convincingly, chemical modification did not destroy the colloidal properties of the organic materials we now know are polymeric. For example, natural rubber was considered by most early 20th century chemists to consist of aggregates of small eight-membered rings (a cyclic isoprene dimer, in modern terminology). It was argued by proponents of this concept that the aggregates of these cyclic species were held together by attractive secondary forces arising from the presence of the unsaturated double bonds. Staudinger poked large holes in the cyclic aggregate argument by cleverly hydrogenating natural rubber, eliminating the double bonds, and showing that the modified natural rubber still had "colloidal" properties.

polymers are indicated (many of the acronyms will be defined later). We will feature a number of these materials and their inventors in our *Polymer Milestones* and *Fascinating Polymers* sections later in this book. There is no question that polymer materials have had a huge impact, but is that necessarily a good thing? Some would argue, not!

POLYMER MATERIALS: BOON OR BANE?

You have all probably seen advertisements on TV, which show a premature baby in a plastic incubator, or an air bag deploying in a car crash, saving a life. They all end with soothing words like, ". . . made possible by a material we call plastics." The chemical industry and plastics manufacturers alike have expended much money and effort to change the negative perception that the general public has of plastics. Let's face it, the vast majority of people in the world are woefully ignorant of chemistry and are not persuaded by scientific explanations. (Just think of the arguments about teaching evolution!)

Such advertisements tend to glorify the role of plastics in society, but they are designed to counterbalance the pervasive unfavorable images that plastics evoke. So, have plastics really gotten a bum rap?

Are Plastics Just Cheap and Nasty Materials?

Let's say you need to buy a plastic hose to water the garden. You go to the store and are confronted with three models of different quality and price: a "cheap and nasty" hose at $6.99 (no warranty), an "up-scale" hose at $13.99 (warranty 1 year) and a "top-of-the-line" hose at $59.97 (warranty 5 years). Which hose are you most likely to buy? You'll probably mutter, "I'm not paying 14 bucks for a damn garden hose and 60 bucks is just ridiculous!" So you settle for the "cheap and nasty" hose! Guess what? The hose works fine for the first year, but as you stagger into the house after a final fall Sunday spent raking leaves, you haven't the energy to put away your cheap hose and you leave it in the garden over the winter. "It'll be OK!" But when you reemerge to use the hose next

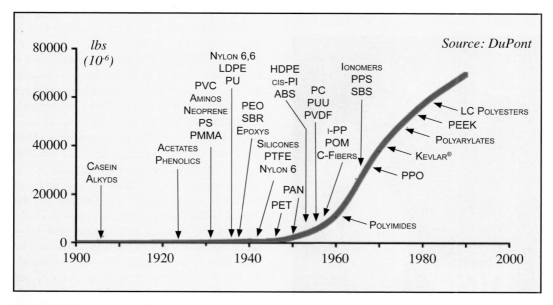

FIGURE 1-9 Growth of US plastics production (with approximate dates of the commercial introduction of different plastics).

FIGURE 1-10 Not really a flattering image!

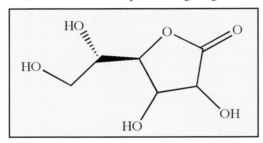

FIGURE 1-11 Chemical structure of vitamin C.

FIGURE 1-12 An old DuPont advertisement (Courtesy: DuPont).

spring, you find it has perished and is full of cracks. "Typical! Useless bloody plastics! Now I'll have to buy a new hose." Which one are you most likely to buy? You've got it! This is called planned obsolescence and Sears and Wal-Mart love you for it! As the old adage states: "You get what you pay for" and the "cheap and nasty" reputation (Figure 1-10) is not necessarily the fault of plastic materials.

Traditional versus Synthetic Materials

There's a common belief that anything "natural" must be superior to something that is man-made or synthesized. You have all heard something like this: "I'm not buying that synthetic vitamin C, it's nowhere near as good as the natural vitamin C you get from rose hips." This, of course, is unmitigated rubbish. It's the same molecule and has the same chemistry! Vitamin C is L-ascorbic acid, a molecule with a specific formula, $C_6H_8O_6$, which has the chemical structure shown in Figure 1-11, no matter where it comes from or how it's synthesized.

In much the same way, "natural" polymeric fibers like wool, cotton, silk, etc., are often touted as superior to anything that is man-made or "synthetic." But is this fair? There is no doubt that natural fibers have a unique set of properties that have withstood the test of time (e.g., it is difficult, but not impossible, to match silk's "feel" or cotton's "ability to breathe"). On the other hand, consider Lycra®, a completely synthetic fiber produced by DuPont (Figure 1-12) that has a truly amazing set of properties and is the major component of Spandex® (a material that keeps string bikinis on!). Or consider the "wrinkle-free" polyester fibers used in clothing and the "stain proof" nylon and polyacrylonitrile polymers used in carpets. The point here is that polymers, be they "natural" or "synthetic," are all macromolecules but with different chemical structures. The challenge is to design polymers that have specific properties that can benefit mankind.

Do Traditional Materials Get a Break?

Now here's something to think about. Glass was discovered centuries ago (see Figure 1-13), but for the sake of argument, let's imagine that it was only invented during the last decade. Do you think it would have been embraced by the public and have passed the scrutiny of the government regulatory agencies? Not bloody likely! Remember glass is relatively heavy (a typical 1.5 L glass bottle is some 5–6 times heavier than its PET plastic counterpart) and it has the unfortunate tendency to shatter into very sharp and dangerous fragments when accidentally dropped. Can't you see the headlines: "New Material Cuts Off Baby's Finger!"; "Shampoo bottle breaks in shower, cutting feet!"; "Glass Should Be Banned!". How does glass get away with what could be an overwhelming negative image? Mainly because it's a useful traditional material, and although there are some risks involved in using glass, these are deemed acceptable, as we are all familiar with them.

And what about wood (Figure 1-14), another traditional material? Wood is an intricate natural composite composed of cellulose fibers embedded in a matrix of another polymer called lignin. There is very little that is more appealing than a polished piece of oak or mahogany. But wood burns, and it burns very well, while giving off lots of toxic fumes! You don't hear too many people suggesting that wood be banned as a construction material! But if a new plastic material is introduced into the market and it has the same flammability characteristics as wood, it is inevitable that it will be subject to far more stringent regulations, and may be deemed not up to code or specification.

So, are plastics being treated unequally or unfairly just because they are relatively new and do not have the long track record of traditional materials? Perhaps, but although we should always err on the side of caution, let's try at least to be objective. There is really no such thing as a material that is perfectly safe. One can easily think of scenarios where any material could kill or maim an individual

FIGURE 1-13 A beautiful stained glass window in a church in Lavenham, England.

FIGURE 1-14 Exposed oak beams in a cottage in Suffolk, England that dates back to the 1400s.

under certain circumstances, be they deliberate or just plain unfortunate. Common sense should rule and not the knee-jerk reactions of individuals who are often scientifically illiterate and cannot make reasonable judgments about the balance between acceptable risk and benefits.

Plastics, Energy and the Environment

If, in this age of supersizing, you go to a restaurant and order a steak, you may find that you're served with what appears to be

FIGURE 1-15 The ubiquitous foamed polystyrene "doggy-bag."

FIGURE 1-16 One of your authors' dog, Winnie, patiently waiting for her "doggy-bag."

the better part of a whole cow. Struggle as you might, you cannot finish the meal. The waiter asks: "Do you want a doggy bag?" Out comes a massive foamed polystyrene (PS) container (Figure 1-15). The Greenpeace activist at the next table screams in horror (he was already upset that you didn't order the veggie-burger). But, these foamed PS containers are very well designed and have properties that are close to ideal for their intended purpose. They are lightweight, inert and have extraordinary thermal insulation properties. So your "leftovers" remain hot and uncontaminated for a considerable time until your dog (Figure 1-16), or some mooching in-law is ready to consume them.

That's the good news! But what do we do with this large container once we have consumed the contents? We throw it in the garbage! Just imagine how many of these containers are thrown away each day, each month or in a year? Why don't we just reuse or recycle these PS containers? The reasons are complex, but can be reduced to two major factors: economics and human foibles. When it becomes significantly more expensive to collect, clean, sterilize and redistribute an item than it is to produce the same item from virgin material, there is a major economic disincentive to reuse or recycle. And, frankly, most of us just don't want to be bothered! Again, this is not the fault of the plastics material per se—it did its job beautifully. Disposing of any material, let alone plastics, is a problem and we will revisit the subject later in this section.

But, what about the energy required to make disposable packaging? Is it a drain on natural resources? Not really, we have to keep things in perspective. Although the vast majority of plastics are derived from oil and natural gas, most of the world's oil production is burnt as fuel and only about 4% or so is used to produce plastics (Figure 1-17). Energy conservation should be a top priority in any environmental or political agenda, but, in truth, it is not. Here in the United States, policy is akin to crisis management and the importance of an energy policy runs "hot and cold." Some of us are old enough to remember the energy crisis of the 1970s, when talk of long-term energy planning was rife and money was freely flowing for alternative energy sources. The subsequent glut of oil on the world market soon killed these initiatives. At the time of writing this text, it seems like "deja vu all over again" (to quote Yogi Berra), as we are once again wrestling with high oil prices. And this time it may be for real, as oil and gas reserves are finite and burning fossil fuels is apparently contributing to global warming. It remains only a matter of time before the perception that plastics materials are using up too much of our valuable resources will resurface. What isn't realized is that plastic packaging actually protects the energy invested in producing, distributing and maintaining goods. For example, it takes about 15.8 MJ to bake and distribute a loaf of bread, but only about 1.7 MJ to produce the packaging used to protect it! Before materials were packaged in this way tremendous quantities were lost to "spoilage." This should be kept in mind when constructing an "energy balance."

Let's get back to plastic waste (Figure 1-18). Where does it go? Most of our garbage is buried in landfills (bloody great holes in the ground!). This would not be so bad if there was an unlimited amount of space and the location of sites was not subject to political pressures and the common refrain of NIMBY (not in my back yard)! In general, landfills are a safe (albeit unsightly) method of disposal, if properly managed. Let's consider our discarded foamed polystyrene (PS) container. In an ideal world, one would like to design a container that is cheap, inert and indestructible during its manufacture, distribution, storage and use. Then, after it has served its useful purpose, it could be thrown away and would immediately and magically degrade into innocuous, useful, life-enhancing solids, liquids and gases. But we live in a real world and these properties are usually mutually exclusive, in a chemical sense, and compromises have to be made. It is perhaps ironic that the very chemical structure of foamed PS that is responsible for its desirable thermal and oxidative stability is the

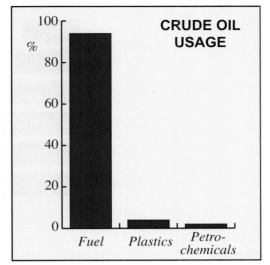

FIGURE 1-17 Chart illustrating approximately where crude oil finds use.

same structure that prevents it from degrading in a timely fashion. And so the PS container just sits in the landfill and if you come back in 50 years or so you could probably dig it up, clean it and reuse it! You may be thinking there is an obvious solution. If PS doesn't significantly degrade in a landfill,

FIGURE 1-18 A plastics dump—not a pretty sight!

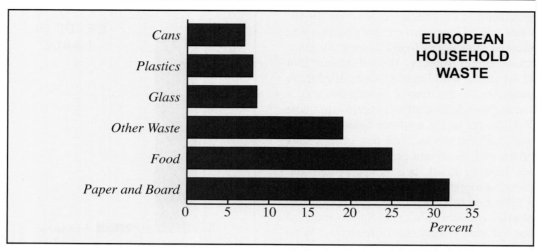

FIGURE 1-19 Chart showing the approximate composition of European household waste (Source: BPF).

why don't we just stop using it? Let's simply substitute "natural" polymers made of cellulose, like paper, cardboard, etc., which are biodegradable. Well, it ain't that simple!

Garbology—the analysis of garbage—has been raised to the level of a legitimate science by anthropology Professor William Rathje of the University of Arizona. He has demonstrated that eliminating plastic waste won't solve the problem. Out of some 160 million tons of waste produced in the United States annually, 20% is food-related, 20–30% is construction debris and 50% is paper products. The chart above (Figure 1-19) shows a slightly different distribution of household waste from European data. Styrofoam, fast food packaging and disposable diapers contribute less than 3%! As Rathje states, "If they [plastics] were banned tomorrow, the people who manage landfills wouldn't know the difference." Moreover, Rathje dispelled the myth that paper products degrade easily in landfills. Using a bucket auger, Rathje and his students showed that some 10–30 readable newspapers were present for every three feet descended into the landfill they studied. Thirty-year-old newspapers have been dug out of landfills that can be dusted off and read! Even an 18-year-old ear of corn and 15-year-old hot dogs were found intact! In other words, it's not just plastics—very little decomposes rapidly in a landfill.

Nevertheless, rapid decomposition would not hurt. So why can't plastic manufacturers simply design polymers that biodegrade easily? Then perhaps they would conveniently "disappear" in a landfill. Again, it's not that simple. Apart from the economic reality that such materials tend to be more expensive and have more limited physical properties than their non-biodegradable counterparts, biodegradability in a typical landfill is not assured. The biodegradation of natural and certain synthetic polymers requires the presence of the correct "bugs" (bacteria) together with the right amount of water and oxygen at an appropriate temperature. Most landfills are not optimized to produce such conditions (economics again). The bottom line: How much are you willing to pay for your "throw-away" container and how much do you care what happens to it?

McDonald's and some other fast-food outlets have adopted a "friend of the environment" strategy by replacing polystyrene foam clam shells with coated paper wrappers. But, what about the trees that get chopped down to produce the paper? A cynic might believe that corporate image and cost might have been overriding factors in this decision! Other companies have introduced starch/limestone-based materials that are deemed "environmentally friendly" because they are also potentially biodegradable. DuPont has recently announced Biomax®, a hydrolytically unstable and biodegradable PET. And keep an eye on Cargill-Dow's poly(lactic acid), a biodegradable plastic made from lactic acid that is derived from renewable natural sugars.

Most polymers are organic, so what about composting (Figure 1-20)? Now there's a "rotten" idea! (Sorry, but you'll have to get used to bad puns in this text.) For an individual who has enough space, maintaining a garden compost heap is arguably one of the more environmentally friendly acts he or she can perform (although the neighbors may not always agree!). Putrescible (look it up!) garden or household waste, through the action of natural microorganisms, transforms biological waste into a valuable soil conditioner, mulch and fertilizer, which can then be used to nourish and promote plant growth. In effect, the compost heap is Mother Nature's recycling plant! And, unlike landfills, a compost heap is optimized for biodegradation. So, wouldn't it be wonderful if plastic waste could be composted? But, as you should know by now, most of it can't. Certain polymers and blends, however, are compostable, and there has been a considerable amount of money and effort expended over the past three decades to develop such materials. The Warner-Lambert company spent a small fortune developing what was essentially a starch-filled polyethylene garbage bag that decomposed in a compost heap. A fine material, but it cost too much and the business was sold off. Polymer blends based on polycaprolactone, by Planet Polymer Technologies, worked beautifully in compostable garbage bags, but its fate was the same: too expensive. Perhaps the large-scale production of Cargill-Dow's polylactic acid compostable plastics will eventually bring the cost down to a point where compostable garbage bags and the like will be the norm. For now, the much cheaper, non-compostable, commodity plastics rule the day. But there are rumblings on the horizon. Recently (2007) the city of San Francisco decided to ban polyethylene grocery bags (or "Witches' Knickers" as they

FIGURE 1-20 One of your authors' rather dilapidated compost heap.

FIGURE 1-21 A large incinerator located in the south of England (Courtesy: British Plastics Federation).

are quaintly called, because that's what they look like when they get caught in trees) in favor of compostable starch-based materials. Many other cities and states across the world are doing the same. Stay tuned!

Burning or incinerating plastics to produce energy seems like a good idea (Figure 1-21). However, incinerators are also subject to the NIMBY syndrome. Approximately 25% of Europe's garbage is incinerated and this can be a very efficient method of disposal, especially in areas which are densely populated and where transportation costs are high. In fact, some of the energy produced from burning plastics can be recovered (Figure 1-22) and used to produce electricity (although many older incinerators do not have this capability). In addition to natural polymers like cellulose, starch, etc., certain plastics, including the high volume commod

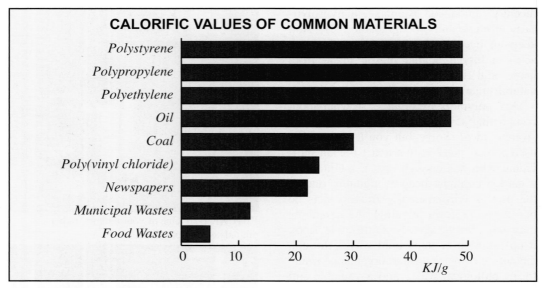

FIGURE 1-22 Chart showing approximate calorific values (Source: BFP).

ity plastics, polyethylene and polypropylene, burn cleanly and easily to yield mainly water and CO/CO_2 gases. This is hardly surprising, as this is akin to burning a high molecular weight candle! For polymers containing aromatic groups and/or atoms other than carbon and hydrogen, like polystyrene, poly(vinyl chloride), polyacrylonitrile etc., the chemistry is more complex and incineration conditions are more stringent. It is also more expensive, because it is necessary to optimize combustion in order to reduce the formation of toxic chemicals and have in place scrubbers to remove them.

Finally, what about recycling? At first glance, this appears to be a fabulous idea, embodying the concept of both conserving energy and resources. Surely, it can only be as a result of some sort of conspiracy by the chemical companies that more of this isn't done. Why can't we take any and all of our plastic waste, grind it into little pieces and mold a useful new article from it? But again, this is not as simple as it sounds. It's been tried and all that is produced is useless rubbish. There are many different types of plastic, having vastly different chemical, physical and mechanical properties. Most of them are fiercely antisocial and do not like to mix with

any other plastic. One can, perhaps, liken it to trying to mix vinegar with oil. They are both technically foods, but they phase-separate! (You'll learn why in the chapter on polymer solutions and blends.) Molded articles made from such mixtures have little cohesive strength and fall apart. Accordingly, different types of plastics must be hand separated (not a cheap process!) before we can recycle them. At the time of writing, only polyethylene terephthalate (PET—coke bottles etc.) and polyethylene (PE—milk containers etc.) are recycled from household waste. And the relative costs of recycled and virgin material is such that the economics of recycling just these two polymers is close to marginal, although that is changing as the costs of feedstocks to synthesize polymers increases. Nevertheless, from 1990 to 1997 there was more than a three-fold increase in both the number of plastic bottles recycled and the number of United States recycling companies (Figure 1-23). In 1998, the amount of recycled plastic bottles in the US rose to 1.45 billion pounds and there was an overall recycling rate of 23.5% for PET and HDPE bottles (which accounted for 99% of all recycled bottles). Clearly, with some 1800 businesses involved in the handling and reclaiming of post-consumer plastics for the purposes of making new products, somebody has to be making money! And this should only be the beginning. With greater awareness and incentives, we should surely be able to increase participation and raise the recycling rate of PET and HDPE to at least 50%. But always remember the Clinton campaign mantra: "It's the economy, stupid"! If recycled material costs more than virgin material, where's the incentive?

Polymers: No Need to Apologize

To wind up these introductory rants, we pinched this section title from an article by Dr. N. C. Billingham. (*Trends in Polymer Science*, **4**, 172, June 1996.) It says it all and we were not smart enough to come up with anything better! To quote Dr. Billingham:

> From packaging materials, through fibers, foams and surface coatings, to continuous extrusions and large scale moldings, plastics have transformed almost every aspect of life. Without them much of modern medicine would be impossible and the consumer electronics and computer industries would disappear. Plastic sewage and water pipes alone [see Figure 1-24] have made an immeasurable contribution to public health worldwide.

We think that industrial and academic scientists and engineers who have worked in the polymer field have much to be proud of. Maybe in the future, biotechnology will be considered the greatest contribution that chemistry has made to mankind, but for now, we believe the synthesis of polymeric materials has to top the list. In the forthcoming chapters and essays, we will try to justify this statement as we introduce you to the science of these materials. But to begin with, in the next chapter, we'll talk about microstructure!

RECOMMENDED READING

N. C. Billingham, "Plastics—no need to apologize." *Trends in Polymer Science*, **4**, 172, June 1996.

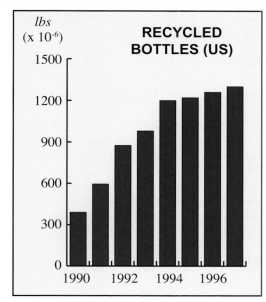

FIGURE 1-23 Chart showing the approximate weight of recycled plastic bottles in the US (Source: www.plastic.com).

FIGURE 1-24 PVC sewage pipes.

S. Fenichell, *Plastic—The Making of a Synthetic Century*, Harper Business, New York, NY, 1996.

R. Hoffmann, "A natural-born fiber," *American Scientist*, **85**, 21, 1997.

M. Kaufman, *The First Century of Plastics—Celluloid and Its Sequel*. The Plastics Institute, London, England, 1963.

S. T. I. Mossman and P. J. T. Morris, Editors, *The Development of Plastics*, The Royal Institute of Chemistry, Cambridge, England, 1994.

W. L. Rathje and C. Murphy, *Rubbish! The Archaeology of Garbage*, HarperCollins Publishers, New York, NY, 1992.

SOME USEFUL WEBSITES

A Brief History of Plastics
www.me.umist.ac.uk/historyp/history.htm

American Plastics Council
www.plastics.org

Association of Plastics Manufacturers in Europe
www.apme.org

British Plastics Federation
www.bpf.co.uk

Cargill-Dow Company
www.cdpoly.com

DuPont Company
www.dupont.com

Industrial Council for Packaging and the Environment
www. incpen.com

National Archives & Records Administration
www.treasurenet.com

National Plastics Center and Museum
www.npcm com

Society of Plastics Engineers
www.4spe.org

STUDY QUESTIONS

In the forthcoming chapters, various study or homework questions will be set forth. These types of questions have been omitted in this chapter, because they would end up being pathetically easy! Instead, you should write a short essay (no more than three typed pages, not including any interesting graphics that you happen upon in your background research). Pick out one of the topics in this chapter, such as natural vs. synthetic materials, the environmental concerns about disposing of plastic waste, or the problems with PVC, etc., and write a critical review. (For example, you may want to consider why more plastics aren't recycled. Just Google "Plastics Recycling" and you'll find a load of stuff!) It does not matter what side of any argument you come down on, as long as your points are critically made and supported by evidence. (For example, if you simply state that "I believe lack of recycling is all a conspiracy by plastics manufacturers" or, more succinctly, "plastics suck,". without some supporting evidence, you won't get many points!)

2

Microstructure and Molecular Weight

POLYMER MICROSTRUCTURE

In an introductory text such as this, it might seem logical to start with describing how polymers are made (synthesis) and proceed on through structure, properties and processing. Indeed, one of the points of including an overview of the early history of polymer materials in the preceding chapter was to show how chemists and entrepreneurs (often one and the same) made incredibly useful materials that had an enormous social impact before having any clear idea of the molecular structure and macromolecular character of their products. However, the development and introduction of analytical techniques capable of probing the fine details of chemical structure, or microstructure, has changed things considerably in the last fifty years or so. We now have a much clearer understanding of the relationship between local structure, molecular weight (chain length) and the mechanism of synthesis, to the point that polymers can be designed to have certain structures. Things rarely come out perfectly, of course, so there are nearly always defects in the materials we try to synthesize, but chemistry and microstructure are so deeply entwined that we believe it is important to have a working knowledge of the latter before dealing with the former. Accordingly, in this chapter we will describe some of the major features of the microstructure of synthetic polymers and how we define averages for describing the distribution of molecular

weights found in these materials. We will deal with some natural polymers and the measurement of molecular weight later in this book.

Linear and Branched Polymers

So far, we've given you a picture representation of a chain as a connected set of beads, like pearls on a string. In chemistry, there are various ways of depicting chemical structures and we will assume that you did not fall asleep in all your high school chemistry classes and know what these are. However, a very, very smart theoretical physicist friend of ours was once asked in a seminar at Penn State, "What is the chemical structure of the polymer you're describing?" He replied, "It is irrelevant." Given that you are not a hard-core theoretician, you should be able to immediately identify the balls and sticks in the model of part of a polyethylene (PE) chain shown in Figure 2.1. (OK, if you do happen to be a theoretical physicist, they are carbon atoms, hydrogen atoms and covalent bonds.) Polyethylene is what we call a simple homopolymer; that is, it is made up of identical units. It is also linear, as opposed to being branched, like the part of a chain shown in Figure 2.2 (the carbon atoms on the branch are shown in red, to distinguish them). This is the type of thing we mean by microstructure. Even seemingly minor differences in chain structure can have a profound effect on things like the ability of a material

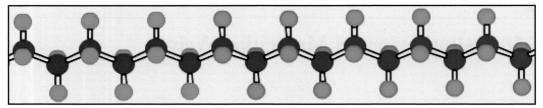

FIGURE 2-1 Part of a linear PE chain.

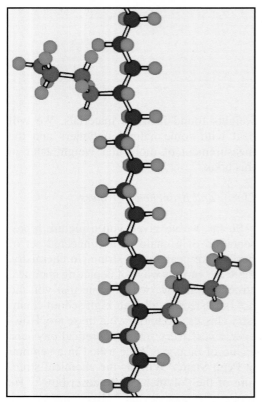

FIGURE 2-2 Part of a branched PE chain.

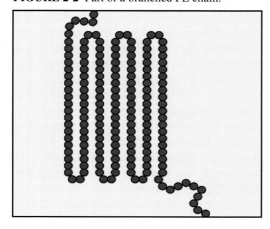

FIGURE 2-3 Packing of linear chains.

to crystallize, hence macroscopic properties such as stiffness and strength. For example, which of the two types of chains, linear or branched, do you think would crystallize most readily (i.e., be capable of arranging itself in a regular, repeating, ordered, three-dimensional array in space)? We have every confidence that your answer is linear (after all, you have chosen to study polymers, so you must be bright!).

Linear chains can stack regularly, as in the two-dimensional representation shown in Figure 2-3. Note how this model has the chain folding back on itself, with bits sticking out of the end. We'll go into this in more detail later. Randomly placed branches of different length would "get in the way" of such regular, close packing. Obviously, if we just have a small number of branches, then crystallinity would be reduced, but not eliminated. Various grades of polyethylene are produced commercially and are often referred to as high density or low density. Which do you think is the high-density polyethylene, the linear, more crystalline stuff, or the (somewhat) branched, less crystalline stuff? Again, this is a rhetorical question, because we are convinced that you have immediately sprung to your feet yelling, "It has to be the linear PE!," startling anybody that happens to be in the same room as you.

If you are a bit slow, or just having a bad day, think about it. If the chains can stack regularly, then they can get closer together than if there are branches that would interfere with this stacking. Chains that cannot crystallize (e.g., highly branched ones), actually look something like cooked spaghetti or random coils. In this state, the chains cannot stack as closely together as in the crystalline lattice, hence, there are fewer chains per unit volume and their density (mass per unit volume) is less.

It may seem like we're getting ahead of ourselves here, talking about crystallinity and properties when we've barely started our discussion of microstructure. This is not because we are absent-minded and have started to digress (although, being professors, we are prone to do this), but because right from the start we want you to think about the relationship between levels of structure and properties. For example, the type of polyethylene that goes into milk jugs is stronger, stiffer, but more opaque (less optically clear) than the type of polyethylene that is used to make film wrap (greater optical clarity, more flexible, but less strong). Can you figure out which type of polyethylene is used to make film wrap? If you are starting to get the hang of this, you should have answered, "It's the low density PE." In general, a more crystalline material is stronger and stiffer, for reasons that will hopefully become clear as you learn more about polymers. But, simplistically, if the forces of cohesion between the molecules are greater, as they generally would be in a crystalline material as opposed to one that is less organized (or amorphous), then one would expect it to be stronger (reality is a lot more complex than this because of the effect of defects, etc.). One would also expect it to melt or soften at a higher temperature, because more heat would be required to overcome these stronger forces and break up the lattice to form a liquid or melt. Similarly, if the molecules of a material are locked in a rigid crystalline lattice, then one would expect it to be stiffer. However, highly crystalline materials are often more brittle, so there is a trade-off.

Finally, a partly crystalline polymer (polymers never crystallize completely) have regions of order and disorder that have different densities and refractive indices, leading to an internal scattering of light that makes the material appear opaque. The degree of optical clarity will vary with crystallinity. If the polymer does not absorb light in the visible part of the spectrum, then one would expect it to be optically clear if it were 100% disordered (cooked spaghetti) or 100% ordered (think of a diamond), because the material would be optically homogeneous. Materials that are in-between scatter light internally to an extent that depends upon the degree of crystallinity, except in the rare case of a material whose amorphous and crystalline domains have the same refractive index. We'll consider the details of polymer structure and morphology later.

Branching Types

So far, all we have considered in terms of chain microstructure is the difference between linear and branched homopolymers. In the rest of this chapter we will introduce other types of microstructures, but as we go along we want you to keep in mind some of the "quick and dirty" arguments we gave concerning the relationship of chain structure to long-range order (crystallinity) and, hence, physical properties. Ask yourself, "Would a chain with this microstructure be capable of crystallizing, and if it did, what would that mean in terms of properties?"

The type of branching we've considered in the preceding section is actually short-chain branching. Long-chain branching can also occur and by doing some clever chemistry one can even make star polymers, where a number of long-chain branches are joined, apparently at a single point, but actually in a well-defined junction zone (Figure 2-4). There is also considerable interest in polymers where branching occurs with wanton profusion, as in so-called dendrimers, an example of which is shown in Figure 2-5. Here you can see that the core of the molecule has four short branches joined at a single point. (We would actually say that we had a core of functionality four.) Each one of these four segments then branches into two additional segments, so that the second layer from the core in the onionskin-like structure has 8 segments. The next layer or generation has 16, and so on. Hyperbranched polymers are similar, but less perfect, and can have different topologies.

FIGURE 2-4 Different types of branching.

Network Formation

Branching can also lead to the formation of densely connected networks. For example, if we start with a Y-shaped molecule, where each arm of the Y has a reactive group that is capable of reacting with and connecting to any other group on another Y, then as the reaction proceeds we would build up molecules of various sizes. The unreacted ends of these larger molecules would also continue to react, ultimately forming a network of interconnected units. We say that the Y-shaped molecule is trifunctional. Tetrafunctional (four reactive groups) molecules (X-shaped molecules) can also react

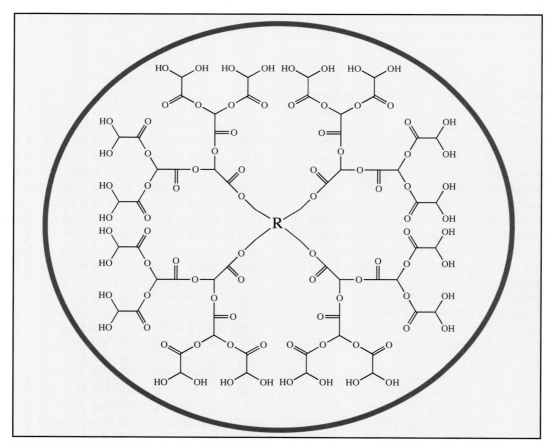

FIGURE 2-5 Schematic representation of a dendrimer.

to form networks. If you think about it, all you would be able to make from bifunctional molecules is linear chains! More on this later when we discuss polymerization. But for now, back to network formation.

An example of forming a network from a trifunctional molecule is the reaction of phenol with formaldehyde. The hydrogens in the ortho and para positions to the OH group can react with fomaldehyde to form (initially) oligomers, like the molecule displayed in Figure 2-6. As the reaction proceeds, molecules like this continue to react, building up an interconnected network. The figure shows what a small part of such a network would look like. Phenolic resins like this were first made around the turn of the 20th century by Leo Baekeland (see *Polymer*

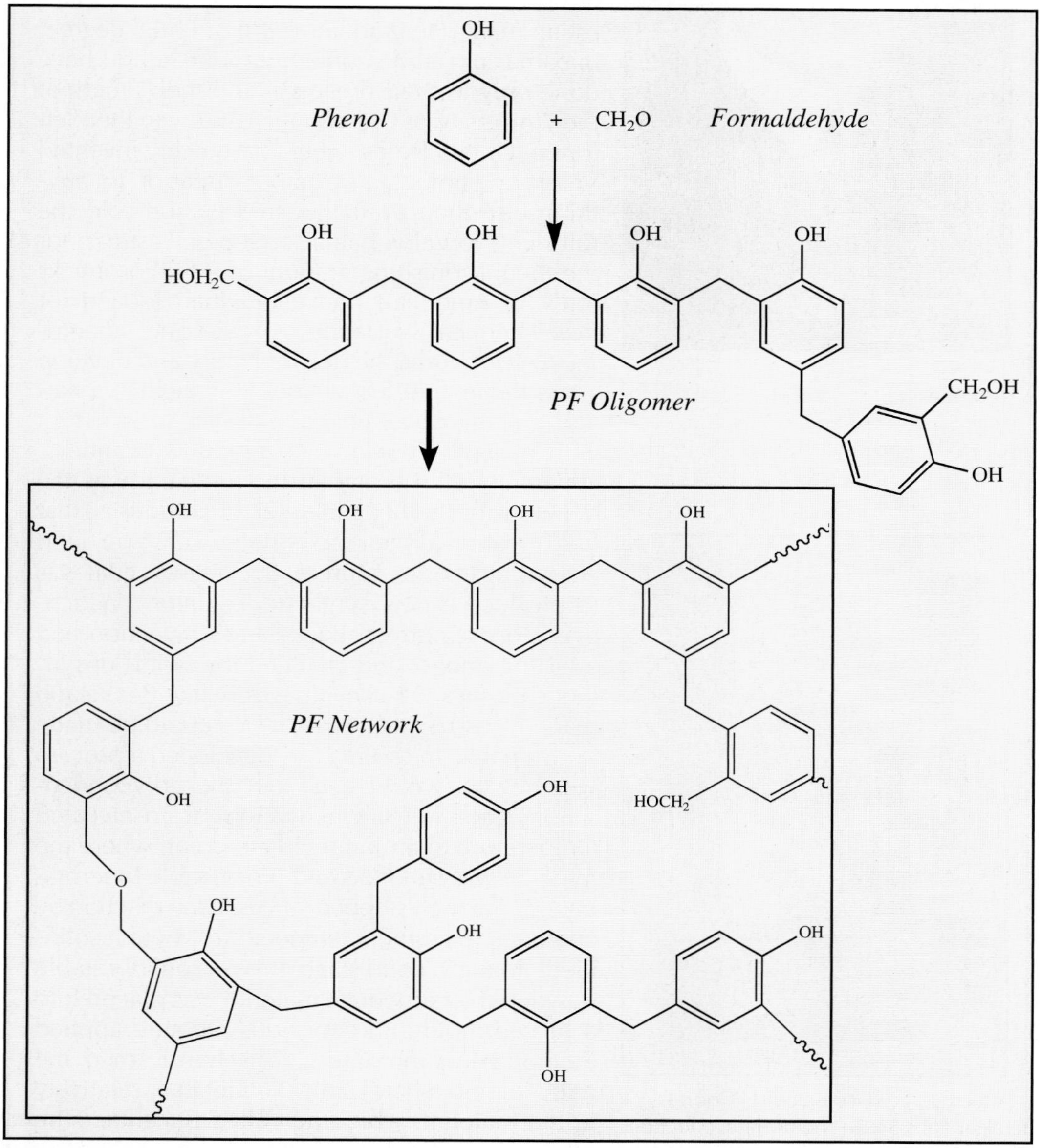

FIGURE 2-6 Representation of the formation of a typical phenolic resin.

POLYMER MILESTONES—LEO BAEKELAND

Leo Baekeland (Courtesy: Edgar Fahs Smith Memorial Collection, University of Pennsylvania Library).

Upon arriving home you find a letter from an old friend inviting you to accompany him on a trip into the North Woods. You stroll into your den to look over your fishing rods and rifles; again you encounter this material, for the reel on the rod and the butt plates on the guns are formed of it.

Returning to the drawing room you join your wife for an evening's radio concert. Should you examine closely you will discover that the radio apparatus is made almost entirely of it.

The Material *of a* Thousand Uses

Advertisement for Bakelite (Courtesy: Edgar Fahs Smith Memorial Collection, University of Pennsylvania Library).

It was actually an Englishman, Arthur Smith, who in 1899 was awarded the first patent for the application of phenol/formaldehyde (PF) resins as electrical insulators. But it was Dr. Leo Hendrik Baekeland, a Belgium chemist born in Ghent and living in the United States, who made the big breakthrough. Baekeland was an interesting man. He graduated with a Ph.D. degree, maxima cum laude (something your authors have done only in their dreams!), and then taught at the University of Ghent until 1889. He then left for the United States, where he quickly invented Velox, a photographic paper superior to anything else then available. In 1899 he sold the full rights to Velox paper to George Eastman for the then astonishing amount of a million bucks (only in America)! Baekeland then looked for new chemical worlds to conquer and, according to one source, he decided to try and develop a less flammable replacement for shellac, a natural coating obtained from beetles, one use of which was as a wood varnish for bowling alleys, apparently all the rage at that time. It was the demands of the fledgling electrical industry that had created a shellac shortage, however, and according to other sources, Baekeland's goal was to produce a new synthetic insulator. Whichever story is correct, Baekeland's invention had a lasting impact and changed the world of synthetic plastics. The key discovery that Baekeland made in 1907 was that PF resins could be made in two parts. In essence, he developed a process whereby he would carry out the polycondensation reaction (polymerization) at an elevated temperature to an intermediate point, where the material was still fluid and processable (uncrosslinked). He then stopped ("froze") the reaction by cooling it to ambient temperature where it solidified. In the second stage, this reasonably stable powdered intermediate material was placed into a heated mold and enormous pressure applied (compression molding). This transformed the material into a hard, shiny, intractable, relatively brittle material, which he called Bakelite, "The Material of a Thousand Uses."

POLYMER MILESTONES—JAMES SWINBURNE

In many texts Baekeland is given the credit for the discovery of phenolic resins, but this is not quite right. The famous German Nobel Prize–winning chemist, Adolf von Baeyer, was the first to report (in 1872) that when phenol is reacted with an aldehyde, resinous materials are formed (which he evidently considered annoying rather than a potentially useful discovery). Moreover, Arthur Smith, was awarded the first patent for the application of phenol/formaldehyde (PF) resins as electrical insulators. Apart from the contributions of Baeyer, Smith and others, Adolf Luft, a scent chemist from Galicia (in the Austro-Hungarian Empire) deserves mention. Luft obtained a patent in England (#10,218) in 1902 for the *"Process for Producing Plastic Compounds"* from phenol, formaldehyde and sulfuric acid. It was an awkward process, but Luft made several samples. Enter now James Swinburne, later Sir James, an Eng-

Sir James Swinburne (Courtesy: London Science Museum).

lishman who, by chance, happened to see a sample of Luft's resin in a patent agent's office, and described it as looking like " . . . half a pint of beer had frozen and come out of its glass." Obviously a man with a sense of perspective! Swinburne saw the potential, formed a company called the Fireproof Celluloid Company, Ltd., and invited Luft to England for discussions. After much experimentation, including the use of base catalyzed reactions and fillers, Swinburne and his coworkers had made rods, sheets and lacquers. Imagine his chagrin, when he found out that Baekeland had filed his patent in England one day before him. Seeking legal council, Swinburne was advised that ". . . as we had sent out numerous samples, though we had not sold any, both patents were invalid." He dropped his patent application, but refused to stop manufacturing. The lacquer business was particularly promising, as brass bedsteads were becoming all the rage and the phenolic lacquers could be used to coat the brass fixtures and prevent tarnishing. The Fireproof Celluloid Company, having used up all its capital, went into liquidation. Undaunted and with an endearing sense of humor, Swinburne and colleagues, formed a new company, the Damard Lacquer Company (the silent "h" is missing!) and moved his manufacturing facility to Birmingham where the majority of the brass beds were being made. (Incidentally, Birmingham, the birthplace of one of your authors, is not on your prime list of hot vacation spots!). Evidently, the relationship between Swinburne and Baekeland was cordial. Swinburne acquired a license from the Baekeland company and has written, "Dr. Baekeland and his people had always treated us, not as rivals, but as friends and collaborators." Rare, indeed!

FIGURE 2-7 Cross-linked polymer chains.

Milestones—page 28), who modestly called the resulting materials Bakelite.

Cross-linking

Networks can also be made by taking linear polymer chains and linking them using covalent bonds (Figure 2-7). We call this cross-linking. An example of cross-linking is the reaction of natural rubber or polyisoprene with sulfur (or, as we prefer, sulphur). The sulfur interconnects the chains by reacting with the double bonds. This process was originally discovered by Charles Goodyear, who called it vulcanization (see *Polymer Milestones*—next page). Note that the linkages shown in Figure 2-7 actually consist of short chains of sulfur atoms. Cross-linking is crucial in making elastomers with useful properties, as it prevents the chains from slipping past one another—see Rubber Elasticity in Chapter 13.

Polymer Isomerism

Now we move on to another topic under the general heading of microstructure, namely *isomerism*. If you've forgotten what isomerism is or never knew in the first place, check out the definition given in the box below.

> **DEFINITION: ISOMERISM**
>
> *Two molecules are said to be isomers if they are made up of the same number and types of atoms, but differ in the arrangement of these atoms.*

There are various types of isomerism that occur in nature and to us they find their most fearsome form in describing the structure of various sugars. Fortunately, the types of isomerism we will consider are not as complex or involved. Specifically, we will consider:

1. Sequence isomerism
2. Stereoisomerism (in vinyl polymers)
3. Structural isomerism (diene polymers)

Sequence Isomerism

When a monomer unit adds to a growing chain it usually does so in a preferred direction (Figure 2-8). Polystyrene, poly(methyl methacrylate) and poly(vinyl chloride) are only a few examples of common polymers where addition is almost exclusively what we call head-to-tail. To illustrate what we mean by this, consider a polymer chain during polymerization. If the mechanism of polymerization is something called chain polymerization, then there will be an active site at the end of this chain to which the next unit will add. We have shown a vinyl polymer with the general structural formula $CH_2=CXY$ (Figure 2-9). If the X = H and the Y = Cl, then this would be vinyl chloride. If we label the CH_2 part the "tail" (T) and the CXY part the "head" (H), then it is easy to see that this monomer can add to the chain in either of two ways, TH or HT. As mentioned above, in many common polymers, such as polystyrene, addition occurs almost exclusively in a head-to-tail fashion. Obviously, steric fac-

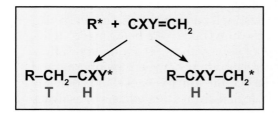

FIGURE 2-8 Sequence isomerism.

POLYMER MILESTONES—CHARLES GOODYEAR

Charles Goodyear was a young bankrupt hardware merchant when he visited the Roxbury India Rubber Co. store, America's first rubber manufacturer, in 1834. He was trying to sell an idea for a new valve to be used in rubber life preservers. But his timing was lousy! The company had just discovered the "downside" of natural rubber. There was a heat wave and thousands of articles manufactured by the company had either "melted" in storage or been returned by outraged customers. The directors evidently "met in the middle of the night to bury $20,000 worth of stinking rejects in a pit". (www.goodyear.com). Rubber was seemingly dead! While in jail for debt and in his wife's kitchen when liberated, Goodyear experimented by kneading and mixing rubber with virtually anything he could get his

Charles Goodyear (Courtesy: Goodyear Tire & Rubber).

hands on, in the hope that he would find the "magic" ingredient that would modify rubber and eliminate its few unfortunate properties. Supposedly, one day in 1840, he accidentally dropped a mixture of rubber and sulfur onto a hot stove. He might have expected that it would become a gooey mess, but instead it "vulcanized" and was still flexible the next day. Vulcanization was the breakthrough everybody was looking for! Goodyear found that when rubber was heated with sulfur it did not flow and it retained much of its elasticity over a much wider range of temperatures. It also had much better solvent resistance. He had actually cross-linked the chains, although he didn't know it. Hancock subsequently also found that masticated natural rubber, when mixed with sulfur, produced a superior elastic product (he evidently had examined some of Goodyear's samples, observed a yellowish sulfur "bloom" on the surface and quickly reinvented vulcanization). A friend of Hancock, Brockedon, is credited with coining the name vulcanization, after Vulcan, the Roman god of fire. Ironically, Goodyear obtained his US patent in 1844, a year later than Hancock's UK patent. Actually, Goodyear filed for a UK patent, but found that Hancock had filed a few weeks earlier. Goodyear sued, was offered a half-share of the Hancock patent to drop his suit, declined, and lost the case! Bloody lawyers! Goodyear was fighting lawyers and people who attempted to circumvent his patents for the rest of his life. When he died, in 1860, he was $200,000 in debt. His family received plenty of royalties, however, and were well taken care of.

POLYMER MILESTONES—THOMAS HANCOCK & CHARLES MACINTOSH

Thomas Hancock (Source: The Malaysian Rubber Board).

An important invention that was to have an enormous impact on the eventual commercialization of rubber and its products was a machine that softened, mixed and shaped rubber. This machine was designed by an Englishman, Thomas Hancock. Natural rubber has an extraordinarily high molecular weight (compared to most synthetic elastomers made today). This renders it essentially insoluble, highly elastic and difficult to shape (a consequence of entanglements or labile cross-links). By subjecting rubber to high shear in his machine, Hancock was literally tearing apart or cleaving the polymer chains and effectively reducing the overall size or average molecular weight of the rubber. This process was given the delightful term, *mastication*. (No, not what you're thinking; behave!) Following mastication the natural rubber is more soluble, less intractable and easier to mold into an article. Hancock, who was also involved in the discovery of vulcanization (see the *Polymer Milestone* concerning Charles Goodyear, page 31), found that masticated natural rubber, when mixed with sulfur, produced a superior elastic product. Another interesting inventor of the period was the Scottish chemist Charles Macintosh. The origin of the word "mac" (often spelled mackintosh), meaning a waterproof raincoat, comes from his name and he is often given credit for inventing the rubberized fabric that was manufactured by cementing two pieces of cloth together using rubber dissolved in coal-tar naphtha. However, it was actually Alexander Parkes, who came up with the idea and he sold the rights to Macintosh, who was awarded the patent for rubberized fabric in 1832. (Remember Parkes? He was the prolific inventor who discovered Parkesine, a modification of cellulose—see page 8—from where we also get the word parka.) By "sandwiching" rubber between two pieces of fabric, some of the unfavorable properties, tackiness and flow, were minimized and the material could be used to make rain wear and other articles that required waterproof fabrics. Still, the material left a lot to be desired and it took the discovery of vulcanization and collaboration with Thomas Hancock before the major problems of gum rubber were overcome and modern rubberized materials were developed.

Charles Macintosh (Courtesy: The London Science Museum).

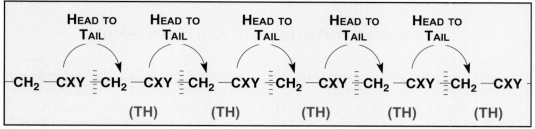

FIGURE 2-9 Head-to-tail microstructure.

tors play a role here; the bloody great benzene rings on adjacent units would strongly repel during the polymerization process if they were head-to-head. In other polymers, particularly those with smaller substituents, head-to-head and tail-to-tail placements can occur. A good example (Figure 2-10) is poly(vinylidene fluoride), $-CH_2-CF_2-$. The yellow balls represent fluorine atoms.

Stereoisomerism in Vinyl Polymers

Polymerization of a vinyl monomer, $CH_2=CHX$, where X may be a halogen, alkyl or other chemical group (anything except hydrogen!) leads to polymers with microstructures that are described in terms of *tacticity*. The substituent placed on every other

carbon atom has two possible arrangements relative to the chain and the next X group along the chain. These arrangements are called *racemic* and *meso*, just to annoy you and make your life difficult (Figure 2-11). Imagine the backbone carbon atoms of the chain to be arranged in the form of a zigzag, all in the plane of the page; we call a pair of adjacent units (diad) where the substituent (X) is in one unit above the plane of the page and in the other below the plane of the page, a *racemic diad*. If in the two units, the substituent groups are both on the same side of the chain, then we call it a *meso diad*. What do we call polymers made up of such regularly arranged units? Not meso and racemic polymers, but *isotactic* and *syndiotactic!*

Parts of polypropylene chains ($-CH_2-$

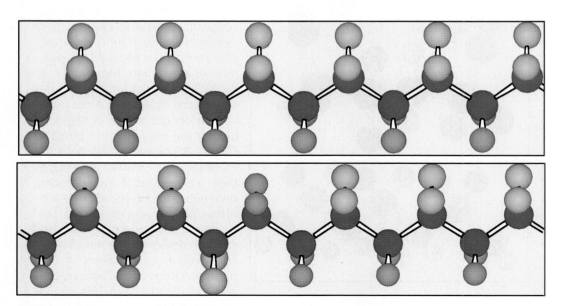

FIGURE 2-10 (Top) head-to-tail PVDF;. (bottom) head-to-head and tail-to-tail sequences in PVDF.

FASCINATING POLYMERS—PVDF & PIEZOELECTRICITY

Piezoelectric materials are fascinating. Under appropriate circumstances, if pressure is applied to a piezoelectric material, a voltage is generated. The converse is also true. The history of piezoelectricity is rich in contributions from famous scientists such as Coulomb, P. and J. Curie, Hankel, Lippmann, Lord Kelvin, Voight, Langevin and Born. Piezoelectric transducers were developed at the beginning of the 20th century, notably by Langevin, and immediately found application in detecting the sound waves emanating from submerged submarines. Langevin was to became the foremost authority on sonar equipment and the field of ultrasonics. Nowadays, piezoelectric materials are ubiquitous and used in all sorts of useful applications, including crystal oscillators, transducers in telephone speakers, headphones, sonar arrays, mechanical actuators, etc. It was the Japanese scientist, Kawai, who discovered a strong piezoelectric effect in PVDF in 1969. To obtain a useful transducer material, PVDF is extruded and drawn (stretched) while being subjected to a strong electric polarization field. The origin of piezoelectricity in PVDF is usually explained in terms of a "dipole model." PVDF is a semicrystalline polymer and the crystalline phase is composed of two major contributions: an α-phase, which consists of a series of antiparallel chains, and a β-phase, consisting of parallel chains where the dipoles align. Under the application of an electric field the polymer chains align themselves in the field, by rotating the dipoles around the chain axis. A net polarization occurs, which is responsible for the piezoelectric effect in PVDF. A polymeric piezoelectric material has many advantages over conventional piezoelectric ceramics. Flexible and large area ultrasonic transducers can be fabricated. Expect to find PVDF used in audio devices like microphones, high frequency speakers, hydrophones, ultrasound scanners, etc., and in pressure switches, actuators and robotics.

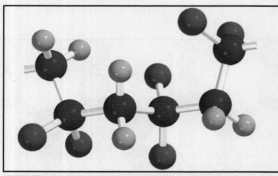

Space filling model of PVDF.

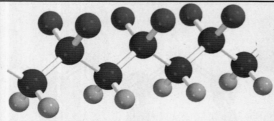

PVDF: Top, α-phase; bottom, β-phase.

CHCH$_3$–)$_n$, where all the methyl (CH$_3$) substituents are on the same side are shown schematically in Figure 2-12. (The methyl carbon atoms are shown in red to make them easier to see.) We call this arrangement *isotactic*, so this polymer is isotactic polypropylene. Because of steric repulsions between CH$_3$ groups on adjacent units, the chain would not want to sit in this planar zigzag shape or conformation, but would fold into a different shape by rotating the polymer backbone bonds. More on this when we discuss conformations. Also shown in Figures 2-13 and 2-14 are two more polypropylene chains. One of these consists of units that are all racemic to one another and is called *syndiotactic*. The other has a random arrangement of units and we call such chains *atactic*.

Structural Isomerism

So far, we've described sequence and stereoisomerism. Now we come to another type of microstructure that some students find difficult, mainly because they hate trying to memorize organic nomenclature (who

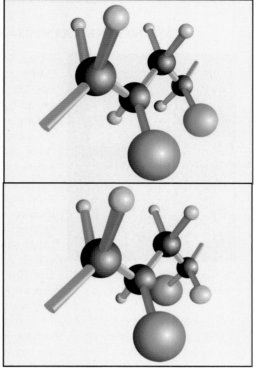

FIGURE 2-11 (Top) meso diad (m); (bottom) racemic diad (r).

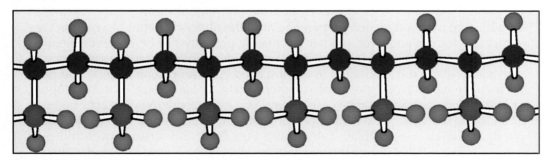

FIGURE 2-12 Isotactic polypropylene (planar zigzag conformation).

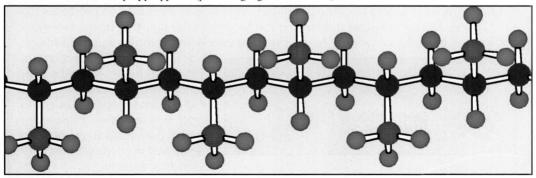

FIGURE 2-13 Syndiotactic polypropylene (planar zigzag conformation).

FASCINATING POLYMERS—POLYPROPYLENE TAPE & HINGE

Package wrapped with oriented polypropylene strapping.

As we will see in Chapter 3, isotactic polypropylene (*i*-PP) is commonly synthesized using a Ziegler-type catalyst. It has a very regular structure and readily crystallizes. At room temperature *i*-PP is a hard, tough plastic with a melting point of about 165°C and a glass transition temperature of about 0°C. So where does polypropylene excel? Well you've all come across the stuff shown in picture on the left. Your mother sends you a "care package" containing "real" food and the parcel is wrapped with this innocuous looking opaque tape (called strapping). You hungrily grab it, thinking, "If I give this a quick jerk it will break and let me get to the goodies." But like hell it does! After much cursing you inevitably have to go and find a sharp knife or pair of scissors to cut the strapping. Chances are, this tape is oriented polypropylene. When *i*-PP is extruded from the melt through a die shaped in the form of a slit, a film or tape is formed. If the film is simply supported and allowed to cool, the tape is unoriented and the polypropylene crystallizes in a spherulitic morphology (more on this in Chapter 8). However, if the *i*-PP tape is rapidly stretched (drawn) during the process of cooling and crystallization, orientation in the direction of extrusion occurs and there is a significant change in the crystalline morphology. This gives oriented *i*-PP its superior strength in the draw direction. You might note, however, that the *i*-PP tape is easily defibrillated (can be pulled apart in a direction perpendicular to the fiber orientation)—you don't get something for nothing! Yet another fascinating property of *i*-PP deserves mention and is based upon similar arguments. It is the extraordinary fatigue resistance of orientated *i*-PP that is embodied in the so-called "polypropylene hinge." You have all seen the typical "one piece" plastic tool, tackle, or lunch box, or the analogous polypropylene suit and attaché cases. Have you ever wondered why, after opening and closing the box or case hundreds of times, the hinge is still intact? Most hard plastic materials would break if they were bent back and forward over and over again. It turns out that during the injection molding process used to manufacture the box or case, the molten *i*-PP is forced, under enormous pressure, through a narrow slit in the mold that is to become the hinge. This orients the *i*-PP and produces the very strong, fatigue-resistant material in the hinge. In the figure opposite we show a clip used to seal a colostomy bag. This is an excellent example of an i-PP hinge.

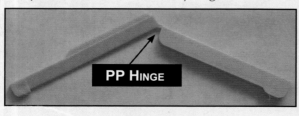

PP Hinge

A polypropylene clip used to seal colostomy bags.

Look at the piffling amount of i-PP in the hinge. One of your authors, who unfortunately has firsthand knowledge, has yet to break one of these clips even though they have been opened and closed numerous times over periods of months.

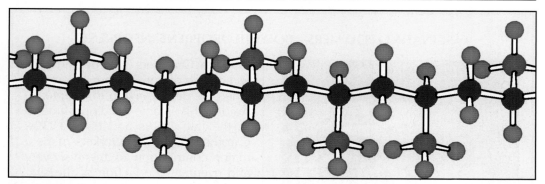

FIGURE 2-14 Atactic polypropylene (planar zigzag conformation).

doesn't?). This is structural isomerism and is seen in the synthesis of polydienes from conjugated dienes (Figure 2-15). The polymers made from these monomers are important because they (and also many copolymers that incorporate these units) are elastomers or rubbers. Polyisoprene, for example, is natural rubber; polychloroprene is Neoprene (you know, what wet suits are made of). However, natural rubber is a particular structural isomer of polyisoprene, *cis*-1,4-polyisoprene. So what on earth does *cis* mean and where do the numbers come from?

Let's start with the numbers. This is pretty easy—we just label the carbon atoms of the monomer in order, as shown in Figure 2-16. Now we have not done any chemistry yet, but hopefully you will recall that a covalent bond can be thought of (very crudely and inaccurately) as a sharing of electrons between atoms. So we can "break" one of the bonds in one of the double bonds and leave the carbon atoms still attached by a single bond (Figure 2-17). This leaves an "unshared" electron on each of the carbon atoms involved. These are represented by dots. Now let's do the same for the other double bond (Figure 2-18). Obviously we can now connect the middle dots to make a new double bond, leaving unshared electrons "hanging out" on carbon atoms 1 and 4 (Figure 2-19). Can you guess what's coming next? Of course, two such units can become attached through their carbon atoms 1 and 4. This can be repeated, building up a chain with 1,4 linkages (Figure 2-20). Naturally,

$$CH_2 = CX - CH = CH_2$$

where, if:

X = H we have butadiene
X = CH$_3$ we have isoprene
X = Cl we have chloroprene

FIGURE 2-15 Diene monomers.

$$CH_2 = CX - CH = CH_2$$
$$ 1 2 3 4$$

FIGURE 2-16 Numbering the carbon atoms.

$$\overset{\bullet}{CH_2} - \overset{\bullet}{CX} - CH = CH_2$$
$$ 1 2 3 4$$

FIGURE 2-17 "Breaking" the first double bond.

$$\overset{\bullet}{CH_2} - \overset{\bullet}{CX} - \overset{\bullet}{CH} - \overset{\bullet}{CH_2}$$
$$ 1 2 3 4$$

FIGURE 2-18 "Breaking" the second double bond.

$$\overset{\bullet}{CH_2} - CX = CH - \overset{\bullet}{CH_2}$$
$$ 1 2 3 4$$

FIGURE 2-19 Reforming the central double bond.

this step-by-step breaking and reforming of bonds is not what happens in a real reaction, we've just constructed this picture to give you an understanding of where the 1,4-

FASCINATING POLYMERS—POLYCHLOROPRENE (NEOPRENE)

Wallace Carothers stretching a Neoprene film (Reproduced with the kind permission of the Hagley Museum and Library).

Wallace Carothers will be the subject of one of our *Polymer Milestones* when we discuss nylon in Chapter 3. Among his many accomplishments in the late 1920s and early 1930s, Carothers and his coworkers made a major contribution to the discovery and eventual production of the synthetic rubber, polychloroprene. It was synthesized from the diene monomer, chloroprene, $CH_2=CCl–CH=CH_2$. Chloroprene, which is a very reactive monomer—it spontaneously polymerizes in the absence of inhibitors—was a product of some classic studies on acetylene chemistry performed by Carothers and coworkers at that time. In common with butadiene and isoprene, in free radical polymerization chloroprene is incorporated into the growing chain as a number of different structural isomers. Elastomeric materials having very different physical and mechanical properties can be made by simply varying the polymerization temperature. In November 1931, DuPont launched this new synthetic elastomer that was given the trade name Neoprene. It was found to have good resistance to oil (far superior to natural rubber), heat and weathering. Commercial production started in 1932 and Neoprene was an important synthetic elastomer used in WWII. Neoprene can be vulcanized and is still produced today, finding applications as hoses, weather strips, gloves, adhesives and corrosion-resistant apparel. As fate would have it, one of your authors worked at DuPont's Experimental Station in the 1970s, in what was then the Elastomers Department. Imagine his chagrin when informed that his first research project involved the characterization of Neoprenes. "Wow, Carothers invented Neoprene some 40 years ago, what can there possibly be that is not known about polychloroprene?" In retrospect, a rather naive response. But, there was also another problem, Neoprene was rumored to be on the way out. Other elastomers that were cheaper and easier to produce were touted to replace it. The industrial emulsion polymerization of chloroprene had always been a pain. The monomer is very reactive, a suspected carcinogen that contains chlorine (an environmental "no-no"), and the polymer has to be isolated after about 60–70% conversion of the monomer (necessitating recovery of the unused monomer), otherwise it forms an insoluble gel. But, every time the demise of Neoprene has been predicted, other applications for the material have surfaced. Look at all the wet suits, boots and swimming gear that is made from Neoprene today. Carothers produced a "keeper"!

$$- CH_2 - CX = CH - CH_2 - CH_2 - CX = CH - CH_2 - CH_2 - CX = CH - CH_2 -$$
$$1 \quad 2 \quad 3 \quad 4 \quad 1 \quad 2 \quad 3 \quad 4 \quad 1 \quad 2 \quad 3 \quad 4$$

FIGURE 2-20 Formation of a 1,4-polydiene microstructure.

part of the nomenclature comes from. This approach is useful, however, in that it should also now be immediately clear that we can add units by just "breaking" (i.e., reacting) one of the double bonds (1,2- or 3,4-) and incorporating those units into the chain. The possibilities are summarized in Figure 2-21, using polyisoprene as an example.

The exact proportion of the various types of units that you get depends upon the method of polymerization and the precise experimental conditions. There is still one thing we haven't explained, however. There are two types of 1,4 units: *cis* and *trans*. This simply refers to the arrangement of carbon atoms around the double bond. If the main-chain (backbone) carbon atoms lie directly across the double bond from one another, then this is what we call *trans*. If these atoms lie on the same side of the double bond, then they are *cis* to one another. Chains that are completely *cis* or completely *trans*, or are some mixture of the two, can be synthesized, depending upon the method of polymeriza-

tion. Most rubber trees make *cis*-1,4-polyisoprene, natural rubber.

Copolymers

So far in our discussion of microstructure, we have considered homopolymers. To some degree, however, there is an element of semantics involved in our definition. Is a branched polyethylene a true "homopolymer" or should it be considered a copolymer of ethylene and whatever units comprise the branches? Here our concern is "real" copolymers, those synthesized from two (or more) distinct monomers. The simplest possible arrangements are shown in Figure 2-22 and are self-explanatory. But, as we will see, real life is more complex. True random copolymers are rare and in most cases there are tendencies to "blockiness" or alternating arrangements. There are also graft copolymers, but we will discuss all this in more detail when we consider copolymerization.

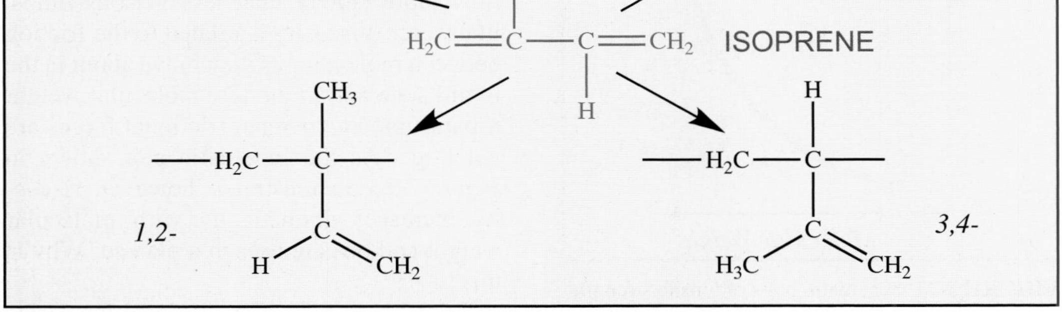

FIGURE 2-21 Structural isomers of polyisoprene.

-A-A-B-A-B-B-B-A-
RANDOM COPOLYMER

-B-A-B-A-B-A-B-A-
ALTERNATING COPOLYMER

-A-A-A-A-B-B-B-B-
BLOCK COPOLYMER

FIGURE 2-22 Copolymer types.

MOLECULAR WEIGHT

Now we are almost ready to discuss polymer synthesis, but first we need to revisit a topic we touched on earlier—molecular weight. We will see that the mechanism of

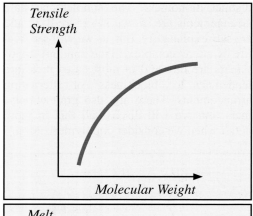

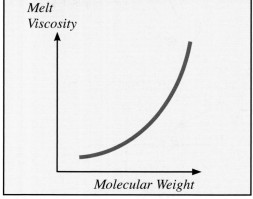

FIGURE 2-23 Schematic plots of tensile strength (top) and melt viscosity (bottom) versus polymer molecular weight.

polymerization can have a profound effect on both molecular weight and molecular weight distributions. It is a topic that some people have trouble with at the start of their studies, because they get confused by the different averages involved. But it is important to persevere, as the graphs shown in Figure 2-23 would indicate. These are schematic plots of strength and viscosity (i.e., the "thickness" of a fluid) as a function of molecular weight or chain length. Let's look at these in a bit more detail.

In Chapter 1 we discussed the effect of molecular weight on the physical properties of a set of n-paraffins. The first few members of the series (ethane, propane, butane) are all gases at ordinary temperatures and pressures. But as they get bigger, the forces of attraction between the molecules increase and we get liquids (e.g., octane, found in gasoline). Larger still, the cohesive forces result in "semi-solids," things like the paraffin waxes. Finally, at high molecular weights, the forces of cohesion (and chain entanglements) are such that we now have a solid—linear polyethylene. Clearly, "strength" in this series goes from zero (gases and liquids cannot hold their own shape) to a value that will level off at high molecular weight.

So, strength plotted as a function of molecular weight will look something like the graph at the top of Figure 2-23. At the upper end there will not be much difference in strength if we, say, double the molecular weight from one million to two million. Viscosity is a different story, however. Viscosity measures how easily fluids flow. Highly viscous (or "thick") fluids, like molasses, flow more slowly than less viscous fluids, like water. Viscosity is related to the friction between molecules as they move about in the liquid state and in the low molecular weight n-paraffins, like octane, frictional forces are not large. As the schematic plot shown in Figure 2-23 demonstrates, however, viscosity increases dramatically with molecular weight and does not reach a plateau. Why is this?

Think about the difference between a nice linguini and spaghettios (or chopped up spaghetti)—you know, the stuff you get out of a can (Figure 2-24). The small stuff can be stirred more easily because the chains are too short to tangle up with one another. However, viscosity does not increase in a nice linear fashion with molecular weight (hence, entanglements), but in what we call a power-law fashion. If you double the molecular weight, viscosity can increase by a factor of about 10! This can make very high molecular weight polymers too viscous to process by ordinary methods. So, molecular weight is really important and we need to know more about it.

Molecular Weight Distributions

The problem with describing the molecular weight of synthetic polymers is that there is always a distribution of chain lengths (although certain polymerizations can give very narrow distributions). The solution would seem to be simple—let's just define an average. For example, let's start with 200 monomers (or, more realistically, 200 moles of monomer) and polymerize these to give 10 chains (or 10 moles of chains). The average "length" of the chain, or degree of polymerization, is then 20: 200 divided by 10. If the weight of each unit in the chain was 100 units, the average molecular weight would be 20,000. But this says nothing about the distribution. Do we have one chain 191 units long and 9 unreacted monomers, or some other, more symmetric distribution about the average, as illustrated in Figure 2-25?

Definition of Number and Weight Average Molecular Weight

Also, what we have calculated is something called a number average, which is defined mathematically below. If you're not used to dealing with summations, this looks horrible. To give you a feel for how it works, and also introduce a different average—the weight average—let's consider a ridiculous example. Behold an elephant with four mosquitoes on its bum (Figure 2-26). If the elephant weighs 10,000 lbs and the

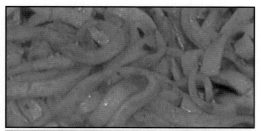

FIGURE 2-24 A plate of linguini (top) and spaghettios (bottom).

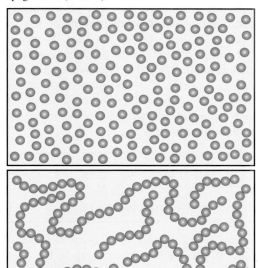

FIGURE 2-25 Schematic diagram of 200 monomers (top) and 10 chains formed from them.

mosquitoes 1 lb each (you know, the type of mosquito you meet every time you go camping), what would be the average weight of each of the things or species present? Now, if we were to calculate the average weight of each thing present in the same way as we calculated the average length of each of 10

FIGURE 2-26 An elephant with four mosquitoes on its bum!

chains that were synthesized from a total of 200 monomers (200/10 = 20), we would say that the total weight of everything present is equal to 10,000 + 1 + 1 + 1 + 1 lbs. There are five things present, so the average weight of each is about 2,000 lbs. We have used Equation 2-1, where N_x is the number of things belonging to species x present (1 elephant and 4 mosquitoes), while M_x is the weight of each of these things.

$$\overline{M}_n = \frac{\sum N_x M_x}{\sum N_x}$$

EQUATION 2-1

This is the *number average* of this "sample." If you were to now go up to this elephant and yanked on his (or her) trunk, he (or she) would probably give you a good stomping. Obviously, the effect of the mosquitoes on this stomping would be negligible compared to that of the elephant, so the number average does not give a good measure of this important physical property (stomping ability!). What if we average a different way, by weight? If W_x is the total weight of species, x, present, we could write Equation 2-2.

$$\overline{M}_w = \frac{\sum W_x M_x}{\sum W_x}$$

EQUATION 2-2

The *weight average* of this collection is about 10,000, a better reflection of stomping ability.

Now let's say we had a sample with 5 (moles of) chains of "length" (degree of polymerization or DP) 100 (i.e., has 100 monomer units in the chain), 5 (moles of) chains of length 150 and 5 (moles of) chains of length 200. If you're quick and intuitive when it comes to numbers, you might immediately grasp that the average DP of this sample is 150. If the molecular weight of each unit in the chain were again 100, the number average molecular weight would then be 15,000. If you're a bit slow, or just having a bad day, it's pretty easy to work this out using the expression for number average shown as Equation 2-1. Now N_x is simply the number (of moles) of chains of the x-"species." We have three species in our sample; chains of DP 100, chains of DP 150 and chains of DP 200, whose weights M_x are therefore 10,000, 15,000, and 20,000, respectively. Before proceeding, see if you can substitute correctly into the Equation 2-1. (For those of you that are mathematically challenged we do this for you in Equation 2-3.)

$$\overline{M}_n = \frac{\sum N_x M_x}{\sum N_x}$$

$$= \frac{5[10,000] + 5[15,000] + 5[20,000]}{5 + 5 + 5}$$

$$= 15,000$$

EQUATION 2-3

What about weight average molecular weight? First, recall the definition of weight average (Equation 2-2). Now note that the total weight of species x present is just the molecular weight of each chain of type x multiplied by the number of chains of this type (e.g., 5 chains, each of weight 10,000 means that W_x is 50,000). Accordingly, we can substitute into Equation 2-2 to obtain a different form of the equation for weight average (Equation 2-4).

$$W_x = N_x M_x$$

EQUATION 2-4

This then yields the expression shown in Equation 2-5.

$$\overline{M}_w = \frac{\sum W_x M_x}{\sum W_x} = \frac{\sum N_x (M_x)^2}{\sum N_x M_x}$$

EQUATION 2-5

We can now calculate the weight average as shown in Equation 2-6. Note that the weight average is larger than the number average. This is always true, except in the case when all the chains are the same length (as in proteins, where the number and sequence of amino acids is specified by the cell's machinery). In this case the number average equals the weight average.

$$\overline{M}_w = \frac{\sum N_x (M_x)^2}{\sum N_x M_x}$$

$$= \frac{5[10,000]^2 + 5[15,000]^2 + 5[20,000]^2}{5[10,000] + 5[15,000] + 5[20,000]}$$

$$= 16,111$$

EQUATION 2-6

We call the ratio of the two averages the polydispersity of the system (Equation 2-7). It is a measure of the breadth of the distribution.

$$Polydispersity = \frac{\overline{M}_w}{\overline{M}_n} \geq 1$$

EQUATION 2-7

Which Average Molecular Weight?

We have seen that average molecular weight is not unique. It turns out that there are more than two ways to define an average. Look at the definitions of number and weight average reproduced in Equations 2-8.

$$\overline{M}_n = \frac{\sum N_x M_x}{\sum N_x}$$

$$\overline{M}_w = \frac{\sum N_x M_x^2}{\sum N_x M_x}$$

$$\overline{M}_z = \frac{\sum N_x M_x^3}{\sum N_x M_x^2}$$

EQUATIONS 2-8

You can see that we can go from number to weight average by multiplying each of the terms inside the summations by M_x. Higher order averages can be constructed in the same way; e.g., the z-average (Equation 2-8). The ratios of these averages can be related to the moments of the molecular weight distribution and tell us about its breadth and "skewedness."

But we're sure you are starting to get sick of molecular weight, so let's leave the definitions there. However, there are methods for measuring number and weight average (also z-average, although this is harder) and you

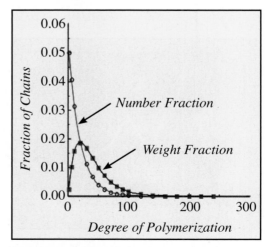

FIGURE 2-27 Graph of the fraction of chains versus the degree of polymerization.

will have to learn about these as you go on in this subject. Also, many students think of distributions solely in terms of "bell curves" or Gaussian distributions. In many polymerizations, the distributions are more complex, as illustrated by the plot of the fraction of chains having a particular chain length versus DP shown in Figure 2-27. It turns out that the entire distribution can be measured by chromatographic methods, but this is a more advanced topic and at this point we will just let sleeping dogs lie.

RECOMMENDED READING

H. R. Allcock, F. W. Lampe and J. E. Mark, *Contemporary Polymer Chemistry*, 3rd. Edition, Prentice Hall, New Jersey, 2003.

J. L. Koenig, *Chemical Microstructure of Polymer Chains*, Wiley, New York, 1982.

G. Odian, *Principles of Polymerization*, 3rd. Edition, Wiley, New York, 1991.

STUDY QUESTIONS

1. What is the difference in the chain structure of atactic, isotactic, and syndiotactic polystyrene. (Hint: a picture is worth 1,000 words!) Which one of these will most likely be incapable of crystallizing? Explain why?

2. Sketch the types of structural isomers that can be formed from the polymerization of chloroprene ($CH_2=CCl-CH=CH_2$).

3. You are given the infrared spectra of two samples of polyethylene (see below) which are known to have approximately the same (high) molecular weight and the same type of end groups. One of these samples, say sample A, displays a band due to the presence of methyl groups, while in the other sample B, this band is extremely weak. What is the difference in microstructure of these two samples? Which one of these do you think would be more crystalline and why?

4. Which of these samples (A or B) would you use to:

   A. Make a flexible, optically clear polymer film.
   B. Make a milk jug, where stiffness and good barrier properties are more important than optical clarity. Explain the basis of your choice (i.e., don't be lazy and just guess!).

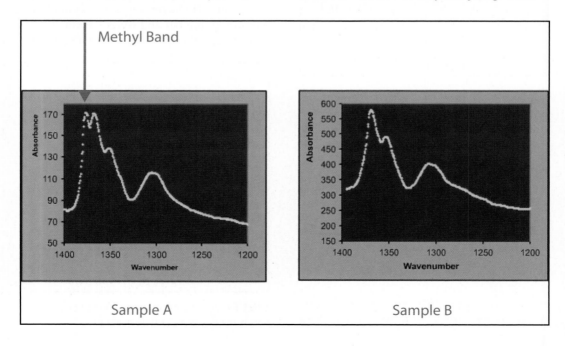

5. You are given the following set of mono-disperse polymer samples:

Sample	Number of moles	Molecular weight
1	10	10,000
2	20	20,000
3	30	20,000
4	40	30,000
5	30	50,000
6	20	60,000
7	10	80,000

Calculate the number and weight average of a mixture of this set of samples.

What is the polydispersity of the mixture?

6. You are asked to identify four samples handed to you by some rotten, sadistic, polymer science professor:

SAMPLE A is an opaque rectangular-shaped film that has some flexibility and appears to be tough. (It did not break when one of the other students in your class suddenly went berserk and pounded it with a hammer.)

SAMPLE B is transparent, fairly rigid and brittle.

SAMPLE C is not transparent because it has been mixed with some gunk (a filler). However it stretches many times its original length, but does not return to its original length if it is held extended for anything more than a few seconds.

SAMPLE D appears identical to sample C but does return to its original length after stretching.

Because you have no idea how to start answering this question, you bother your professor with incessant e-mails and you whine a lot during class. Finally, he or she gives you some clues, just to get rid of you. You discover that one of the samples is poly-isoprene, one is atactic polystyrene, one is high-density polyethylene and the last is also one of these three polymers, but its micro-structure has been altered by some chemical treatment. Identify which sample is which polymer, giving reasons for each choice (zero points for just guessing and not giving a reason). Explain how the microstructure of sample C was altered to give sample D.

7. You are an engineer working for a company that synthesizes and uses polymers for a range of products. A competitor makes an identical polymer, polypropylene, but its properties are superior. One of the organic chemists in your company comes up to you one day and says he has reproduced the competitor's product exactly, because his polymer has the same number average molecular weight as your competitor's. Your response is:

A. He must be demented because he is wearing a Notre Dame football T-shirt. He is also an organic chemist and cannot possibly be right.

B. He is right, even if he is demented. An average is an average and the samples must be equivalent.

C. He is wrong because the distribution of molecular weights could be very different.

D. He is right because if the distributions are different then the average would also be different. Also, he would only be truly demented if he was wearing a Notre Dame tee-shirt.

Explain which of your highly biased, unprofessional responses is probably correct, giving a reasoned argument rather than a blindly prejudicial statement as an explanation.

8. Polymers can be classified as thermoplastic or thermosetting. Explain the difference.

3

Polymer Synthesis

INTRODUCTION

Now that you have learned something about polymer microstructure, the multifarious ways monomer units can be arranged in chains, it is time to move on and look at some of the details of how these units are linked together in the first place—the science (and sometimes art) of polymer synthesis

Major Classifications

Polymerization reactions can be classified in various ways. Wallace Carothers, a great and tragic figure in the history of polymer science (see *Polymer Milestones* on the next page), suggested that most polymerizations fall into one of two broad categories, condensation or addition. For reasons that will become obvious, the terms step-growth and chain polymerization provide a more accurate and complete description, but dinosaurs like your aging authors have a lingering nostalgia for these older terms and still consider them powerfully descriptive, if not always accurate. You'll see why as you work your way through this chapter. Before getting to the nitty-gritty, however, we want to use the rest of this section to review some basic chemistry. Those of you with a chemical background won't need this, but we also teach engineers, many of whom (in our experience) are deeply allergic to organic chemistry.

Condensation Reactions

Figure 3-1 shows a simple condensation reaction. This is aptly named because it involves the splitting out of a molecule of water, as in the reaction of acetic acid, the active ingredient of vinegar (the stuff many Brits still put on their chips) and ethanol (or ethyl alcohol), the active ingredient of beer. You can see that the molecules combine to form something called ethyl acetate and also a molecule of water. The acetate molecule contains an ester group and you may, if you are as bright as we hope you are, immediately wonder if reactions like this lead to polyesters. Yes they do, but before getting to that, let's say a little more about the nature of reactions.

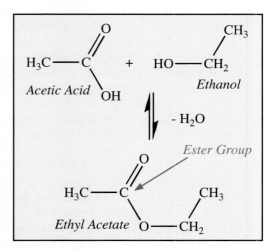

FIGURE 3-1 Formation of an ester.

POLYMER MILESTONES—WALLACE CAROTHERS

Wallace Carothers in his laboratory (reproduced with the kind permission of the Hagley Museum and Library).

Wallace ·Hume Carothers appears to have been born with chemistry in his blood! After studying and teaching chemistry in Missouri, where his enormous potential was plainly evident, he obtained his doctorate from the University of Illinois, and then took up a teaching position at Harvard. Charles Stine, who in 1928 was the research director and originator of the new fundamental research program at the E. I. du Pont de Nemours Company (now called simply DuPont), pulled off the recruiting coup of the decade, if not the century. He persuaded Carothers, who had only been at Harvard for some three semesters, to leave his teaching post and join DuPont. Carothers, who was to head DuPont's newly formed fundamental research laboratory, was given wide discretion in the research he was to perform. In fact, he clearly stated at the outset that his goal was to tackle the controversy concerning the existence of macromolecules from the synthetic side, by synthesizing compounds of high molecular weight and known constitution. He is best known for the discoveries of nylon and neoprene, but his general contributions to polymer synthesis and polymer structure/property relationships were truly seminal. Wallace Carothers was probably the first to appreciate the enormous importance of monomer functionality in the production of condensation or step-growth polymers. Functionality is simply the number of reacting groups per reacting monomer. He demonstrated that polymers can only be formed if the functionality of the monomer is two or greater; that a functionality of two yields linear chains and that if the functionality is more than two, branched chains and 3-dimensional networks are formed. Moreover, he recognized that to obtain linear polymers of high molecular weight, very pure monomers had to be used and that stoichiometric equivalence of reactive groups was essential. This insight resulted in the polymerization of some simple polyesters and then nylons. Carothers is also credited with categorizing the two major classes of polymerization: condensation or step-growth and addition. Nobody says that life's fair, but in the opinion of your authors, Carothers deserved to win the Nobel Prize in chemistry. But he didn't. He was, however, the first industrial scientist elected to the National Academy of Sciences and to all those who have studied polymer science, his place in history is secure. By all accounts, Carothers was a quiet, unassuming man, absorbed in his studies, but also interested in music, art, sports and politics. His achievements have stood the test of time, but the tragedy is that he never lived to see the extraordinary success of nylon—the subject of a forthcoming *Fascinating Polymers*. The bouts of depression to which he had long been subject became increasingly frequent after 1935 and he began to believe he was a failure as a scientist. He left his home in Wilmington on April 28, 1937 and checked into a Philadelphia hotel room, where he swallowed lemon juice laced with potassium cyanide.

Why Do Molecules React?

So, why should acetic acid and ethanol react in the first place? Why aren't they happy just the way they are? You might at first think it's just a matter of energy. In other words, if you add up all the energies of the bonds and you find that the products, ethyl acetate and water, have a lower energy than the reactants, acetic acid and ethyl alcohol, then the reaction will go, just like water runs down hill (going from a position of higher potential energy to one of lower potential energy). The excess energy is then released as heat (it's what we call an exothermic reaction). It's more complicated than this, however. It turns out that some reactions are endothermic (they take in heat from the surroundings) and many reactions are also reversible, including this one! In other words, while some of the acetic acid molecules are busy reacting with alcohol molecules to make acetate molecules and water, some of the acetate molecules and water are busy making ethanol and acetic acid! The system comes to a state of dynamic equilibrium. "But wait a minute," you might add, "I'll do an experiment and waste a good beer by mixing it with vinegar." You find nothing happens. So, what's going on?

There are a number of factors that determine if molecules will react under a given set of conditions. We found a beautiful way of understanding this on the web.[1] Think of a chemical reaction as a crime. In order for it to occur, there has to be both motive and opportunity. The motive is provided by thermodynamics. The products formed in the reaction need to be more stable than the reactants. There are two parts to this. Some reactions occur because of energetic factors: the molecules formed in the reaction have stronger bonds (and are then at a lower energy) than the ones you started with. Other reactions go because of entropy, the final state of the system is more disordered than the initial state. This can happen if a number of smaller molecules are formed from fewer

larger ones. It is the balance of energetic and entropic factors that determine the position of equilibrium, whether or not you form a significant amount of products. Entropy is, of course, a concept that strikes fear and loathing into the hearts of many science and engineering majors, so we will come back and discuss this in more depth later in this book. For now, just accept that some reactions go essentially to completion, some don't seem to go at all, while others dither about in the middle!

"Fine," you might say, "but I still want to know why I wasted a good beer on an experiment that apparently didn't work." There's one more factor. The chemical bonds in ethyl alcohol and acetic acid are quite stable and so the molecules sit around happily at room temperature doing nothing in particular (except moving about). The problem is that there is an energy barrier that the reactants have to get over to form products, illustrated schematically below (Figure 3-2). At room temperature, the alcohol and acid molecules don't have enough energy to react. (It's like the mornings you don't have enough energy to get out of bed and go to class.)

How do molecules get over this barrier? One important way is through collisions. The molecules have kinetic energy associated with their motion. The higher the temperature the faster they move. If you heat the mixture, a certain fraction of collisions will then involve enough energy to get the molecules "over the hump." Essentially,

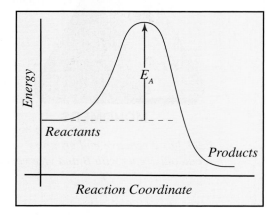

FIGURE 3-2 Energy barrier diagram.

[1] http://neon.cm.utexas.edu/academic/courses/Spring2001/CH610B/Iverson/).

there is a rearrangement of electrons. Just as in our discussion of diene polymers (Chapter 2), we show this simplistically in Figure 3-3, depicting a covalent bond as a sharing of electrons instead of the "probability cloud" orbital description that comes out of quantum mechanics. The rearrangement we have shown in this figure is not what really happens at all. These reactions are actually acid-catalyzed and involve an intermediate protonated state that breaks apart to give the final products. Nevertheless, if your grasp of organic chemistry is weak, this simple picture should give you a feel for the ultimate rearrangements that occur in these types of chemical reactions.

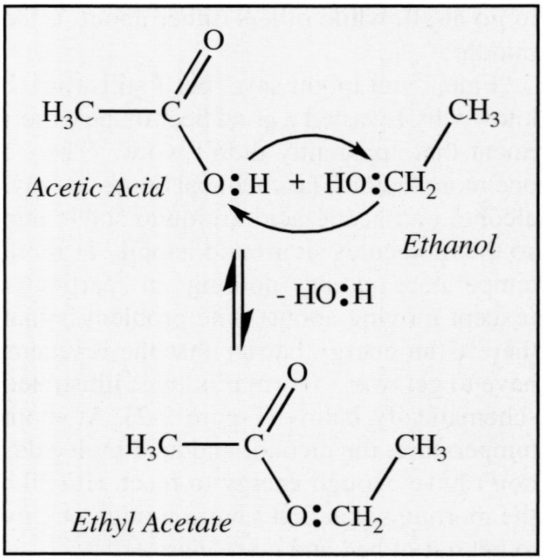

Basic Features of Step-Growth Polymerization

Having given you a very crude picture of what happens in a condensation reaction, let us now turn our attention to making a polymer, a polyester. The question is this: if we heat acetic acid and ethanol up to just over 100°C, to get the reaction going and drive off water, why don't we form polymer? The answer is simple: once you react each functional group, the acid and the alcohol, that's the end of it, because the methyl groups don't react under these conditions, nor (under other conditions) in such a way as to lead to the formation of a polymer. Obviously, to make linear chains we need bifunctional molecules, as illustrated in Figure 3-4, where we use the letters A and B to represent

FIGURE 3-3 Formation of an ester (shared electron approximation).

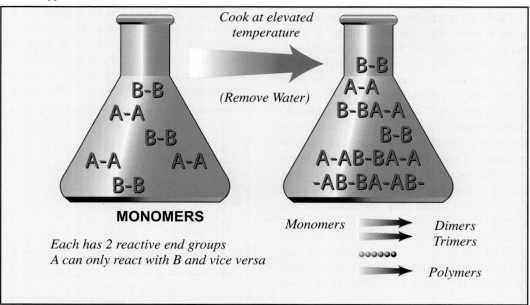

FIGURE 3-4 Schematic representation of the early stages of step-growth polymerization.

the reactive functional groups at each end of the molecule. Note that the reaction doesn't happen all in one go, all the monomers suddenly connecting to make chains, but in a step-growth fashion. Before getting to this, however, let's continue our simple review of some of the chemistry that is important in synthesis by considering how double bonds react.

Basic Features of Addition Polymerization

Again, we will show you a simple "shared electrons" model of the covalent bond, just so that you can get a feel for what's going on (Figure 3-5). Just keep in mind that as you learn more chemistry you will have to start thinking in terms of orbitals and such-like. Anyway, it should be immediately apparent that we can "break" one of the covalent bonds in a molecule containing double bonds, "localizing" the electrons on each of the carbon atoms. If we do this to a bunch of such molecules, we can then link them together by forming new bonds between unshared electrons on adjacent molecules. Of course, it doesn't happen like this at all. By giving you this picture, however, we hope you immediately pick up on the fact that molecules with double (or triple) bonds are capable of polymerizing. So are certain types of ring molecules. Opening up the ring by breaking a bond would also give a molecule with unshared electrons on each end (only certain types of bonds in certain ring molecules will do this). The actual polymerization process is more like a chain reaction, with the addition of monomers one at a time to an active site at the end of a growing chain (Figure 3-6). This picture of the polymerization process only represents one of the steps in a chain polymerization, however, one that we call propagation. How do these polymerizations get started—a process we call initiation? Also, how do these reactions finish or terminate to give us a fully formed and no longer reacting chain? Having covered some basic chemistry, let's now reconsider step-growth polymerization in some greater detail.

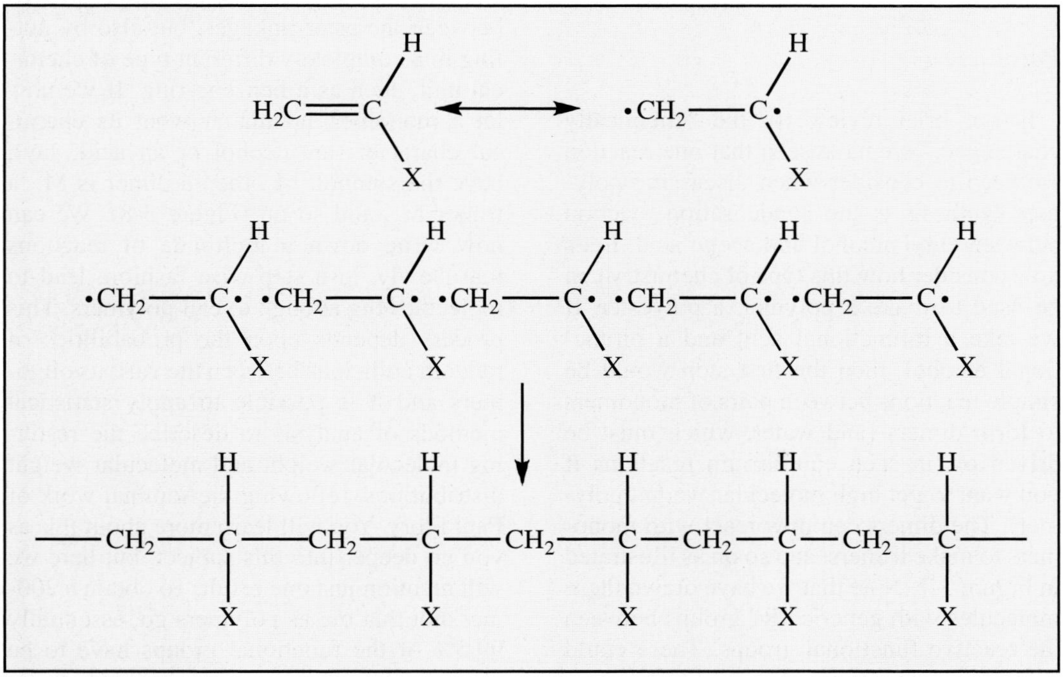

FIGURE 3-5 Formation of an addition polymer from an olefin (shared electron approximation).

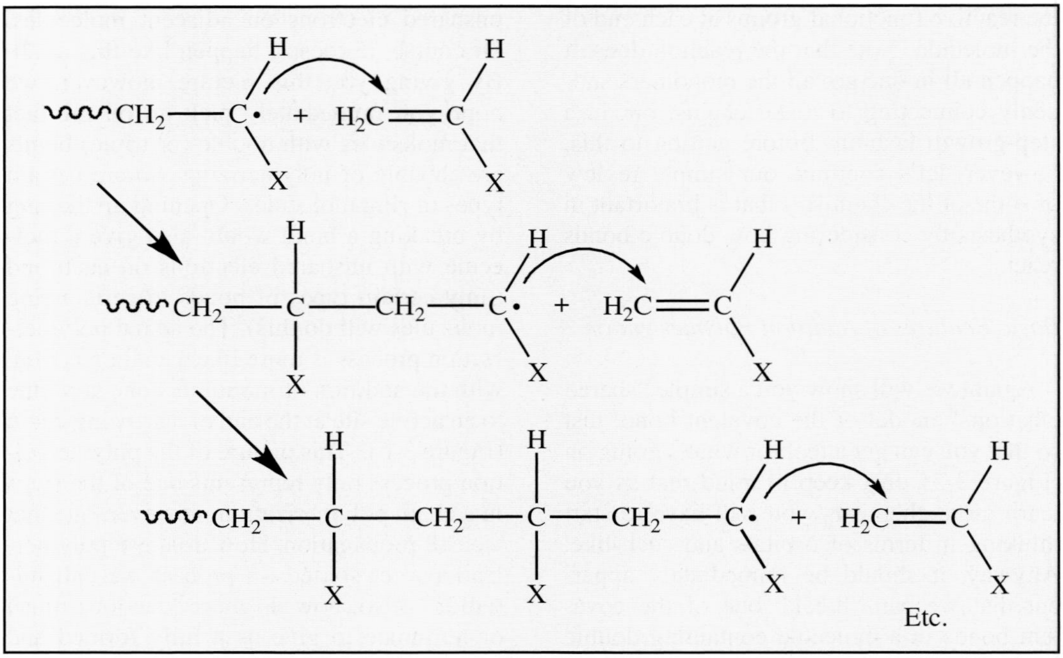

FIGURE 3-6 Depiction of a polymerization chain-like reaction.

STEP-GROWTH POLYMERIZATIONS

Polyesters

In our brief review for the "chemically challenged," we have seen that one reaction we need to consider when discussing polymer synthesis is the condensation reaction between ethyl alcohol and acetic acid. Let's now consider how this type of chemistry can be used to make a polymer, a polyester. If we take a bifunctional acid and a bifunctional alcohol, then the first step would be simple reactions between pairs of monomers to form dimers (and water, which must be driven off in such equilibrium reactions if you want to get high molecular weight polymer). The dimers can now react with monomers to make trimers, and so on as illustrated in Figure 3-7. Note that we have drawn these molecules with generic "R" groups between the reactive functional groups. These could consist of different numbers, say n and m, of CH_2 groups. Polyesters are a class of poly-

mers, and you can make different types by changing not only the number of CH_2 groups between the ester linkages, but also by putting in a completely different type of chemical unit, such as a benzene ring. If we now let a monomer, no matter what its chemical character (an alcohol or an acid, say), have the symbol, M_1, then a dimer is M_2, a trimer M_3, and so on (Figure 3-8). We can now write down a multitude of reactions that slowly, in a step-wise fashion, lead to molecules big enough to call polymers. This process depends upon the probabilities of random collisions between the various oligomers and it is possible to apply statistical methods of analysis to describe the resulting molecular weight and molecular weight distributions, following the seminal work of Paul Flory. You will learn more about this as you go deeper into this subject, but here we will mention just one result. To obtain a 200-mer, not that big as polymers go, essentially 99.5% of the functional groups have to be reacted. Obtaining a yield of 99.5% on an industrial scale can be a challenge!

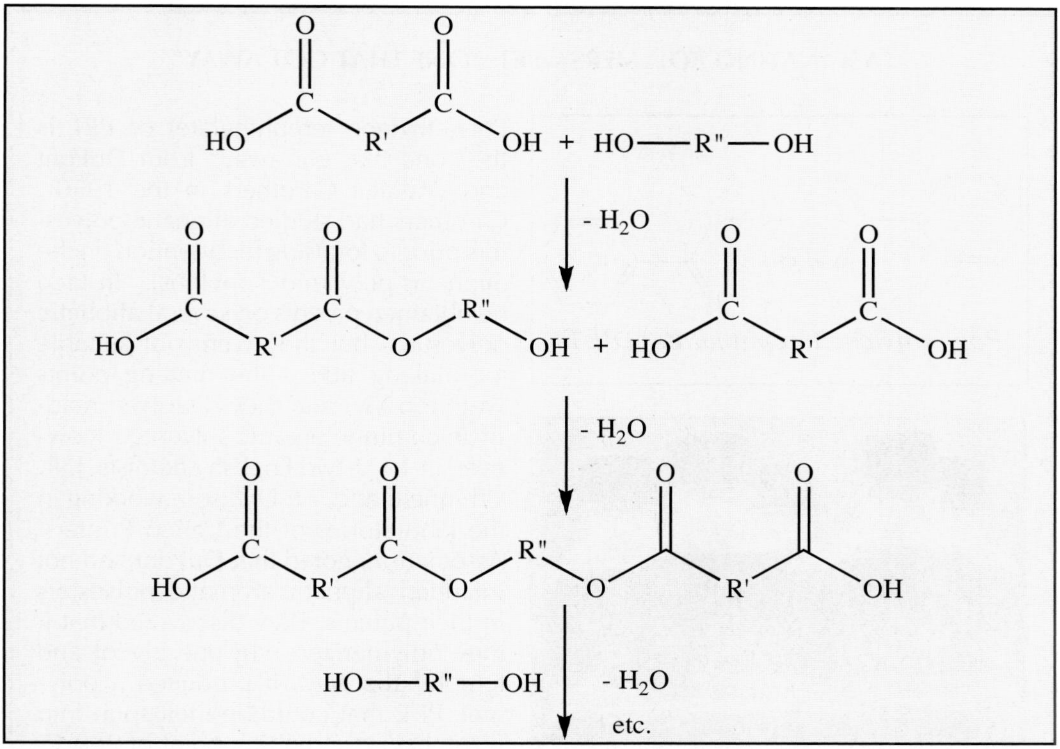

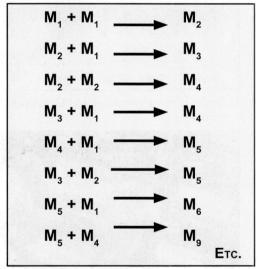

FIGURE 3-7 Formation of dimers, trimers, etc., in the step-growth polymerization of polyesters.

Polyamides (Nylons)

Hopefully you now have the basic idea that linear chains can be synthesized by reacting appropriate bifunctional molecules. In the example we just gave, a molecule of water was split out. Similar step-wise reactions lead to the formation of nylons, again a class of polymers rather than an individual type, this time characterized by an amide linkage between the functional groups, illustrated in Figure 3-9. One common nylon is nylon 6,6. So, you're saying to yourself, where do these 6's come from all of a sudden? The polymer, whose repeat unit is shown in the figure, is called nylon 6,6, because of the number of carbon atoms in each of its constituent units. So what would nylon 6,10 look like? It could be synthesized from sebacic acid (which has 8 CH_2 groups and a total of 10 carbon atoms) and hexamethylene diamine—the number of carbon atoms in the diamine is, by convention, the first quoted in the numbering

$M_1 + M_1$	$\longrightarrow$	M_2
$M_2 + M_1$	$\longrightarrow$	M_3
$M_2 + M_2$	$\longrightarrow$	M_4
$M_3 + M_1$	$\longrightarrow$	M_4
$M_4 + M_1$	$\longrightarrow$	M_5
$M_3 + M_2$	$\longrightarrow$	M_5
$M_5 + M_1$	$\longrightarrow$	M_6
$M_5 + M_4$	$\longrightarrow$	M_9

Etc.

FIGURE 3-8 Depiction of the formation of multimers in step-growth polymerization.

scheme (Figure 3-10).

There are also nylons that are designated by a single number, e.g., nylon 6 (Figure 3-11), nylon 12, etc. The synthesis of such nylons

FASCINATING POLYMERS—PET: "ONE THAT GOT AWAY"

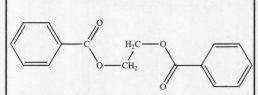

Poly(ethylene terephthalate) (PET)

Examples of the now ubiquitous PET mineral water bottles.

Polester fiber production (Courtesy: Hoechst/Celanese).

Poly(ethylene terephthalate) or PET is the "one that got away" from DuPont and Wallace Carothers in the 1930s. Carothers had studied aliphatic polyesters prior to focusing his attention on the aliphatic polyamides (nylons). In fact, he obtained patents on several aliphatic polyesters, but they were not suitable for making fibers (the melting points were too low and they dissolved readily in common organic solvents). However, in 1941 two English chemists, J. R. Whinfield and J. T. Dickson, working in the laboratories of the Calico Printers' Association, noted that DuPont had not included aliphatic/aromatic polyesters in their patents. They discovered that if they polymerized ethylene glycol and terephthalic acid, it produced a polymer, PET, that could be melt-spun into fibers having a melting point of 240°C. Although WWII delayed development for some 5 years, Whinfield and Dickson obtained patents for PET in 1946. Calico Printers licensed the product to ICI, who, in 1950, started production of Terylene®. DuPont, who had now realized the potential of PET fibers, obtained exclusive rights to produce it in the United States, calling it Dacron®. PET has since become the dominant polyester produced and it finds application in clothing (the dreaded Dacron® polyester leisure suits made infamous in the "disco" years), in Mylar® films (overheads) and in the myriad bottles that contain soda, mineral water and other liquid refreshments. PET is manufactured by condensation polymerization using a process of ester interchange from dimethylterephthalate and ethylene glycol. It is a glassy polymer at room temperature (T_g about 80°C) and, under favorable conditions, it can crystallize (T_m about 265°C).

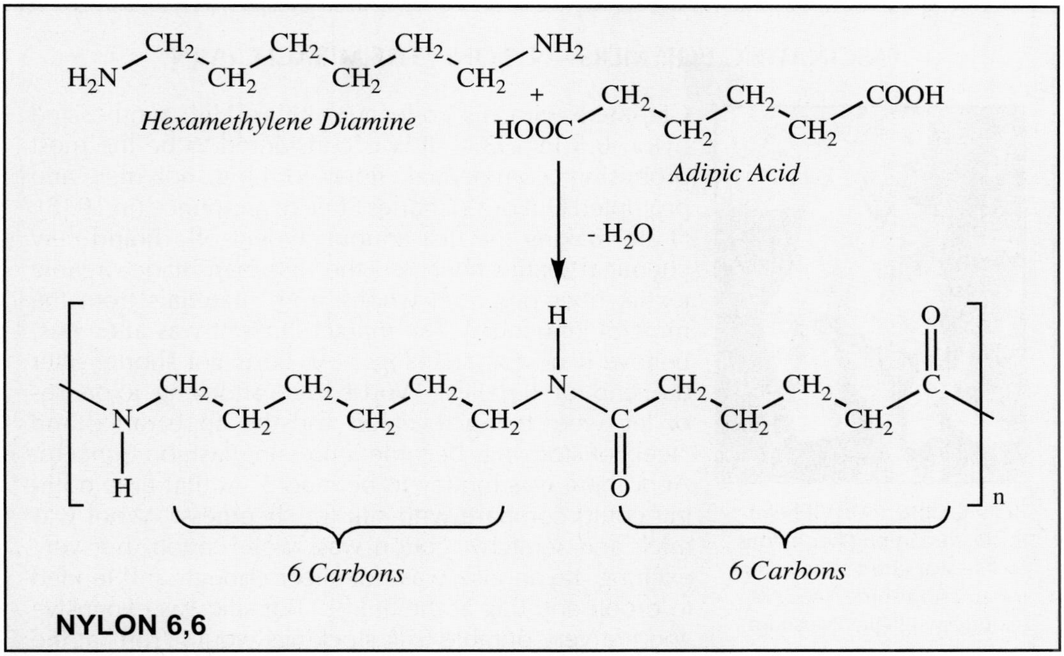

FIGURE 3-9 Formation of nylon 6,6.

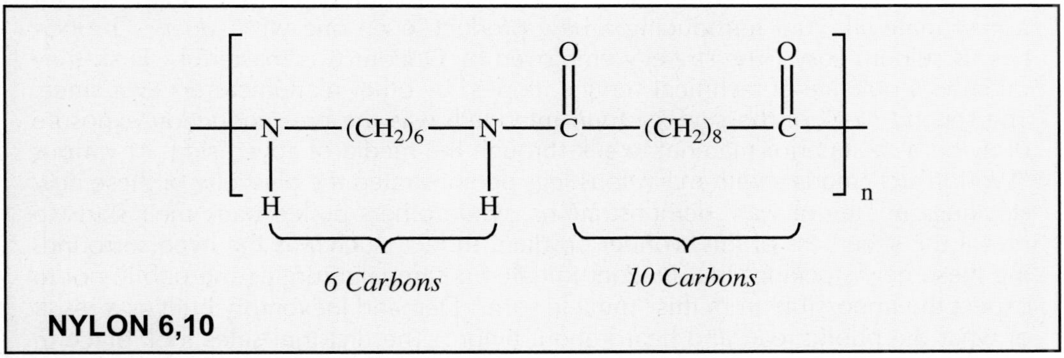

FIGURE 3-10 Chemical repeating unit of nylon 6,10.

may be achieved by step-growth polymerization from an appropriate A-B amino-acid, NH_2-R-COOH. However, nylon 6, a commercially significant material, is usually polymerized from the cyclic monomer caprolactam by ring-opening polymerization.

Another important aspect of step-growth polymerizations is that, in addition to driving the reaction to high degree of conversions, it is also necessary to have precisely equivalent amounts of the monomers (you will see why when you study the statistics of these reactions). This is not always easily accom-

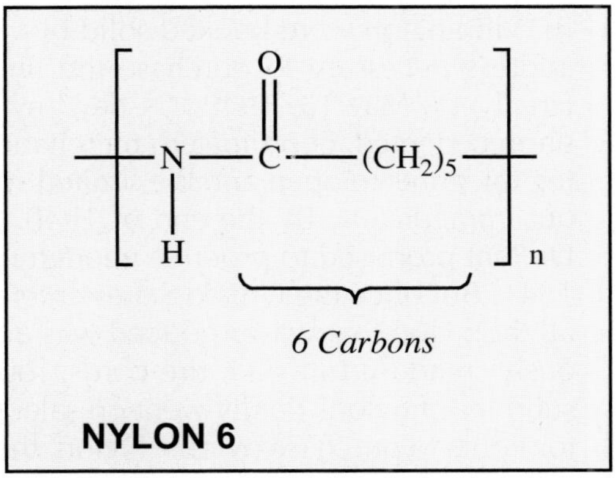

FIGURE 3-11 Chemical repeating unit of nylon 6.

FASCINATING POLYMERS—NYLON: "THE MIRACLE YARN"

Betty Grable, a WWII heart-throb, auctioning her nylons for the war effort—they went for an astonishing $40,000 (Courtesy: Hagley Museum).

Carothers with his colleague, Julian Hill, synthesized nylon 6,6 in 1935. It was considered to be the most promising commercial fiber forming polymer and prompted DuPont's Charles Stine to announce (in 1938): "I am making the first announcement of a brand new chemical textile fiber . . . the first man-made organic textile fiber prepared wholly from materials from the mineral kingdom." The market DuPont was after was, believe it or not, stockings. As skirts got shorter after the end of the First World War, shocking expanses of leg were being revealed and the appearance and "feel" of stockings became a pressing fashion concern. And, there was money to be made! At that time nothing could compare with silk for sheerness. Wool was thick and scratchy; cotton was, well, cotton, not very exciting; rayon also was not sheer enough and tended to droop and bag at the ankles. But silk was expensive and not very durable (silk stockings would "run" at the slightest provocation). Nevertheless, about 1.6 million pairs of silk stockings were being purchased a day in the United States alone. The market was therefore ripe for a new material. But introducing a new product, even one with superior properties, is seldom easy. The strategy employed by DuPont was masterful. First, they initiated a practice of technical service, copied by other manufacturers ever since. The second "leg" of the strategy (pun intended) was the pre-production exposure of nylon as a superior material to silk through the media of advertising. At various "World Fairs" models with marvelous legs demonstrated the elasticity of these new stockings in "tug-of war" demonstrations, while others pulled back their skirts to reveal the sheerness of this wonder product. In fact, such was the hype surrounding these new stockings that DuPont initiated a campaign urging the public not to expect the impossible from this "miracle yarn." Demand inexorably built as a result of what the public read and heard about nylon. The first trial sales took place in Wilmington, Delaware, in 1939. At Braunstein's department store the crowds were so large they pushed back the counter, pinning saleswomen to the wall. Hotel rooms in Wilmington were booked solid by wealthy women seeking to establish the local address necessary to purchase the limit of three pairs of stockings per customer. Finally, on May 15, 1940, "N-day," nylon stockings went on sale nationwide in an unprecedented, coordinated, merchandising event. Crowds lined up for hours waiting for stores to open and the limited supply of three-quarters of a million pairs sold out immediately. By the end of 1940, 36 million pairs of "nylons" had been sold. DuPont promised to produce enough nylon to supply 10% of the hosiery market in 1941. But once the United States entered World War II, all this changed. Practically all the nylon that was produced was acquired by the military for parachutes (previously made from silk), tire cords, ropes, and so on. After the war, when limited supplies of nylons finally went on sale, pandemonium ensued with people lining up for hours to purchase nylons. Nylon, the first true synthetic fiber, was just in the right place at just the right time with just the right properties.

plished on an industrial scale. One trick that is possible with nylons is to convert the acid and amine (which is a base) to salts, which will precipitate out of solution in the form of 1:1 complexes (Figure 3-12).

"Condensation" reactions can involve the splitting out of molecules other than water. For example, nylon 6,6 can be made from hexamethylene diamine and adipoyl chloride (instead of adipic acid). In this case a molecule of HCl is split or "condensed" out. There is a neat trick you can perform with this system that is commonly called "The Nylon Rope Trick." The acid chloride will dissolve in an organic solvent, such as chloroform, while the diamine will dissolve in water. These two solutions do not want to mix and when carefully added to a beaker they form a phase-separated system. Polymerization can then occur at the interface between the phases (an interfacial polymerization), as illustrated in Figure 3-13.

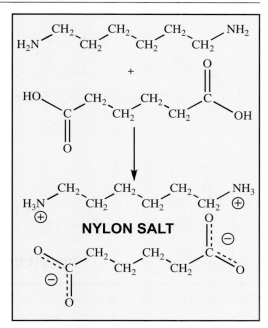

FIGURE 3-12 Nylon salt formation.

Polyurethanes

Some step-growth polymerizations involve reactions where no small molecules (like HCl or H_2O) are split out at all, as in the synthesis of polyurethanes, shown schematically in Figure 3-14. Isocyanates are very reactive and the reaction between isocyanates and aliphatic hydroxyl groups is extraordinarily fast, as it is with amines, carboxylic acids and many other groups, including water. This latter reaction is particularly convenient, as CO_2 can be produced in-situ (Figure 3-15) and used to produce polyure-

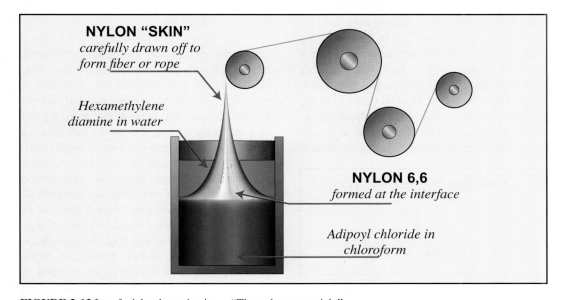

FIGURE 3-13 Interfacial polymerization—"The nylon rope trick."

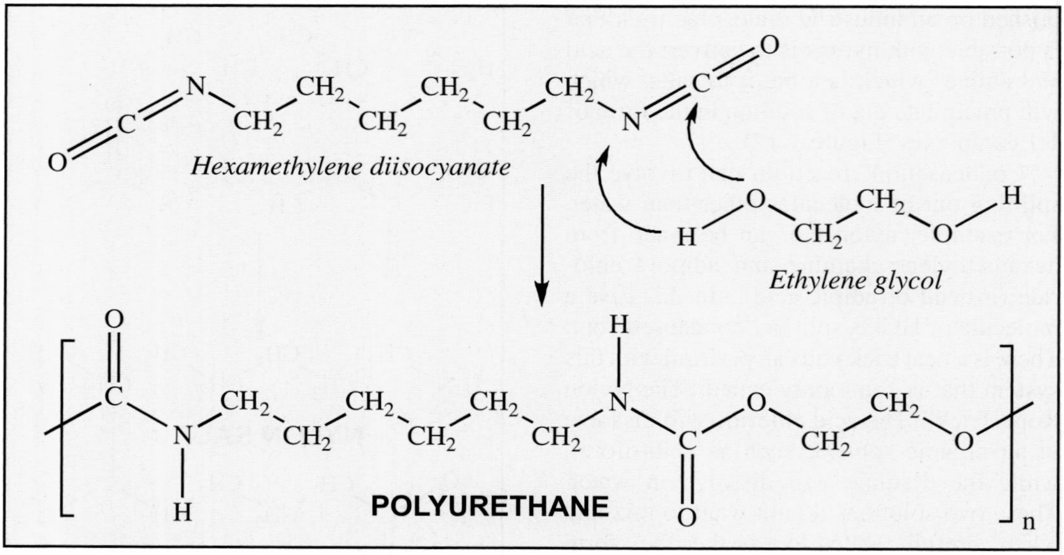

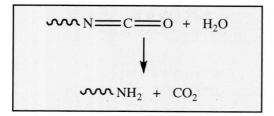

FIGURE 3-14 Formation of polyurethane.

thane foams. The rapid reaction rate is the key to the versatility of polyurethane chemistry and serves to distinguish it from other conventional step-growth polymerizations.

Isocyanates will aggressively seek out hydroxyl groups, even if they are present in only small concentrations, such as end groups of a "prepolymer." Thus block-type copolymers can be made. Think, for example, of what type of PU foam you would get if you started with a system consisting of a diisocyanate and a poly(propylene oxide) prepolymer or "polyol" that had, say, an average degree of polymerization of 40 ($n = 40$)—see Figure 3-16. Now compare this with one made from the simple propylene glycol $HO–CH(CH_3)CH_2–OH$ ($n = 1$). In the former, each relatively stiff urethane group would be separated by a chain of about 40 very flexible propylene oxide segments and

the foam would be quite rubbery. On the other hand, in the latter foam the urethane groups are separated by only one propylene segment, and the resultant foam would be much more rigid. This is an over-simplification, but it should be obvious that one can "dial in" desired mechanical properties by changing the chemistry and size of the blocks.

Transesterification

Clearly, classifying the above reactions as step-growth polymerizations rather than condensation polymerizations provides a more apt description. It turns out that even in those reactions that involve the splitting out of a molecule of water, as in the synthesis of polyesters, things can get more complicated, in this case, because of transesterification (or ester interchange) reactions. An example is the synthesis of polyethylene terephthalate, or PET, the stuff most people think of as "polyester," because it is used to make things one comes across every day, such as those ubiquitous plastic water bottles or items of clothing. The first step in the reaction is shown in Figure 3-17 which is the simple esterification of dimethyl terephthalate using

$$\sim\!\!\sim\!\!N\!\!=\!\!C\!\!=\!\!O \;+\; H_2O$$

$$\downarrow$$

$$\sim\!\!\sim\!\!NH_2 \;+\; CO_2$$

FIGURE 3-15 Reaction of isocyanate with water to yield an amine and CO_2.

POLYMER MILESTONES—OTTO BAYER

At the ripe old age of 31, Otto Bayer, who just happened to have the same surname as the famous German company that he worked for, Farbenfabriken Bayer A.G., became the head of their Central Scientific Laboratory in Leverkusen. This was in 1934 and Otto Bayer's initial focus was on dyestuff chemistry, a field in which German chemists excelled. But he realized that the future of the company required diversification into new areas of industrial chemistry. Crop protection and the emerging field of macromolecular science appeared to offer considerable potential. Otto Bayer was aware of the work of Staudinger and Carothers. He was particularly impressed with the latter's discovery of polyamides and its implications for the fiber and textile industries. He knew that in the condensation (linear step-growth) polymerization used to prepare nylon and polyesters a small molecule, usually water, is formed and has to be removed.

Otto Bayer (Courtesy: Bayer Corporation).

Bayer reasoned that a superior reaction would be one in which there was no small molecule produced. During his Ph.D. organic chemistry studies, he had used a test for amines that involved forming urea derivatives with phenyl isocyanate. Could he form polymers using similar reactions with compounds that contained two isocyanate functionalities? When he suggested this to his superior, he elicited the response, "You don't seem to be the right man to run the laboratory!" First of all, simple monofunctional isocyanates were prepared at that time by the reaction of an amine with phosgene. (Phosgene is that miserable poisonous gas, $COCl_2$, used as a weapon in the trenches during WWI. So, to state the obvious, it was "environmentally challenging.") One of Bayer's closest colleagues remarked: "If you had ever tried to make a mono-isocyanate, you wouldn't have come up with the mad idea of trying to produce diisocyanates. The reaction of phosgene on diamines would result in almost anything except diisocyanate, and would probably be thoroughly unstable, anyway." Others retorted: "It will explode!" To say the least, management was not impressed. The project, with such obvious insuperable difficulties, appeared futile. But Otto Bayer was to prove them wrong! Notwithstanding the reservations that Otto Bayer's colleagues had about diisocyanates, Heinrich Rinke, who worked with Bayer, succeeded in synthesizing them and also a viscous polymeric material from which filaments could be drawn—an encouraging development that led to the original polyurethane patent registered in 1937. The Bayer scientists discovered that toluene, for example, could first be nitrated to yield a mixture of 2,4- and 2,6-dinitrotoluenes. In a subsequent reduction, the dinitrotoluenes were then converted into the corresponding diaminotoluenes. The final step was the reaction of phosgene with the diaminotoluenes to yield the 2,4- and 2,6-toluene diisocyanates (TDI). So it turns out that making diisocyanates, while still formidable on an industrial scale, was not as difficult as originally surmised (although the efficiency with which the plant and production engineers at Bayer were able to do this still amazes your authors).

FASCINATING POLYMERS—SPANDEX ELASTOMERIC FIBERS

Lycra® in action! (Courtesy: DuPont)

While we are on the subject of segmented polyurethanes, we would be remiss if we did not mention the incredible elastomeric fibers that are produced under the generic name, spandex. These are the elastic fibers that are responsible for "comfort stretch" garments like panty hose, lingerie, girdles, hosiery, form-fitting tights and skimpy bathing suits. In the early 1950s, both the DuPont and Bayer companies were actively involved in research designed to commercialize an elastomeric fiber based upon urethane chemistry. First, it was necessary to define the chemical, physical and mechanical properties that would be required of the segmented polyurethane elastomer, optimize the composition to obtain these properties, then develop an appropriate fiber spinning process. DuPont's research led to the development of a polyether/polyurea urethane which they called Lycra®. Bayer's research was along very similar lines and they developed a polyester/polyurea urethane named Dorlastan®. These two products hit the market in the early 1960s and the world hasn't looked the same since. Let's briefly see what was involved in producing the original Lycra® and Dorlastan® polymers. New developments are not created in a vacuum. Basic knowledge drawn from related fields is often crucial and the development of spandex was no exception to this rule. The key ingredients were in place in the early 1950s: the contributions of Otto Bayer and his colleagues to the chemistry of diisocyanates and polyurethanes, the contributions of Wallace Carothers and others to polymer chemistry and the advances in rubber (especially vulcanization) and synthetic fiber spinning technologies. Scientists at DuPont and Bayer arrived at similar ideas and focused on making a flexible prepolymer (the soft segment) with isocyanate end groups that could be chain-extended using a multifunctional amine (the hard segment). DuPont settled on a soft segment of poly(tetramethylene oxide glycol) which they end-capped with a diisocyanate. Bayer's soft segment was based upon a diisocyanate-capped aliphatic polyester. Chain extension and the development of the hard segment was originally achieved using hydrazine (DuPont) or a dihydrazide (Bayer). Many improvements have been made since the introduction of the original Lycra® and Dorlastan® polymers and many other manufacturers have since entered the market. One of your authors has had the opportunity to visit DuPont's Lycra® plant in Waynesboro, VA, and observe close-up how spandex fibers are produced industrially. The simple description of segmented polyurethane/ureas doesn't even come close to describing the complexity of the industrial processes involved in forming elastomeric spandex fibers. It is truly impressive how the scientists, technologists and engineers have overcome numerous problems and can rapidly spin complex spandex fibers of different morphologies and deniers. With so many variables, the efficient spinning of elastomeric fibers is as much an art as it is a precise science.

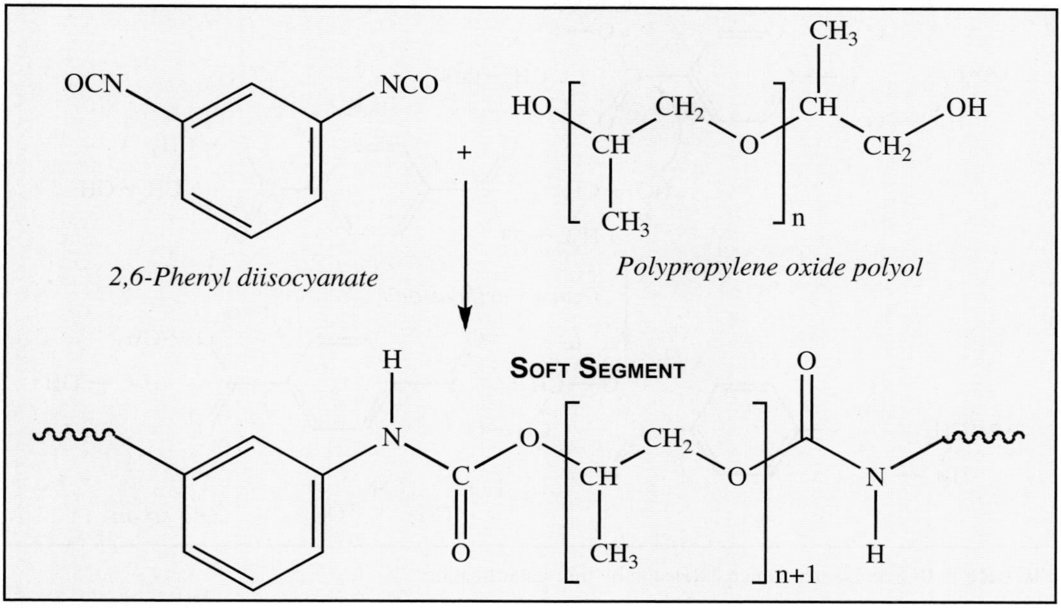

FIGURE 3.16 Formation of a block polyurethane copolymer.

two equivalents of ethylene glycol, with the elimination of methanol. The next chain building steps involve a successive process of transesterification with the elimination of ethylene glycol as illustrated in Figure 3-18.

Network Formation

We have seen how monofunctional molecules react once to give a slightly larger molecule and bifunctional molecules react

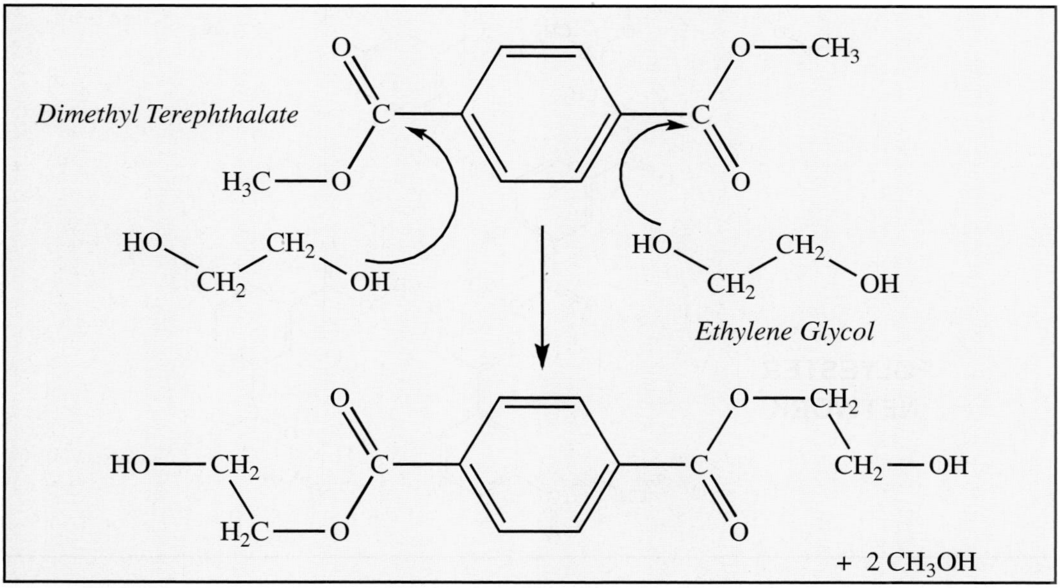

FIGURE 3.17 Initial step: esterification of dimethylterephalate.

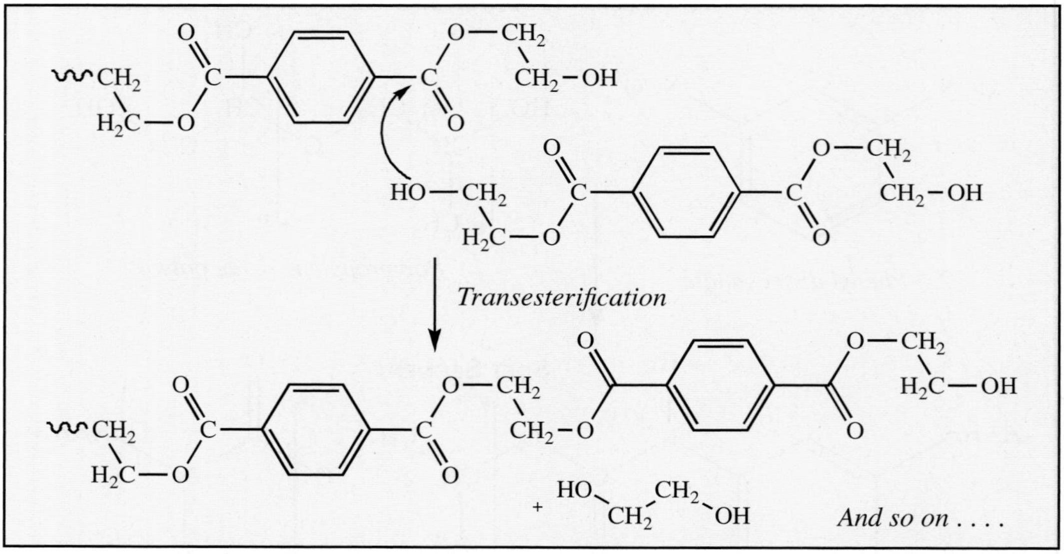

FIGURE 3-18 Second step: chain extension by transesterification.

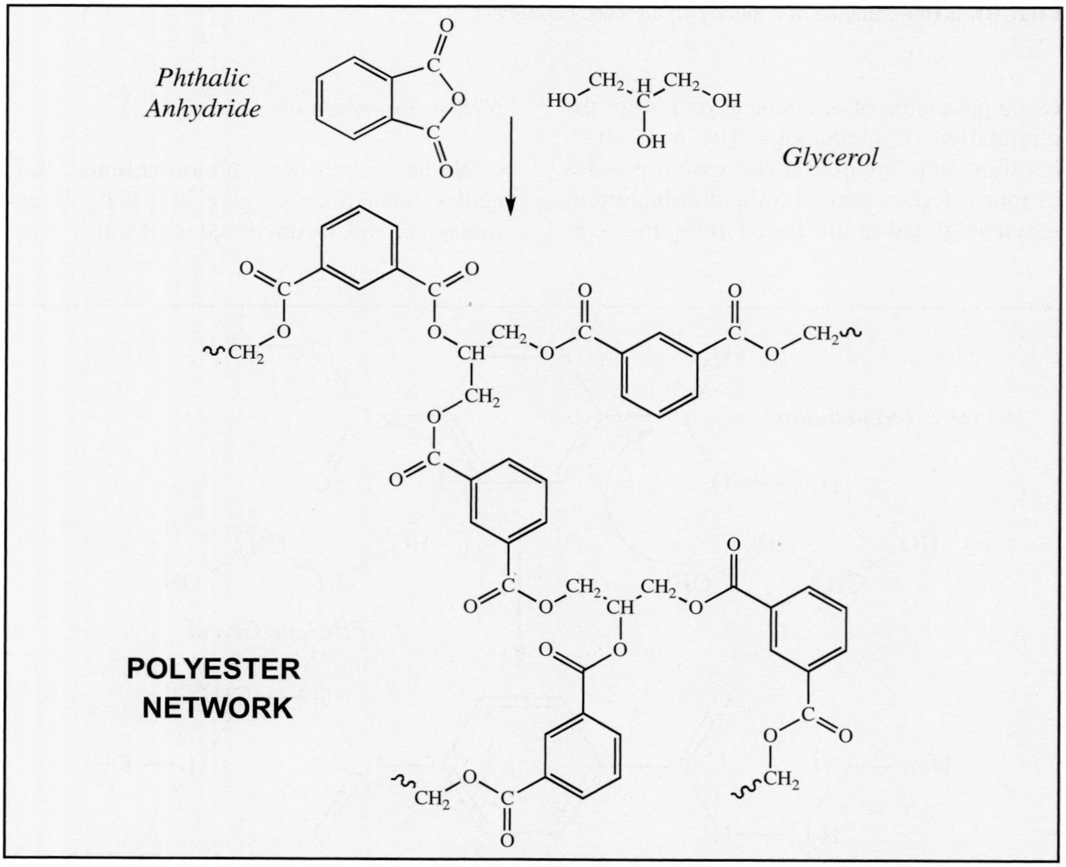

FIGURE 3-19 Typical part of a polyester network formed from phthalic anhydride (difunctional) and glycerol (trifunctional).

to give chains. How would you make chains that branch and then, perhaps, interconnect to form networks? The answer of course is to use multifunctional ($f > 2$) monomers. We actually gave an example in our preliminary discussion of chain architecture, when we discussed the reaction of phenol with formaldehyde (see Chapter 2)—a reaction that Baekeland took advantage of in the development of Bakelite materials. Here we will use as an example (Figure 3-19) a polyester network, such as that formed by the reaction of glycerol and phthalic anhydride (PA) that is the basis of alkyd resins that have been used extensively in the surface coatings industry.

That's our quick and dirty look at step-growth polymerization. The crucial feature of this type of reaction is the slow build-up of chains in a step-wise process. Now let's take a closer look at addition polymerization.

CHAIN OR ADDITION POLYMERS

The central feature of chain or addition polymerization is that there is an active site at the end of a growing chain where monomers are added sequentially, one by one. At any particular time in the reaction vessel, there are usually some fully formed and no longer reacting polymer chains, unreacted monomer and, finally, a very small number of chains "caught" in the process of growing or polymerizing, as depicted in Figure 3-20.

The types of monomers that can be polymerized by a chain mechanism are illustrated in Table 3-1. These either have "unsaturated" (double or triple) bonds or are cyclic. The letters "X" and "Y" in some of these structures are used to represent various types of atoms or functional groups (e.g., $-CH_3$, $-Cl$, a benzene ring, etc.)

In our simplistic review of some of the chemistry associated with chain polymerization (Figure 3-6), we observed that these types of molecules were capable of polymerizing because you can "open up" one of the bonds on, say, a vinyl monomer, and link the monomers to form a chain. There are two things you should note about this. First, this represents just one type of chain polymerization. Second, there are other parts to the reaction we have to consider: how it gets started (initiation), how it finishes (termination) and what side reactions can occur (chain transfer). Let's start by illustrating two other types of chain polymerizations.

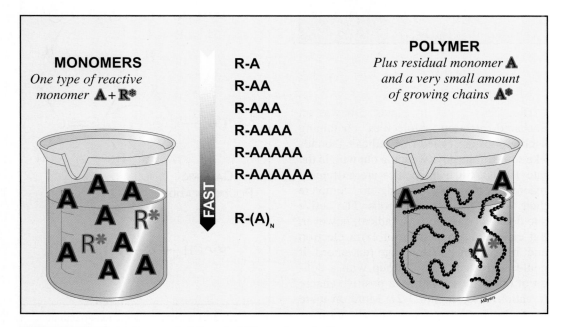

FIGURE 3-20 Schematic representation of addition polymerization.

TABLE 3-1 MONOMERS THAT CAN BE POLYMERIZED BY CHAIN POLYMERIZATION

MONOMER TYPES
Olefin, Vinyl & Vinylidine
Various Dienes
Acetylene
H—C≡C—H
Cyclic

Free Radical Polymerization

A free radical polymerization can be initiated in a number of ways. One of the most common is to use an initiator such as a peroxide (a molecule containing O–O single bonds) that is relatively stable at room temperature but readily cleaves to give radicals when heated or irradiated (Figure 3-22). This is called homolytic scission, although, if you're an engineer, that is probably information overload. *Initiation* consists of two steps, the formation of the primary radical followed by the addition of this radical to a monomer.

Then we're off to the races! *Propagation* proceeds like a chain reaction, with the

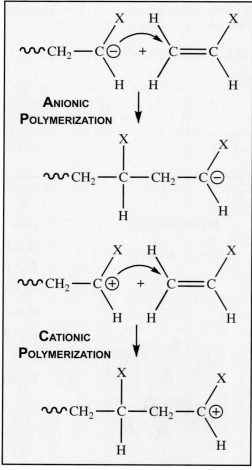

ANIONIC POLYMERIZATION

CATIONIC POLYMERIZATION

FIGURE 3-21 Schematic depiction of anionic (top) and cationic (bottom) polymerization.

The "active site" in Figure 3-6 was an unshared electron. Molecules containing such groups are called free radicals (sounds like some of the guys we hung out with in the late 1960s). Such molecules are extremely reactive, constantly seeking to complete their "outer shell" of electrons. The active site does not have to be a radical, however, but can be a group with an extra electron and, hence, a negative charge (an anion). It could equivalently be a group with a deficit of one electron, hence, a positive charge (a cation)—see Figure 3-21. More on these later. But let's first focus our attention on free radical polymerization.

addition of monomers one at a time as they collide and react with the free radical site at the end of the growing chain. The reaction mechanism essentially involves opening the bond of the monomer and the formation of a new radical species at the end of the chain (Figure 3-23).

So, we have a bunch of growing chains reacting with monomers which (in a batch reaction) gradually get depleted. Obviously, the ends of two growing chains could also meet as a result of their random motions. What do you think would happen then? Yes, they could react with one another, terminating each chain. Termination can occur by two mechanisms, however (Figure 3-24). The first is simple combination, where the radicals collide and form a new covalent bond, joining the two original chains to make one larger one. The second mechanism is called disproportionation (who thinks up this stuff?) and involves the transfer of a proton from one chain to another with the corresponding rearrangement of electrons.

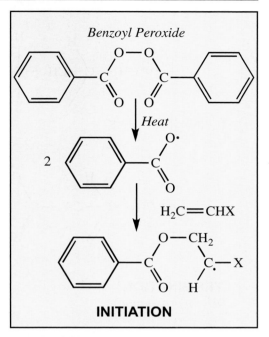

INITIATION

FIGURE 3-22 Generation of free radicals from benzoyl peroxide followed by reaction with the first monomer (initiation).

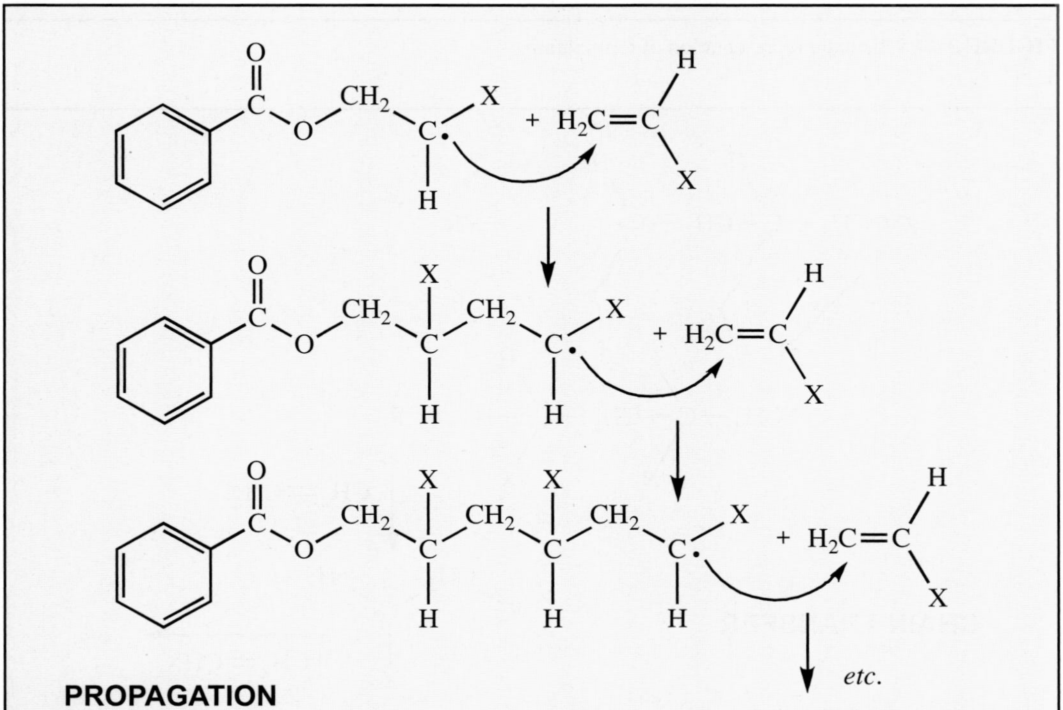

PROPAGATION

FIGURE 3-23 Schematic representation of propagation.

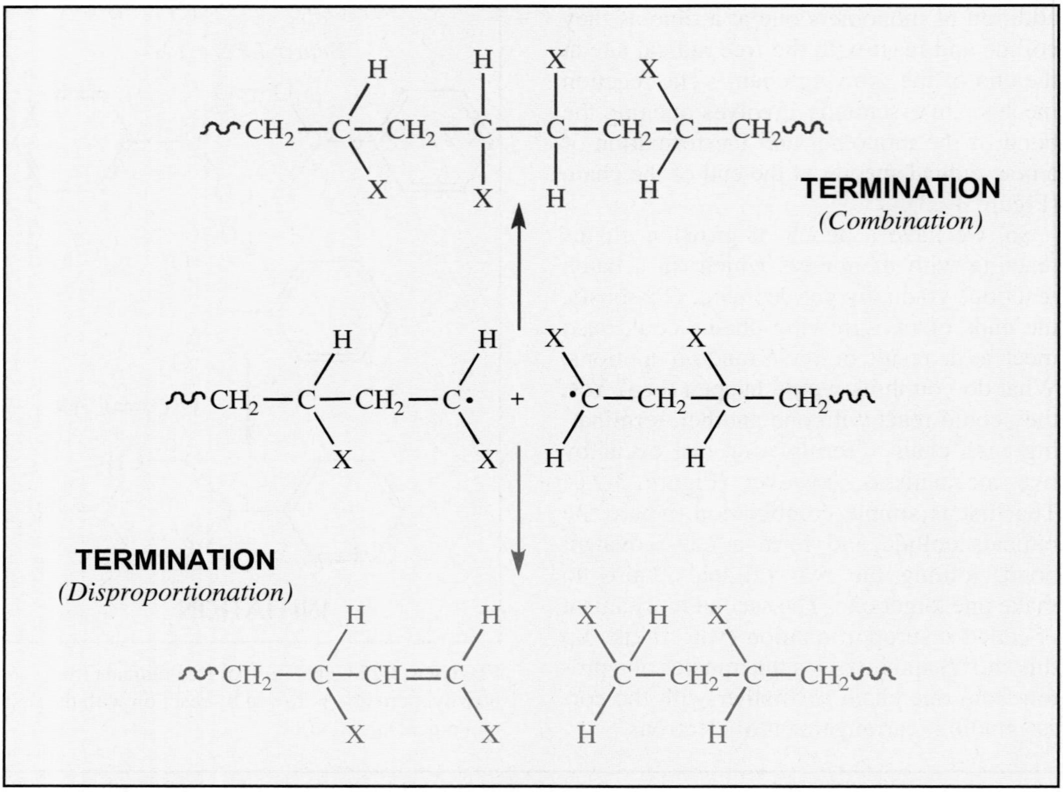

FIGURE 3-24 Schematic representation of termination.

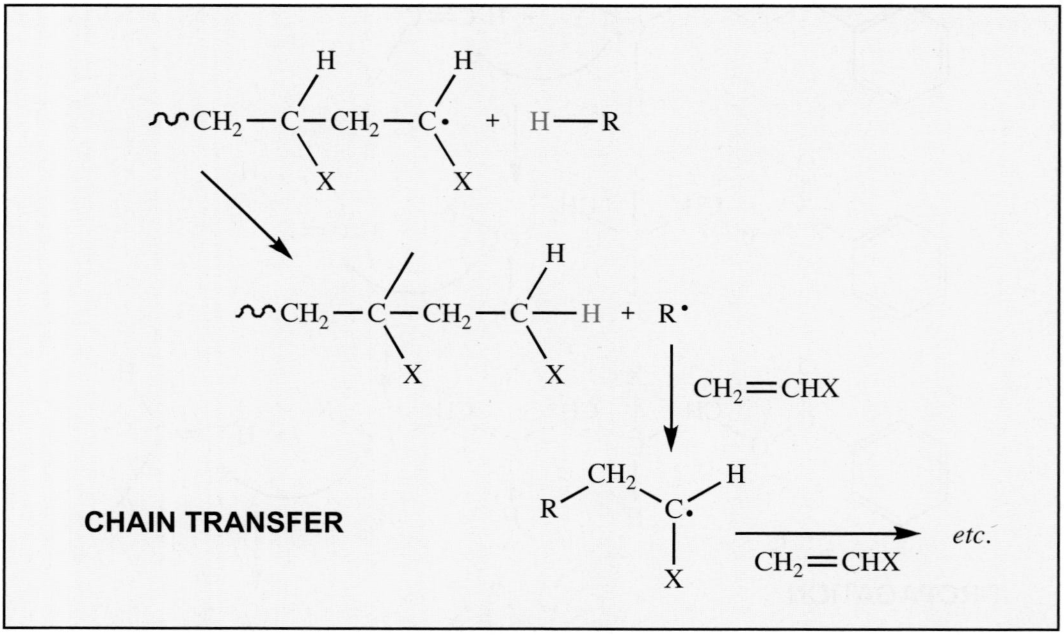

FIGURE 3-25 Schematic representation of termination.

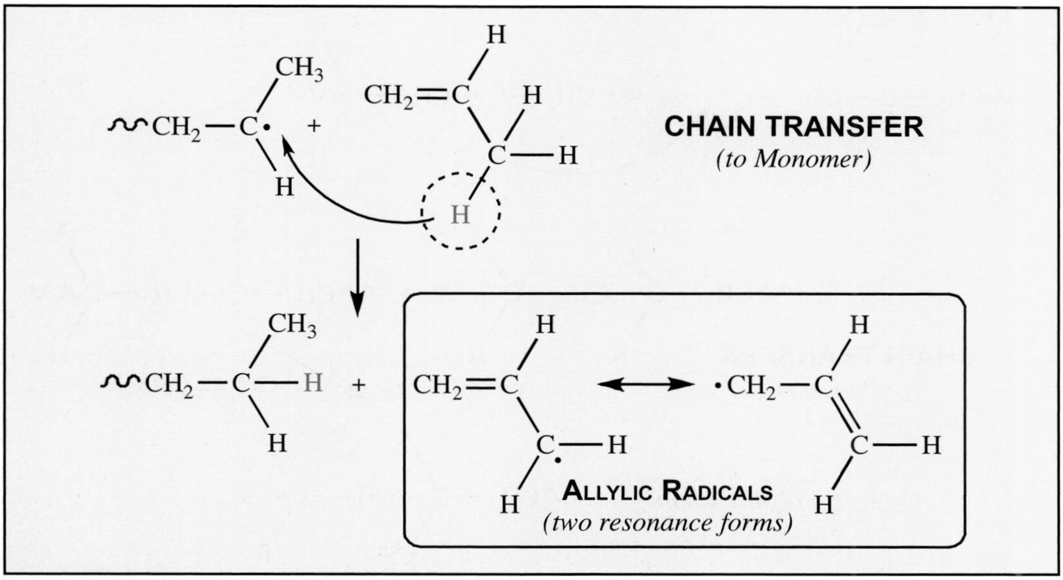

FIGURE 3-26 Formation of allylic propylene radicals.

Radicals are reactive beasts and in addition to propagating and terminating the active site at the chain end they can engage in reactions with solvent, monomer, initiator, or even other chains. This process is called *chain transfer*, because the process "caps" the growing chain, usually (but not always) by saturating it with a proton, but at the same time generates another radical that, if it is capable of adding a monomer, can start a new chain growing (Figure 3-25). A particularly interesting case of chain transfer to monomer involves so-called allylic monomers (i.e., those with the structure $CH_2=CH–CH_2X$) such as propylene (where $X = H$). The hydrogens α to the double bond (those on the next carbon atom along) readily transfer to a radical species, forming an allylic radical that has a high resonance sta-

bility (Figure 3-26). These radicals, in turn, can terminate by reacting with each other or a propagating chain (Figure 3-27). The rate of chain transfer relative to propagation is such that chains of any length never have a chance to form (we call this *inhibition*). As a result, α-olefins, such as propylene, cannot be produced by conventional free radical polymerization.

Chain transfer can also occur to polymer molecules, resulting in the formation of a branched molecule. *Long-chain branches* (Figure 3-28), as opposed to the short branches, will be considered next. This type of chain transfer becomes significant when the concentration of polymer is high (i.e., in the later stages of a batch polymerization), particularly when the polymer has a very reactive propagating radical (e.g., polyeth-

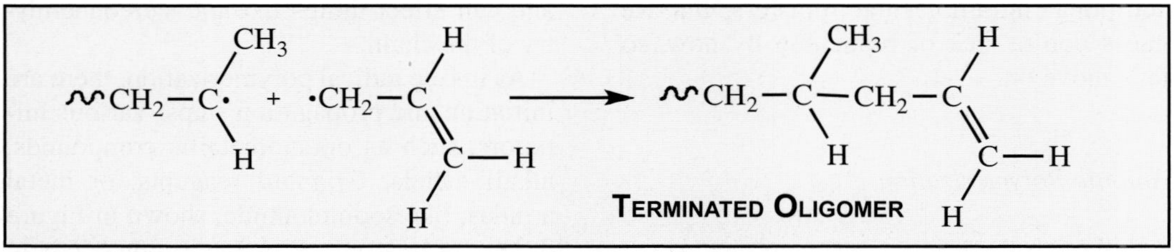

FIGURE 3-27 Formation of terminated propylene oligomer.

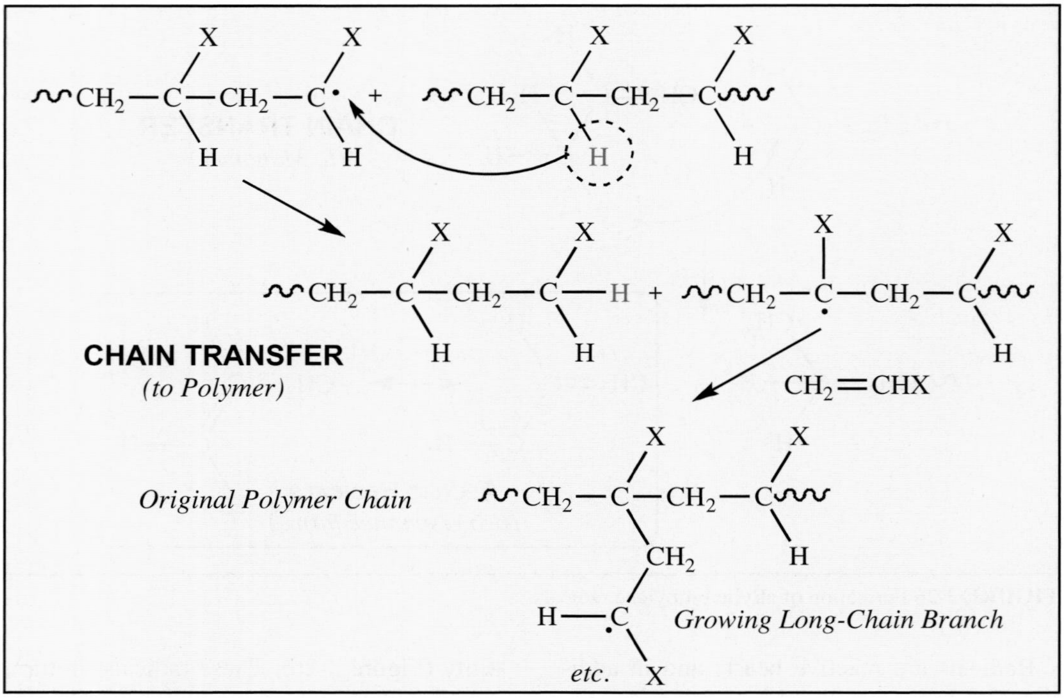

FIGURE 3-28 Formation of long-chain branches.

ylene, poly(vinyl acetate), poly(vinyl chloride), etc.).

In addition to long branches, free radically polymerized polyethylene also has *short-chain branches* produced by a process called backbiting. This is an intramolecular chain transfer reaction that occurs locally among groups near the end of the chain. The formation of a butyl branch is illustrated in Figure 3-29. Butyl branches are thought to predominate because of a (stable) six–membered ring transition state, but amyl and hexyl branches are also formed, where hydrogens are abstracted from the sixth and seventh methylene groups from the radical end. Ethyl branches are also formed from additional intramolecular transfers, but we guess you're sick of branching by now, so let's move on.

Anionic Polymerization

Although there is an unshared electron at the end of a growing chain in free radical polymerization, there is still an equal number of protons and electrons, hence, overall charge neutrality. Certain monomers can be polymerized anionically, however, such that there is an excess of one electron at the active site, hence, a negative charge (see the definition in the box opposite if your confused about anodes and anions). If the active site is a carbon atom, as shown previously, then this charged species is called a carbanion. Shown in Figure 3-30 is an oxanion, formed in this case during an anionic ring-opening polymerization. Of course, there is a positively charged counterion hanging around somewhere and the separation between the two depends upon the nature of the solvent and can affect things like the stereochemistry of the chain.

As in free radical polymerization, there are initiation and propagation steps. Various initiators, such as organometallic compounds, alkali metals, Grignard reagents, or metal amides, like sodium amide, shown in Figure 3-31, can be used. Propagation proceeds in the usual manner, but there is no termination

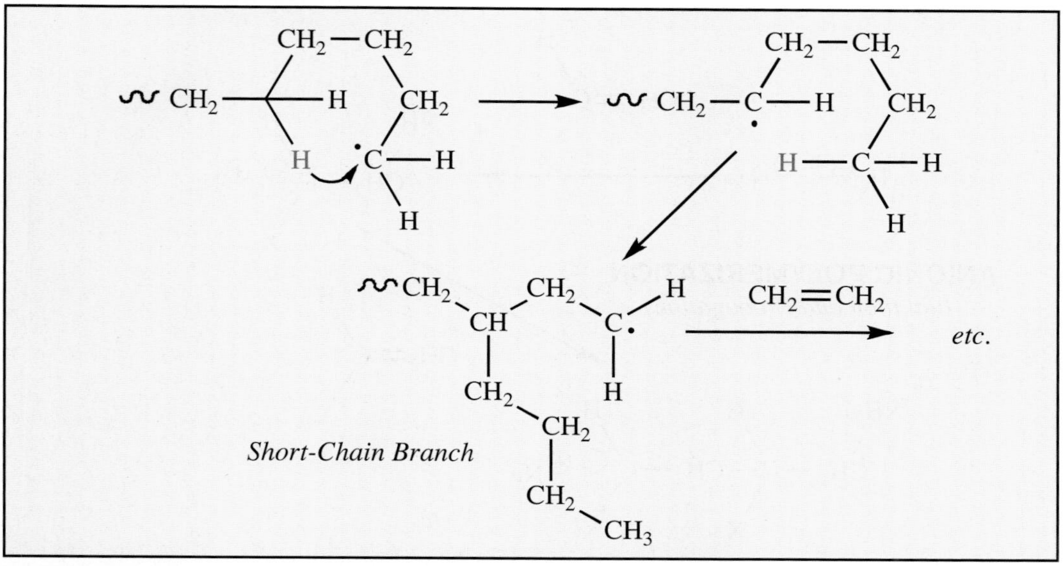

FIGURE 3-29 Backbiting—formation of short-chain branches.

of the type that occurs when free radicals collide. (Why not?)

If a solvent that is able to release a proton is used, however, it can react with the active site. Ammonia is an example of such a protic solvent and the reaction results in the formation of a negatively charged NH_2 ion, which can initiate the polymerization of a new chain. In other words, we have chain transfer to solvent (Figure 3-32). What do you think would happen if we used an inert or non-protic solvent (one that does not readily release a proton)?

To answer this question, let's consider the polymerization of styrene initiated by metallic sodium in an "inert" solvent in which there are no contaminants (i.e., there are no molecules with active hydrogens around). This polymerization is also interesting because anion radicals are initially formed. These quickly combine to form a dimer that propagates from both ends (Figure 3-33). (Do you get a glimmer of an idea of how it might be possible to make a "star" polymer?) Then, if there is nothing for the anion to react with, there is no termination. Combination with the counterion occurs in only a few instances; the ions hang around one another and their attractions are mediated by

On Anodes and Anions

Those of you who have done some electro-chemistry may be confused by the fact that there is a negative charge on an anion, because in an electrolytic cell the anode is positively charged. But an anion is then a species that migrates towards the anode and must therefore be negatively charged. (Confusingly, in other types of cells the anode can be negative, the definition of an anode is actually in terms of an oxidation site—aren't you glad we told you that?)

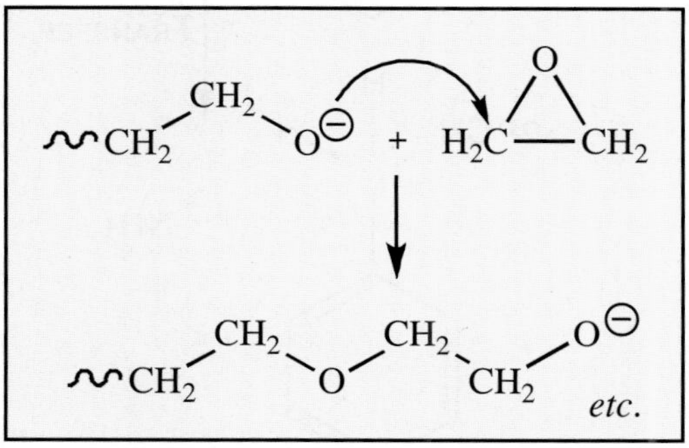

FIGURE 3-30 Schematic representation of the anionic ring opening polymerization of ethylene oxide.

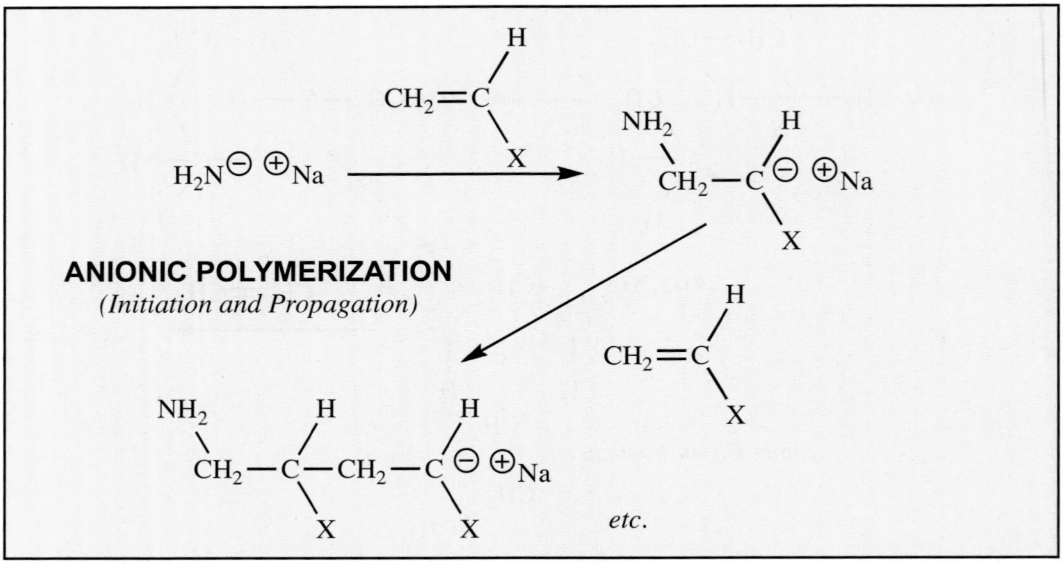

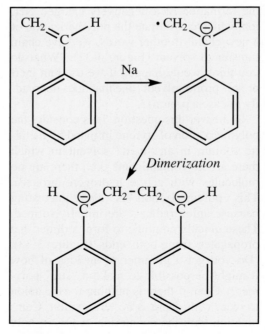

FIGURE 3-31 Schematic representation of initiation by sodium amide and subsequent propagation for anionic polymerization.

solvent. This allows the synthesis of block copolymers. Because the active site stays "alive,",one can first polymerize styrene, for example. Then, when the styrene has been effectively used up, butadiene can be added and the polymerization will restart. A third block of styrene can then be polymerized, making the type of triblock copolymer that is used as thermoplastic rubber (Figure 3-34). Living polymers are not eternal, however, and one cannot keep adding blocks forever.

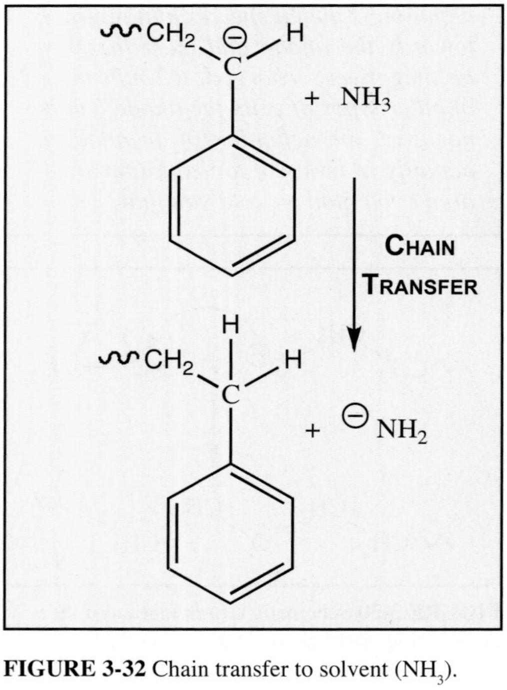

FIGURE 3-32 Chain transfer to solvent (NH_3).

FIGURE 3-33 Formation of a styryl anion radical and subsequent dimerization.

POLYMER MILESTONES—MICHAEL SZWARC

Michael Szwarc was born (1909) and educated in Warsaw, Poland and received a chemical engineering degree at the Warsaw Polytechnic in 1932. He subsequently emigrated to Israel and obtained a Ph.D. in organic chemistry from Hebrew University in 1945. Then he moved to England and obtained Ph.D. (physical chemistry) and D.Sc. (1949) degrees from Manchester University in 1947 and 1949, respectively. He lectured at Manchester University and performed research primarily on the bond dissociation energies of polyatomic molecules. It was during this time that he discovered the spontaneous polymerization of $CH_2=C_6H_4=CH_2$ and this was the catalyst that led to his lifelong passion for polymer chemistry. In 1952, he accepted an appointment in

Michael Szwarc (Courtesy: SUNY College of Environmental Science and Forestry).

the United States as a Professor of Physical and Polymer Chemistry at the State University of New York (SUNY) in Syracuse. He initiated what was to become the Polymer Research Institute at SUNY and served as its Director until his retirement in 1979. While Professor Szwarc made many seminal contributions to the field of chemistry and will long be remembered for his contributions to the modern kinetics of polymerization, it was his discovery of living polymerization that elevates him into that special class of truly outstanding polymer scientists. In 1956, Szwarc announced his discovery of living anionic polymerization and, as they say in the classics, things were never the same. The kinetics of living polymerization were novel, and it was now feasible to produce polymers and oligomers with controlled narrow molecular weight distributions and well defined end-groups. Moreover, it was also possible to use living polymerization to produce a wide variety of unique, well controlled, block and star copolymers, including the highly successful thermoplastic elastomers that we now take for granted and use in everyday life. Szwarc's ideas made polymer chemists take a new look at the fundamentals of polymerization and it was not long before living polymerization was extended first to cationic polymerization and then later even to free radical polymerization. Michael Szwarc published three books that were concerned with carbanions and living polymers, some 500 papers and won many prestigious awards (too numerous to mention here). He died in 2000 at the age of 91 in San Diego, California.

A knife handle made of Kraton® which is a block copolymer of styrene and butadiene that is made by living anionic polymerization (Source: www.knifeoutlet.com).

FASCINATING POLYMERS—SUPER-GLUES

Harry Coover (Source: National Inventors Hall of Fame).

Super-glues, the superstar divas of modern adhesives, are currently manufactured by a number of different companies and marketed under such trendy trade names as Krazy Glue®, Black Max®, and many others. So what is super-glue and how was it discovered? The common ingredient in all super-glues is an alkyl cyanoacrylate monomer. The discovery (and rediscovery) of so-called super-glue is yet another case of the intervention of serendipity and the presence of a very observant scientist. Dr. Harry Coover, working in the Eastman Kodak Research Laboratories during WWII, was investigating the polymerization of various types of acrylic monomers, including alkyl cyanoacrylates, with the aim (pun intended!) of making an optically clear plastic for gun-sights. Cyanoacrylates showed some promise, but were abandoned because they had the unfortunate propensity to stick to everything! In 1950, Dr. Coover transferred to the Tennessee Eastman Company, supervising research on tougher, more heat-resistant acrylics that were being designed for jet canopies. Evidently a coworker, Dr. Fred Joyner (an appropriate name if ever we've heard one), wanted to determine the refractive index of ethyl cyanoacrylate and smeared some of this monomer between the two glass prisms that are put into the refractometer. You've guessed it; he glued them together and couldn't pry them apart! Bummer! But this was to lead Coover and his coworkers to realize that they may have a super-glue on their hands and developed what was called Eastman #910 adhesive.

Vinyl monomers that have electron-withdrawing substituent groups capable of stabilizing a carbanion through resonance or induction usually undergo facile anionic polymerization when initiated by a base. The actual structure of the monomer dictates the minimum strength of the base required to initiate polymerization. Most commercial anionic polymerization is performed using alkali metal initiators. However, vinylidene cyanide, $CH_2=C(CN)_2$, for example, polymerizes explosively when exposed to even a very weak base like water! This is the basic idea behind super-glue. By backing off a bit, replacing one –CN group with the less reactive –C(O)O– group, which produces a cyanoacrylate, we can moderate the polymerization reaction with water. The polymerization is still incredibly fast, but not explosive. So it is moisture in the air or on the surface of the item that you wish to glue together that initiates rapid polymerization of cyanoacrylates and super-glues your fingers together!

Alkyl cyanoacrylate

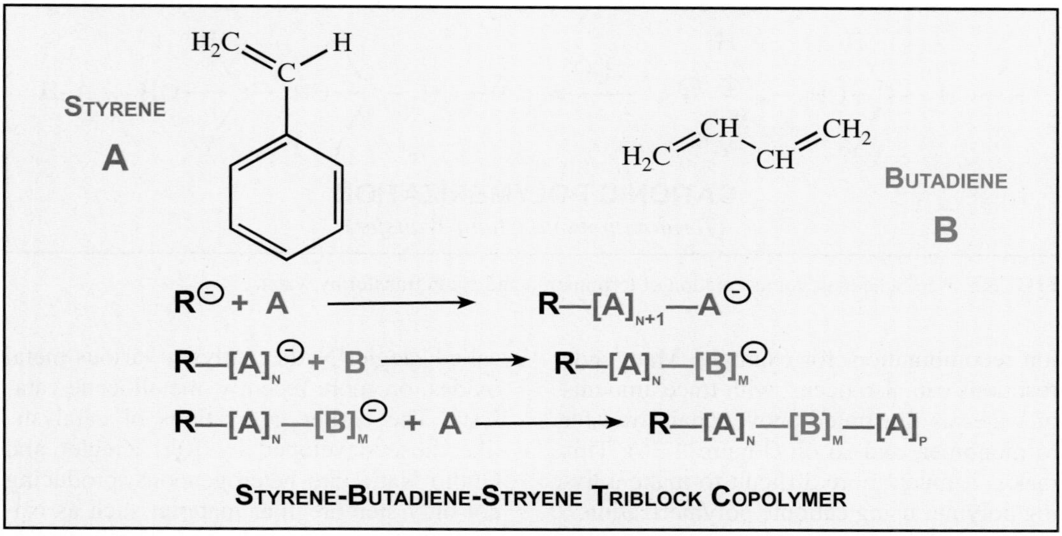

FIGURE 3-34 Schematic representation of the formation of a styrene/butadiene (SBR) triblock copolymer by living anionic polymerization.

Eventually, the anion will spontaneously terminate by mechanisms that are apparently not yet completely established. There are a lot more interesting things about anionic polymerization—the effect of polar groups, the fact that not all monomers can be used to make block copolymers, the ability to make certain polymers with very narrow molecular weight distributions, and so on—but these topics are for more advanced treatments, so now we will turn our attention to cationic polymerization.

Cationic Polymerization

As you, by now, have doubtless anticipated, cationic polymerizations involve an active site where there is a positive charge because, in effect, there is a deficit of one electron at the active site (Figure 3-21). Cationic polymerizations can be initiated by protonic acids (Figure 3-35) or Lewis acids (the latter sometimes combined with certain halogens). Unlike anionic polymerization, termination can occur, by anion–cat-

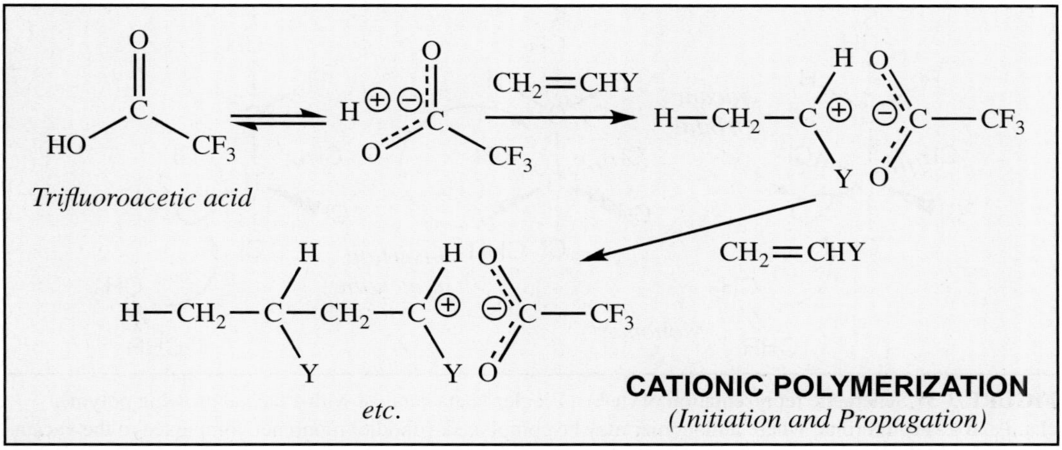

FIGURE 3-35 Schematic representation of initiation and propagation for cationic polymerization.

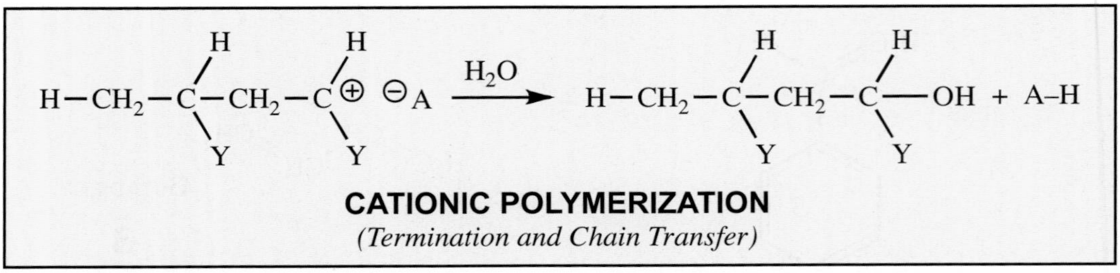

$$H-CH_2-\overset{\overset{\displaystyle H}{|}}{\underset{\underset{\displaystyle Y}{|}}{C}}-CH_2-\overset{\overset{\displaystyle H}{|}}{\underset{\underset{\displaystyle Y}{|}}{C}}\overset{\oplus}{} \ominus A \xrightarrow{H_2O} H-CH_2-\overset{\overset{\displaystyle H}{|}}{\underset{\underset{\displaystyle Y}{|}}{C}}-CH_2-\overset{\overset{\displaystyle H}{|}}{\underset{\underset{\displaystyle Y}{|}}{C}}-OH + A-H$$

CATIONIC POLYMERIZATION
(Termination and Chain Transfer)

FIGURE 3-36 Schematic representation of termination and chain transfer by water.

ion recombination, for example. Many side reactions can also occur; with trace amounts of water as illustrated above, chain transfer to monomer, and so on (Figure 3-36). This makes it much more difficult to make a living polymer using cationic polymerization.

Coordination Polymerization

Some reactions are best described as coordination polymerizations, since they usually involve complexes formed between a transition metal and the π electrons of the monomer. Many of these reactions are similar to anionic polymerizations and could be considered under that category. These types of polymerizations usually lead to linear and stereoregular chains and often use so-

called Ziegler-Natta catalysts, various metal oxides, or, more recently, metallocene catalysts. The earlier generations of catalysts, like those developed by Karl Ziegler and Giulio Natta, are heterogeneous, producing not only stereoregular material such as isotactic polypropylene, but for this polymer, atactic material and stereoblock polymers as well. The degree of stereoregularity depends on the nature of the catalyst, the nature of the monomer, temperature, the nature of the solvent; and so on.

Ziegler-Natta catalysts generally consist of a metal organic compound involving a metal from groups I–III of the periodic table, such as triethyl aluminium, and a transition metal compound (from groups IV–VIII), such as titanium tetrachloride (Figure 3-37).

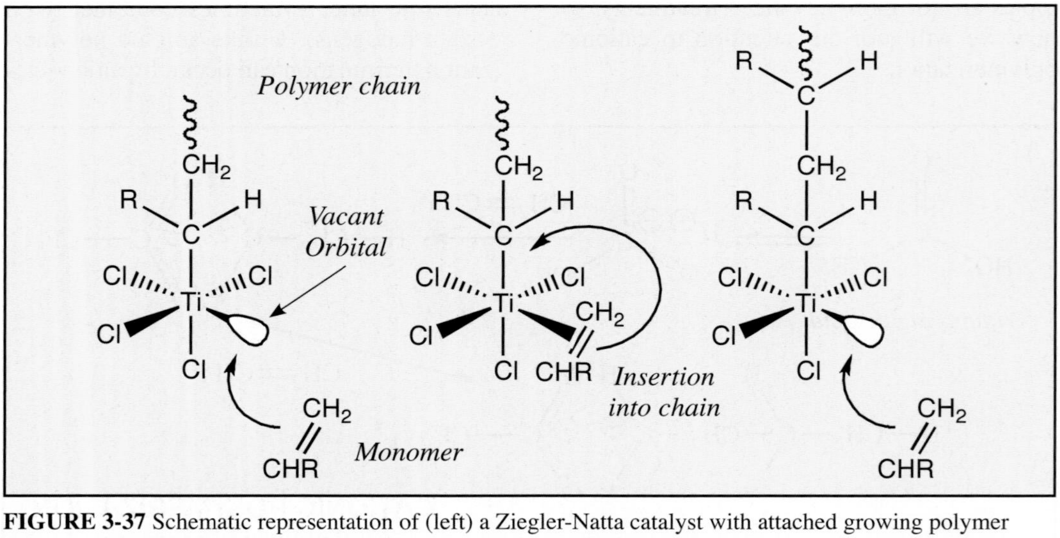

FIGURE 3-37 Schematic representation of (left) a Ziegler-Natta catalyst with attached growing polymer chain and a vacant orbital where a monomer may be complexed: (middle) monomer complexed to the vacant orbital and (right) insertion into the polymer chain leaving the vacant orbital ready for complexation with another monomer.

POLYMER MILESTONES—KARL ZIEGLER

In 1943, two years before the end of WWII, Karl Ziegler accepted the position as Head of the Kaiser-Wilhelm-Institut für Kohlenforschrung (Institute for Coal Research) in Mülheim an der Ruhr. Ziegler, who was not a coal chemist, accepted the position only on the condition that, "I must, so I maintained, have complete freedom to work in the whole area of the chemistry of carbon compounds (organic chemistry) without having to make allowance for whether my work had anything directly to do with coal or not." There was a commonly held belief after WWII that the golden age of polymer chemistry had come and gone; that all useful polymers had been discovered. This prediction has "legs" and seems to surface every ten years or so. Ziegler certainly dispelled this idea in the 1950s, with the discovery of the catalysts that bear his name. It doesn't take much imagination to realize that things must have been very tough

Karl Ziegler (Courtesy: Max Plank Institute, Mulheim an der Ruhr).

for Ziegler and the Coal Institute during the period from 1943 (while the war was in progress) to the end of the decade. But it was during this time that Ziegler and his coworkers laid the foundation for what was to culminate in the discovery of organo-metallic polymerization catalysts. Ziegler and H. G. Gellert observed that ethylene reacts to produce chain molecules (which, at most, had wax-like qualities) when certain organo-aluminum compounds are present. But attempts to increase the chain size were not successful. After going down various blind alleys, Ziegler and Holzkamp discovered the reason for this. Traces of metals, such as nickel, had an enormous effect on these reactions and the molecular weight of the final product. This appears to be yet another example of the intervention of Lady Luck. It is perhaps somewhat ironic that it was probably a contaminated or unclean autoclave that was responsible for the discovery of the so-called Ziegler catalysts. Ziegler and his coworkers, most notably Holzkamp, Breil and Martin, then initiated a systematic research program to find other metal compounds that might do the same thing. Serendipity struck and during this search they hit the mother lode. Ethylene, in the presence of a combination of an organo-aluminum and a zirconium compound, at relatively low pressure, produced polyethylene. In his lecture at the Nobel Prize award ceremony, Ziegler described his accomplishment thus, "The development began at the end of 1953 when Holzkamp, Breil, Martin and myself, in a few almost dramatic days, observed that the gas ethylene could be polymerized at 100, 20, or 5 atmospheres and finally even at atmospheric pressure very rapidly in the presence of certain, easily prepared, catalysts to give a high molecular weight plastic." Ziegler's discovery of mixed organo-metallic catalysts was to reverberate around the world of polymer chemistry and many other polymers were developed using this technology—including the so-called tactic polymers.

The metal organic compound acts as a weak anionic initiator, first forming a complex whose nature is still open to debate. Polymerization proceeds by a process of insertion. The transition metal ion (Ti in this example) is connected to the end of the growing chain and simultaneously coordinates the incoming monomer at a vacant orbital site. Two general mechanisms have been proposed: monometallic and bimetallic. For simplicity, we have only illustrated one of these, the monometallic mechanism. Isotactic placement can then occur if the coordinated monomer is inserted into the chain in such a way that the growing chain remains attached to the transition metal ion in the same position. Or, if the chain becomes attached to the transition metal ion in the position of the orbital that was initially vacant, syndiotactic addition will occur. This becomes more favored at lower temperatures, but vinyl monomers usually form isotactic chains with these catalysts. Because of the heterogeneous nature of the geometry of the catalyst surface, atactic and stereoblock polymers can also be formed.

Metallocene Catalysts

Every 25 years or so, there appears to be a "quantum jump" in polyolefin technology. Metallocene catalysts are considered by many to be one such revolution in polymer science. They have been referred to as "designer" catalysts, because of their ability to polymerize an incredible variety of olefin monomers to form copolymers with different microstructures, narrow (uniform) molecular weights and reproducible distributions of composition and branching. Metallocenes are not new and were evidently known to Natta as early as 1957. They are based upon metal complexes of cyclopentadienyl—you are possibly familiar with ferrocene (Figure 3-38), for example, the iron "sandwich," that was discovered in 1951—but the kinetics of polymerization using these catalysts was deemed too slow for them to be of any commercial interest.

The cyclopentadienyl ligand was of great interest to inorganic chemists, however, and hundreds of metallocene molecules were characterized. Nevertheless, some 20 years were to pass before Professor Walter Kaminsky, now at the University of Hamburg, discovered a cocatalyst, methylaluminoxane (MAO), that was to spark a revolution. In very simple terms, Kaminsky discovered that if a large excess of the cocatalyst MAO to metallocene is used, the metallocene molecule is modified and the resultant complex becomes a highly reactive polymerization catalyst for olefins and other specific monomers. Following Kaminsky's discovery of the cocatalyst MAO, there was a flurry of activity, most notably in the laboratories of the Exxon and Dow companies, that led to further developments in this field.

Metallocenes used for the polymerization of olefins are complex catalysts composed of a positively charged metal ion, typically zirconium or titanium, sandwiched between one or two electron-donating cyclopentadienyl ligands. Usually there is a chemical bridge attaching the two rings, which could be an ethylene group or a silicon atom. Thus, the metallocene resembles a half-opened clamshell (Figure 3-39). We will not go into mechanistic details, but, very simply, the methyl groups of the cocatalyst MAO displace two of the initial ligands (e.g., chlorine) of the metal complex. Subsequently, one of these methyl groups is eliminated and the active polymerization catalyst is actually an alpha-agostic π-complex into which the olefin is inserted.

There are many variations of metallocenes that have been developed and patented by Exxon, Dow and many other companies. By

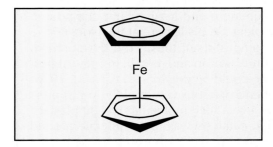

FIGURE 3-38 Schematic representation of ferrocene.

varying the bridging groups and attaching other functionalities to the cyclopentadienyl rings (e.g., an indenyl ligand), the steric and electronic properties of the catalysts can be fine-tuned to give the desired selectivity. For the purposes of this broad sweep, however, this is more than enough detail.

Living Free Radical Polymerization

If you had taken our introductory polymer science course about a decade ago, we would have emphatically stated that living polymerizations are restricted to anionic and a few rare cationic processes. However, just to prove us wrong, a group of scientists, including, most notably, Professor Matyjaszewski at Carnegie Mellon University, have upset the apple cart. With Jin-Shan Wang he published in 1995 a seminal paper entitled *Controlled "Living" Radical Polymerization*. In it, the authors show how a transition-metal species, M_t^n (such as $CuCl/2,2'$-bipyridine), can abstract a halide, X, from an initiator (e.g., phenyl ethyl chloride) to give a carbon-based radical, R•, and an oxidized species $M_t^{n+1}X$. This radical can now react with a vinyl monomer, Y, (such as styrene) to give the growing polymer chain radical species RY•, which, in turn, reacts with $M_t^{n+1}X$ to give RYX and regenerate the original reduced catalyst species, M_t^n (Figure 3-40). This process continues and in contrast to conventional free radical polymer-

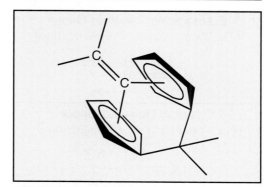

FIGURE 3-39 Schematic representation of a metallocene catalyst showing the insertion of the monomer.

ization, the molecular weight of the polymer increases linearly as a function of monomer conversion. This type of living free radical polymerization therefore has characteristics of both conventional free radical addition and step-growth processes. This is a fascinating and active field of polymer chemistry that has many important ramifications. But for our purposes, in this introductory text, this is hopefully enough to whet your appetite to seek out more information from other sources.

Monomer Types for the Different Polymerization Methods

As you might guess, not all monomers can be polymerized by a given chain polymerization method. There is a selectivity involved

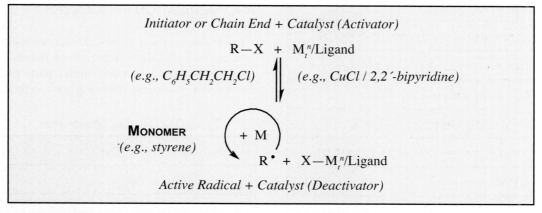

FIGURE 3-40 Schematic representation of living free radical polymerization.

ELECTRON WITHDRAWING GROUP
(Facilitates Anionic Polymerization)

$$\overset{\delta\oplus}{\underset{}{H_2C}} = \overset{\delta\ominus}{\underset{}{CH}} - X$$
→

ELECTRON DONATING GROUP
(Facilitates Cationic Polymerization)

$$\overset{\delta\ominus}{\underset{}{H_2C}} = \overset{\delta\oplus}{\underset{}{CH}} - X$$
←

FIGURE 3-41 Electron donating and withdrawing substituents.

that depends upon chemical structure (i.e., the inductive and resonance characteristics of the group X in the vinyl monomer shown in Figure 3-41). With the exception of α-olefins like propylene, most monomers with C=C double bonds can be polymerized free radically, although at varying rates (Table 3-2). Monomers are much more selective with respect to ionic initiators. Electron-withdrawing substitutents, such as cyano, acid or ester, facilitate anionic polymeriza-

TABLE 3-2 FREE RADICAL MONOMERS

ETHYLENE $CH_2=CH_2$
TETRAFLUOROETHYLENE $CF_2=CF_2$
BUTADIENE $CH_2=CH–CH=CH_2$
ISOPRENE $CH_2=C(CH_3)–CH=CH_2$
CHLOROPRENE $CH_2=C(Cl)–CH=CH_2$
STYRENE $CH_2=CH(C_6H_5)$
VINYL CHLORIDE $CH_2=CHCl$
VINYLIDINE CHLORIDE $CH_2=CCl_2$
VINYL ACETATE $CH_2=CH–OC(O)CH_3$
METHYL ACRYLATE $CH_2=CH–C(O)OCH_3$
ACRYLONITRILE $CH_2=CHCN$

TABLE 3-3 ANIONIC MONOMERS

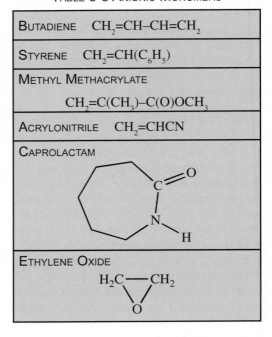

BUTADIENE $CH_2=CH–CH=CH_2$
STYRENE $CH_2=CH(C_6H_5)$
METHYL METHACRYLATE $CH_2=C(CH_3)–C(O)OCH_3$
ACRYLONITRILE $CH_2=CHCN$
CAPROLACTAM
ETHYLENE OXIDE

tion (Table 3-3). On the other hand, electron-donating substituents, such as alkyl, alkoxy and phenyl groups, increase the electron density on the C=C double bond and facilitate cationic polymerization (Table 3-4). Finally, Ziegler-Natta (and other) catalysts can be used to polymerize a variety of α-olefins (e.g., ethylene and propylene) and styrene, but many polar monomers cannot be polymerized this way, as they inactivate the initiator, either through complexation or reaction with the metal components.

POLYMERIZATION PROCESSES

In the final section of this chapter we will give a quick and dirty overview of polymerization processes. As with other parts of this subject, you could devote a book to this

TABLE 3-4 CATIONIC MONOMERS

ISOBUTYLENE $CH_2=C(CH_3)_2$
STYRENE $CH_2=CH(C_6H_5)$
VINYL METHYL ETHER $CH_2=CH(OCH_3)$

topic alone, so we cannot do it justice here. However, you should know something of the different methods that can be used, so that you do not appear totally ignorant if you're forced to talk to a synthetic chemist.

First, we have to distinguish between making polymers in the laboratory and on the industrial scale. A picture of Wallace Carothers in his lab, synthesizing (we think) nylon, is shown in the *Polymer Milestone* at the beginning of this chapter. At this scale, polymerizations need not be driven to high conversions, unless the chemistry demands it. Furthermore, subsequent separations (from solvent, for example) are not usually a problem. You just tell your graduate student to get on with it and try not to blow up the lab!

At the industrial scale (e.g., Figures 3-42 and 3-43), things that are minor problems in the laboratory can become big headaches. The viscosity of the polymerization mass can become unmanageably large, heat transfer from a large vessel can be a big problem (nearly all polymerizations are exothermic), and so on. Furthermore, designing a process to handle one of these problems might result in changes in microstructure and molecular weight relative to what was originally obtained in laboratory preparations. But this is the sort of thing you have to get into if you're a process engineer.

Most laboratory polymerizations use one of the following methods: bulk, solution, suspension or emulsion. We will describe these shortly, but if we are talking about large-scale polymerizations it is useful to first make a distinction between batch and continuous processes. The latter are more cost efficient, but cannot be applied to every type of polymerization. As we will see in Chapter 4, step-growth polymerizations are generally slow (hours), so that continuous reactors with inordinately long residence times would be required to make polymers with acceptable molecular weights. A second distinction that can be made is between single-phase processes and multiphase processes. Single-phase processes include bulk (or mass), melt and solution polymerizations.

FIGURE 3-42 Manufacture of neoprene at DuPont circa 1940 (reproduced with the kind permission of the Hagley Museum and Library).

In a bulk polymerization only reactants (and added catalyst, if necessary) are included in the reaction vessel. This type of process is widely used for step-growth polymerizations where high molecular weight polymer is only produced in the last stages of the polymerization. As a result, the viscosity of the reaction medium stays low throughout most of the reaction. However, crystallization of the polymer that is formed can lead to all sorts of

FIGURE 3-43 A modern polymerization plant (Courtesy: Exxon/Mobil).

FIGURE 3-44 Lucite® [poly(methyl methacrylate)] manufacture at DuPont circa 1940s (reproduced with the kind permission of the Hagley Museum and Library).

vent can occur i.e., the solvent can take part in the reaction. In addition, the problem of removing and recycling the solvent is introduced, which is expensive and not a trivial factor in today's environmentally conscious society. Even trace amounts of potentially toxic or carcinogenic solvents can pose a major predicament.

Obviously, one solvent, ideal from many points of view, is water. Nobody cares about trace amounts of water in polymer films that might be used to wrap food, for example. Trace amounts of benzene (a grade A carcinogen) would be unacceptable, however. The problem is, of course, that most polymers (or monomers for that matter) do not dissolve in water. This brings us to the topic of multiphase processes where polymerizations are performed in water with the monomer or polymer suspended in the form of droplets or dispersed in the form of an emulsion.

Suspension Polymerization

With the exception of a few polar molecules, such as acrylic acid, ethylene oxide, vinyl pyrrolidone and the like, most monomers are largely insoluble in water. It is important to emphasize the word, largely, because even when you see a grossly phase-separated system, like oil and water, a small amount of oil has actually dissolved in the water and vice versa. This will be important when we consider emulsion polymerization. A mixture of the two will phase separate into layers, usually a less dense "oily" layer of monomer sitting on top of the water layer. Polymerization in this state would obviously not have much of an advantage over bulk polymerization. Continuous rapid stirring can produce smaller globs or spherical beads of monomer suspended in the water, however, and each bead then becomes a miniature reaction vessel, as depicted schematically in Figure 3-45. The initiator used must also be essentially water insoluble, so that it prefers to reside in the monomer phase where it can initiate polymerization. Problems of heat transfer and viscosity are overcome by this method. In addition, the

problems, so that bulk polymerizations are carried out at temperatures higher than the melting point of the polymer and are often called melt polymerizations.

Chain polymerizations are less often performed in the bulk, because of problems with the control of the reaction. [An interesting exception is poly(methyl methacrylate), a polymer that is soluble in its own monomer (not all polymers are), and which is synthesized commercially by chain (free radical) polymerization very slowly in bulk (Figure 3-44). The resulting polymer has outstanding optical properties (clarity) because there are very few impurities.] In bulk polymerizations there is a tendency for the reaction mass to form a gel (i.e., have an extraordinarily high viscosity) and "hot spots" can develop. At the extreme, the reaction rate can accelerate to "runaway" proportions (for reasons we will discuss when we consider kinetics) with potentially disastrous (explosive) consequences. Viscosity and heat control can be achieved, if necessary, by carrying out the polymerizations to a relatively low conversion, with the unreacted monomer being separated and recycled. Another way to control the viscosity and heat transfer problems of chain polymerizations is to perform the polymerization in solution. A major concern with this method is that chain transfer to sol-

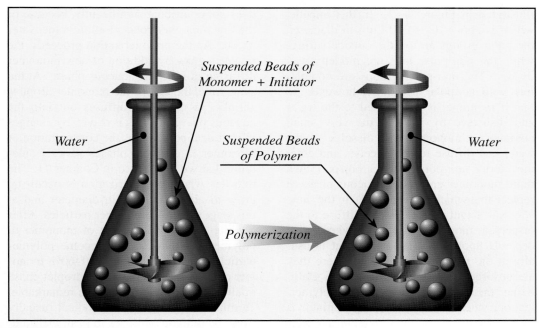

FIGURE 3-45 Schematic representation of suspension polymerization.

final product is in the form of "beads," which is convenient for subsequent processing. There are difficulties involving coalescence of particles, however, so a variety of additives (protective colloids, etc.) are used to stabilize the droplets. The beads or monomer droplets in such a suspension polymerization (sometimes also called bead or pearl polymerization) are usually about 5 mm in diameter and require the mechanical energy of stirring to maintain their integrity. If the stirring is stopped, a gross phase separation into two layers occurs.

Emulsion Polymerization

There is a way to suspend even smaller monomer particles in water such that the monomer droplets are stable and do not aggregate to form a separate layer. Essentially, a surfactant (soap) is used to form an emulsion. Surfactant molecules consist of a polar head (hydrophilic) group attached to a non-polar (hydrophobic) tail, such that it looks something like a tadpole, as depicted in Figure 3-46.

In water, soap molecules arrange them-

selves so as to keep the polar groups in contact with water molecules, but the non-polar tails as far from the water as possible (at concentrations above a certain level, called the critical micelle concentration). One way of doing this is to form a micelle, which usually has a spherical or rod-like shape, as

FIGURE 3-46 Schematic representation of a surfactant molecule with a hydrophilic head and a hydrophobic tail.

FIGURE 3-47 Two-dimensional schematic representation of a spherical micelle.

illustrated in Figure 3-47. In this micelle, which is about 10^{-3} to 10^{-4} μm in diameter, the polar groups are on the outside surface while the non-polar tails are hidden away inside. The non-polar groups are compatible with nonpolar monomer, however, so that if monomer is now added to the water and dispersed by stirring, the very small amounts of monomer that dissolve in the water can diffuse to the micelles and enter the interior, non-polar hydrocarbon part (then more monomer enters the aqueous phase to replace that which has departed). In the same way, surfactant molecules can diffuse to the dispersed monomer droplets (whose size depends upon the stirring rate, but is usually in the range of 1–10 mm), where they are absorbed onto their surface, thus stabilizing them. Unlike suspension polymerization, where a water-insoluble initiator is used, in emulsion polymerization a water-soluble initiator is added. The polymerization, for the most part, occurs in the swollen micelles, which can be thought of as a meeting place for the water-soluble initiator and the (largely) water-insoluble monomer (Figure 3-48). A small amount of polymerization sometimes occurs in the monomer droplets and almost certainly in solution, but the latter does not contribute significantly, because of the low monomer concentration in the water phase. As the polymerization proceeds, the micelles grow by addition of new monomer diffusing in from the aqueous phase. At the same time, the size of the monomer droplets shrinks, as monomer diffuses out into the aqueous phase. All this is driven by changes in chemical potential as the concentration of monomer in different phases varies, something that we will get to in Chapter 11. The micelles where polymerization is occurring grow to about 0.5 mm in diameter and at this stage are called polymer particles. After a while (~15% conversion of monomer to polymer) all the micelles become polymer particles and a while later (40–60% monomer conversion) the monomer droplet phase finally disappears. It is all quite remarkable. Termination occurs when a radical (usually from the initiator) diffuses in from the aqueous phase. This is a major advantage of the emulsion technique for the polymerization of monomers such as butadiene, which cannot be polymerized easily by free radical means using homogeneous (single-phase) polymerization, because it has a fairly high rate of termination. The termination step in an emulsion polymerization is controlled by

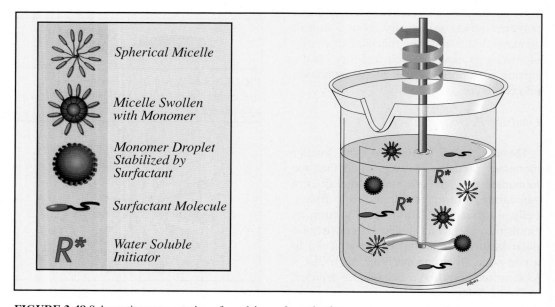

FIGURE 3-48 Schematic representation of emulsion polymerization.

FASCINATING POLYMERS—LATEX PAINTS

Water-based latex paints first appeared on the scene in the late 1940s and started a mini-revolution. They were more environmentally friendly and did not have the high viscosity problems associated with their solvent-based cousins. Actually, we have

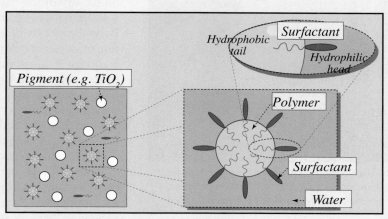

Schematic representation of an emulsion paint.

to go back to the early 1930s to the seminal work of Bayer's chemists for the genesis of emulsion paints, for it was they who first synthesized styrene/butadiene copolymers by emulsion polymerization. The cartoon above represents, in a very simple fashion, a typical can of a latex paint. Essentially water-insoluble monomers like vinyl acetate are polymerized in very small particles (about 1 μm) that are stabilized in water by surfactants. This is the binder, and although the polymer is generally of high molecular weight, the overall viscosity of the emulsion is relatively low and paints containing high amounts of solids that can be easily applied can be

prepared. To make a paint, pigments are added, along with several other additives. Remember, in a latex paint, the tiny water-insoluble polymer (and pigment) particles are stabilized and suspended in water. When you wash your paint brush under the tap you are, in effect, flushing these very small particles down the drain. On the wall, the latex paint is drying by evaporation of water into the atmosphere. As the majority of the water evaporates the polymer (and pigment), particles pack closer and closer together. The small gaps between these particles act as capillaries and squeeze the surfactants to the surface. Now the polymer particles fuse together in a process known as coalescence, which results in a coherent insoluble filled plastic film. This is why you can't wash the paint off the wall!

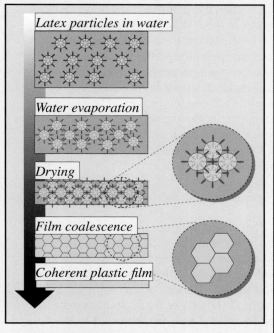

Schematic representation of film formation.

FIGURE 3-49 BUNA rubber, a styrene/butadiene random copolymer made by emulsion polymerization manufacture, circa 1940s (Courtesy: Bayer).

the rate of arrival of radicals at the polymer particles, which depends only on the concentration of initiator and the concentration of surfactant (hence, initial micelle concentration). There are some obvious additional advantages to emulsion polymerization; as in suspension polymerization, the viscosity is always low and heat control is relatively straightforward (which, for example, was put to good use in the manufacture of BUNA rubber in the 1940s—see Figure 3-49). Also, the final product is an emulsion of ~0.1 μm diameter polymer particles that is typically 50% by volume polymer and 50% water. This makes it almost immediately applicable as a surface coating (paint). The major disadvantage is the presence of the surfactant, which is difficult to completely remove even if the polymer product is precipitated and washed.

Multiphase Processes

There are various types of multiphase processes that are widely used in the mass production of polymers. The two phases can both be liquids, as in suspension and emulsion polymerization, or can be a gas/solid, gas/melt (liquid) or liquid/solid system. In the interfacial polymerization of nylon 6,6, for example, the two monomers are initially dissolved in different solvents, hexamethylene diamine in water and adipoyl chloride in chloroform, as we mentioned previously. In this system, the aqueous phase sits on the top of the chloroform solution and a solid polymer is formed at the interface. Polymerizations of an initial miscible mixture of, say, monomer + initiator or monomer + solvent + initiator also become multiphase if the polymer formed is insoluble in its own monomer e.g., poly(vinyl chloride) or poly(acrylonitrile), or the monomer solvent mixture (e.g., the cationic polymerization of isobutylene in methylene chloride.

Finally, multiphase processes, where the monomer is a gas, are important methods for polymerizing monomers such as ethylene and propylene. Some low density polyethylenes, as well as several well-known ethylene copolymers such as ethylene-co-vinyl acetate copolymers, are still made by high-pressure free radical polymerization methods. The process is usually continuous, with compressed ethylene, for example, at about 200°C flowing through a heavy-walled tubular reactor. Initiator is injected into the ethylene stream as it enters the vessel. About 15% of the entering gas is converted to molten polymer which is carried out in the gas stream (there is often a polymer build-up on the lower surface of the reactor which must be periodically removed by a process called "blowing down"). These days, however, polymers like high density polyethylene are often produced in a gas/solid multiphase process. Catalyst can be suspended in an upward flowing stream of monomer gas in a fluidized bed reactor, for example, and crystalline polyethylene grows on the catalyst surface. Polyethylene and polypropylene are also polymerized by forming a slurry of monomer, insoluble catalyst and a solvent and again the polymer "grows" on the catalyst surface. But this is enough; you should have the general picture by now. There is a range of industrially important methods for synthesizing polymers and if your interest is in the large-scale production of these materials, God bless your engineering heart, you will have to study the methods we have mentioned (and others) in much greater depth.

RECOMMENDED READING

H. R. Allcock, F. W. Lampe and J. E. Mark, *Contemporary Polymer Chemistry*, 3rd. Edition, Prentice Hall, New Jersey, 2003.

P. J. Flory, *Principles of Polymer Chemistry*, Cornell University Press, Ithaca, New York, 1953.

G. Odian, *Principles of Polymerization*, 3rd. Edition, Wiley, New York, 1991.

K. J. Saunders, *Organic Polymer Chemistry*, Chapman and Hall, London, 1973.

STUDY QUESTIONS

1. In study questions for Chapter 2 you were given the infrared spectra of two samples. These samples had approximately the same (high) molecular weight and the same type of end groups. One of these samples, sample A, displayed a prominent band due to the presence of methyl groups, while in the other sample, B, this band is extremely weak. Explain how sample A was probably synthesized and how this leads to short-chain branching. Explain how sample B was probably polymerized to give an essentially linear chain.

2. What are the major differences between step-growth and chain (or addition) polymerization?

3. How does "living" polymerization differ from ordinary addition polymerization? Show how the former can be used to synthesize styrene-butadiene-styrene block copolymers.

4. Nylon 6 and nylon 6,6 are two commercially produced polyamides. How are they synthesized and how do their chemical repeat units differ? (Note: you won't find a description of the synthesis of nylon 6 in the book. Check other sources!)

5. What is the difference between anionic, cationic, and free radical polymerization?

6. Can propylene by synthesized free radically? If not, explain why not.

7. Describe how urethane chemistry can be used to make both flexible and rigid foams.

8. What is transesterification? Use the synthesis of poly(ethylene terephthalate) as an example in giving your answer.

9. Polystyrene foams can be made by a suspension polymerization method (not mentioned in this book—check other sources). Describe this process.

10. Compare and contrast suspension and emulsion polymerization methods.

4

Polymerization Kinetics

INTRODUCTION

The study of the chemical kinetics of polymerization is important at all sorts of levels and the ramifications of the results range across the subject, from the process engineering problems of large-scale synthesis to an understanding of the microstructure of copolymers. First, kinetics tells us how long a reaction takes. We will see that step-growth polymerizations generally take a long time, but some chain polymerizations can proceed at (literally) explosive speeds (Figure 4-1). Second, kinetics affects microstructure and chain length. For example, the rate of adding different monomers affects copolymer composition. However, here we will initially confine our attention to the kinetics of homopolymerization. We will consider copolymerization separately, as it is useful to consider some aspects of the statistics of polymerization before getting to that subject.

CHEMICAL KINETICS

First things first. Do you recall anything that you did in your studies of chemical kinetics? Do the names van't Hoff and Arrhenius mean anything to you? If they don't, because studying physical chemistry gave you brain lesions, or you contracted some dreaded, horrible disease like Congo's Wallop that erased your memory banks, then you should read our brief review of kinetics.

But, if you're comfortable with the concept of rate constants, jump to page 92!

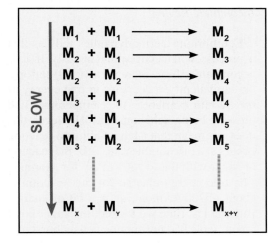

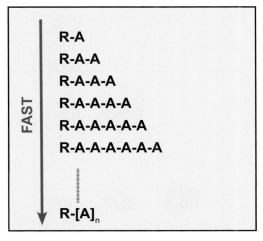

FIGURE 4-1 Relative reaction rates.

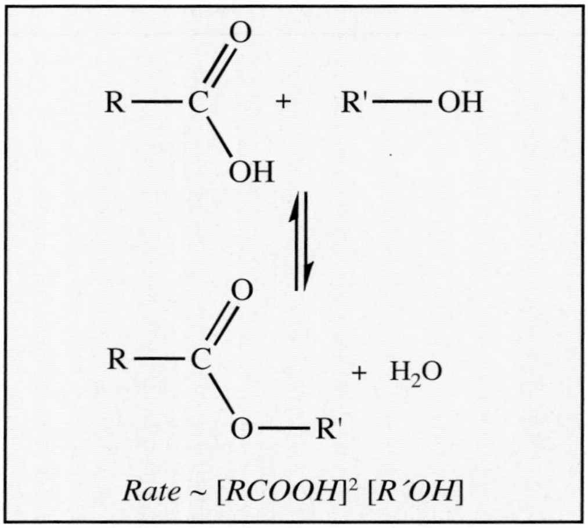

$$Rate \sim [RCOOH]^2 [R'OH]$$

FIGURE 4-2 Ester formation.

Kinetics—A Superficial and Thoroughly Incomplete Review

Equilibrium thermodynamics tells us that a process will occur spontaneously if it is accompanied by an increase in (total) entropy, or, equivalently, a decrease in free energy of the system (variables like temperature and pressure being held constant). However, this does not mean that such an allowed reaction occurs at a measurable rate. Take the reaction of sugar with the oxygen in air, for example. The free energy change for this reaction is about −5700 kJ/mole, quite a bit of energy. But, the last time we looked, the sugar bowl on our table had not spontaneously burst into flames. The rate of this reaction is negligible at ordinary temperatures and pressures.

Kinetics is obviously important in polymer science, not just in the study of things like the rate of polymerization (where the danger is not that the reaction rate will be too slow, but too fast, thus blowing up the lab or the plant where you happen to be working), but also in dealing with subjects such as crystallization. It is also one of the topics that you should bring with you from your studies of chemistry to the study of polymers. Except that your love life was in one of its periodic bouts of upheaval when you took the courses and you're a bit confused by the whole subject, particularly chemical kinetics. So, as with thermodynamics later on, we will review some basic concepts here.

As we note in our *Science Milestone*, opposite, van't Hoff classified reactions as unimolecular, bimolecular, etc., according to the relationship between their rates and the concentration of reactants. It did not escape van't Hoff that the power to which a concentration was raised in his equations did not necessarily correspond to the number of molecules involved in a particular reaction. For example, when we consider the formation of esters from acids and alcohols, we will find that if there is no added catalyst the rate of the reaction is proportional to the product of the concentration of alcohol groups and the square of the concentration of acid groups (Figure 4-2). We now talk about the order of a reaction, of course, a clarification in terminology that was introduced by Ostwald in 1887. A first-order reaction is what van't Hoff would have referred to as unimolecular, and depends on the concentration of a single reactant.

A great example of a first-order reaction is the radioactive decay of unstable nuclei, where we have, for example, atoms of "A" decaying to certain products (Figure 4-3). If the rate of this decay is simply proportional to how much "A" is present, we can write a simple differential equation. We will let N be the number of unstable nuclei present at any moment (Equation 4-1), and k is, in this case, a decay constant.

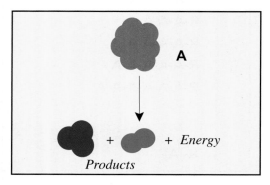

FIGURE 4-3 Depiction of a first-order reaction.

$$-dN/dt \sim N \quad or \quad -dN/dt = kN$$
EQUATION 4-1

A MILESTONE IN SCIENCE—VAN'T HOFF

As we have mentioned, the study of the chemical kinetics of polymerization is extremely important. The foundations of this subject were established by van't Hoff, who, in the 1870s, had turned his attention from organic to physical chemistry, writing a text that an English obituarist unkindly described as "unreadable."* The basic theory of reaction kinetics was described in a later, apparently more readable text, which built on the insights of earlier workers who had observed a relationship between the rate of a reaction and the concentration of reactants present at any given moment. He classified reactions as *unimolecular* if their rate was proportional to the first power of the reactant's concentration; *bimolecular*, if it depended on the product of the concentration of two reactants (or the square of a single reactant); and so on. This, in turn, allowed van't Hoff to express the rate of a reaction in terms of a differential equation, thus laying the foundation for the formalism that we use today.

Van't Hoff (Courtesy: The Edgar Fahs Smith Collection of The University of Pennsylvania Library).

*Quote from *The Norton History of Chemistry,* W. H. Brock, W. W. Norton & Co. New York, 1993.

Note that we use a minus sign, because we are describing something that is decreasing with time. You will see a lot of equations of this form, the derivative of something being proportional to itself, and you should get into the habit of recognizing that the integral is an exponential (Equation 4-2) where N_0 is the amount of "A" at $t = 0$.

$$\int_{N_0}^{N} dN/N = \int_{t_0}^{t} kt$$

Hence:

$$N = N_0 \exp(-kt)$$

EQUATION 4-2

Thus, if we start with one mole of the radioactive isotope of carbon, ^{14}C, then its decay has the exponential form shown in Figure 4-4. The decay of initiator in a batch free

radical polymerization has the same form, as you will see. But just to finish up on this little aside on radioactive decay, you may know that the amount of residual ^{14}C is used to date archaeological remains and other objects, but of course these materials never start out with a mole of ^{14}C, so you are usually trying to detect very small amounts. In practice,

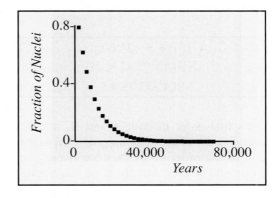

FIGURE 4-4 Radioactive decay.

this limits the accuracy of the technique to remains that are less than 30,000 years old.

But, enough of this dalliance in other fields, let's get back to Ostwald and the order of a reaction. We've illustrated a first-order decay process, but if we were talking about a chemical reaction, rather than radioactive decay, we would use concentration in moles per liter (mol L^{-1}) rather than using the number of molecules or moles of a material in our differential equation. This is usually indicated by putting square brackets around the symbol for the reacting group, where k is now called the rate constant (Equation 4-3).

$$A \longrightarrow Products$$
$$-d[A]/dt = k[A]$$

EQUATION 4-3

Keep in mind what we've said before, if we are describing a species (reactant) that is decreasing with time we put in a minus sign. So far so good. It should now be obvious that if the rate of the reaction depends on the square of the concentration of "A" (Equation 4-4), then we have what we call a second-order reaction.

$$-d[A]/dt = k[A]^2$$

EQUATION 4-4

But, what if we now have a reaction where A and B produce products (Equation 4-5), what is the order of this reaction?

$$A + B \longrightarrow Products$$
$$d[P]/dt = k[A][B]$$

EQUATION 4-5

It's first-order in both A and B, but the overall order of the reaction is defined as the sum of the order of all the components, so this would be two. Finally, if we had a complex reaction like that depicted in Equation 4-6 then the overall order would be $x + y$.

$$-dP/dt = k[A]^x[B]^y$$

EQUATION 4-6

This raises an important question: how do you know what the order of a reaction happens to be? The answer is: you usually don't! There is not (yet?) a universal molecular theory that allows one to predict the order of a reaction. You have to do a lot of careful experiments to measure these quantities and also the rate constants, which, as we will see when we get to the kinetics of polymerization, will allow us to obtain lots of important information about molecular weight, copolymer composition, and so on, as well as information about the speed of a reaction. So, one of the problems with the study of kinetics is that, unlike thermodynamics, where you can start with essentially three laws (one of the four laws is not really a law) and then logically construct the whole edifice, with kinetics you essentially have a phenomenological science, a fancy way of saying that we simply describe the formal structure of the subject in terms of experimentally observed quantities. This is not to say that there are not useful theories, but none, alas, seem to be perfect, just like the theories of the golf swing that we have diligently studied and applied (imperfectly) over the years. Nevertheless, we will mention a couple: the first relating to the important question of the temperature dependence of the rate constant.

You probably already know from your elementary chemistry labs that to get a reaction going or to speed it up you raise the temperature. The relationship between reaction rates and temperature was first established empirically by Arrhenius around the end of the nineteenth century, when data on reaction rates were first being accumulated. Arrhenius actually used van't Hoff's derivation for the rate constant, so it is perhaps somewhat unfair that the equation below is simply called the Arrhenius equation. He noticed that the log of the rate constant is proportional to $1/T$, as illustrated in Figure 4-5, and the constant of proportionality (slope) was characteristic of the reaction. For reasons we will shortly make clear we will let this constant be $-E_a/R$. If we let the intercept of the line be A, then we obtain Equation 4-7.

$$\ln k = \ln A - E_a/RT$$
or
$$k = Ae^{-E_a/RT}$$
EQUATION 4-7

For gas phase reactions, this can be understood in terms of collision theory. Imagine molecules of types A and B bouncing around in a container. These molecules can react to form a product P, but, first, they have to come in contact. Of course, through their random motions they will collide, but this does not mean they will necessarily react. They may just bounce apart (top of Figure 4-6). Collision theory says that they will react if their energy of collision is more than a certain minimum. If you remember your basic physical chemistry, you should recall that the fraction of collisions that involve an energy greater than this minimum, which we will call the activation energy, E_a, goes as $\exp(-E_a/RT)$. (If you've forgotten all this stuff, check our review of thermodynamics in Chapter 10.)

Let's look at this in a bit more detail. First of all, the rate at which molecules of A and B collide must simply be proportional to their concentration—if we double the number of A and B molecules we would expect that their rate of collision would also double (Equation 4-8).

$$Reaction\ Rate \sim [A][B]$$
EQUATION 4-8

However, only that fraction of collisions with sufficient energy can react (Equation 4-9).

$$Reaction\ Rate \sim [A][B]e^{-E_a/RT}$$
EQUATION 4-9

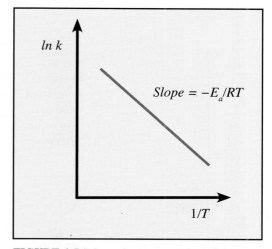

FIGURE 4-5 Schematic representation of an Arrhenius plot.

Comparing this to our second-order rate law equation we obtain Equation 4-10, which is the empirical relationship established by Arrhenius.

$$Reaction\ Rate = k[A][B]$$
so that:
$$k \sim e^{-E_a/RT} = Ae^{-E_a/RT}$$
EQUATION 4-10

Using the kinetic theory of gases, it is also possible to obtain an expression for the constant of proportionality, A, but the

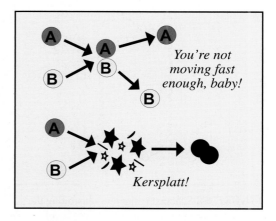

FIGURE 4-6 Schematic of collision theory.

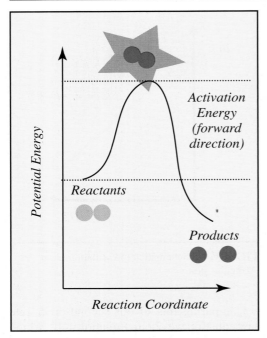

FIGURE 4-7 Schematic of activation energy.

experimental value is always less than the one calculated from theory. This is presumably because, for a reaction to occur, the molecules also have to be aligned correctly when they collide. This is wonderful stuff, in that first and foremost, it does what all good theories should do, provide insight and understanding. But it does not apply to the liquid state where molecules are essentially in contact with their neighbors all the time. A more sophisticated theory that applies to reactions in both the gaseous and liquid state is activated complex theory, where an unstable reaction intermediate either rearranges to form products or collapses to give back the reactants (Figure 4-7). By assuming that the activated complex is in equilibrium with the reactants, an equation of the Arrhenius form can be obtained. But you don't need this for what we will cover in polymerization kinetics, so we will call it quits here and get back to polymers.

THE REACTIVITY OF CHAIN MOLECULES

When Carothers was first attempting to synthesize large molecules using condensation reactions, focusing initially on polyesters, he ran into a roadblock. At first it seemed that a molecular weight of about 6000 was the upper limit of what could be obtained by these reactions. At that time, the consensus of opinion was that the reactivity of the end group decreased as the chain length got larger. It turns out that these esterifications were reversible reactions and the polyesters were being hydrolyzed back to acids and alcohols. But before that was fully accepted, a young Paul Flory made a bold and controversial assumption—that the intrinsic reactivity of a functional group is independent of the molecule to which it is attached (Figure 4-8). He then proceeded to prove his point with a detailed study of the kinetics of polyesterification. So that seems like a good point to start our studies of polymerization kinetics.

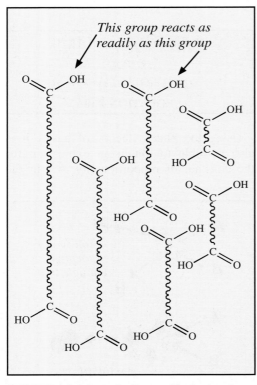

FIGURE 4-8 Schematic diagram illustrating the independence of chain length on reactivity of functional end groups.

THE KINETICS OF STEP-GROWTH POLYMERIZATION

We will start our discussion by discussing polyesterifications of the type illustrated in Figure 4-9. Here A–A is a diacid and B–B is a dialcohol (diol). One's first guess would be that this bimolecular reaction would be second-order with a rate constant k_2 (Equation 4-11).

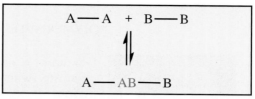

FIGURE 4-9 Representation of A-A, B-B polyesterification.

$$Rate = -d[A]/dt = k_2[A][B]$$

EQUATION 4-11

But it isn't! Experimental measurements show that the (overall) reaction is actually third-order (rate constant = k_3) because the reaction is catalyzed by acids (so one of the reacting components also acts as a catalyst— Equation 4-12).

$$Rate = -d[A]/dt = k_3[A]^2[B]$$

EQUATION 4-12

Note two important things. First, the rate of the reaction is described in terms of the disappearance of one of the functional groups, in this case the A's or acids. Because A's only react with B's and the stoichiometry is 1:1, we could have just as easily chosen the B's to follow. Second, the quantities [A] and [B] are the concentrations of functional groups, not monomers or molecules. In this reaction, there are two functional groups per monomer, so if some nasty, sadistic professor was to set you a homework question where the concentration of monomers was given, you would have to multiply these numbers by two to get the concentration of functional groups.

Now let's consider the important case where we have exactly equal concentrations of functional groups, so we can put $c = [A] = [B]$. Instead of Equation 4-12 we can write Equation 4-13.

$$Rate = -dc/dt = k_3 c^3$$

EQUATION 4-13

And if the initial concentration (time, $t = 0$) of monomer is c_0 then we can integrate this equation as shown in Equations 4-14.

$$\int_{c_0}^{c} dc/c^3 = \int_{t_0}^{t} dt$$

Hence:

$$2k_3 t = 1/c^2 - 1/c_0^2$$

EQUATIONS 4-14

The Extent of Reaction (p)

If this reaction is indeed third-order, *and if Flory's assumption that the intrinsic reactivity of a functional group is independent of chain length is correct,* then a plot of $1/c^2$ versus t should be linear. Because it provides a direct link to the statistics of polymerization, however, it is useful to first follow Flory and define a new parameter, p, the extent of reaction Equation 4-15.

$$p = \frac{Number\ of\ COOH\ groups\ reacted}{Total\ \#\ of\ original\ COOH\ groups}$$

EQUATION 4-15

The fraction of unreacted groups must then, by definition, be $(1 - p)$, so that c, the concentration of groups remaining at time t, must be this fraction times the initial concentration, c_0.

The Kinetics of Uncatalyzed Polyesterifications

Substituting $c = c_0(1 - p)$ into Equation 4-14 gives the Equation 4-16.

POLYMER MILESTONES—PAUL FLORY

Paul Flory

It is hard to find a topic in the field of polymer physical chemistry where Paul Flory has not made seminal contributions, if not the seminal contribution. Not only are his theories and models the starting point for scientists entering this field, he was also an experimentalist, always looking to test the validity of his ideas and insights. Paul Flory was born in Sterling, Illinois, on June 19th, 1910. After receiving his undergraduate degree in chemistry from Manchester College in Illinois, he became a graduate student at The Ohio State University in 1931 (nobody's perfect). This is where he grew to love physical chemistry and in 1934 he received his doctoral degree, specializing in photochemistry and spectroscopy. Carothers was looking for a physical chemist to join his group and DuPont hired the newly graduated Paul Flory the same year. Flory later remarked that he was not a synthetic organic chemist and Carothers didn't try to make him one, but instead encouraged him to carry out "mathematical investigations of polymerization." This resulted in his lifelong interest in the fundamental science of polymer materials. But, as Flory later remarked, Carothers death "was one of the most profoundly shocking events of my life," and he left DuPont for a series of academic and industrial appointments, including stints at Standard Oil and Goodyear because of the "urgency of research and development on synthetic rubber." In 1948, Flory was invited by the famous chemist, Peter Debye, to Cornell and some six months later Flory was offered a professorship. What followed was a most productive period of research and teaching, culminating in the 1953 publication of the now classic text, *Principles of Polymer Chemistry*. We tell all of our students that if you are serious in making polymer science the focus of your career, buy this book and absorb it! Your authors are still in awe of the depth of understanding that Flory possessed at that time. On a number of occasions we have thought that we had discovered some fundamental insight about polymers, only to find that Flory, in an aside or footnote in his book or published works, had obviously considered and recognized it! In 1957, Flory moved to the Mellon Institute in Pittsburgh, but was not able to establish the broad program of basic research that he desired and in 1961 he accepted a Professorship in Chemistry at Stanford University. As we have mentioned, Flory's contributions span the field of polymer physical chemistry. Starting with his work on the kinetics and statistics of polymerization initiated at DuPont, he went on to lay the foundations of polymer solution theory. His work, and the independent contributions of Huggins, culminated in the Flory-Huggins equation, still the starting point for discussing the phase behavior of polymer solutions and blends. There followed investigations of the frictional properties of polymer solutions, the excluded volume effect, theories of rubber elasticity, liquid crystalline polymers and an extensive body of work concerned with the configurations of polymer chains. In 1974, Flory's accomplishments were finally recognized by his peers and he was awarded the Nobel Prize in chemistry. He died quite unexpectedly in 1985.

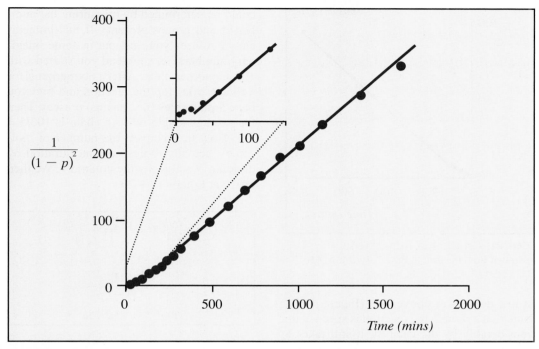

FIGURE 4-10 Graph of $1/(1 - p)^2$ versus t [redrawn from the data of P. J. Flory, *JACS*, **61**, 3334 (1939)].

$$2c_0 k_3 t = 1/(1 - p)^2 - 1$$

EQUATION 4-16

So now a plot of $1/(1 - p)^2$ versus t should be linear. And it is! Except right at the beginning, in the initial stages of the polymerization (Figure 4-10). The curvature at low concentrations is typical of simple (i.e., non-polymer forming) esterifications, however, and can be attributed to the large changes in character of the medium in the initial stages of the reaction. Clearly, over most of the reaction, the plot is linear and the reactivity of the groups would seem to be independent of chain length. Note also how long the reaction takes to reach high degrees of conversion. We'll come back to this shortly.

The Kinetics of Catalyzed Polyesterifications

Further evidence supporting Flory's assumption and analysis came from studies of the same reaction, catalyzed by the addi-

tion of small amounts of a strong acid (*p*-toluene sulfonic acid). Because the amount of this acid catalyst remains constant throughout the reaction, its concentration can be folded into the definition of a rate constant, $k' = k_2[\text{Acid}]$ (Equation 4-17).

$$Rate = -d[A]/dt = k'[A][B]$$

EQUATION 4-17

Again assuming equal amounts of reacting functional groups, we get Equation 4-18.

$$Rate = -dc/dt = k'c^2$$

EQUATION 4-18

Finally, upon integrating we arrive at Equation 4-19.

$$c_0 k' t = 1/(1 - p) - 1$$

EQUATION 4-19

A plot of $1/(1 - p)$ is linear again, except

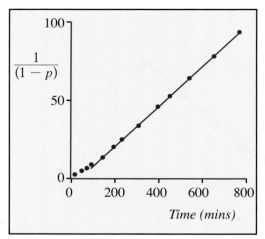

FIGURE 4-11 Graph of $1/(1 - p)$ versus t [redrawn from the data of P. J. Flory, *JACS*, **61**, 3334 (1939)].

at low degrees of conversion (Figure 4-11). Note that although the reaction is speeded up considerably by the catalyst, it still takes a long time to get to high degrees of conversion. So, is it important to carry the reaction to high degrees of conversion? For step-growth polymerizations, absolutely! To understand why, just recall the definition of number average degree of polymerization. This is simply the number of monomer molecules you started with divided by the number of molecules present at a given moment in the polymerization (which, in principle,

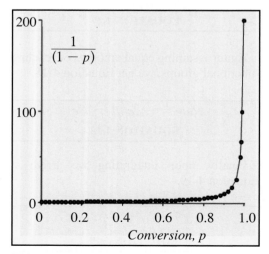

FIGURE 4-13 Graph of $1/(1 - p)$ versus conversion, p.

could be determined by measuring the number of end groups present). If this immediately confuses you, just put in some imaginary numbers; let's pretend you started with 100 monomer units. After polymerizing for a time t, you stop the reaction and find you have 5 molecules (i.e., chains) present. Then the average length of each chain is 100/5 = 20. Now let's convert the number of molecules, actually moles, to concentration by dividing N and N_0 by the volume V. We then arrive at Equations 4-20.

$$\overline{x_n} = \frac{N_0/V}{N/V} = \frac{c_0}{c} = \frac{c_0}{c_0(1 - p)}$$

or:

$$\overline{x_n} = \frac{1}{(1 - p)}$$

EQUATIONS 4-20

Accordingly, you don't get to decent degrees of polymerization, let's say 200, unless you have high degrees of conversion. For this example, $p = 0.995$, or a conversion of 99.5%—and you want to do this on an industrial scale! Organic chemist friends of ours jump up and down in glee if they get reaction yields of 80%, which, for our reaction, would give us a number average degree of polymerization of 5; not good. The plot shown in Figure 4-12 demonstrates how you only get decent degrees of polymerization at high values of conversion. We'll revisit this when we talk about the statistics of polymerization.

THE KINETICS OF CHAIN POLYMERIZATIONS

We have seen that the rate of step-growth or condensation polymerization is relatively slow and macromolecules are only produced at high degrees of conversion. In contrast, chain or addition polymerizations occur rapidly and polymer is produced in the initial stages of the reaction. Instead of having monomers going to oligomers and then to polymers, with essentially all the molecules taking part in the reaction at the same time, a chain is initiated, propagates quickly, and then (usually, but not always) terminates. At any instant of time in the reaction pot there is monomer, fully formed polymer chains, and just a (relatively) few molecules that are in the process of growing. We'll now look at the kinetics of this process, focusing mainly on free radical polymerization, as this will give you a feel for the general approach. We'll then indicate how the kinetics can be modified to account for other types of chain polymerizations.

In our discussion of polymer synthesis, which you have presumably studied first before getting to kinetics, we saw that the process of free radical polymerization can be broken down into four steps (Figure 4-13). The simplest approach is to consider initiation, propagation and termination first, then see how the equations are modified to account for chain transfer. We have three goals. First, obtain an expression for the rate of polymerization in terms of the concentration of things like monomer and initiator and a set of rate constants; second, obtain an expression for the degree of conversion; finally, obtain an expression for the average degree of polymerization.

Initiation

Initiation of free radical polymerizations can be achieved in a number of ways and for some monomers simply heating them will often do the trick. Most commonly, however, free radicals are generated by adding an initiator, which forms radicals upon being heated or irradiated. There are several groups of such initiators and a common example of one type is benzoyl peroxide, shown in Figure 4-14. The cleavage of the unstable O–O single bond upon heating gives two fragments or radicals, each with an unshared electron. We can describe this decomposition process using a rate constant k_d (Equation 4-21).

$$I_2 \xrightarrow{\ k_d\ } 2RO^\bullet$$
EQUATION 4-21

- **INITIATION**
- **PROPAGATION**
- **CHAIN TRANSFER**
- **TERMINATION**

FIGURE 4-13 The main steps of free radical polymerization.

In some treatments of the kinetics of polymerization this step alone is considered initiation. Others consider initiation to also include a second step: the reaction of this very reactive radical (because of the unshared electron left "hanging out") with the first monomer (Equation 4-22).

$$RO^\bullet + M \xrightarrow{\ k_i\ } M_1^\bullet$$
EQUATION 4-22

This is the way we were taught, so that's what you're going to get! However, it doesn't make a lot of difference, because the decomposition of the initiator is much slower than

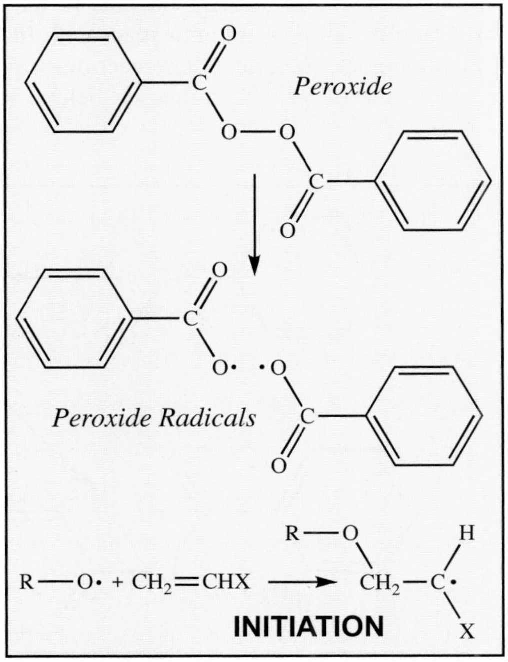

FIGURE 4-14 Initiation.

the addition of the first monomer and is, thus, the rate-limiting step. We can therefore write Equations 4-23 for the rate of initiation in terms of the rate of formation of the radical species $M_1^\bullet$.

$$r_i = \frac{[dM_1^\bullet]}{dt} = 2k_d[I]$$

and

$$r_i = -2\frac{d[I]}{dt}$$

Hence:

$$-\frac{d[I]}{dt} = \frac{1}{2}\frac{d[M_I^\bullet]}{dt} = k_d[I]$$

EQUATIONS 4-23

Note that we have written r_i in terms of both the rate of formation of $M_1^\bullet$ species and rate of disappearance of monomer and then substituted to get the final equation. The factor 2 appears because each initiator molecule (in this case peroxide) gives two radicals and can start two chains. One could also fold the constant factor 2 into the rate constant, if one were of a mind to do so. However, not all of the so-called primary radicals formed by decomposition of initiator react with the first monomer. Several other reactions can occur, some of which are shown below in

Figure 4-15, so only a fraction, f, of initially formed radicals start chains. Therefore, the kinetic equation for the rate of initiation has to be modified to include this "fudge factor" (Equation 4-24).

$$r_i = \frac{[dM_I^\bullet]}{dt} = 2fk_d[I]$$

EQUATION 4-24

Propagation

Propagation consists of the successive addition of monomers. If we again make the assumption that reactivity is independent of chain length, so that each successive addition can be described by the same rate constant, k_p, then we obtain Equations 4-25.

$$M_1^\bullet + M \xrightarrow{k_p} M_2^\bullet$$
$$M_2^\bullet + M \xrightarrow{k_p} M_3^\bullet$$
$$\vdots \quad \vdots \quad \vdots$$
$$M_x^\bullet + M \xrightarrow{k_p} M_{x+1}^\bullet$$

$$r_p = -\frac{d[M]}{dt} = -k_p[M_1^\bullet][M]$$

EQUATIONS 4-25

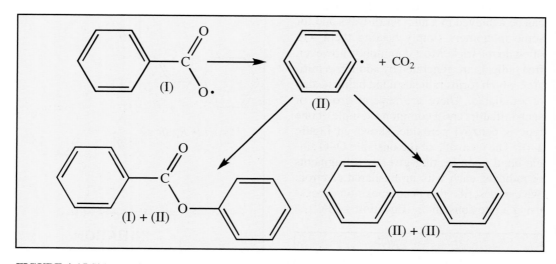

FIGURE 4-15 Side reactions of the decomposition of benzoyl peroxide.

Termination

Termination, you will no doubt recall, can occur by two mechanisms: first, combination, where the radicals collide and form a new covalent bond, joining the two original chains to make one larger one; second, disproportion, which involves the transfer of a proton from one chain to another with the corresponding rearrangement of electrons (Equations 4-26).

$$M_x^{\bullet} + M_y^{\bullet} \xrightarrow{\;k_{tc}\;} M_{x+y}$$
$$M_x^{\bullet} + M_y^{\bullet} \xrightarrow{\;k_{td}\;} M_x + M_y$$

EQUATIONS 4-26

Both types of termination reactions involve two radicals and are kinetically identical, so they can be combined to give an equation in terms of an overall termination rate constant, k_t (Equations 4-27).

$$r_t = -\frac{d[M^{\bullet}]}{dt}$$
$$r_t = 2k_{tc}[M^{\bullet}][M^{\bullet}] + 2k_{td}[M^{\bullet}][M^{\bullet}]$$
$$k_t = k_{tc} + k_{td}$$
$$r_t = 2k_t[M^{\bullet}]^2$$

EQUATIONS 4-27

Again, the factor 2 is included explicitly, this time because two radicals are consumed in each termination reaction.

Summary of Kinetic Equations

With all the equations that have now been thrown at you, you're probably a bit confused about where we are and where we're going. So, first of all, we've summarized our three kinetic equations, for initiation, propagation and termination, in Figure 4-16. Now, the first thing we want to obtain is an expression for the rate of polymerization, R_p. This is simply the rate of formation of polymer, which, if you think about it, is the same as the rate of disappearance of

$$r_i = \frac{d[M_i^{\bullet}]}{dt} = 2fk_d[I]$$
$$r_p = -\frac{d[M]}{dt} = k_p[M_1^{\bullet}][M]$$
$$r_t = 2k_t[M^{\bullet}]^2$$

FIGURE 4-16 Summary of kinetic equations.

monomer (it is not evaporating into a black hole in space, after all), which, if you think about it some more, is the same as the rate of propagation! Well, we have an equation for that. But, even though it is expressed in terms of the concentration of monomer and a rate constant, things that we usually know or can measure—it also contains a term in the concentration of radicals—a transient species whose concentration we don't know and generally can't measure.

The Steady-State Assumption

What you have to do is make an assumption to obtain an expression for this quantity in terms of things we know or can measure. We use something called the *steady-state assumption*, which says that the concentration of the transient species is constant or "steady." These little blighters are reacting all the time, of course, and for their concentration to be steady state, radicals must be generated at the same rate as they are being consumed. Radicals are generated by initiation and consumed during termination, so we can write $r_i = r_t$. This, in turn, means that we can express monomer radical concentration in terms of measurable quantities (Equations 4-28).

$$r_i = r_t$$
$$2fk_d[I] = 2k_t[M^{\bullet}]^2$$

Hence:

$$[M^{\bullet}] = \left[\frac{fk_d[I]}{k_t}\right]^{1/2}$$

EQUATIONS 4-28

POLYMER MILESTONES—FAWCETT AND GIBSON

ICI's high pressure lab after the explosion (Courtesy: ICI).

The discovery of polyethylene is often told as though it was purely an accident; and, indeed there were elements of luck associated with its discovery. But, if you delve a little deeper, you will find that "lucky" scientific discoveries are usually made by intelligent, hard working, well-read, inquisitive people, who, above all, recognize something important and unusual when they see it. E. W. Fawcett and R. O. Gibson of the ICI Chemical Company were such individuals and in 1931 they initiated a study of chemical reactions at high pressures. Fawcett, an organic chemist, was well-versed in the ideas of Staudinger, while Gibson was a physical chemist who had ample experience in special high-pressure techniques. They were to form a synergistic partnership. In some high-pressure experiments involving ethylene, Fawcett noted the presence of a small quantity of white particles floating about, but he did not think they were of any significance. But on March 27, 1933, upon dismantling the apparatus following an experiment involving ethylene and benzaldehyde, he found a white solid had coated the wall. It was polyethylene. Analysis of the white solid confirmed that it was indeed a polymer of ethylene, but, in those days there was no accurate method of determining its molecular weight. It was logical to now repeat the experiment, but use pure ethylene and omit the benzaldehyde. Surely this would result in a greater yield of polyethylene. But no, only a few small flakes were obtained. Why not increase the pressure? This appeared to be a reasonable approach, but when it was tried, the result was an explosion that demolished the laboratory! Management was not amused! All this occurred during the Great Depression, and the now familiar refrain, "Can we really afford this academic exercise?", was heard. It is amazing that the whole high-pressure program was not simply disbanded. Although, ICI banned high pressure studies until stronger reaction vessels could be developed. Fawcett and Gibson were urged to find work in other areas and Fawcett left for America to join Carothers' group at DuPont. Gibson, however, had not given up, and, risking both his life and his job, continued experiments in secret after normal working hours. In one of the experiments, (almost) pure ethylene (96%) was again tried at high pressure (200 atmospheres) and high temperature (170°C) in the new facilities and this time a white cloud of polyethylene was prepared in seemingly reasonable amounts. Success at last! But, not so fast. The yield was disappointing. What had happened to the rest of the ethylene? Enter Lady Luck! Oxygen impurities present in trace amounts in the ethylene catalyzed the reaction, while a leak lowered the pressure sufficiently to prevent them from being blown to kingdom-come! Initiation was the key! It took ICI another two years to develop a high-pressure manufacturing process, but by then they knew the polyethylene material was chemically inert, had extraordinary electrical insulating properties, was thermoplastic and could be molded into solid articles, films and fibers.

The Rate of Polymerization

Making this type of assumption may seem a bit dubious to you, but it's often the type of thing you have to do in science. In the end, the validity of any assumption you make must stand the test of experimental confirmation. So let's see how well this and also Flory's assumption, that reactivity is independent of chain length, works. The equation for the rate of polymerization is obtained by substituting the expression for the concentration of radical species into the equation for the rate of polymerization (Equations 4-29).

$$R_p = r_p = -\frac{d[M]}{dt} = k_p \left[\frac{fk_d[I]}{k_t} \right]^{1/2} [M]$$

Keep in mind that:

$$-\frac{d[I]}{dt} = k_d[I]$$

Hence:

$$[I] = [I]_0 e^{-k_d t}$$

so that:

$$R_p = \left\{ k_p \left[\frac{fk_d[I]}{k_t} \right]^{1/2} \right\} \{ [M][I]_0^{1/2} \} \{ e^{-k_d t/2} \}$$

EQUATIONS 4-29

These equations also contain a term in the concentration of initiator, which, in a batch reaction, decreases with time. We've already looked at such a first-order decay process in our brief review of chemical kinetics, so you should find the integration and substitution described in Equations 4-29 straightforward. This equation looks complicated and the first thing you're probably thinking is: "I hope my rotten, sadistic, polymers professor does not ask me to memorize that damn thing!" Heaven forbid! However, you should notice that we have broken the equation up into three bits, each in its own squiggly brackets, for the purposes of our discussion and as an aid in seeing some of the consequences of this result. Let's start with the last term, because it's straightforward. It simply tells us that the rate of reaction will slow down in an exponential fashion as the initiator is used up. The second to last term is more interesting and allows an experimental test of the assumptions that went into the derivation. It says that $R_p \sim [M][I]_0^{1/2}$. In the initial stages of the reaction, it can be assumed that $[I] \sim [I]_0$. Experiments confirm that in the initial stages, R_p indeed varies as $[I]^{1/2}$ at constant $[M]$, or if both are allowed to vary, a plot of R_p vs $[M][I]^{1/2}$ is indeed linear (Figure 4-17).

The final term we want to consider contains all the rate constants. In particular, you

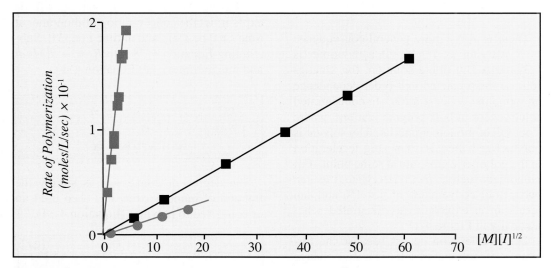

FIGURE 4-17 Graph of R_p versus $[M][I]^{1/2}$ (plotted from the data listed in P. J. Flory's book, *Principles of Polymer Chemistry*, for the polymerizations involving methyl methacrylate and styrene).

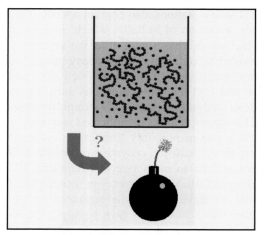

FIGURE 4-18 Schematic representation of the Trommsdorff effect.

should notice that $R_p \sim k_p/k_t^{1/2}$. This tells us a couple of things. First, for the free radical polymerization of ethylene at 130°C and a pressure of 1 bar, the value of this ratio is only 0.05. In other words, termination is much faster than propagation and you don't get polymer! At a temperature of 200°C and a pressure of 2500 bar, $k_p/k_t^{1/2}$ is about 3, enough to give polymer. That's why (in the absence of catalyst) ethylene cannot be polymerized at low pressures.

The Trommsdorff Effect

There is also a more general consequence of R_p varying as $k_p/k_t^{1/2}$. If a monomer is polymerized in dilute solution, the kinetics follows the predicted first-order dependence on monomer concentration. In concentrated solutions, or in bulk polymerizations, where a pot full of monomer undiluted by solvent is polymerized, there is often a big acceleration in the polymerization rate at some point. For monomers such as methyl acrylate, this can occur after conversions of just 1% and can result in an explosion (as illustrated schematically in Figure 4-18)!

This acceleration occurs because the formation of polymer results in a big change in the viscosity of the solution. This does not affect the diffusion of small molecules

like monomers very much, so they get to the growing chain ends readily enough and k_p stays about the same. Termination, however, involves the much slower diffusion of two large molecules, so as the viscosity of the solution increases, the rate of termination decreases, which would be reflected in an apparent decrease in k_t. The factor $k_p/k_t^{1/2}$ can then increase dramatically, resulting in a correspondingly large increase in the rate of polymerization.

The heat evolved from these exothermic reactions also increases significantly and if steps are not taken to dissipate it, there can be unfortunate consequences (you can blow up the lab, not to mention yourself)! This is called the Trommsdorff effect, even though it was apparently originally discovered by two guys named Norrish and Smith. This effect can be explored further by considering another useful quantity we can obtain from kinetics, an equation for the degree of conversion as a function of time.

Conversion

Let's go back to our expression for the rate of polymerization, $R_p = -d[M]/dt$. Integrating this will give us an expression for the change in monomer concentration with time. In case you've just crawled out of bed and haven't had your coffee yet, we will be explicit; let the concentration of monomer at time $t = 0$ be $[M]_0$, and at time t be $[M]$. Substituting Equation 4-28 into $R_p = -d[M]/dt$ and integrating we get Equation 4-30.

$$\ln \frac{[M]}{[M]_0} = 2k_p \left(\frac{f}{k_d k_t} \right)^{1/2} [I]_0^{1/2} \{1 - e^{-k_d t/2}\}$$

EQUATION 4-30

The degree of conversion is simply the fraction of monomer that has been used up or converted to polymer (Equation 4-31).

$$Conversion = \frac{[M]_0 - [M]}{[M]_0} = 1 - \frac{[M]}{[M]_0}$$

EQUATION 4-31

FASCINATING POLYMERS—THE PLASTIC SODA BOTTLE

The poly(ethylene terephthalate) (PET) soda bottle is dominant and ubiquitous. PET is a good barrier to carbon dioxide (the fizz in soda), is relatively light (compared to glass) and the bottles perform their function admirably. But, decades ago, PET had serious competition from a random SAN copolymer composed of styrene (S) and acrylonitrile (AN). A major fight ensued between the various manufacturers, who were attempting to have their own materials chosen for this lucrative market. Responsible scientists are always concerned about the safety of such materials, especially if they are to be used in a situation where they come in contact with food or beverages. In the case of the soda bottle, one major concern was whether or not any of the ingredients of the plastic, or impurities in it, could leach into the soda over time and be detrimental to people's health. In fact, in mice studies, AN was shown to produce rare brain cancers. It also didn't help that another name for acrylonitrile is vinyl cyanide! (This produces visions of "megabucks" in the eyes of an unscrupulous lawyer!) It was established that the monomer, AN, was only present in soda bottles in infinitesimally small amounts (one can never state that there is absolutely nothing of something present in anything!), so that the probability that acrylonitrile could leach into the soda was incredibly small. Accordingly, over the lifetime of the product the bottle was effectively "safe." Nevertheless, the SAN

A nice collection of PET bottles.

bottle was condemned to death! Would you let your baby drink the soda or suck on the bottle if there were any possibility that some acrylonitrile, even if undetectable, could be imbibed? It's cyanide for Heaven's sake! Only it really isn't. Although somewhat of an exaggeration, this is analogous to saying that there is chlorine, which was used as a poisonous gas in the trenches of WWI, in common table salt! This type of fallacious argument, which falls under the general heading of "junk science," is not uncommon in the courtroom. Perception, not scientific logic, is often the rule of the day. The PET bottle was clearly an appropriate choice for soda bottles given that AN was a known carcinogen. But true scientists can never say that there are no monomers (terephthalate esters and ethylene glycol) present in PET that could be leached out into your coke. And, in fact, when you study the statistics of linear polycondensation, you'll discover that the presence of monomers in PET is inevitable! The appropriate questions are at what level are the low molecular weight species present, at what rate do they diffuse into the contents of the bottle and at these concentrations do they pose a threat to humans, big and small? There is no credible evidence that PET bottles are harmful. But remember, common sense is paramount and *nothing* we do as humans is absolutely risk free.

$$Conversion = 1 - exp\left(-\left\{2k_p\left(\frac{f}{k_d k_t}\right)^{1/2}[I]_0^{1/2}(1 - e^{-k_d t/2})\right\}\right)$$

EQUATION 4-32

$$Maximum\ Conversion = 1 - exp\left(-\left\{2k_p\left(\frac{f}{k_d k_t}\right)^{1/2}[I]_0^{1/2}\right\}\right)$$

EQUATION 4-33

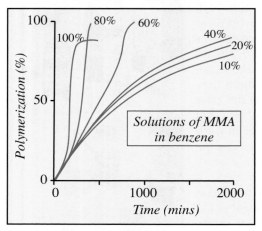

FIGURE 4-19 Graph of conversion (%) versus time [redrawn from the data of G. V. Schultz and G. Harborth, *Macromol. Chem.*, **1**, 106, (1967)].

Substituting for the last term (Equation 4-30), we then get a rather horrible-looking expression for conversion as a function of time (Equation 4-32). Note that because the concentration of initiator decreases exponentially with time, we get the last term in brackets, which tells us conversion never gets to be complete but is always less than

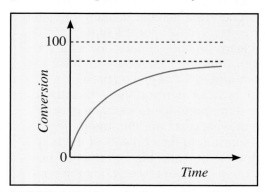

FIGURE 4-20 Schematic of conversion (%) versus time.

1. We'll come back to that in a minute. First let's look at a typical example of experimental data, such as that obtained for the polymerization of methyl methacrylate. Plots of conversion versus time for solutions of up to 40% in benzene follow the predicted relationship nicely. But, at higher concentrations and in the bulk (i.e., 100% methyl methacrylate), the Trommsdorff or auto-acceleration effect can clearly be seen (Figure 4-19).

Finally, an expression for the maximum conversion can be obtained by allowing t to go to infinity in Equation 4-32, which produces Equation 4-33. This is important, in that it tells us that in these types of batch reactions there is always residual monomer (Figure 4-20). If the monomer is nasty and/or carcinogenic, like vinyl chloride or acrylonitrile (which you could call vinyl cyanide if you really want to make it sound bad—see *Fascinating Polymers* on the previous page), then your product might be unacceptable for certain applications, like beverage bottles, where unreacted monomer might leach out into the stuff that people drink.

Average Chain Length

It was a relatively trivial matter to write an equation for the number average degree of polymerization in terms of degree of conversion for a step-growth or condensation polymerization, because in the reaction pot there is a continuous distribution of species present and all these species are taking part in the polymerization. High molecular weight material is not formed until high degree of conversions are reached. In chain polymerizations, chains are initiated, grow quickly and (usually) terminate, forming polymer in the first few seconds. There is no continuous distribution; but, depending on the polymerization conditions, at any instant of time there is a mixture of monomer, fully formed and no longer polymerizing polymer, and just a few growing chains (Figure 4-21).

Accordingly, unlike step-growth polymerizations, where it makes sense to define an average over the whole reaction pot, including unreacted monomers, oligomers, and so on, for chain polymerizations it makes more sense to define an average in terms of the reaction product, leaving out unreacted monomer. But we also have to account for the fact that we don't have steady-state conditions and the nature of the product (the length of the polymer chains) changes with time, as the concentration of monomer, unreacted initiator, etc., decreases as the reaction proceeds. We therefore define a kinetic chain length, v, which is the average number of monomers polymerized per chain radical at a particular instant of time during the polymerization. This is given by the rate of monomer addition to growing chains divided by the rate at which chains are started. In other words, the rate of propagation divided by the rate of initiation. If you find this confusing or incomprehensible, go back to the old trick of putting in some arbitrary numbers. Let's say that during a time period, t, 100 chains are started and 1,000,000 monomers are reacted or used up. The average degree of polymerization of the chains that were initiated in this time period would be 1,000,000/100, or 10,000.

Obviously, the longer the time period we choose, the more things will be different at the end than at the beginning. What we really want to do is consider shorter and shorter time periods. In the limit we then get Equation 4-34, which will be the number average length of the chains being formed at the chosen instant of time.

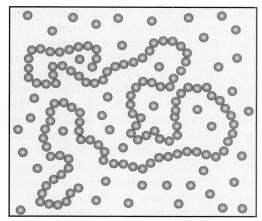

FIGURE 4-21 Schematic diagram of a terminated polymer chain in a sea of monomers.

It is then a trivial matter to substitute our kinetic equations for r_p and r_i (Equation 4-35).

$$v = \frac{r_p}{r_i} = \frac{k_p [M^\bullet][M]}{2fk_d [I]}$$

EQUATION 4-35

We can then make the usual steady-state assumption and substitute for the concentration of chain radicals (Equation 4-36).

$$v = \frac{k_p}{2(fk_d k_t)^{1/2}} \frac{[M]}{[I]^{1/2}}$$

EQUATION 4-36

Note that unlike the rate of polymerization, where R_p goes as $[M][I]^{1/2}$, the kinetic chain length goes as $[M]/[I]^{1/2}$. This makes intuitive sense if you think about it; the fewer chains you start, the smaller the number of radical species at any instant of time, hence the probability of chains meeting to terminate is less and the chains on average are longer. It also means that you cannot arbitrarily change conditions to affect the rate of polymerization without also changing the molecular weight of the product.

The kinetic chain length is clearly equal to the number average degree of polymeriza-

$$v = \frac{Rate\ of\ Propagation}{Rate\ of\ Initiation} = \frac{r_p}{r_i}$$

EQUATION 4-34

tion, if termination is by disproportionation. But what if termination is by combination? Does the number average degree of polymerization equal v or $2v$? The answer is so trivial we'll just observe that you should give up polymer science and join a flaky internet startup if you don't know the answer. The real problem is that termination often occurs by both mechanisms, but not in equal proportions. To account for this, a new parameter ξ is defined, equal to the average number of "dead" chains formed per termination. Because termination by disproportionation gives two dead chains, while termination by combination only gives one, we can write Equation 4-37 for the rate of dead chain formation (i.e., disappearance of chain radicals, $M^{\bullet}$).

$$-\frac{d[M^{\bullet}]}{dt} = (2k_{td} + k_{tc})[M^{\bullet}]^{2}$$

EQUATION 4-37

Note that disproportionation results in the formation of two dead chains, hence the factor 2 in the equation. Dividing this term by the expression for the rate of termination we get Equation 4-38.

$$\xi = \frac{(2k_{td} + k_{tc})[M^{\bullet}]^{2}}{(k_{td} + k_{tc})[M^{\bullet}]^{2}} = \frac{(2k_{td} + k_{tc})}{k_{t}}$$

EQUATION 4-38

It is then easy to figure out that the instantaneous number average degree of polymerization is given by Equation 4-39.

$$\overline{x_n} = \frac{2v}{\xi} = 2v\left(\frac{k_{tc} + k_{td}}{k_{tc} + 2k_{td}}\right)$$

EQUATION 4-39

So, if termination is predominantly by combination ($k_{tc} \gg k_{td}$) the average degree of polymerization is equal to $2v$, while if ter-

mination is predominantly by disproportionation ($k_{td} \gg k_{tc}$) it is simply equal to v.

A more formal derivation of Equation 4-39, useful when we get to chain transfer, is obtained starting from $v = r_p/r_t$. (Remember the steady-state assumption, $r_i = r_t$?) To obtain the number average degree of polymerization instead of the kinetic chain length we then use the rate of dead chain formation instead of r_t (Equations 4-40).

$$\overline{x_n} = \frac{k_p[M][M^{\bullet}]}{(2k_{td} + k_{tc})[M^{\bullet}]^{2}}$$

$$\overline{x_n} = \left(\frac{k_p}{\xi(fk_d k_t)^{1/2}}\right)\frac{[M]}{[I]^{1/2}}$$

$$\overline{x_n} = 2v\left(\frac{k_{tc} + k_{td}}{k_{tc} + 2k_{td}}\right)$$

EQUATIONS 4-40

Chain Transfer

Right at the beginning of our discussion of the kinetics of free radical polymerization, we mentioned chain transfer, then proceeded to ignore it! It's time to go back and account for this factor. Chain transfer is simply the process by which a growing chain radical is terminated and a new one started. Usually, the process occurs by proton transfer (but note that it is not always a proton that is transferred). A lot of components in a polymerization can act as so-called chain transfer agents: monomer, solvent, the polymer chains themselves (giving long chain branching), etc. Specific chain transfer agents are also often added to a polymerization to control the molecular weight. In general (although not always), chain transfer does not affect the rate of polymerization, because as soon as one chain is terminated by chain transfer another is started by the newly formed radical. So all we will consider here is how the instantaneous average molecular weight is affected.

The effect of chain transfer can easily be

accounted for by realizing that this process simply provides an additional mechanism for the rate of dead chain formation. If $[T]$ is the concentration of the chain transfer agent, then we can write Equation 4-41.

$$-\frac{d[M^{\bullet}]}{dt} = (2k_{td} + k_{tc})[M^{\bullet}]^2 + k_{tr}[T][M^{\bullet}]$$

EQUATION 4-41

Substituting as before leads to Equation 4-42.

$$\overline{x}_n = \frac{k_p}{\xi(fk_d k_t[I])^{1/2} + k_{tr}[T]}$$

EQUATION 4-42

It is convenient to rearrange this by inverting the equation (Equation 4-43), as this gives the instantaneous number average molecular weight in terms of the average chain length that would be obtained in the absence of chain transfer and terms describing the effect of chain transfer ($C = k_{tr}/k_p$).

$$\frac{1}{\overline{x}_n} = \frac{1}{(\overline{x}_n)_0} + C\frac{[T]}{[M]}$$

EQUATION 4-43

A plot of the reciprocal of the number average degree of polymerization versus $[T]/[M]$ should then be linear, if the analysis is correct. And indeed it is, as illustrated for the polymerization of styrene in various solvents, which act as chain transfer agents (hence the use of the symbol $[S]$ instead of $[T]$ in Figure 4-22).

Anionic Polymerization

Now let's consider the kinetics of other types of chain polymerizations. Except that now you're going to do the work! But it's not that bad, you just follow the general scheme that we set up for free radical polymeriza-

tion. We'll start with anionic polymerization in a protic solvent. First you write down the chemical equations for each of the steps (Figure 4.23). We'll assume $NaNH_2$ is the initiator. Remember (see Chapter 3) that there is no termination step, but there is chain transfer to solvent. [N.B.: we're assuming there are no impurities present (like water) which can terminate the polymerization by chain transfer.]

Now write down the kinetic equations for each of the steps (Equations 4-44).

$$R_i = \frac{d[M^{\ominus}]}{dt} = k_i[NH_2^{\ominus}][M]$$

$$R_p = -\frac{d[M]}{dt} = k_p[M^{\ominus}][M]$$

$$R_{tr} = -\frac{d[M^{\ominus}]}{dt} = k_{tr}[M^{\ominus}][NH_3]$$

$$k_d = \frac{[NH_2^{\ominus}][Na^{\oplus}]}{[NaNH_2]}$$

EQUATIONS 4-44

What we want you to do is show that the rate of polymerization is second-order in

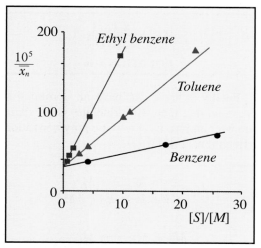

FIGURE 4-22 Graph of $1/x_n$ versus $[S]/[M]$ [redrawn from the data of R. A. Gregg and F. R. Mayo, *Faraday Soc. Discussions*, **2**, 328 (1947)].

$$NaNH_2 \underset{}{\overset{k_d}{\rightleftharpoons}} Na^{\oplus} + H_2N^{\ominus}$$

$$H_2N^{\ominus} + M \xrightarrow{k_i} H_2N{-}M^{\ominus}$$

$$H_2N{-}M_i^{\ominus} + M \xrightarrow{k_p} H_2N{-}M_{i+1}^{\ominus}$$

$$H_2N{-}M_i^{\ominus} + NH_3 \xrightarrow{k_{tr}} H_2N{-}M_iH + H_2N^{\ominus}$$

FIGURE 4-23 Chemical equations for anionic polymerization in a protic solvent.

monomer concentration—you should get Equation 4-45.

$$R_p = k_p \frac{k_i}{k_{tr}} k_d^{1/2} \frac{[NaNH_2]^{1/2}}{[NH_3]}[M]^2$$

EQUATION 4-45

You will need to make a steady-state assumption that the concentration of anion radicals is constant (and you should also note that $[Na^+] = [NH_2^-]$). But you're not done yet! Now if you assume that the number average degree of polymerization is equal to the average number of monomers added to an anion before chain transfer to NH_3, derive Equation 4-46.

$$\bar{x}_n = \frac{k_p[M]}{k_{tr}[NH_3]}$$

EQUATION 4-46

Finally, what if we have an aprotic solvent, so that there (in principle) is no chain transfer and we have a living polymerization (Equations 4-47)?

$$I^{\ominus} + M \xrightarrow{k_i} M_1^{\ominus}$$
$$-d[I^{\ominus}]/dt = k_i[I^{\ominus}][M]$$

$$M_1^{\ominus} + M \xrightarrow{k_p} M_{1+1}^{\ominus}$$
$$-d[M]/dt = k_p[M^{\ominus}][M]$$

EQUATIONS 4-47

We want you to show that the rate of polymerization is first-order in monomer concentration (assume that initiation is fast, so that $[M^{\ominus}] = [I]$).

Cationic Polymerization

In cationic polymerization, all sorts of things can occur. For this problem we will again make things easy for you and provide the equations for initiation and propagation (Figure 4-24). What we want you to do is derive an equation for the rate of propagation for two limiting conditions.

a) Assume termination occurs solely by combination of the cation chains with their counterions (Equation 4-48):

$$M_i^{\oplus}A^{\ominus} \overset{k_t}{\rightleftharpoons} M_iA$$

EQUATION 4-48

b) Assume no termination, but there is chain transfer to solvent (Equation 4-49):

$$M_i^{\oplus} + S \xrightarrow{k_{tr}} M_iS_1 + S_2^{\oplus}$$

EQUATION 4-49

Now you have to work out the rest of the equations and the appropriate assumptions yourself! (You should find that the rate of polymerization is second-order in monomer concentration.)

INITIATION

$$C + AH \rightleftharpoons \{H^{\oplus}\ CA^{\ominus}\} \qquad K = \frac{[H^+CA^-]}{[C][AH]}$$

$$\{H^{\oplus}\ CA^{\ominus}\} + M \xrightarrow{k_i} HM^{\oplus}CA^{\ominus} \qquad R_i = k_i[H^+CA^-][M]$$

PROPAGATION

$$M_i^{\oplus} + M \xrightarrow{k_p} M_{i+1}^{\oplus} \qquad R_p = -\frac{d[M]}{dt} = k_p[M^+][M]$$

FIGURE 4-24 Chemical and kinetic equations for cationic polymerization.

RECOMMENDED READING

H. R. Allcock, F. W. Lampe and J. E. Mark, *Contemporary Polymer Chemistry*, 3rd. Edition, Prentice Hall, New Jersey, 2003.

P. J. Flory, *Principles of Polymer Chemistry*, Cornell University Press, Ithaca, New York, 1953.

G. Odian, *Principles of Polymerization*, 3rd. Edition, Wiley, New York, 1991.

STUDY QUESTIONS

1. Derive an expression for the relationship between the extent of reaction, p, and time, t, in an uncatalyzed polyesterification. Show all the integrations explicitly!

2. You are working for a company that many years ago conducted a number of polycondensation reaction experiments between diol and diacid monomers. You need data on the kinetics of one of these reactions from the archives, but much was lost in a flood. All you find are the results of the original experiments, where small aliquots of sample were withdrawn from the reaction vessel, quenched to low temperature to stop the reaction and titrated to determine the amount of unreacted acid functional groups. The results given below show how much of a standard NaOH solution (in cc/gm) was necessary to neutralize the acid functional groups. Using these results, determine if the reaction was catalyzed or uncatalyzed.

Time (mins)	Titer/g (cm^3/g)	Time (mins)	Titer/g (cm^3/g)
0	136.9	300	4.107
25	68.47	400	3.013
50	34.23	500	2.163
100	17.12	600	1.918
150	10.55	700	1.644
200	6.848		

Hint: calculate the extent of reaction, p, from this data and proceed from there.

3. Derive the expression for the rate of a free radical polymerization. Using this expression, account for the Trommsdorff effect and the inability of ethylene to polymerize free radically at ordinary temperatures and pressures.

4. In a reaction vessel containing pure monomer and no initiator, free radicals were generated thermally and this process involved two molecules of monomer

$$M + M \xrightarrow{\Delta} R^{\bullet}$$

A. Show that if a simple assumption is made concerning the order of this reaction, then the rate of polymerization is proportional to the square of the monomer concentration. Hint: although there is no separately added initiator, you can obtain an expression for the rate of initiation by simply writing down an expression for the rate of formation of $R^{\cdot}$,

$$r_i = \frac{d[R^{\cdot}]}{dt} = ?$$

B. Derive an expression for the kinetic chain length.

C. How does $\bar{x}_n$ change with an increase in the monomer concentration?

5. If equal concentrations of acrylonitrile and methyl methacrylate were each polymerized at 60°C with equal concentrations of the same initiator, which polymer would have the higher DP and by how much? Assume polyacrylonitrile undergoes termination only by radical combination and poly(methyl methacrylate) by disproportionation, that no chain transfer occurs, and that initiator efficiencies are the same in both reactions. (Use Table 4-1 below.)

6. What concentration of benzoyl peroxide

($k_d = 1.45 \times 10^{-6}$ L/mol-s at 60°C) would be needed to polymerize a 1.00 M solution of styrene to a molecular weight of 2500 (number average). Assume that termination occurs only by radical combination and that initiator efficiency is 100%. (Use Table 4-1 below.)

7. How much isopropyl alcohol, chain transfer constant ($C = k_{tr}/k_p$) equals 3.1×10^{-4} (deg. of polym.)$^{-1}$ would have to be added to the polymerization to lower the degree of polymerization to 1250?

8. In the section on anionic polymerization, we mentioned that it was time for you to do some work! So here is the homework question you have been dreading. Show that in an anionic polymerization of an appropriate monomer in a protic solvent, NH_3, initiated using $NaNH_2$, the rate of polymerization is second-order in monomer concentration. Also, obtain an expression for the kinetic chain length. Try not to throw up on your answer sheet!

9. Similarly for cationic polymerization (Figure 4.24), derive an equation for the rate of propagation for two limiting conditions:

TABLE 4-1 REPRESENTATIVE PROPAGATION AND TERMINATION RATE CONSTANTS

MONOMER	TEMPERATURE (°C)	$k_p \times 10^{-3}$ (L/mol-s)	$k_t \times 10^{-7}$ (L/mol-s)
Acrylonitrile	60	1.96	78.2
Chloroprene	40	0.220	9.7
Ethylene	83	0.242	54.0
Methyl acrylate	60	2.09	0.95
Methyl methacrylate	60	0.515	2.55
Styrene	60	0.176	7.2
Vinyl acetate	50	2.64	11.7
Vinyl chloride	50	1.10	21.0
Tetrafluoroethylene	40	7.40	7.4

Note: $k_p \times 10^{-3}$ in the heading of the column means that all the numbers have been multiplied by 10^{-3}. In other words, the value of k_p for acrylonitrile should be 1.96×10^3.

A. When termination occurs solely by combination of the cation chains and their counterions.

B. When there is no termination, but there is chain transfer to solvent (Equation 4-48).

5

Polymerization: Probability and Statistics

INTRODUCTION

Probability and statistics are powerful tools in the study of polymerization and can be used to describe things like molecular weight distributions in homopolymerizations and sequence distributions in copolymerization. In an ideal world, most students should have at least a passing acquaintance with the fundamentals of simple probability theory before getting to study polymers in detail (even if it is only to ask the questions posed in Figures 5-1 and 5-2.) But, just as with kinetics and thermodynamics, it is our experience that for many most of this stuff hasn't quite set in and remains a confused jumble occupying parts of the brain that remain curiously unconnected. Fortunately, most of what we'll cover here is very simple; but just in case you've skillfully avoided all courses involving basic probability and statistics up to this point, or if you've destroyed the part of your brain that carries this information as a result of some wicked over-indulgence, we'll try to give you some of the basics.

Probability of an Event

Actually, for the subject we will cover first, the statistics of linear polycondensation, all you really need is a simple definition, some common sense and the ability to reason. OK, in the common sense department some of you are already in trouble! All we can do is

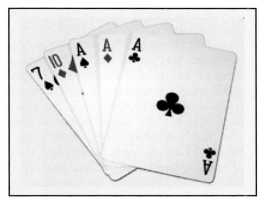

FIGURE 5-1 What is the probability of getting three aces in a poker hand?

remind you of the definition (Equation 5-1):

$$P\{E\} = \frac{N_E}{N}$$

EQUATION 5-1

This simply states that the probability of an event, E, written as $P\{E\}$, is equal to N_E, the number of E events occurring, divided by

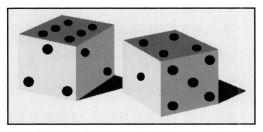

FIGURE 5-2 How many ways are there of throwing a ten?

A-B + A-B ⇌ A-BA-B

TYPE I

A-A + B-B ⇌ A-AB-B

TYPE II

FIGURE 5-3 Types of step-growth polymerization.

N, the total number of events. For example, just imagine that you toss a coin one hundred times and measure the number of heads. You expect N_E would be about 50 and $P\{E\}$ would be about one-half. Most often, N_E wouldn't be exactly 50, you would have to repeat the 100 coin tosses a large number of times and average all the values to get $P\{E\}$ = 0.5, but you get the idea. The important thing at this point is to associate the probability of an event occurring with the number fraction of such events, given that you are measuring a really large number of events. That won't, in general, be a problem for us, because if we are considering, say, a mole of reacting molecules, we are dealing with billions and billions of species. Also, note that the probability of an event is always between 0 and 1. So, if you have a homework problem where your answer is outside these limits, you screwed up! Moreover, if there are only two possible events, as in our coin toss (heads or tails), then the probability of the other event occurring (tossing a tail) is $1 - P\{E\}$. Just keep all this stuff in the back of your mind as you go on to our first topic, step-growth polymerization.

STATISTICS OF LINEAR POLYCONDENSATION

As we mentioned in the introduction to this chapter, we will start by considering the statistics of linear step-growth polymerization. Remember that there are two types of such reactions: in the first, each bifunctional monomer has different but complementary functional groups, an acid, A, and an alcohol, B, for example (i.e., A–B); in the second type, each monomer only has one type of functional group (i.e., A–A and B–B). In each case an "A" can only react with a "B,",in this example to give an ester, which we've labeled either AB or BA in Figure 5-3 (think about it—they are equivalent and only differ in direction along the chain).

In reactions of type I, we can count the number of molecules by measuring the concentration (i.e., number/unit volume) of one of the end groups, say, the acid, by titration (in the good old days) or by spectroscopic methods.

It's easy to see that counting the unreacted A's counts the molecules, but let's see if you can remember something about averages. If, in making this bunch of oligomers, we had started with 48 monomers, what would be the number average degree of polymerization of the molecules in the box (Figure 5-4)? The answer of course is 3 (48/16), which you hopefully didn't get by guessing, but by remembering that the number average degree of polymerization is simply the number of molecules (monomers) you started with divided by the number of molecules you have at the time you stop the reaction and count the end groups (e.g., the A's).

It is easy to verify that measuring the number of end groups also gives a measure of the number of molecules in a type II polycondensation, providing that you start with exactly equal numbers of molecules. Check it out by drawing pictures—you'll find that the number of molecules that end up with A's on both ends is equal to the number with B's

A-BA-BA-BA-BA-B   A-B

A-BA-BA-B   A-B   A-BA-BA-B

A-BA-BA-BA-B

A-BA-BA-BA-BA-BA-BA-BA-B

A-B   A-BA-B   A-BA-BA-B   A-B

A-B

A-BA-BA-BA-BA-BA-BA-BA-B

A-BA-B   A-BA-BA-B   A-BA-BA-B

FIGURE 5-4 A bunch of oligomers.

$$p = \frac{\textit{Number of COOH groups reacted}}{\textit{Number of COOH groups originally present}}$$

EQUATION 5-2

on both ends—so you get an accurate count. We've covered this already in our discussion of kinetics. What we want to do here is show how we can use probability and statistics to determine what happens if we don't have an equal number of reacting functional groups, or we have some monofunctional additive or contaminant. We can also use these tools to calculate distributions, which give us much more information than simple averages.

To do all this, we first have to revisit an old friend, the parameter, p, which we defined in our discussion of kinetics as the extent of reaction, equal to the fraction of functional groups of a particular type that have reacted. For example, if we are measuring acid groups we would use Equation 5-2. We also saw that the number average degree of polymerization was given by (Equation 5-3):

$$\overline{x}_n = \frac{N_0}{N} = \frac{1}{(1-p)}$$

EQUATION 5-3

so that "high polymer" (an old term that we still like) is only produced at high conversions (see previous Chapter—Figure 4-12). For example, if you wanted to obtain a degree of polymerization of 200, what degree of conversion would you need to achieve in your reaction?

Non-Stoichiometric Equivalence

The relationship between number average degree of polymerization and p (Equation 5-3) only applies to polymerizations where there are exactly equal numbers of A and B functional groups. Equivalence is obtained directly in step-growth polymerizations of type I (A–B), but is more difficult to achieve for polymerizations of type II (A–A and B–B). However, if the number of functional groups is not exactly the same, we still

have the same definition of number average degree of polymerization (Equation 5-4);

$$\overline{x}_n = \frac{N_0}{N}$$

EQUATION 5-4

but the relationship to p is different. What we have to do is relate N and N_0 to the concentration of monomers and the extent of reaction.

We will start by defining a new quantity, r, equal to the ratio of the number of functional groups (not monomers)—Equation 5-5.

$$r = \frac{N_A}{N_B}$$

EQUATION 5-5

The parameter r is defined so as to always be a fraction, so that the subscripts A and B no longer refer to a particular monomer (we've been pretty consistent in letting A represent carboxylic acid functional groups in polyesterifications). In other words, B must always represent the functional groups that are present in excess. Keep in mind that N_A and N_B represent the number of functional groups present, so that the number of monomers, N_0, is given by Equation 5-6.

$$N_0 = \frac{N_A + N_B}{2} = \frac{N_A}{2}\left(\frac{1+r}{r}\right)$$

EQUATION 5-6

This assumes each is bifunctional. We have N_0, now all we need is N. Remember, we can count N by counting the number of end groups. Because we have unequal numbers of functional groups, we need to count all the end groups, not just those of one type, and then divide by 2. The number of chain ends, or end groups, after a fraction p have reacted is given by (Equation 5-7):

$$N_A(1 - p) + (N_B - pN_A)$$

EQUATION 5-7

Your first reaction on seeing this might be: huh? Remember, A represents the minority species. After a fraction p have reacted, there are $1 - p$ left. The number of A groups that have reacted is pN_A. But, A only reacts with B, so the number of B groups that have reacted must also be pN_A. The number of molecules, N, left after a degree of conversion, p, is then obtained by dividing by 2 and the equation can be rearranged by substituting $N_B = N_A r$ to yield Equations 5-8.

$$N = \frac{1}{2}[N_A(1 - p) + (N_B - pN_A)]$$

$$N = \frac{N_A}{2}\left[(1 - p) + \frac{(1 - rp)}{r}\right]$$

EQUATIONS 5-8

Now we have expressions for both N and N_0 and we can substitute to obtain the Equation 5-9.

$$\overline{x}_n = \frac{N_0}{N} = \frac{(1 + r)}{(1 + r) - 2rp}$$

EQUATION 5-9

Note that we recover our original expression (Equation 5-3) if $r = 1$ (it's always good to check these things). Also, this equation tells us that there is a theoretical upper limit that can be achieved, even if the conversion of the minor component (A) is pushed to the limit ($p \longrightarrow 1$).

$$\overline{x}_n = \frac{1 + r}{1 - r} \ (as \ p \to 1)$$

EQUATION 5-10

For example, if we start with 100 moles of

A and 99 moles of B, what is the maximum number average degree of polymerization?

Chain Stoppers

Finally, small amounts of monofunctional groups (e.g., acetic acid, CH_3COOH) can be added to a polymerization to control molecular weight, or can be present as an impurity in the monomer "feed."

A-BA-BA-BA-BA-BA-BA-R

I'm unreactive! ➚

It turns out (see Flory) that we can use exactly the same equation to describe the degree of polymerization (Equation 5-9), as long as r is redefined (Equation 5-11) as:

$$r = \frac{N_A}{N_B + 2N_B^M}$$

EQUATION 5-11

This is the basis for some mean and nasty homework and test questions, so keep it in mind! Note that N_B^M is the number of *monofunctional* molecules.

The Probability That a Group Has Reacted

So far we haven't used probabilities in any meaningful way. Brace yourself, because now we start! Probability has actually entered our equations implicitly through the parameter, p, *the fraction of functional groups that have reacted*. This is also equal to the probability of finding that one such group, chosen at random from the reaction vessel, has reacted. Once again, your reaction might be: huh? Think of it this way. Let's say you start with a bowl of apples. Then you sprinkle some magic foo-foo dust on them and half turn to oranges (Figure 5-5). What would be the probability that blindfold you randomly choose an item from the bowl that has reacted? You would guess 0.5, and you would be right. (Be careful here, we are assuming that you are choosing according to number, not size. If the oranges are twice as big as the apples and you choose by asking the probability that an element of volume is occupied by an orange, you would get a different answer. This distinction is important when we discuss polymer solutions.)

The Probability of Finding an X-mer

What we are trying to do is determine the distribution of species in the pot at some point in time defined by the degree of conversion. In other words, what would be the fraction, by number, of monomers, dimers, trimers, etc., when a fraction p groups (e.g., $p = 0.8$) have reacted? What would be the distribution by weight? Let's start with the distribution by number and follow the original analysis of Flory. We ask what would be the probability that a molecule, taken at random from the reaction pot after a conversion p, is an *x-mer*, i.e., has x units in the chain. For example, the oligomer above is made up of five monomers and is a "*5-mer.*" The way you tackle this is to say: well, if I have a 5-mer then this means that 4 of my groups have reacted. Because A's only react with B's and vice-versa, we only have to focus on one of the functional groups. We'll choose the B's. So, what is the probability that the first B group in the chain, going from left to right, has reacted?

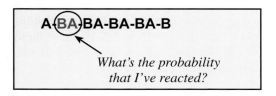

What's the probability that I've reacted?

If you didn't say p, then you're really lost when it comes to probabilities and you should give up polymer science and become a sociologist. Moving on, what is the individual probability that the second group has reacted? Another p, of course. But, and this is the crux, what is the probability that both groups taken together have reacted?

FIGURE 5-5 A nice bowl of apples and oranges.

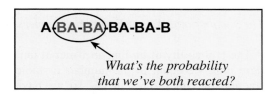

What's the probability that we've both reacted?

It's like asking what is the probability that on tossing a coin you get two heads in a row: we know that the individual probability of getting a head each time is 0.5, but now we're asking something different, what is the probability that in two throws we get two heads? Or in this case, what is the probability that two groups have reacted? It's p^2, of course. We're getting there. In our 5-mer, four B groups have reacted, and the probability that all four taken together have reacted must therefore be p^4.

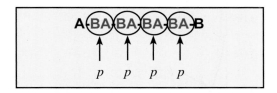

$$p \quad p \quad p \quad p$$

Do you think we've finished? No, there's still one B left unaccounted for. We've got to the end of the chain and there's still one B group hanging out, unreacted. We can't ignore it. The probability that this group has not reacted is $1 - p$.

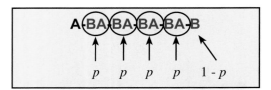

So the overall probability of finding a 5-mer after p groups have reacted is $p^4(1 - p)$. Now we can simply generalize this to the probability of finding an x-mer. Piece of cake. If the probability of finding an x-mer is P_x, then we obtain Equation 5-12.

$$P_x = p^{x-1}(1 - p)$$

EQUATION 5-12

The Number and Weight Fraction of X-mers

The probability of finding an x-mer at random must be equal to the number or mole fraction of x-mers present, X_x, so we have our first answer (Equation 5-13).

$$P_x = \frac{N_x}{N} = X_x = (1 - p)p^{(x-1)}$$

EQUATION 5-13

So the mole fraction of monomers present

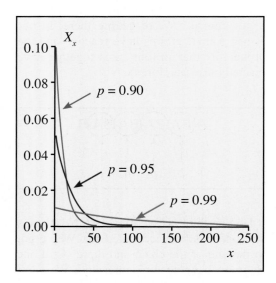

FIGURE 5-6 Number or mole fraction of x-mers for different values of p.

when $p = 0.5$ is simply $(1 - 0.5)0.5^0 = 0.5$. The mole fraction of dimers is $(1 - 0.5)0.5^1 = 0.25$. And so on. You can obtain your own plots of X_x versus x for various values of p (Figure 5-6). No matter what the value of p, one species always predominates by number. What species is that? You should have found that there is always a larger number of monomers present than any other species at any degree of conversion. Accordingly, for certain applications, like food wrap or beverage bottles, this residual monomer better not be toxic or carcinogenic!

Although, by number, we always have more monomers than anything else, by weight, the proportion can be small. You can have a mixture consisting of 10 moles of monomer and 1 mole of 200-mer, for example, and the latter will out outweigh the former by a factor of 20. So, we would also like to calculate the distribution by weight. The weight of an x-mer is simply x times the weight of each unit, xM_0, so the weight of all x-mers present must be the number of x-mers, N_x, multiplied by this factor, $N_x x M_0$. The weight fraction of x-mers is then given by the Equation 5-14.

$$w_x = xN_xM_0$$

EQUATION 5-14

The Number and Weight Distribution Functions

We need to express N_x in terms of x and p and to do so we simply use a set of substitutions. Equations 5-15 and 5-16 show these for the number and weight distributions, respectively.

$$\overline{x}_n = \frac{N_0}{N} = \frac{1}{(1 - p)}$$

We have:

$$N = N_0(1 - p)$$

Hence:

$$N_x = N_0(1 - p)^2 p^{(x-1)}$$

EQUATIONS 5-15

$$w_x = \frac{xN_xM_0}{N_0M_0} = \frac{wt \ of \ all \ x\text{--}mers}{wt \ of \ all \ units}$$

$$N = N_0(1 - p)$$

$$N_x = N_0(1 - p)^2 p^{(x-1)}$$

Hence:

$$w_x = x(1 - p)^2 p^{(x-1)}$$

EQUATIONS 5-16

(Note that we have assumed that the molecular weight of a monomer is the same as that of a repeat unit—this often isn't true, but the correction is easily made and this way we don't clutter things up with too much algebra.)

Values of the weight fraction of x-mer, w_x, plotted as a function of x, are shown in Figure 5-7 for the same values of p that were used to plot values of the number or mole fraction earlier. Note that there is now a maximum that shifts to higher values of x as p increases. The distribution also broadens considerably at high values of conversion.

The Schulz-Flory Distribution

The distribution described by Equations 5-15 and 5-16 is called the *most probable*, or *Schulz-Flory* distribution. It rests on the assumption that we made in our discussion of the kinetics of these reactions, but not explicitly in our treatment of their statistics. This assumption, of course, is that reactivity is independent of chain length. This allows us to assume that the probability that a group has reacted is p, no matter what the length of the chain in which it is found. How good is this assumption and, hence, the equations? We assumed that this could be easily checked using size exclusion chromatography, which was not available when these equations were first derived. To our surprise, we have not been able to find a paper containing such data. A painstaking fractionation study by Taylor was performed many years ago and the results are shown in Figure 5-8. Fractions obtained by precipitation from solution almost invariably contain a distribution of

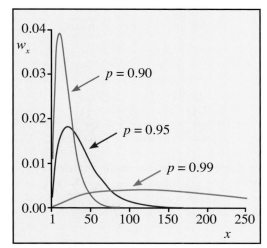

FIGURE 5-7 Weight fraction of x-mers for different values of p.

species, as opposed to a single type of x-mer. However, both Taylor's experimental results and their agreement with the experiment shown in Figure 5-8 should be considered excellent and a vindication of the validity of the assumptions made in the derivation.

Finally, the equations for the most probable distribution can be used to calculate various average degrees of polymerization: number, weight, z, etc. Earlier, we derived an equation for the number average in a very simple way. The method shown in Figures 5-9 and 5-10 is more complicated. Its value is that, unlike the simple approach, it

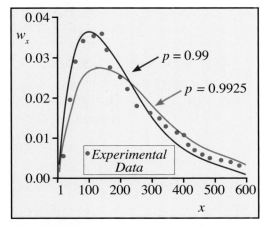

FIGURE 5-8 Experimental results redrawn from the data of G. B. Taylor, *JACS*, **69**, 638 (1947).

Definition of number average:

$$\overline{x}_n = \frac{\sum x N_x}{\sum N_x}$$

Which can be written:

$$\overline{x}_n = \frac{N_1}{\sum N_x} + \frac{2N_2}{\sum N_x} + \frac{3N_3}{\sum N_x} + \cdots$$

Each term is in the mole fraction of its respective species $\left(N_x / \sum N_x\right)$ *so we can write:*

$$\overline{x}_n = \sum x X_x$$

Substituting for X_x, *we get:*

$$\overline{x}_n = \sum x p^{(x-1)}(1-p)$$

If $p < 1$, *which it is, then the summation term is part of a well known series that converges:*

$$\sum x p^{(x-1)} = \frac{1}{(1-p)^2}$$

Hence:

$$\overline{x}_n = \frac{1}{(1-p)}$$

FIGURE 5-9 Derivation of the number average degree of polymerization in terms of *p*.

From the definition of weight average:

$$\overline{x}_w = \sum x w_x = \sum x^2 p^{(x-1)}(1-p)^2$$

Again the summation is in terms of a series that converges:

$$\sum x^2 p^{(x-1)} = \frac{(1+p)}{(1-p)^3}$$

Hence:

$$\overline{x}_w = \frac{(1+p)}{(1-p)}$$

FIGURE 5-10 Derivation of the weight average degree of polymerization in terms of *p*.

is easily extended to the calculation of the higher averages. It is straightforward, but you should work through the algebra yourself with a pencil and paper. Maybe we're optimists; we may just have to set this as a test question!

Polydispersity

As we have mentioned previously, the weight average is always larger than the number average, except for monodisperse samples where they are exactly equal. Also, if you delve deeper into probability and statistics, you will find that the averages contain terms in the moments of the distribution and their ratios are related to things like the breadth of the distribution, its skewedness, and so on. In polymer science, we generally only use the ratio of the weight average to number average, which we call the polydispersity, and which is a measure of the breadth of the molecular weight distribution. For the most probable distribution in a step-growth polymerization, we can substitute to obtain Equation 5-17. Note that at high degrees of conversion the polydispersity tends to 2.

$$\frac{\overline{x}_w}{\overline{x}_n} = (1-p)$$

$$As\ p \to 1; \quad \frac{\overline{x}_w}{\overline{x}_n} \to 2$$

EQUATION 5-17

STATISTICS OF CHAIN POLYMERIZATIONS

Molecular weight averages and distributions can also be calculated for chain polymerizations using similar arguments to those used in dealing with step-growth polymerization. However, because we are dealing with a kinetic chain length we'll see that in polymerizations that are not "living," the results are less useful, in the sense that these quantities change over the course of the polymerization.

Free Radical Polymerization

First we need to redefine the parameter, p. In a step-growth polymerization, this was defined by the fraction of groups that have reacted, which is equal to the probability that a particular group has reacted. The equivalent definition here is in terms of the probability that a given polymer chain radical will react with monomer and continue to grow, as opposed to terminate. To keep things simple we will assume that termination proceeds solely by disproportionation. We then get Equation 5-18 for p, equal to the rate of growth divided by the rate of growth plus the rate of termination (think about it: we want the fraction of reactions that lead to growth). The probability of termination is then simply $1 - p$.

$$p = \frac{k_p[M][M^{\cdot}]}{k_p[M][M^{\cdot}] + 2k_{td}[M^{\cdot}]^2}$$

$$= \frac{k_p[M]}{k_p[M] + 2k_{td}[M^{\cdot}]}$$

EQUATION 5-18

The probability of finding an x-mer, or, equivalently, the mole fraction of x-mers polymerized at some instant of time, is then given by exactly the same expression as that obtained in dealing with condensation polymerizations (Equation 5-19).

$$P_x = p^{(x-1)}(1 - p)$$

EQUATION 5-19

We showed how this equation could be used to obtain expressions for the number and weight average molecular weight in the preceding section. As in step-growth polymerization, we obtain for the polydispersity (Equation 5-20),

$$\frac{\overline{x_w}}{\overline{x_n}} = (1 + p)$$

EQUATION 5-20

which is close to 2 for values of p approaching unity.

But, once again, its time for you to do some work. By substituting for p and using the steady-state assumption (if necessary, go back to Chapter 4), show that you obtain the expression for the number average degree of polymerization given by Equation 5-21.

$$\overline{x_n} = 1 + \frac{k_p[M]}{2(k_d k_{td} f[I])^{1/2}} = 1 + \upsilon$$

EQUATION 5-21

Also note that this is equal to the kinetic chain length, υ, as p tends to 1 (i.e., when the rate of propagation is much larger than the rate of termination, so that n is large). Also, show that the molecular weight distribution, as described by the polydispersity, becomes sharper ($= 1 + p/2$) if termination is by combination rather than disproportionation. Of course, we could go on, obtaining equations for the situation where termination is by both mechanisms, looking at the effect of chain transfer, and so on, but what we have given you here should be enough to give you a feel for how to approach these problems. One final point: unless you have very low degrees of conversion, you won't get these distributions. This is because the concentration of monomer (and initiator) changes as the polymerization proceeds. As a result, you get a superposition of all the instantaneous distributions and the overall distribution becomes much broader.

Ionic Polymerizations

For anionic polymerizations in protic media, you get the same expressions as those obtained for free radical polymerization with termination by disproportionation (p is still the probability of chain growth, but now $1 - p$ is the probability of chain transfer). Again, the averages and distributions you can obtain are only valid for low degrees of conversion. For cationic polymerization, there are several types of transfer and termination reactions that occur in most reactions, so you would have to be stark raving nuts to want to derive the equations. Instead, we will close this section by mentioning the special case of living polymerization.

Living Polymerizations

Although we want to say something about living polymerizations, we figure you've had your fill of probability theory at this point, so we will just give you the result. Essentially, you have to figure out how to distribute the monomers among the growing chains. You obtain a Poisson distribution in the number fraction of *x*-mers (Equation 5-22).

$$X_x = \frac{e^{-\nu}\nu^{x-1}}{(x-1)!}$$

EQUATION 5-22

Here ν is the number of molecules reacted per growing chain. After obtaining an equivalent expression for the weight fraction of *x*-mers and the number and weight average

degree of polymerization, you get the following expression for the polydispersity (Equation 5-23).

$$\frac{\overline{x_n}}{\overline{x_w}} = 1 + \frac{\nu}{(\nu+1)^2}$$

EQUATION 5-23

So, as ν gets larger, the molecular weight distribution narrows and approaches values of 1. (Of course, this assumes that the rate of initiation is fast and there is an absence of chain transfer reactions, so, in practice, the molecular weight distributions are a little broader than this.)

BRANCHING AND GELATION

So far, we have only considered linear polymerization reactions, as in condensation reactions involving bifunctional monomers (A–B or A–A/B–B pairs). Obviously, incorporating multifunctional monomers into this type of polymerization results in the synthesis of highly branched polymers and can lead to the formation of very large interconnected molecules. These can have macroscopic dimensions and are considered to be "infinite networks". The formation of such gels does not follow automatically from the incorporation of multifunctional monomers into the reaction pot, however. So, keeping in mind that an A can only react with a B and vice-versa, which of the reactions in Figure 5-11 do you think would lead to network formation?

I: (i) and (ii)
II: (iii) and (iv)
III: (v) and (vi)

If you answered "III" you're doing well. Let's look at this in a little more detail. First of all, bifunctional monomers can only give linear polymers, so you should have dismissed schemes (i) and (ii) out of hand; self-condensation of units in scheme (iii) leads to randomly branched structures. This is because each molecule present in the reaction pot, from monomer through oligomers

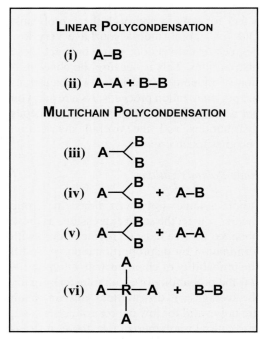

FIGURE 5-11 Multifunctional step-growth polymerization schemes.

FASCINATING POLYMERS—FORMICA® LAMINATES

This fascinating polymer-based material, Formica®, involves the formation of highly cross-linked polymer networks. The Formica Company, originally an electrical insulator business, was founded in 1913 by Daniel J. O'Conor and Herbert A. Faber, two friends, both of whom had been working for the Westinghouse Company. Although severely under-funded, the business grew and eventually attracted the attention of the "big boys." The Bakelite Company was supplying O'Conor and Faber with the resin to produce their plastic laminates. Westinghouse started to make similar laminates and negotiated an exclusive deal with Bakelite for their resin, which effectively cut off the supply to our two fledgling entrepreneurs. They had to scramble to find another supplier and came across "Redmanol," a resin produced by L. V. Redman, a Canadian chemist. With additional financial backing, the first Formica®-brand sheet laminate came off the press in July of 1914. The company grew during the next five years and then the lawyers got into the act! A bitter series of lawsuits involving Formica, Westinghouse, General Electric, Bakelite, Redman and others ensued. Formica survived and there was a consolidation of resin manufacturers into the new Bakelite Corporation, which sold to all. Formica is one of those words cherished by industrial moguls; a namebrand that is universally recognized. To make a Formica®-like laminate it is first necessary to start with an absorbent kraft paper (which looks rather similar to the generic brown-looking toilet paper that one might find in a public lavatory!). The kraft paper is then soaked in a solution of a PF resin and left to dry. Industrially, this is done by a continuous process where a massive roll of kraft paper is passed through a tank containing the PF resin solution, through two rollers that resemble an old fashioned wringer, through a heated drying oven, and finally through a knife that cuts the impregnated paper into standard sheets. PF resins are always an amber or brown color, which limits their applications. Over time two other resins, urea/formaldehyde (UF) and melamine/formaldehyde (MF), were developed, which are clear and more stable. They are, however, more expensive. These resins are used to impregnate high quality paper for the decorative surface sheets. Dried PF resin soaked kraft paper sheets are stacked one upon another. A pattern or veneer (wood grain) printed on high grade paper is placed next upon the stack. Then a translucent sheet (urea or melamine/formaldehyde resin) is added, followed by a texture sheet (if a matte finish is required). The stack is then compression molded into the familiar Formica® sheet.

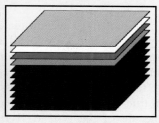

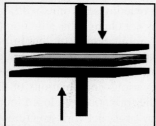

Schematic diagrams of (top) stacked dried PF resin-soaked kraft paper (brown) plus a metal foil (optional), a pattern or veneer sheet (red) and a translucent sheet (blue); (middle) compression molding between highly polished heavy steel platens under high temperature and pressure; (bottom) a sheet of Formica® laminate.

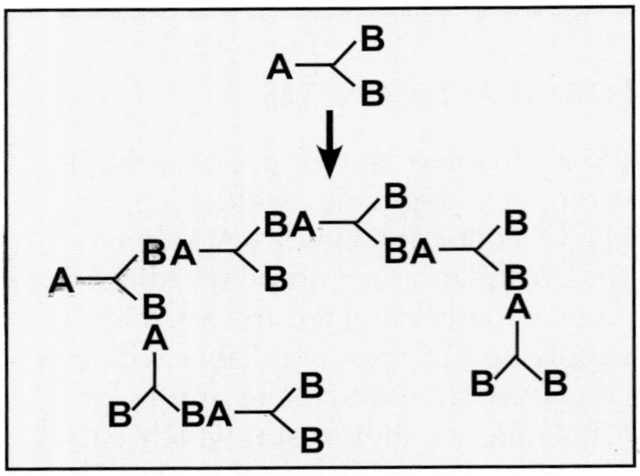

FIGURE 5-12 Polymerization of A–B$_2$.

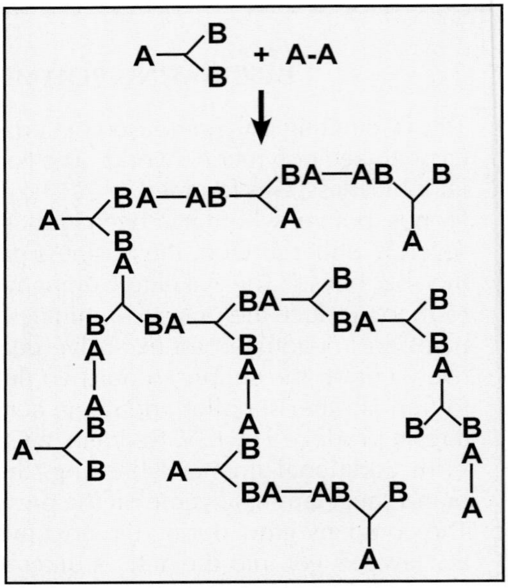

FIGURE 5-14 Polymerization of A–B$_2$ with A–B.

to high molecular weight material, contains only one reactive A group; all the rest are B's (Figure 5-12). Hence, networks are never formed, the molecule just keeps on branching, forming no "closed loops." This is not just true of trifunctional molecules, but any molecule of the type A–R–B$_{f-1}$. Note also that reacting such units with A–B molecules, does not change this situation (Figure 5-13).

Combinations of multifunctional and bifunctional monomers, such as those depicted in (v) and (vi) of Figure 5-11 (there are various combinations of this type), lead to the formation of closed loops and infinite (very large) network structures (Figure 5-14). This is because the initial oligomers formed have more than one "A" or "B" group and this allows for the formation of a large interconnected structure. Networks formed by step-growth polymerizations are industrially important (e.g., phenolic resins, polyurethanes), so we will discuss them here. Highly branched materials formed by condensation reactions are also interesting, so we will consider the statistics of their polymerization.

Statistical arguments can also be applied to the cross-linking of addition polymers,

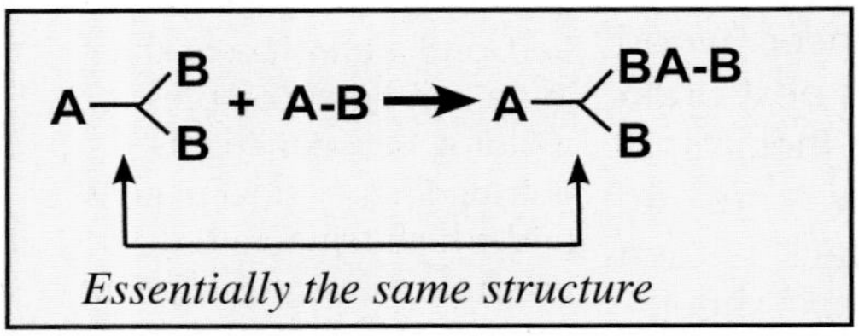

such as the dienes, and the networks formed by including small amounts of units like divinylbenzene in a chain polymerization, but not in as complete a manner. We will therefore focus our attention on the statistics of branching and network formation in step-growth polymerizations.

Theory of Gelation

We will start by considering the critical conditions for the formation of infinite networks and the molecular weight distributions characteristic of these types of non-linear polymerizations. We will take as an example the polymerization of two difunctional molecules with a trifunctional molecule of the type illustrated in Figure 5-15. As before, we assume that A's can only react with B's and vice-versa. In the early stages of the polymerization, oligomeric structures similar to those shown in Figure 5-15 are formed. Flory defined the conditions under which an infinite network will be formed by making two assumptions. First, the usual one of equal reactivity of all functional groups, regardless of where they are located. Second, he assumed that intramolecular condensation does not occur. In other words, the functional

Essentially the same structure

FIGURE 5-13 Polymerization of A–B$_2$ with A–B.

groups on the oligomer do not bend back and react with each other, they only react with groups on other molecules. We will see that this is not a good assumption.

The α Parameter

Flory defined a parameter, α, as *the probability that a given functional group that is part of a branch point leads via a chain of bifunctional units to another branch unit*, as illustrated in Figure 5-16.

To illustrate the methodology, let's take the specific case of a pot containing known amounts of a trifunctional monomer, $R-A_3$, and bifunctional monomers A–A and B–B. We now ask the question: "What is the probability that a molecule selected at random from the reaction pot has the structure shown in Figure 5-16?" This might at first seem a pointless question, but, hang in there, it will get us where we want to go, after we define some terms. These are given in the box opposite, together with some probabilities that follow from these definitions (Figure 5-17).

If we now examine the chain connecting two branch points (i.e., the $R-A_3$ trifunctional units), we can now go from one end to the other and write down the probability that each unit, in turn, has reacted. Let's start at the left-hand branch point. The probability that an A group on this unit has reacted with a B unit is simply p_A, the probability that any A group has reacted. Next, we need the probability that the other B unit on this B–B molecule has reacted with the A of an A–A unit (not the A of an $R-A_3$ unit). So, as we illustrate in Figure 5-18, the fraction of A's in A–A units is $(1 - \varrho)$, and the probability that a B group has reacted is just p_B, so this probability is simply $p_B(1 - \varrho)$. Proceeding down the chain, the probability that the other end of this A–A molecule has reacted with a B–B unit is p_B. Now, if the linear part of the chain has i A–A and B–B units, the probability that all the groups have reacted is simply $[p_B(1 - \varrho)p_A]^i$. Finally, we need the probability that the end B group on this chain has reacted with the A group of an $R-A_3$ unit. This is ϱp_B. If you didn't get that, look at the

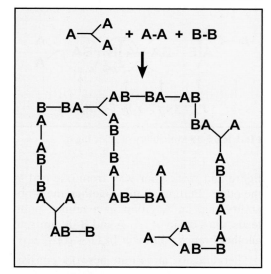

FIGURE 5-15 Polymerization of $R-A_3$, A–A and B–B.

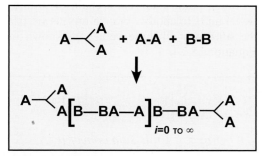

FIGURE 5-16 Structure used in the definition of α.

DEFINITIONS

p_A is the probability that an A group has reacted

p_B is the probability that an B group has reacted

ϱ is the ratio of A's (both reacted and unreacted) belonging to branch units to the total number of A's

$p_B\varrho$ is the probability that a B group has reacted with a branch unit

$p_B(1 - \varrho)$ is the probability that a B group is connected to an A–A unit

FIGURE 5-17 Summary of definitions.

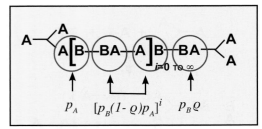

FIGURE 5-18 Probability of selecting α.

figure and work your way from one end to the other. Putting all this together, the probability that an A group in a branch point leads, via a chain of i A–A and B–B units, to another branch point is $p_A[p_B(1 - \varrho)p_A]^i\, p_B\varrho$.

Theoretically, an infinite network consisting of chains of all lengths, from $i = 0$ to $i = \infty$, can be present, so α, the probability that a given functional group that is part of a branch point leads via a chain of bifunctional units to another branch unit, is simply the sum over all these values, as shown in Equations 5-24.

$$\alpha = \sum_{i=0}^{\infty} [p_A p_B (1 - \rho)]^i p_A p_B \rho$$

For x < 1 this sum converges:

$$\sum_{i=0}^{\infty} x^i = \sum_{i=1}^{\infty} x^{i-1} = \frac{1}{1 - x}$$

Hence:

$$\alpha = \frac{p_A p_B \rho}{1 - p_A p_B (1 - \rho)}$$

EQUATIONS 5-24

As in the case of linear polycondensation, we again let the ratio of A to B groups initially present be $r = [A]_0/[B]_0$, it then follows that $r = p_B/p_A$. This is important. Check for yourself: start with 2 moles of A and 1 mole of B, then let 0.5 moles of each react. If there are more A than B groups, the probability that a B group has reacted (equal to the fraction of B groups that have reacted) is greater than the probability that an A group

has reacted. Using this definition of r, we then get Equations 5-25.

$$\alpha = \frac{rp_A^2\rho}{1 - rp_A^2(1 - \rho)}$$

or

$$\alpha = \frac{p_B^2\rho}{r - p_B^2(1 - \rho)}$$

EQUATIONS 5-25

There are two special cases where these equations are even simpler, summarized in Equations 5-26.

If there no A–A groups, $\rho = 1$; *and*

$$\alpha = rp_A^2 = \frac{p_B^2}{r}$$

If $p_A = p_B = p$, $r = 1$; *and*

$$\alpha = \frac{p^2\rho}{1 - p^2(1 - \rho)}$$

EQUATION 5-26

Also, because of the way Flory defined the parameters r and ϱ, the equations are independent of the functionality of the branch unit, R–A$_f$. However, the derivation scheme is not entirely general; it does not apply to situations where the multifunctional monomers contain both A and B groups, or to the condensation of two types of multifunctional units, one bearing A groups, the other having B groups. Furthermore, it is often the case that the reactivities of the functional groups on the branching unit are not identical, as in the case of glycerol (Figure 5-19), where the secondary alcohol is less reactive than the primary alcohol. Nevertheless, it is usually possible to calculate α using a different methodology or a modification to the procedure described above. What is of more interest to us here is the fact that there is a critical value of α where the incipient formation of

FASCINATING POLYMERS—ALKYD RESINS

Cross-linked or 3-dimensional network polyesters, such as those based upon the reaction of glycerol and phthalic anhydride (PA), have been used extensively as surface coatings. Not usually by themselves, however. The network polymer formed by glycerol and PA was described by Smith as early as 1901, but was not a useful product, because it resembled glass and was far too brittle. In the mid 1920s, Kienle, working for the General Electric Company, made an important discovery. These network polyesters could be modified significantly by so-called drying oils. The presence of drying oils, usually natural products, made the initial resin soluble in aliphatic solvents. Upon drying and exposure to oxygen, these modified resins rapidly cross-linked and the resultant films were flexible and durable. Oil-modified polyesters, or alkyd resins, were born and still account for a significant fraction of the polymeric resins used in the surface coating industry. Alkyd resins are made principally from polyols (multifunctional alcohols), dibasic acids (or anhydrides) and modifying oils (or their corresponding acids). For you who enjoy cooking, alkyds are right up Chef Emeril Lagasse's alley, as the resins are literally cooked up from a myriad of complex recipes, often containing combinations of "natural" oils, such as those obtained from sunflowers and soybeans! There is really no such thing as a typical alkyd. Most alkyds contain phthalic acid, ethylene glycol, perhaps a tetrafunctional alcohol like pentaerythritol, and then various combinations of drying and non-drying oils, such as hydrolyzed linseed, tung, safflower, soybean or coconut oil. Common or garden vegetable oils are esters of saturated and unsaturated monocarboxylic acids of glycerol. For example, the main ingredient of a drying oil derived from hydrolyzed sunflower seeds is linoleic acid, which is an unsaturated fatty acid that contains two unconjugated double bonds. Different oils contain different distributions of saturated and unsaturated fatty acids. In the so-called fatty acid process, modifying oils are first hydrolyzed to the corresponding

Sunflowers—a source of natural modifying oils.

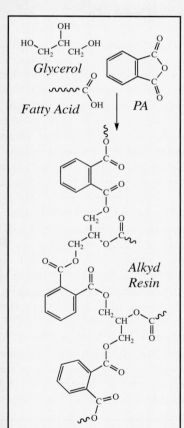

Formation of a typical alkyd resin.

fatty acids and then the polymerization is performed at 200–240°C in the presence of polyol (e.g., glycerol, pentaerythritol) and dibasic acid (e.g., phthalic anhydride). A typical polymer chain is depicted above. The properties of alkyds are dictated primarily by the amount and nature of the modifying oils incorporated. Alkyds are categorized as drying oil, semi-drying oil and non-drying oil resins.

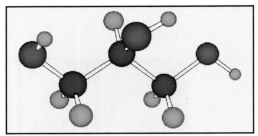

FIGURE 5-19 Glycerol.

an infinite network occurs and this depends only on the functionality of the network.

Critical Value of α

The argument Flory used was beautiful in its simplicity. Let's take as an example network formation using a trifunctional branching unit. A chain that reacts with this unit leads to a branch point that can react with two more chains. In turn, these two branch points can react with four chains that can lead to four more branch points, and so on (Figure 5-20). Now, if $\alpha < 0.5$, there is less than an even chance that each chain will lead to a branch point, thus to two more chains, and so on. In other words, there is greater than an even chance that it will end in an unreacted functional group, so that eventually termination must outweigh continuation of the chain through branching. Hence, for a trifunctionally branched system, when $\alpha < 0.5$, the size of all molecules formed must be limited. Conversely, when $\alpha > 0.5$, structures of an infinite size are possible. This does not mean that all the original monomers become incorporated into one huge molecule: both gel (insoluble infinite network) and sol (soluble molecules of finite size) will exist

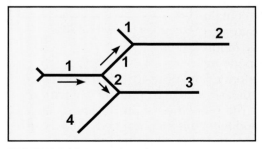

FIGURE 5-20 Generation of chains.

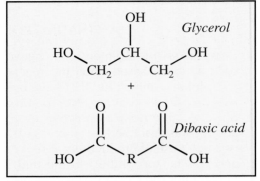

FIGURE 5-21 Kienle's experimental reaction.

in various amounts. This argument can be generalized to a network of any functionality, f, where the critical value of α is given by Equation 5-27.

$$\alpha_c = \frac{1}{f-1}$$

EQUATION 5-27

Gelation: Theory versus Experiment

It has been found experimentally that the incipient gel point occurs at somewhat larger values of α than predicted from the relationship for α_c derived by Flory. Kienle et al.,[2] for example, studied the reaction between glycerol and equivalent amounts of a dibasic acid (Figure 5-21).

In this case, $r = 1$ and $\alpha = p^2$ (see Equation 5-26). They experimentally determined the value of p at which gelation occurred. The onset of gelation can be determined by nitrogen bubbles bled from a capillary into the reaction pot. At the incipient gel point, there is an abrupt loss of fluidity and the bubbles suddenly stop rising. Their experimentally determined value of α was 0.58, significantly higher than the theoretical value of $\alpha_c = 0.5$ and far too large a discrepancy to be accounted for by the difference in reactivity of the secondary alcohol of glycerol relative to the primary OH groups.

[2] R. H. Kienle et al. *J.A.C.S.*, **61**, 2258, 1939; *ibid.* **61**, 2268, 1939; *ibid.* **62**, 1053, 1940; *ibid.* **63**, 481, 1941.

Similarly, Flory determined a value of $\alpha_c = 0.60 \pm 0.02$ for polycondensations of ethylene glycol with various amounts of bifunctional and trifunctional carboxylic acid-containing molecules (Figure 5-22 and see Flory's book *Principles of Polymer Chemistry*). He also experimentally determined (Figure 5-23) p_A, hence α, measured the viscosity, which increases dramatically near the gel point, and calculated the number average degree of polymerization using Equation 5-28.

$$\bar{x}_n = \frac{f(1 - \rho + 1/r) + 2\rho}{f(1 - \rho + 1/r - 2p_A) + 2\rho}$$

EQUATION 5-28

Note that $\bar{x}_n$ is not very large, nor is it increasing rapidly at the gel point. This means that the proportion of large molecules present at this stage is small, but profoundly affects the viscosity.

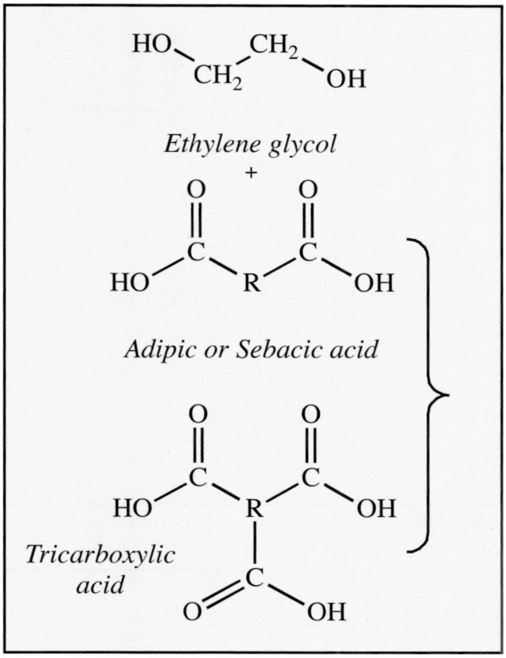

FIGURE 5-22 Flory's experimental recipe.

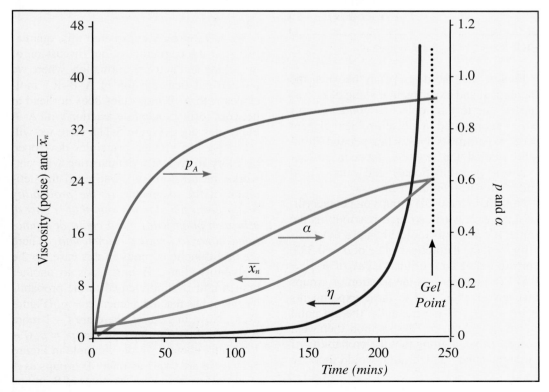

FIGURE 5-23 Flory's experimental results [replotted from the data of P. J. Flory, *Principles of Polymer Chemistry*, Cornell University Press (1953)].

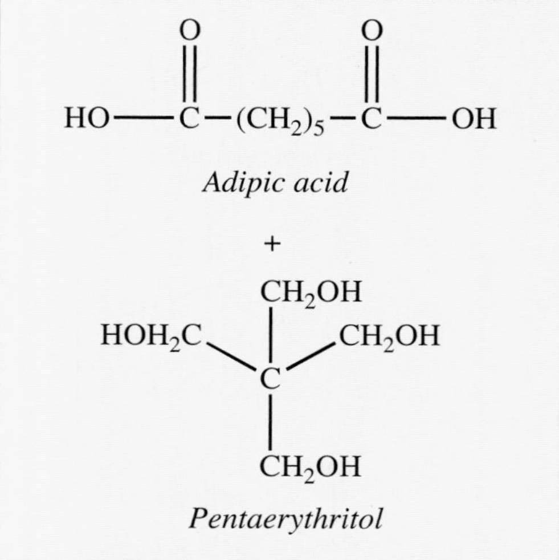

FIGURE 5-24 Stockmayer's experimental reaction.

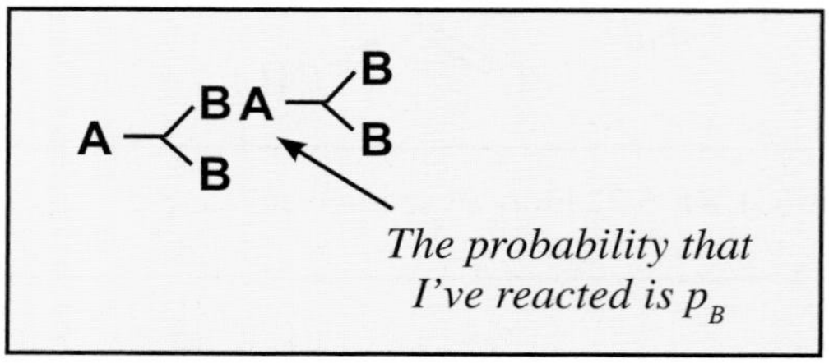

FIGURE 5-25 AB_2 dimer formation.

Finally, that the discrepancy between the theoretical and experimental value of α_c was indeed due to intramolecular condensations was demonstrated in a beautiful and elegant set of experiments by Stockmayer and Weil,[3] who studied the reaction between pentaerythritol (tetrafunctional) and adipic acid (bifunctional)—Figure 5-24.

Note that all the OH groups on pentaerythritol are equivalent, so their reactivity is the same. Because this molecule has a functionality $f = 4$, the theoretical value of $\alpha_c = 0.333$, corresponding to a critical value of p, $p_c = 0.577$ (remember p is the fraction of groups that have reacted and in these experiments $p = p_A = p_B$). Experimentally, the gel point occurred at $p = 0.63$. These authors then went a step further, arguing that the probability of intramolecular condensation would increase

[3] W. H. Stockmayer and L. L. Weil, Chapter 6 in *Advancing Fronts in Chemistry*, S. B. Twiss Ed., Reinhold, New York, 1945.

if the system was diluted with an inert solvent. They then performed a set of experiments as a function of solvent concentration, plotting the value of p_c vs the inverse of this concentration, thus obtaining an extrapolated value of p_c corresponding to an infinite concentration, where intramolecular condensation should be completely eliminated. This extrapolated value was determined to be $p_c = 0.578 \pm 0.005$, in remarkable agreement with theory.

Random Branching Without Network Formation

To complete our discussion of branching, we are going to consider two special cases. The first of these involves random branching without network formation. There is presently (circa. 2007) considerable interest in the potential uses of hyperbranched polymers, and this is what you get if you perform a polycondensation on an $A–R–B_{f-1}$ molecule, where, again, an A can only react with a B.

We have already considered this, qualitatively, at the beginning of our discussion of branching and network formation, where we pointed out that reaction of $A–B–R_{f-1}$ molecules with $A–B$ molecules does not lead to network formation, while reaction with $A–A$ molecules does (Figure 5-11). Here we will see if we can use the principles developed by Flory to formally demonstrate why networks are not formed. First recall the definition of the parameter, α: *the probability that a given functional group that is part of a branch point leads via a chain of bifunctional units to another branch unit*. There are no bifunctional units in this case, so the probability that a B unit leads to another branch unit is simply equal to the probability that it has reacted. Hence, $\alpha = p_B$ (Figure 5-25). Now, in general, there are $f - 1$ more B groups than A groups, so that $p_A = p_B(f - 1)$. Think about it. In the example in Figure 5-25 there are twice as many B groups as A groups. If we start with 20 moles of B groups and 10 moles of A groups and 5 moles of each react, then the fraction of A groups that have reacted (equal to the probability that an A group has reacted) is 5/10. The fraction of B groups that have reacted is 5/20. Then $p_A = (3 - 1)p_B$. Substituting, we then get the expression (Equation 5-29):

$$\alpha = \frac{p_A}{(f-1)}$$

EQUATION 5-29

which can be compared to the expression for the critical value of α, α_c (Equation 5-30):

$$\alpha_c = \frac{1}{(f-1)}$$

EQUATION 5-30

The magnitude of α must always be less than α_c, because $p_A < 1$. Thus gelation cannot occur. At high degrees of conversion, very large hyperbranched molecules may be produced, but not a network.

Flory also derived equations for the number and weight average degree of polymerization, the equations for which, along with that for the polydispersity, we have reproduced in Equations 5-31.

$$\overline{x}_n = \frac{1}{\left[1 - \alpha(f-1)\right]}$$

$$\overline{x}_w = \frac{\left[1 - \alpha^2(f-1)\right]}{\left[1 - \alpha(f-1)\right]^2}$$

$$\frac{\overline{x}_w}{\overline{x}_n} = \frac{\left[1 - \alpha^2(f-1)\right]}{\left[1 - \alpha(f-1)\right]}$$

EQUATIONS 5-31

Remember that in the polycondensation of bifunctional monomers to give linear chains, the polydispersity approaches values of 2 at high degrees of conversion. Generally, if branching occurs the polydispersity index gets much larger than this. For example, in

the polycondensation of A–R–B$_2$, where f = 3, the polydispersity has values of about 6, 11 and 51 for p_A values of 0.90, 0.95, and 0.99, respectively. We mention this because it leads us to the final special case we wish to consider, where the polydispersity actually narrows with increasing conversion.

This occurs if small amounts of a multifunctional monomer, R–A$_f$, are reacted with A–B bifunctional monomers (Figure 5-26). In the initial stages of the polymerization, linear chains of A–B units are the main product. But, as the reaction approaches completion, these linear species disappear in favor of multi-chain star-like polymer molecules.

Why would this lead to a narrower molecular weight distribution? Look at it this way: let's imagine that the reaction proceeds in two separate steps; first linear chains are polymerized almost to complete conversion, so that the polydispersity is close to 2 (Figure 5-27).

Now, take these chains at random and attach them to R–A$_f$ molecules until they all have attached chains. Because large, small and intermediate size chains are combined into larger molecules at random, the molecular weight distribution *narrows*. This can

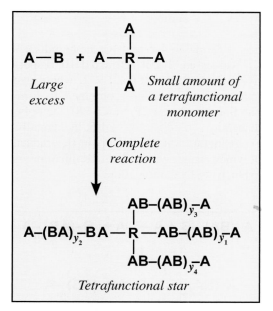

FIGURE 5-26 Polymerization of A–B with a small amount of R–A$_4$.

A—B $\longrightarrow$ A–(BA)$_{y-1}$B

FIGURE 5-27 Initial polymerization of A–B.

be shown formally using combinatorics. You have to describe the number of ways chains of different length can be distributed amongst the *f* arms of the star. If you're used to this stuff, the exercise is straightforward, but if your grasp of statistics is less than firm, the results can appear quite alarming. We will content ourselves with just the result obtained by Stockmayer, who used a number of simplifying assumptions to obtain the equation reproduced in Equation 5-32.

$$\frac{\overline{x_w}}{\overline{x_n}} \approx \left[1 + \frac{1}{f} \right]$$

EQUATION 5-32

As in so many things in this field, if you want to work through the arguments yourself, you cannot do better than go to Flory—see *Principles of Polymer Chemistry*, Chapter IX. Stockmayer's equation illustrates the point we wish to make with dazzling simplicity: as *f*, the number of branches, increases, the polydispersity decreases. Thus for values of *f* equal to 4, 5 and 10, the polydispersity values are 1.25, 1.20 and 1.10, respectively. Note also that for $f = 2$, where two independent chains are combined to form one linear molecule (Figure 5-28), the polydispersity is predicted to be 1.5. Incidentally, an analogous situation occurs in free radical polymerization when chain termination is exclusively by combination.

A–(BA)$_n$–B + A–R–A + B–(AB)$_m$–A

$\downarrow$

A–(BA)$_n$–BA–R–AB–(AB)$_m$–A

FIGURE 5-28 Polymerization of A–B + RA$_2$.

RECOMMENDED READING

P. J. Flory, *Principles of Polymer Chemistry*, Cornell University Press, Ithaca, New York, 1953.

J. L. Koenig, *Chemical Microstructure of Polymer Chains*, J. Wiley & Sons, New York, 1982.

STUDY QUESTIONS

1. Calculate and plot the number and weight distributions of *x*-mers found in a step-growth polymerization for conversions *p* of 0.90, 0.95, and 0.99 for $x = 1$ to 300. Comment on what species predominate by weight and by number. (Hint: you may want to use *Excel* or *Mathematica*!)

2. Derive explicitly (i.e., show every step) equations for the number and weight average molecular weight in a linear polycondensation reaction of A–B-type monomers.

3. Derive an expression for the polydispersity found in a linear polycondensation of equimolar amounts of A–A- and B–B-type monomers. What is the limiting value? How does this limiting compare to that calculated for a free radical polymerization? Explain why the polydispersity of a sample of (say) polystyrene polymerized to a high degree of conversion differs from this.

4. For a linear condensation reaction of A–A- and B–B-type *monomers*

A. Calculate the extent of reaction (*p*) necessary to achieve a number average degree of polymerization of 200, assuming you start with 50 moles of A and 50 moles of B monomers.

B. Calculate the weight average degree of polymerization and polydispersity of this sample.

C. Calculate the degree of polymerization that could be achieved at this degree of conversion (i.e., the value of *p* calculated in part A) if there were an impurity present, such that you not only had 50 moles of B *mono-*

mers, but also 1 mole of a monofunctional (R–B) molecule.

D. Calculate the maximum degree of polymerization that could be achieved if there is a 2% excess of B units present in the polymerization described in part A (i.e., $r = 100/102$).

5. Consider the reaction of adipic acid with pentaerythritol described in the text (Figure 5-25). In a reaction that started with 1 mole of pentaerythritol and 2 moles of adipic acid, it was found (by titrating an aliquot) that at the incipient gel point there were 1.48 moles of *acid groups* that were unreacted. Calculate the critical value of the parameter, α. What is the theoretical value of α_c? What is the reason for any difference in the two values?

6. Describe what types of trifunctional molecules you would use to make a hyper-branched polymer. Derive an expression for α and α_c demonstrating that a network structure can never be formed. Briefly discuss how the polydispersity changes with conversion compared to what is found in linear polycondensations.

6

Copolymerization

INTRODUCTION

Now we come to a subject where we will employ what we have learned in applying both kinetics and statistics to polymerizations: copolymerization. In our discussion of chain microstructure, we have already discussed the various types of copolymers that can be formed, and these are illustrated in Figure 6-1. We'll start by briefly considering some examples, to give you a feel for the range of materials produced commercially.

Copolymers are ubiquitous and important because they allow monomers to be combined in such a way so as to provide useful and sometimes unique properties. For example, linear polyethylene (PE) and isotactic polypropylene (*i*-PP) are both semi-crystalline plastics, but copolymers of ethylene and propylene (EPR) (usually with other comonomers) are rubbers at room temperature (depending on composition). The homopolymers are shown in the top two figures in Figure 6-2, and if you don't know which one's which by now you should collapse in deepest humiliation. A section of an EPR copolymer chain is shown at the bottom.

Block copolymers of styrene and butadiene are used to make thermoplastic rubbers, or impact-resistant glassy polymers, depending on the length, arrangement and relative proportions of the blocks. Polyurethanes are an example of what are called segmented block copolymers and these we have mentioned in Chapter 3.

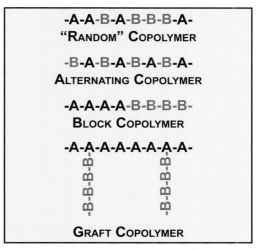

FIGURE 6-1 Major copolymer types.

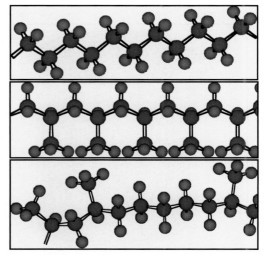

FIGURE 6-2 Chain structures of linear PE (top), *i*-PP (middle) and EPR (bottom).

Graft copolymers, such as "HIPS" (high impact polystyrene) or ABS plastics (acrylonitrile/butadiene/styrene) can be made by dissolving a polymer (A) in a solvent containing a different monomer (B), generating an active site such as a free radical on this polymer (e.g., by irradiation or reaction with an initiator), thus starting a polymerization of the new monomer on this site (Figure 6-3).

Ionomers consist of statistical copolymers of a non-polar monomer, such as ethylene, with (usually) a small proportion of ionizable units, like methacrylic acid. Ethylene-co-methacrylic acid copolymers (~5% methacrylic acid) are used to make cut-proof golf balls (see *Fascinating Polymers* opposite). The protons on the carboxylic acid groups are exchanged with metal ions to form salts. These ionic species phase-separate into microdomains or clusters which act as cross-links, or, more accurately, junction zones (Figure 6-4). (We discuss interactions in a little more detail in Chapter 8.)

Copolymers are synthesized using exactly the same type of chemistry as homopolymers, of course, but everything depends on the way you do it. Step-growth polymerizations of two different monomers often give you truly random copolymers, because rearrangement reactions like transesterification can scramble any initial non-random sequence distribution imposed by the kinet-

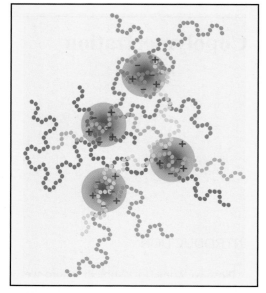

FIGURE 6-4 Schematic diagram of an ionomer.

ics of the reaction. In a chain polymerization of two monomers in the same pot or reaction vessel, this doesn't occur and you get a statistical or random copolymer. The actual microstructure will depend on the rate at which one monomer will add to the chain relative to the other, as we will see. If you want to get a block copolymer, you need to add the monomers sequentially and use a living polymerization technique, usually anionic, and nowadays some ingenious free

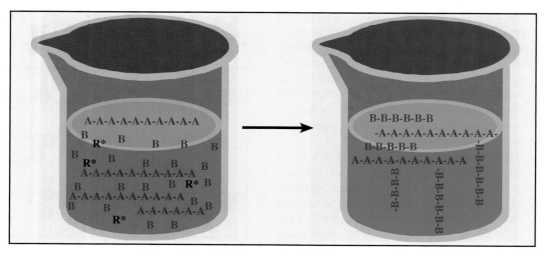

FIGURE 6-3 Schematic diagram illustrating the formation of a graft copolymer.`

FASCINATING POLYMERS—SURLYN® IONOMERS

There is a superb little article, which we highly recommend, that can be found on DuPont's website entitled *"Discontinuous Innovation: How It Really Works"* by Parry Norling and, a good friend of ours, Bob Statz. It concerns the sudden appearance of a major technological breakthrough that can yield entirely new products, processes or services. The development of Surlyn® ionomers is a classic example of this phenomenon. During the 1940s and 50s, after ICI's discovery of low density polyethylene, DuPont and others investigated various ethylene copolymers with the goal of producing materials with particular and useful properties. Ethylene-co-vinyl acetate (EVA), for example, was one such material introduced during that period that is still manufactured in large quantities today. In the early 1960s DuPont was investigating ethyl-

Dick Rees (Source: DuPont).

ene copolymers of methacrylic acid (EMAA). Ethylene-co-methacrylic acid (EMAA) copolymers are produced commercially by a high-temperature/high-pressure free radical polymerization process similar to that used to originally produce LDPE. These materials, which fall today under the umbrella of DuPont's Nucrel® family, had unique flow properties and outstanding adhesion to aluminum foil. Enter Dick Rees, who was attempting to find ways of cross-linking EMAA copolymers. In one experiment, Rees decided to try epichlorohydrin as a potential cross-linking agent and for this proposed reaction he first needed to make the sodium salt of the EMAA copolymer. To his surprise, he obtained a gel-like material. This simple sodium salt of EMAA was transparent, tough, stiff and resilient. Rees had discovered ionomers. So why did Rees obtain a gel-like material? It turns out that the highly polar ionic moieties like to segregate from their non-polar ethylene counterparts of the copolymer chains and form ionic domains (often referred to as multiplets or clusters). If you can imagine something like this occurring on all or most of the copolymer chains, then you should see why Rees formed a 3-dimensional network or gel. As the gel-like material that Rees formed has labile (non-permanent) ionic cross-links, it is possible to heat the material, overcome these intermolecular ionic forces, and mold the material in conventional thermoplastic processing equipment. Then, upon cooling, the ionic cross-links (domains) reform and the material acts like a thermoset. Of course, *the* most important use of Surlyn® ionomers is for cut-proof golf ball covers! Unlike your authors, who have perfect, fluid-like golf swings that are the envy of our colleagues and PGA professionals alike, the vast majority of golfers have really awful swings and deserve the derogatory description, "hacker!" Until the late 60s, golf balls had covers that were predominantly made of Balata. When a hacker attacks a Balata-covered ball, he or she is often rewarded with a large cut or "smile" on the ball which renders it useless. Of course, this built-in obsolescence was advantageous to golf ball manufacturers—they could sell more golf balls! Accordingly, when DuPont came along extolling the abrasion resistance and "cut-proof" virtues of Surlyn® ionomers, they were met with polite disinterest. But this changed when RAM, a company with only a very small market share, worked with DuPont and produced a superb cut-resistant Surlyn® covered ball.

$$F_1 = \frac{(r_1 - 1)f_1^2 + f_1}{(r_1 + r_2 - 2)f_1^2 + 2(1 - r_2)f_1 + r_2}$$

Bet you can't wait to derive this sucker!

FIGURE 6-5 One form of the copolymer equation.

radical methods. Our initial focus here will be the kinetics of chain-growth copolymerization, particularly free radical, as this is a commercially important area which has been extensively studied and a knowledge of the kinetics allows a prediction of the instantaneous copolymer composition. Then we will turn our attention to the application of probability theory to sequence distribution and the characterization of copolymer chains.

KINETICS OF CHAIN-GROWTH COPOLYMERIZATION

Free Radical Copolymerization

We should point out that the equation we will derive, the copolymer equation (Figure 6-5), should be applicable to other types of polymerizations, such as those utilizing catalysts. Many commercially used catalysts are heterogeneous, however, meaning that we get polymers with different characteristics (sequence distributions) produced at different sites. The copolymer equation should apply to the polymers produced at each site, but the final product contains all these jumbled up together and there is no way to judge what

FIGURE 6-6 Propagation steps.

came from where. So for simplicity, we'll just stick to free radical copolymerization. Similarly, we will just consider the copolymerization of two monomers, M_1 and M_2. The methodology for copolymers with three or more monomers would be the same, it's just that the algebra gets out of hand. Here, you just want to learn the general approach.

Free radical copolymerization involves initiation, propagation and termination, just like homopolymerization. With one or two exceptions, initiation and termination are pretty much the same in these polymerizations and it is the propagation step that gives copolymerization its special character. For the copolymerization of two monomers, there are four possible propagation steps (Figure 6-6). If the monomer at the terminal position of the growing chain radical happens to be a monomer of type 1, M_1, then it can either add another M_1 monomer, or it can add an M_2 monomer. The same two possibilities apply if the chain end is an M_2.

Each of these propagation steps can proceed at different rates, so we need four rate constants to describe the kinetics of the propagation step. We will use a double subscript where the first number refers to the monomer at the terminal position of the chain radical while the second number refers to the type of monomer that is being added. The rate constant k_{11}, for example, refers to the rate of adding a monomer of type 1 to a growing chain whose terminal group also happens to be a monomer of type 1. The rate constant k_{21} would describe the rate of adding a 1 to an end of type 2. And so on.

Terminal Model

Note that here we are actually making an assumption: that the rate constants depend only on the nature of the terminal group—hence, the *terminal* model. There is also something called the *penultimate* model, where there is now a dependence on the character of the final two units in the chain radical (Figure 6-7). This obviously complicates the algebra, because for each reaction in the terminal model we now have two in the penultimate model, giving us eight rate constants describing propagation! However, the terminal model appears to describe the majority of copolymerizations very well and it is certainly all we need to show how you approach a problem like this, so we will stick to the terminal model. We will briefly come back to the penultimate model when we consider the statistics of copolymerization, to show how you could distinguish between the two using knowledge of sequence distributions.

If the terminal model applies to the copolymerization that interests us and we happen to know the rate constants (we'll discuss how to measure these later), then we can immediately describe some interesting limiting conditions.

1. If $k_{11} \gg k_{12}$ and $k_{22} \gg k_{21}$

If the rate of adding a monomer of type 1 to a terminal chain radical of type 1 is much greater than adding a monomer of type 2 to this same radical, then obviously there will be a tendency to get sequences or blocks of monomers of type 1 (assuming, of course, that we don't have big disparities in the concentration of the monomers)—as illustrated in Figure 6-8. If k_{22} is also much greater than k_{21}, there would be a tendency to form block copolymers and even some homopolymers, as long as the difference in the rate constants is large enough.

2. If $k_{12} \gg k_{11}$ and $k_{21} \gg k_{22}$

Similarly, if the rate of adding a M_2 to a M_1 and a M_1 to a M_2 is much greater than the rate of adding monomers of the same type, then there is a tendency to form alternating copolymers (Figure 6-9).

3. If $k_{12} = k_{11}$ and $k_{21} = k_{22}$

Finally, if the rate of adding either M_1 or M_2 to a M_1 radical is the same, and the rate of adding either M_1 or M_2 to a M_2 radical is equivalent, then the copolymers are truly random. More on this shortly!

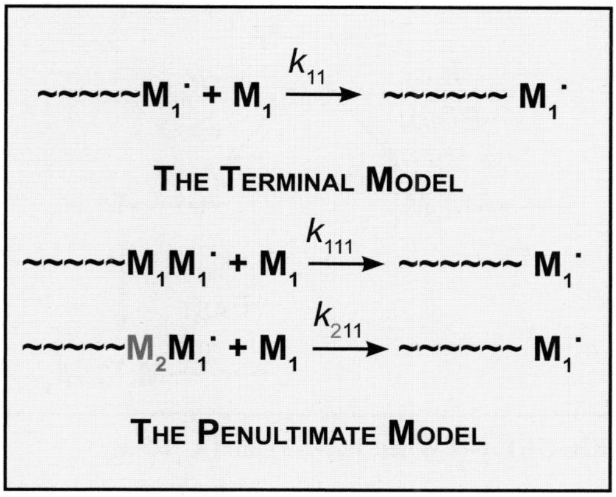

FIGURE 6-7 Two models of free radical copolymerization.

Deriving the Copolymer Equation

Obviously, what we would really like to do is not just have a feel for tendencies, useful as this is, but also calculate copolymer composition and sequence distributions, things that can also be measured by spectroscopic methods. We will start by using kinetics to obtain an equation for the *instantaneous* copolymer composition (it changes as the copolymerization proceeds). Later we will use statistical methods to describe and calculate sequence distributions. In deriving the copolymer equation, we only have to consider the propagation step and apply our old friend, the steady-state assumption, to the radical species present in the polymerization, $\sim\sim\sim M_1^{\bullet}$ and $\sim\sim\sim M_2^{\bullet}$.

Note that the first and third reactions in the four possible propagation steps shown in

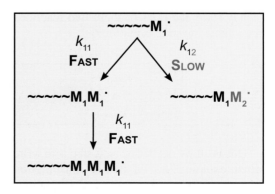

FIGURE 6-8 Situation when $k_{11} \gg k_{12}$.

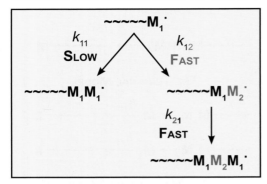

FIGURE 6-9 When $k_{12} \gg k_{11}$ and $k_{21} \gg k_{22}$.

Figure 6-6 do not result in a change in the character of the radical species. Only reactions 2 and 4 change the nature of the radical chain end group.

The Steady-State Assumption

When we were discussing homopolymerization, we used the steady-state assumption to obtain an equation for the concentration of the transient species, the radical chains, which were all of identical character. Here we define the character of the radical chain according to the nature of the terminal group, so we have two types of transient species and all we can get from the steady-state assumption is an equation for the ratio of their concentrations (Equation 6-1). To be explicit, if we consider radical chain species $\sim\sim\sim M_1^{\bullet}$ (we could just as well consider $\sim\sim\sim M_2^{\bullet}$), then the only reaction in Figure 6-6 that leads to a new $\sim\sim\sim M_1^{\bullet}$ radical is reaction 3. The only reaction that leads to a removal of $\sim\sim\sim M_1^{\bullet}$ (by conversion to $\sim\sim\sim M_2^{\bullet}$) is reaction 2. At steady state, the rates of these two reactions are equated to give Equation 6-1.

$$\frac{[M_1^{\bullet}]}{[M_2^{\bullet}]} = \frac{k_{21}[M_1]}{k_{12}[M_2]}$$

EQUATION 6-1

This is all we need. The next step is to write down equations for the rate of consumption of monomer 1 and monomer 2 (Equations 6-2).

$$-\frac{d[M_1]}{dt} = k_{11}[M_1^{\bullet}][M_1] + k_{21}[M_2^{\bullet}][M_1]$$

$$-\frac{d[M_2]}{dt} = k_{22}[M_2^{\bullet}][M_2] + k_{12}[M_1^{\bullet}][M_2]$$

EQUATIONS 6-2

These monomers are not just disappearing off the face of the earth, into the black hole in our clothes dryers that manages to selectively consume just one of the members of each pair of socks we own. (We now buy identical socks to thwart this incarnation of Maxwell's demon!) They are actually going into the polymer chains that are being formed at that instant of time. The ratio of these two equations must therefore represent the ratio of monomer 1 to monomer 2 incorporated into the chains at that moment.

Equations 6-3 contain the expression obtained by dividing Equations 6-2 and rearranging the result so as to be in terms of the ratio of the radical chain concentrations. Equation 6-1 from the steady-state assumption is then substituted.

Note we have defined two new parameters, the reactivity ratios r_1 and r_2.

$$\frac{d[M_1]}{d[M_2]} = \frac{k_{11}[M_1]\dfrac{[M_1^{\bullet}]}{[M_2^{\bullet}]} + k_{21}[M_1]}{k_{22}[M_2] + k_{12}[M_2]\dfrac{[M_1^{\bullet}]}{[M_2^{\bullet}]}}$$

$$\frac{[M_1^{\bullet}]}{[M_2^{\bullet}]} = \frac{k_{21}[M_1]}{k_{12}[M_2]}$$

$$r_1 = \frac{k_{11}}{k_{12}} \qquad r_2 = \frac{k_{22}}{k_{21}}$$

$$\frac{d[M_1]}{d[M_2]} = \frac{r_1\dfrac{[M_1]}{[M_2]} + 1}{r_2\dfrac{[M_2]}{[M_1]} + 1}$$

EQUATIONS 6-3

These are the ratios of the rate constants describing the addition of a like monomer (e.g., k_{11}) relative to an unlike monomer (e.g., k_{12}). Substituting, we get one form of the instantaneous copolymer composition equation, describing the ratio of monomer 1 to monomer 2 that is being incorporated into the polymer at a given instant of time.

This equation is often expressed in alternative forms. One such form merely substitutes new variables for the ratio of monomer "feed," $x = [M_1]/[M_2]$, and the instantaneous copolymer composition, $y = d[M_1]/d[M_2]$. Hence we obtain Equation 6-4.

$$y = \frac{1 + r_1 x}{1 + \frac{r_2}{x}}$$

EQUATION 6-4

Another widely used form expresses the instantaneous monomer and copolymer concentrations in terms of *mole fractions*. For monomer 1 this would be defined as:

$$f_1 = [M_1]/([M_1] + [M_2])$$

and

$$F_1 = d[M_1]/(d[M_1] + d[M_2])$$

(Of course, $f_2 = 1 - f_1$ and $F_2 = 1 - F_1$.) This yields the instantaneous copolymer composition, as given in Equation 6-5.

$$F_1 = \frac{(r_1 - 1)f_1^2 + f_1}{(r_1 + r_2 - 2)f_1^2 + 2(1 - r_2)f_1 + r_2}$$

EQUATION 6-5

Reactivity Ratios

The reactivity ratios, $r_1 = k_{11}/k_{12}$ and $r_2 = k_{22}/k_{21}$, are extremely important quantities, expressing the relative preference of the radical species for the monomers. If r_1 is greater than 1, for example, it means that a chain with a terminal radical of type 1 would rather add another monomer of type 1 than a monomer of type 2. Furthermore, the reactivity ratios are the only two independent rate variables that we need to know or measure, as opposed to the four individual rate constants.

We'll discuss the measurement of reactivity ratios shortly. First, we'll have a look at what types of copolymers we would expect to get for certain limiting values of the reactivity ratios and then have an initial look at the problem of composition drift.

When we first introduced the rate constants, we took a quick look at what types of copolymers we would expect to get under certain limiting conditions. It's useful to repeat this exercise in terms of the reactivity ratios, going into a little more detail for certain limiting cases. This time, however, you should figure it out for yourself first!

SPECIAL CASE 1. If $r_1 = r_2 = 0$

What type of polymer would you expect to get—a random, copolymer, an alternating copolymer, a block copolymer and/or homopolymers? Think about it: if $r_1 = r_2 = 0$ this means that a radical chain whose terminal group is of type 1 never wants to add a monomer of type 1, and a radical of type 2 never wants to add one of its own kind, either. (For $r_1 \sim 0$ and $r_2 \sim 0$, k_{11} and $k_{22} \sim 0$.) For values of $r_1 = 0$, $r_2 = 0$, we would get a perfectly alternating copolymer (see Figure 6-1). In the more probable case, where r_1 and r_2 are small, we would get a strong *tendency to alternation*. Note that the copolymer composition depends not only on the reactivity ratios, but also on the monomer "feed" composition. Also, in discussing these tendencies, we are assuming that the monomer concentrations are not widely different at the moment we are considering the copolymer composition.

SPECIAL CASE 2. If $r_1 = r_2 \sim \infty$

If $r_1 = \infty$, then k_{12} must equal zero. Similarly, if $r_2 = \infty$, then k_{21} must equal zero. Accordingly, chain radicals with a terminal group of type 1 will only add monomers of type 1, while M_2 chain radicals will show equivalent discrimination, only wishing to add monomers of their own type. Hence, only *homopolymers* are produced.

If r_1 and r_2 are both large, but not infinite, then *block or blocky copolymers* will be produced, perhaps with some homopolymer, depending on how large the reactivity ratios are and the relative concentration of the monomers in the feed.

SPECIAL CASE 3. If $r_1 = r_2 = 1$

Under these conditions, monomers of type 1 and type 2 add with equal facility to the growing chain, regardless of the nature of the radical species in the terminal position. The resulting copolymer is *truly random* and its composition is exactly the same as the "feed," or the reaction mass (i.e., $F_1 = f_1$).

SPECIAL CASE 4. If $r_1 r_2 = 1$

This special case is called an *ideal copolymerization*. As in special case 3, the distribution of monomers in the copolymer is random, but the copolymer composition is not usually the same as that of the monomers in the reaction mass. It has been our experience that many students have a bit of trouble seeing why this condition should give a random copolymer. A lot of students tend to associate the word random with a 50:50 composition where the monomers are distributed according to coin toss statistics. However, a 90:10 distribution of, say, monomer 1 and monomer 2 can be random. Certainly, there will be "blocks" of M_1 units, just because there are so many of them relative to M_2's, but the copolymer is random if the probability of finding a 1 at any point in the chain is 0.9, while the probability of finding a 2 is 0.1.

Now, the condition $r_1 = r_2 = 1$ means that $k_{11}/k_{12} = k_{21}/k_{22}$ and the rate of adding a 1 to the terminal group on the chain relative to a 2 does not depend on the nature of the terminal group. For example, if the terminal group is a 1 and the ratio of the rate constants describing the addition of a 1 relative to a 2 is 10, then the same ratio of the rate constants is found if the terminal group is a 2. This means that the placement of units will be random, even though, in this example, there will be a lot more M_1 units being incorporated into the copolymer chain than M_2 units (depending on the composition of the feed).

Composition Drift

If one monomer is being incorporated into the polymer at a faster rate than the other, then the composition of the "feed" or monomer reaction mass is not constant, but changes with time. This, in turn, means that the composition of the copolymer produced at, say, the start of the reaction is very different to that which is being produced at the end. Getting back to the instantaneous copolymer composition, the special cases we have discussed are just that: in practice, special case 2 ($r_1 = r_2 = \infty$) has not been obtained, but there are examples where special case 1 ($r_1 = r_2 = 0$) has been approached ($r_1 = r_2$ = very small). We will consider an example of this after considering the statistics of copolymer composition. Similarly, certain systems are almost ideal ($r_1 r_2 = 1$), but most often $r_1 r_2 < 1$. A better feel for how reactivity ratios affect copolymer composition can be obtained by making plots of the instantaneous copolymer composition as a function of feed concentration (e.g., F_1 versus f_1), as shown in Figure 6-10. Keep in mind that this does not account for composition drift, but tells us what instantaneous copolymer concentration we would get for a given concentration of monomers. The special case $r_1 = r_2$ = 1 gives the straight line along the diagonal of this plot (copolymer composition equals feed composition). The curve above this diagonal ($r_1 > 1$; $r_2 < 1$) represents the situation where the copolymer formed instantaneously is always richer in M_1's than M_2's. If both reactivity ratios are smaller than 1 ($r_1 < 1$; $r_2 < 1$), then curves crossing the diagonal, at a point called the *azeotrope*, can be obtained (by analogy to the situation where a vapor has the same composition as a liquid with which it is in equilibrium). At the azeotrope the copolymer has exactly the same composition as the monomer feed, hence:

$$d[M_1]/d[M_2] = [M_1]/[M_2]$$

FASCINATING POLYMERS—BUNA RUBBER

The IG Farben Company was formed in 1925 by the merging of Bayer with the BASF and Hoechst companies. Research initially focused on the polymerization of butadiene, a compound for which an economic synthesis had recently been discovered. Chemists at IG Farben succeeded in producing a polybutadiene rubber using sodium (Na) as a catalyst. The word Buna was derived from a combination of Bu, for butadiene, and na for sodium. But this Buna rubber was "sticky" and had poor tear and tensile strengths. Not an ideal material for a tire! In attempting to reduce the "stickiness" of the rubber, chemists at Bayer's laboratory in Leverkusen made some important discoveries. They turned their attention to emulsion polymerization. In this process, the water insoluble butadiene (oil) is emulsified into fine particles that are stabilized in water using a soap or surfactant. Different water soluble catalysts were employed, as sodium could not be

Buna rubber sheets (Source: Bayer).

used. (For those of you who are chemically challenged, sodium reacts violently with water!) But the polybutadiene produced was still too "tacky" and difficult to handle. Some improvements were made by adding drying oils, like linseed oil, at the start of the polymerization, but the breakthrough came from another direction. Kurt Meisenburg, also working in Leverkusen, was studying the polymerization of styrene, and discovered that a small amount of styrene added to the butadiene before polymerization had the same effect as the drying oils. A little later, in June 1929, a colleague, Walter Bock, polymerized butadiene and styrene (2:1 ratio) and produced a styrene-co-butadiene copolymer, that, when compounded with carbon black, had excellent wear resistance. In fact, in some ways it was better than natural rubber! A key patent was filed. This material was called Buna S. As we see it, IG Farben registered Buna as a trademark in 1930 to confuse young students. We say this because the word Buna was derived from butadiene and sodium and there was no sodium used to produce Buna S. In fact, the Buna S rubber synthesized by Walter Bock (S for Styrol, the German equivalent of styrene) was a 2:1 copolymer of butadiene and styrene. What was not known at the time was that, under the polymerization conditions used, butadiene and styrene fortuitously add to the growing chain in an essentially random or statistical fashion. Composition drift was not a problem. Moreover, the butadiene was incorporated into the polymer chain as three distinct structural isomers, referred to as *cis*-1,4-, *trans*-1,4- and 1,2- placements. Stereoisomerism was not a factor in the case of styrene incorporation. Thus, Buna S is an irregular amorphous material, incapable of crystallization. The copolymer microstructure is complex and made up of four structural units. Changing polymerization conditions, e.g., polymerization temperature, changes the concentration and distribution of the butadiene structural isomers. This affects the physical, chemical and mechanical properties of the rubber and this became important to the development of the synthetic rubber industry.

POLYMER MILESTONES—THE AMERICAN SYNTHETIC RUBBER PROGRAM

Modern truck tire prior to curing (Courtesy: Bridgestone Americas).

"War! What's it good for?" As the popular rock ballad of the Vietnam Era responds: *"Absolutely nothing!"* But there are times when you have to fight and, throughout history, advances in materials have been stimulated by war. The development and production of synthetic rubber was no exception. Until Japan attacked Pearl Harbor in December of 1941, the United States bungled along, with Standard Oil and the major rubber and chemical companies mainly embroiled in arguments about patent rights and infringements involving Buna S-type rubber, of which only a paltry 231 tons was actually produced in the United States in 1941. It's remarkable, but war and a common enemy quickly resolve irreconcilable differences, and within one week of the declaration of war with Japan, the major US oil, chemical and rubber industries had agreed to share patents and information under the auspices of a government-owned agency called the Rubber Reserve Company. As the vast majority of natural rubber production was now under the control of the Japanese, production of major quantities of synthetic rubber was crucial. Just to give you a feel for how essential rubber was for the war effort, a Sherman tank used about a ton of rubber, a battleship, 75 tons, and a B-17 Flying Fortress, half a ton. In 1943, US military use alone was greater than 500,000 tons. By the early 1940s the US stockpile of natural rubber was disappearing rapidly. Edward Weidlein, who at the time was the Director of Research at the Mellon Institute in Pittsburgh, was cajoled into taking control of the government's Synthetic Rubber Program. This required the patience of Job, because he had to get the scientific managers and research chemists of the major chemical and rubber companies to decide on a common recipe for making a version of Buna S—a process akin to herding cats! They finally settled on a recipe for GR-S (Government Rubber-Styrene). Waldo Serman played a major role in the GR-S program and pointed out that the ease with which the rubber can be processed before curing was as important as the properties of the final vulcanized material. One essential ingredient, which was referred to, whimsically, as OEI, was the chain transfer agent, dodecyl mercaptan ($CH_3(CH_2)_{11}SH$). (This is a repulsive chemical that smells like rotten garlic.) OEI acts to reduce the average molecular weight of the elastomer and improves "processability." The US "mutual recipe" was hardly changed throughout the duration of the war. It was an emulsion copolymerization of styrene and butadiene using potassium persulfate initiator and OEI modifier. And it was superior to the German product! Impressively, in the space of a couple of years the United States went from an insignificant amount of synthetic GR-S rubber production to some 730,000 tons in 1945 (6.5 times the maximum German production rate).

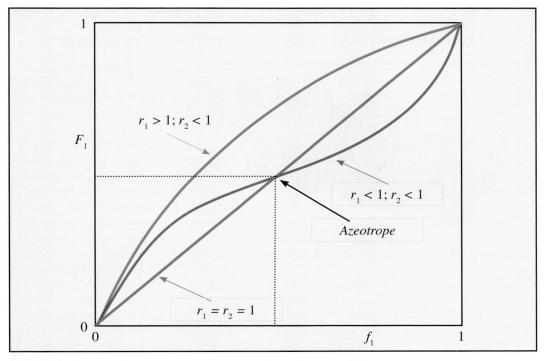

FIGURE 6-10 Effect of reactivity ratios on the copolymer composition as a function of feed concentration.

and there is no composition drift, because monomer is incorporated into the copolymer chains in proportion to the composition of the feed. (Yet another starting point for a sickening homework question!)

Determination of Reactivity Ratios

So far we have discussed reactivity ratios as if they are known quantities. And many of them are (you can find their values in the Polymer Handbook), thanks to sterling work by many polymer chemists over the years. But what if you're confronted with a situation where you don't have this information – how would you determine the reactivity ratios of a given pair of monomers? Essentially, there are two sets of approaches, both of which depend upon using the copolymer equation in one form or another, hence, the assumption that the terminal model applies to the copolymerization we are considering. A form we will use as a starting point was

given in Equation 6-4.

The older set of methodologies relies on preparing a number of copolymers as a function of monomer composition, measuring the resulting copolymer composition and obtaining the reactivity ratios using various plots. Because of composition drift, each copolymerization must be stopped at low degrees of conversion (<10%) and it is assumed that x in Equation 6-4 is accurately given by the initial monomer concentration, while y is given by the integral or average value of the copolymer composition over the range of composition chosen. This latter quantity is usually measured spectroscopically. Modern methods can, in principle, use the ability of NMR spectroscopy to measure sequence distributions in a single copolymer sample and calculate reactivity ratios from this data. This is not always easy, however, because of the complexity of the spectra, as we will discuss in Chapter 7. So the older methods discussed here are still used and you should know something about them.

FASCINATING POLYMERS—VINYLIDENE CHLORIDE COPOLYMERS

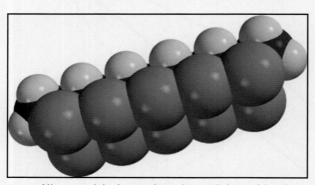

Space filling model of part of a poly(vinylidene chloride) chain (the chlorine atoms are shaded brown).

When asked to name the greatest invention in the history of the world, one of our favorite satirists, Mel Brooks, as the 2000-year-old man, replied to Carl Reiner: "*Saran Wrap.*" You've all used this stuff, even if only for purposes for which it was not intended! Remember the advertisement showing the raw onion wrapped in Saran film, which even a bloodhound couldn't smell? Here's yet another tale of serendipitous discovery. It appears that in 1933, Ralph Wiley, then a college student working in Dow Chemical's laboratories, accidentally discovered a crude form of poly(vinylidene chloride) (PVDC). Following further development, Dow produced "Saran," which then had the appearance of a greasy, dark green film. Saran was sprayed on fighter planes to guard against salty sea spray and also found application in car upholstery. It was approved, after its green color and unpleasant odor were eliminated, for food packaging post-WWII. Poly(vinylidene chloride) is a polymer that has a simple and regular repeating unit containing a methylene (CH_2) and a dichloromethylene (CCl_2) group. In the solid state, it is a highly crystalline plastic with extraordinary barrier properties to oxygen, water and odors. However, the pure homopolymer is almost impossible to process, as it degrades rapidly at melt processing temperatures. Saran® films that are used for flexible packaging materials, especially those that come in contact with food, are typically free radically polymerized copolymers that contain varying amounts of methyl acrylate or vinyl chloride as comonomers. This is a compromise, as the barrier properties decrease a little, but the upside is that these copolymers can be processed readily into films. The comonomers actually disrupt the regularity of the PVDC chain, thereby reducing the melt temperature. Saran resins can be extruded, coextruded or coated to meet specific packaging needs. About 85% of PVDC is used as a thin layer between cellophane, paper and plastic packaging to improve barrier performance. Saran Wrap® film, the first cling wrap designed for household (1953) and commercial use (1949), was introduced by Dow and is currently marketed by S. C. Johnson.

A typical roll of Saran® wrap that one can readily pick up in most supermarkets (Source: www.saranbrands.com).

The Mayo-Lewis Plot

The first method we will consider involves the rearrangement of the copolymer equation (Equation 6-4). The final equation expresses one of the reactivity ratios in terms of the other and the experimental quantities x and y (Equation 6-6).

$$r_2 = \frac{x(1 + r_1 x)}{y} - x$$

EQUATION 6-6

An intersection method is employed, where essentially a first guess or estimate of the value of, say, r_1 is made and the rearranged copolymer equation used to calculate values of r_2 for each value of x and y. You end up with a set of straight lines which should intersect at a point defining the actual values of r_1 and r_2, but generally don't. You have to use some method for picking the best point within an area of intersection (Figure 6-11).

The Fineman-Ross Plot

The Fineman-Ross method uses a more conventional plotting procedure, rearranging the copolymer equation into the following form (Equation 6-7),

$$x\left(1 - \frac{1}{y}\right) = r_1\left(\frac{x^2}{y}\right) - r_2$$

EQUATION 6-7

so that a plot of the left-hand side of the equation versus x^2/y should give a straight line with an intercept $-r_2$ and a slope r_1 as illustrated schematically in Figure 6-12.

The Kelen-Tüdõs[4] Plot

Both the Mayo-Lewis and the Fineman-Ross methods rely on linearizing the copolymer equation. It has been shown that

[4] T. Kelen and F. Tüdõs, *J. Macromol. Sci. - Chem.*, **A9(1)**, 1, 1975.

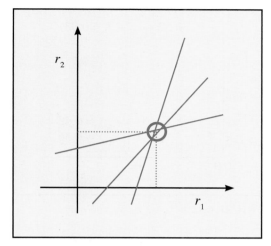

FIGURE 6-11 Mayo-Lewis plot.

this transforms the error structure in such a way that linear least squares procedures are inappropriate. An alternative method by Kelen and Tüdõs, designed to overcome these shortcomings, has gained popularity in recent years. This method has the set of rather awful-looking substitutions shown in Equations 6-8, with the parameter, α, serving to distribute the experimental data uniformly and symmetrically between the limits 0 and 1.

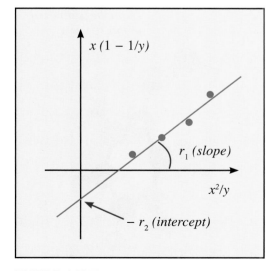

FIGURE 6-12 Fineman-Ross plot.

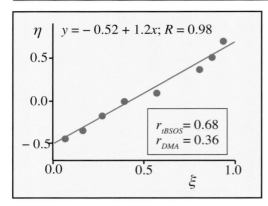

$$\eta \quad y = -0.52 + 1.2x; R = 0.98$$

$$r_{tBSOS} = 0.68$$
$$r_{DMA} = 0.36$$

FIGURE 6-13 A typical Kelen-Tüdõs plot.

$$\eta = r_1\xi - \frac{r_2}{\alpha}(1 - \xi)$$

where:

$$\eta = \frac{x(y - 1)/y}{\alpha + x^2/y} \qquad \xi = \frac{x^2/y}{\alpha + x^2/y}$$

and:

$$\alpha = \sqrt{(x^2/y)_{low} \, (x^2/y)_{high}}$$

EQUATIONS 6-8

The subscripts *low* and *high* in the equation for α represent the lowest and highest values of x^2/y in the experimental data set.

A typical Kelen-Tüdõs plot is shown in

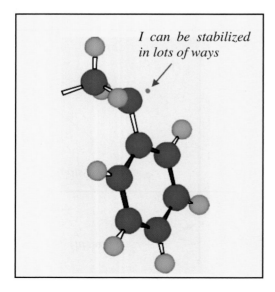

I can be stabilized in lots of ways

FIGURE 6-14 Schematic of styrene stabilization.

Figure 6-13 for data obtained in our laboratories[5] for the system *n*-decyl methacrylate (DMA) copolymerized with *p*(*t*-butyldi-methyl-silyloxy)styrene (*t*-BSOS) — we're interested in some weird stuff! The reactivity ratios are calculated from the intercepts at 0 and 1. Of course, this and the preceding plots are not very interesting to look at, so we normally give our students some data and ask them to compare the values of the reactivity ratios obtained from all three methods. We're nasty!

Monomer Reactivity and the Q-e Scheme

We've talked about reactivity ratios and how they can be measured, but really haven't addressed the question of why one monomer should prefer to react with one radical (or other reactive species) rather than another. As you might guess, there are some obvious factors, like steric hindrance. That is why styrene, with a bloody great benzene ring stuck on one end, polymerizes (free radically) in a head-to-tail fashion, while in poly(vinylidene fluoride) the smaller fluorine atoms allow the formation of some head-to-head and tail-to-tail sequences. We won't go into detail, but two other important factors that affect reactivity are resonance stabilization and polarity. For example, if a monomer has a structure that allows a radical site to be resonance stabilized, as styrene does (Figure 6-14), there is, in general, a decrease in the reactivity of this monomer relative to when there is no resonance stabilization. Polarity comes into play when there are electron-withdrawing or electron-donating groups, which can affect the polarity of the double bond in the monomer. Obviously, a monomer of one type (say, with an electron-withdrawing group) would preferentially add a monomer of the other and in certain circumstances the effect can be so marked that charge transfer complexes are produced. Alfrey and Price attempted to account for these latter two factors (resonance stabilization and polar-

[5] Y. Xu, P. C. Painter, M. M. Coleman, *Polymer*, **34**, 3010, (1993).

ity) with parameters labeled Q and e. This provided a good semi-quantitative way to estimate reactivity ratios at a time when their experimental determination was more difficult. If this interests you, you should go to some standard polymer chemistry books and the original literature and check this out.

COPOLYMER SEQUENCE DISTRIBUTIONS

So far, we have used kinetics to describe the relationship of monomer feed concentration and reactivity ratios to copolymer composition. Now we will show how probability theory can be used to describe sequence distributions. Try to contain your excitement.

We have already seen that, depending on the values of the reactivity ratios, there is a tendency to get random, alternating, blocky, etc., types of copolymers. Probability theory allows us to quantify this in terms of the frequency of occurrence of various sequences, like the triads AAA or ABA in a copolymerization of A and B monomers. The value of this information is that such sequence distributions can be measured directly by NMR spectroscopy, thus allowing a direct probe of copolymer structure and an alternative method for measuring reactivity ratios. As mentioned above, there are problems, as some spectra can be too complex and rich for easy analysis, as we will see in Chapter 7.

General Statistical Relationships

Equation 6-9 was introduced previously as Equation 5-1 in Chapter 5, but when dealing with probabilities we have found that it is never a waste of time repeating the basics. The first thing you must do is remember that the probability of an event occurring, $P\{E\}$, is equal to the number fraction of such events, given that you are measuring a really large number of events.

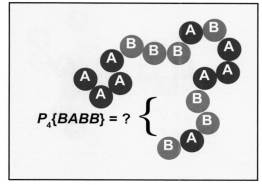

FIGURE 6-15 What is the probability of finding a BABB sequence?

Recall, for example, if you toss a coin one hundred times and measure the number of heads, you might expect that N_E would be about 50 and $P\{E\}$ would be about one-half. Most often N_E wouldn't be exactly 50; you would have to repeat the 100 coin tosses a large number of times and average all the values to get $P\{E\} = 0.5$.

Now let's introduce another definition, the probability of obtaining a particular sequence of n units, $P_n\{x_1 x_2 x_3 x_4, \cdots, x_n\}$, equal to the number fraction of such sequences present. For example, $P_4\{BABB\}$ is the probability that the sequence BABB occurs in a copolymer of just A and B units (Figure 6-15, reading the sequence back from the end of the chain).

The next step is to consider a simple copolymerization of A and B units and then write probability expressions for the monomers, diads, triads, etc., and show how these are interrelated. After a while, the number of such equations looks dauntingly large, but the relationships are actually pretty simple and easy to work out. Let's start with the obvious: if there are just A and B units in the chain then the probability of finding an A, $P_1\{A\}$, is simply the number of A's divided by the total number of A's and B's (Figure 6-16). Similarly for $P_1\{B\}$, which leads to Equation 6-10):

$$P\{E\} = \frac{N_E}{N}$$

EQUATION 6-9

$$P_1\{A\} + P_1\{B\} = 1$$

EQUATION 6-10

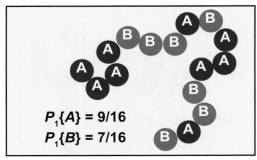

$$P_1\{A\} = 9/16$$
$$P_1\{B\} = 7/16$$

FIGURE 6-16 The probability of finding an A or B in the above sequence.

Now let's consider diad sequences. If we start from one end of the chain and go a monomer at a time and ask what is the probability that this monomer and the next one are both A's, or an A and a B, and so on, you find there are four possibilities (AA, AB, BA, BB) and we get Equation 6-11:

$$P_2\{AA\} + P_2\{AB\} +$$
$$P_2\{BA\} + P_2\{BB\} = 1$$
EQUATION 6-11

The next step is to consider the concentrations or probabilities of lower-order placements ($P\{A\}$ and $P\{B\}$) in terms of the higher-order placements. If we take an A unit, for example, there are two possibilities for the unit that comes next along the chain—it can either be another A or it can be a B. So, we can write Equation 6-12:

$$P_1\{A\} = P_2\{AA\} + P_2\{AB\}$$
EQUATION 6-12

where we have colored the *successor* unit in red to indicate the two possibilities. We can also write an equivalent expression in terms of *predecessor* units (Equation 6-13):

$$P_1\{A\} = P_2\{AA\} + P_2\{BA\}$$
EQUATION 6-13

From these two equations we obtain the important result (Equation 6-14):

$$P_2\{AB\} = P_2\{BA\}$$
EQUATION 6-14

You should have been able to guess that, because the number of AB *diads* you count, working from one end of the chain, must be the same as the number of BA diads you count if you start from the other end of the chain and work backwards. Of course, we can write equivalent expressions in terms of B units and get the same result (Equations 6-15):

$$P_1\{B\} = P_2\{BB\} + P_2\{BA\}$$
$$= P_2\{BB\} + P_2\{AB\}$$
Hence:
$$P_2\{BA\} = P_2\{AB\}$$
EQUATIONS 6-15

As you might guess, we can now go "up to the next level" and express diad sequences in terms of *triad* sequences. If we consider successor and predecessor units to an AA diad, for example, we get Equations 6-16:

$$P_2\{AA\} = P_3\{AAA\} + P_3\{AAB\}$$
$$= P_3\{AAA\} + P_3\{BAA\}$$
Hence:
$$P_3\{AAB\} = P_3\{BAA\}$$
EQUATIONS 6-16

Now you do the same for the other three diad sequences. However, we leave this for you to do as an exercise, hoping it will keep you out of mischief for a while. When you examine your results, you should find that you get Equations 6-17:

$$P_3\{BBA\} = P_3\{ABB\}$$
$$P_3\{ABA\} = P_3\{ABA\}$$
$$P_3\{BAB\} = P_3\{BAB\}$$
EQUATIONS 6-17

Note the *reversibility* of these sequences. As you might guess from this and the qualitative arguments we have made already, the probability or number fraction of a sequence is always equal to that of its mirror image. An example is shown in Equation 6-18:

$$P_6\{ABABBA\} = P_6\{ABBABA\}$$

EQUATION 6-18

Conditional Probability

We're part of the way home, but in order to calculate some of the things we want to know we also need to consider *conditional probabilities*. In this approach we say: "Given something, what is the probability of something else"? You know, given that you engaged in some wicked overindulgence one Friday night, what is the probability that you wake up Saturday morning vowing that next time you'll show some restraint. Except that here we are not concerned with the vagaries of human behavior, but the far less interesting but more exact probabilities of the arrangement of units in a chain. We thus define $P\{x_{n+1}/x_1, x_2, x_3, \ldots, x_n\}$ as the conditional probability that a particular sequence $(x_1, x_2, x_3 \ldots, x_n)$ of units is followed by a particular unit, x_{n+1}. For example, $P\{A/B\}$ is the probability of finding an A unit *given* that the previous unit is a B. By definition, this is equal to the number fraction of A's that follow B's, so we can write (Equations 6-19):

$$P_2\{BA\} = P_1\{B\}P\{A/B\}$$

Similarly:

$$P_2\{AA\} = P_1\{A\}P\{A/A\}$$

EQUATIONS 6-19

And so on. These equations can then be rearranged to give the relationships shown in Equations 6-20.

Obviously, the next step is to express the probability of getting a triad unit in terms of such conditional probabilities. Examples are shown in Equations 6-21:

$$P\{A/A\} = \frac{P_2\{AA\}}{P_1\{A\}}$$

$$P\{A/B\} = \frac{P_2\{BA\}}{P_1\{B\}}$$

$$P\{B/B\} = \frac{P_2\{BB\}}{P_1\{B\}}$$

$$P\{B/A\} = \frac{P_2\{AB\}}{P_1\{A\}}$$

EQUATIONS 6-20

$$P_3\{BAA\} = P_1\{B\}P\{A/B\}P\{A/BA\}$$

$$P_3\{AAA\} = P_1\{B\}P\{A/A\}P\{A/AA\}$$

EQUATIONS 6-21

This starts to look complicated, but all we're really saying is that the probability of getting a BAA sequence is equal to first, the probability of getting a B as the first unit, followed by the probability of getting an A given that the first unit is a B, then the probability of getting another A given that the first two units are a BA sequence. The overall probability of getting a BAA sequence is then these three probabilities multiplied together. These equations and others like them can then be rearranged to give the horrible-looking collection shown in Equations 6-22.

$$P\{A/AA\} = \frac{P_3\{AAA\}}{P_2\{AA\}}$$

$$P\{A/BA\} = \frac{P_3\{BAA\}}{P_2\{BA\}}$$

$$P\{A/AB\} = \frac{P_3\{ABA\}}{P_2\{AB\}}$$

$$P\{A/BB\} = \frac{P_3\{BBA\}}{P_2\{BB\}}$$

$$P\{B/AA\} = \frac{P_3\{AAB\}}{P_2\{AA\}}$$

$$P\{B/BA\} = \frac{P_3\{BAB\}}{P_2\{BA\}}$$

$$P\{B/AB\} = \frac{P_3\{ABB\}}{P_2\{AB\}}$$

$$P\{B/BB\} = \frac{P_3\{BBB\}}{P_2\{BB\}}$$

EQUATIONS 6-22

From all of this, the relationships shown in Equations 6-23 can then be formally obtained.

$$P\{A/A\} + P\{B/A\} = 1$$
$$P\{A/B\} + P\{B/B\} = 1$$
$$P\{A/AA\} + P\{B/AA\} = 1$$
$$P\{A/BA\} + P\{B/BA\} = 1$$
$$P\{A/AB\} + P\{B/AB\} = 1$$
$$P\{A/BB\} + P\{B/BB\} = 1$$

EQUATIONS 6-23

Or you can just write them down using common sense. If you're given an A, then the next unit can only be another A or a B. You may now be saying to yourself, "We are surely done with all these definitions." But no-o-o-o-o! You can have conditional probabilities of different orders! Sometimes this subject shows no mercy.

Conditional Probabilities of Different Orders

We have already shown in Equations 6-21 how you can break down a compound event, like the number fraction of (or probability of finding) a triad sequence such as AAA, into conditional probabilities of different order. Let's first write this in a slightly different form to give you a feel for the nomenclature (Equation 6-24):

$$P\{A^3\} = P_1\{A\}P\{A/A\}P\{A/AA\}$$

EQUATION 6-24

TERMINAL MODEL

$$\sim\sim\sim\sim\sim M_1^\bullet + M_1 \xrightarrow{k_{11}} \sim\sim\sim\sim\sim\sim M_1^\bullet$$

PENULTIMATE MODEL

$$\sim\sim\sim\sim M_1 M_1^\bullet + M_1 \xrightarrow{k_{111}} \sim\sim\sim\sim\sim M_1^\bullet$$
$$\sim\sim\sim\sim M_2 M_1^\bullet + M_1 \xrightarrow{k_{211}} \sim\sim\sim\sim\sim M_1^\bullet$$

FIGURE 6-17 The terminal and penultimate models.

Then generalize this to a consideration of a sequence of n units (Equation 6-25):

$$P\{A^n\} = P_1\{A\}P\{A/A\}P\{A/AA\}\cdots$$
$$\cdots P\{A/A^{n-1}\}$$

EQUATION 6-25

You have no doubt been asking yourself: "Why are these dirty, rotten swine doing this to me?" Step back and think about what we've said about copolymerization. There is something called the *terminal model*, where the rate constants depend only on the nature of the terminal group, and the *penultimate model*, where there is a dependence on the character of the final two units in the chain (Figure 6-17). The use of conditional probabilities is the starting point for distinguishing between these experimentally, as we will see.

For the *terminal* model, then, we can write Equation 6-26 for the number fraction of AAA triads:

$$P_3\{AAA\} = P_1\{A\}P\{A/A\}P\{A/AA\}$$

EQUATION 6-26

But, if the addition of the next unit to a chain depends only on the nature of the end group, we have (Equation 6-27):

$$P\{A/AA\} = P\{A/A\}$$

EQUATION 6-27

and we can write Equation 6-28:

$$P\{A^3\} = P_1\{A\}(P\{A/A\})^2$$

EQUATION 6-28

If the addition of the next group was *independent* of the chain end, Equations 6-29 would be applicable:

$$P\{A/A\} = P_1\{A\}$$

and:

$$P\{A/AA\} = P_1\{A\}$$

Hence:

$$P\{A^3\} = \left(P_1\{A\}\right)^3$$

EQUATIONS 6-29

For the *penultimate* model, we just have the triad conditional probabilities we started with, of course. What we're actually doing is using the theory of Markov chains, named after a Russian mathematician who studied the probability of mutually dependent events. In this general approach we would write (Equation 6-30):

$$P_n\{A^n\} = \left(P_1\{A^k\}P\{A/A\}^k\right)^{n-k}$$

EQUATION 6-30

The terminal and penultimate models then correspond to first- and second-order Markovian statistics, respectively. But you don't actually have to know this, in the sense that we can just proceed using common sense. For example, the probability of finding the sequence ABABA in a system obeying first-order Markovian statistics (i.e., copolymerization where the terminal model applies) is given in Equations 6-31.

$$P_5\{ABABA\} =$$
$$P_1\{A\}P\{B/A\}P\{A/B\}\cdots$$
$$\cdots P\{B/A\}P\{A/B\}$$

or:

$$P_5\{ABABA\} =$$
$$P_1\{A\}(P\{B/A\})^2(P\{A/B\})^2$$

EQUATIONS 6-31

We'll leave it to you to show what the same probability would be if second-order Markovian statistics applied (i.e., the penultimate model).

Some Useful Parameters

These definitions of general statistical relationships and conditional probabilities have given us all the tools we need to describe copolymer microstructure. In the rest of this chapter, we will show how we can relate these probabilities to reactivity ratios, then in the chapter on spectroscopy we can draw a more direct relationship to NMR measurements of sequence distributions. Before getting to that, however, we need to introduce some parameters that are useful in describing copolymers (Figure 6-18). Be grateful for small mercies, because these won't involve the conditional probabilities we have just discussed; those suckers won't rear their ugly heads again until we get to the specifics of the terminal and penultimate models.

The Number Fraction Sequence of Units

Our first definition is the number fraction sequence of (say) A units of length n, which is the number of such sequences divided by the number of all possible sequences of A units (i.e., from $n = 1$ to ∞)—Equation 6-32:

$$N_A(n) = \frac{P_{n+2}\{BA_nB\}}{\sum_1^{\infty} P_{n+2}\{BA_nB\}}$$

EQUATION 6-32

Now, any given sequence of A units must end (and begin) with a B unit (Equation 6-33):

USEFUL PARAMETERS

Number fraction sequence of units

Number average length of runs

Departure from randomness

Run fraction or number

FIGURE 6-18 Useful parameters used to describe copolymer microstructure.

$$\sum_1^\infty P_{n+2}\{BA_nB\} = P_3\{BAB\} + P_4\{BAAB\} + P_5\{BAAAB\} + \cdots$$

EQUATION 6-33

Therefore, the summation reduces to the simple Equation 6-34:

$$\sum_1^\infty P_{n+2}\{BA_nB\} = P_2\{AB\}$$

EQUATION 6-34

Hence, we arrive at the final equation for the number fraction sequence of units (Equation 6-35):

$$N_A(n) = \frac{P_{n+2}\{BA_nB\}}{P_2\{AB\}}$$

EQUATION 6-35

For example, the fraction of AAA units is (Equation 6-36):

$$N_A(3) = \frac{P_5\{BAAAB\}}{P_2\{AB\}}$$

EQUATION 6-36

The Number Average Length of A (and B) Runs

The second definition we wish to introduce is the number average length of A (and B) runs. This is simply defined in Equation 6-37.

$$\bar{l}_A = \frac{\sum_1^\infty n N_A(n)}{\sum_1^\infty N_A(n)}$$

EQUATION 6-37

You might want to compare Equation 6-37 to that of the definition of the number average molecular weight. $N_A(n)$ is the number fraction of sequences of n A's, which we just considered. Now, we know that (Equations 6-38):

$$\sum_1^\infty N_A(n) = 1$$

and:

$$N_A(n) = \frac{P_{n+2}\{BA_nB\}}{P_2\{AB\}}$$

EQUATIONS 6-38

which yields Equation 6-39:

$$\bar{l}_A(n) = \frac{\sum_1^\infty n P_{n+2}\{BA_nB\}}{P_2\{AB\}}$$

EQUATION 6-39

Now we do some substituting. First, writing out the summation expression term by term we have (Equation 6-40):

$$\sum_1^\infty n P_{n+2}\{BA_nB\} = 1P_3\{BAB\} + 2P_4\{BAAB\} + 3P_5\{BAAAB\} + \cdots$$

EQUATION 6-40

If you think about it you will realize that the terms on the right hand side of Equation 6-40 simply add up to all the A's that are present, i.e., (Equation 6-41):

$$\sum_1^\infty n P_{n+2}\{BA_nB\} = P_1\{A\}$$

EQUATION 6-41

And we end up with our expression for the number average length of A and (by analogy) B runs (Equations 6-42):

$$\bar{l}_A = \frac{P_1\{A\}}{P_2\{AB\}} \qquad \bar{l}_B = \frac{P_1\{B\}}{P_2\{BA\}}$$

EQUATIONS 6-42

Accordingly, all we need to calculate these parameters is copolymer composition and diad sequence data.

The Run Fraction or Number

There is also something called the run fraction or number, which we won't use but is commonly found in the NMR literature, so for completeness we will define it here. It is just the fraction of A (and B) sequences that occur in a polymer chain. To illustrate what we mean by this, consider the sequence:

ABAABABAAABBABBBBAAB

Now let's break this up into "runs" of segments of the same type by inserting spaces:

A B AA B A B AAA BB A BBBB AA B

You can see that this chain of 20 units is arranged in 12 alternating runs and the run number, R, is then 12/20 (note that in general the chain has to be long enough that end effects can be ignored). Also note that every run of A units is terminated by an AB link and every run of B units is terminated by a BA link. Thus R is defined in Equations 6-43:

$$R = \text{Fraction of } (AB + BA) \text{ links}$$
$$= P_2\{AB\} + P_2\{BA\}$$
$$= 2P_2\{AB\}$$

EQUATIONS 6-43

This gives us the following useful relationships (Equations 6-44):

$$P_2\{AA\} = P_1\{A\} - \frac{R}{2}$$
and:
$$P_2\{BB\} = P_1\{B\} - \frac{R}{2}$$

EQUATIONS 6-44

A Measure of the Departure from Randomness

Last, but certainly not least, we need to define a parameter that measures the deviation from random statistics. This is given the symbol χ, not to be confused with the Flory-Huggins interaction parameter χ (Chapter 11). The χ we will use here is defined in Equation 6-45:

$$\chi = \frac{P_2\{AB\}}{P_1\{A\}P_1\{B\}}$$

EQUATION 6-45

If the copolymerization were *completely random* $P_2\{AB\} = P_1\{A\}P_1\{B\}$ then χ would equal 1. If there was an *alternating tendency* imposed by the kinetics, then the number fraction of AB diads would exceed $P_1\{A\}P_1\{B\}$ and χ would be >1. If you don't see why this is so, think about the limiting case of a perfectly alternating copolymer. For such a polymer we naturally have exactly equal numbers of A and B segments, so that $P_1\{A\} = P_1\{B\} = 0.5$. Also, there are no AA or BB sequences so $P_2\{AB\} = P_2\{BA\} = 0.5$. Thus the value of χ is 2. Conversely, if the value of χ is less than 1, then there is a *tendency to blockiness*. Again, think about the limiting case. For a diblock copolymer, there is only one AB linkage and χ would have a value that approaches zero for high molecular weight copolymers. A summary of χ values for the different copolymers is presented in Figure 6-19.

SUMMARY OF χ VALUES

$\chi = 1$	*Completely Random*
$\chi > 1$	*Alternating Tendency*
$\chi = 2$	*Completely Alternating*
$\chi < 1$	*Blocky Tendency*
$\chi \approx 0$	*Block Copolymer*

FIGURE 6-19 Values of χ for different types of copolymers.

Terminal Model Revisited

We've now defined all the stuff we want to define and we can now apply this to copolymers and copolymerization. However, some of the applications of probability theory won't be apparent until we discuss spectroscopy and the characterization of chain microstructure. Here we will start with the terminal model, then we will see how composition and things like the parameters $\bar{l}_A$ and χ vary with conversion. First, remind yourself of the four possibilities for chain addition that occur in the copolymerization of A and B units when the rate of addition of a monomer depends only on the nature of the terminal group on a growing chain (Figure 6-20).

Let's start by considering the probability of adding an A to a growing chain end that also has an A as its terminal group. This is simply (Equation 6-46):

$$P\{A/A\} = $$

$$\frac{\text{Rate of producing } AA *}{\text{Sum of all reactions involving } A *}$$

EQUATION 6-46

which in terms of rate constants and reactant concentrations is (Equation 6-47):

$$P\{A/A\} = \frac{k_{AA}[A*][A]}{k_{AA}[A*][A] + k_{AB}[A*][B]}$$

EQUATION 6-47

Dividing through by $k_{AB}[A*][B]$ yields Equation 6-48:

$$P\{A/A\} = \frac{\dfrac{k_{AA}[A]}{k_{AB}[B]}}{\dfrac{k_{AA}[A]}{k_{AB}[B]} + 1} = \frac{r_A x}{r_A x + 1}$$

EQUATION 6-48

where x is once again the monomer feed ratio, $[A]/[B]$. In a similar fashion we can obtain Equation 6-49:

$$P\{B/B\} = \frac{r_B}{r_B + x}$$

EQUATION 6-49

The other two independent conditional probabilities, $P\{A/B\}$ and $P\{B/A\}$ are now readily obtained from (Equations 6-50):

$$P\{A/B\} = 1 - P\{B/B\} = \frac{x}{r_B + x}$$

$$P\{B/A\} = 1 - P\{A/A\} = \frac{1}{r_A x + 1}$$

EQUATIONS 6-50

We can now derive the copolymer equation from these simple conditional probabilities.

TERMINAL GROUP	ADDED GROUP	RATE	FINAL PRODUCT
- - - -A*	A	$k_{AA}[A*][A]$	- - - - AA*
- - - -B*	A	$k_{BA}[B*][A]$	- - - - BA*
- - - -A*	B	$k_{AB}[A*][B]$	- - - - AB*
- - - -B*	B	$k_{BB}[B*][B]$	- - - - BB*

FIGURE 6-20 Terminal model.

The ratio of A to B units in the copolymer is simply $y = P_1\{A\}/P_1\{B\}$ and can be obtained from Equations 6-51:

$$P_2\{AB\} = P_2\{BA\}$$

which means that:

$$P_1\{A\}P\{B/A\} = P_1\{B\}P\{A/B\}$$

EQUATIONS 6-51

Rearranging and substituting we get the Mayo and Lewis form of the copolymer equation (Equation 6-52):

$$y = \frac{P_1\{A\}}{P_1\{B\}} = \frac{P\{A/B\}}{P\{B/A\}} = \frac{1 + r_A x}{1 + r_B/x}$$

EQUATION 6-52

We can express χ (a measure of the departure from randomness) and the other parameters we have defined in terms of both conditional probabilities and then reactivity ratios and molar feed ratios. For example, χ is given by Equation 6-53:

$$\chi = P\{A/B\} + P\{B/A\}$$

EQUATION 6-53

Get the heck out of here, you might say, how did you get that? Figure it out! But, this relationship is certainly correct. Just use Equations 6-10 and 6-51, substitute into Equation 6-53 and you will get the definition of χ (Equation 6-45). Going back to the expression in Equation 6-53 and substituting for the conditional probabilities (Equations 6-48 to 6-50), we then obtain an expression for χ in terms of the reactivity ratios and the molar feed ratio (Equation 6-54):

$$\chi = \frac{r_A x + 2 + r_B/x}{r_A x + 1 + r_A r_B + r_B/x}$$

EQUATION 6-54

Similarly, we leave as a homework question the derivation for the run fraction or number, R from Equations 6-55:

$$R = 2P_1\{B\}P\{A/B\}$$
$$= \frac{2P\{B/A\}}{P\{A/B\} + P\{B/A\}} P\{A/B\}$$

EQUATIONS 6-55

You should get Equation 6-56 for R in terms of the reactivity ratios and the molar feed ratio.

$$R = \frac{2}{r_A x + 2 + r_B/x}$$

EQUATION 6-56

Not to mention the expressions for the number average length of A and B runs (Equations 6-57),

$$\bar{l}_A = \frac{1}{P\{B/A\}} = 1 + r_A x$$

$$\bar{l}_B = \frac{1}{P\{A/B\}} = 1 + r_B/x$$

EQUATIONS 6-57

the number fraction of A and B sequences (Equations 6-58),

$$N_A(n) = \left(\frac{r_A x}{1 + r_A x}\right)^{n-1}\left(1 - \frac{r_A x}{1 + r_A x}\right)$$

$$N_B(n) = \left(\frac{r_B/x}{1 + r_B/x}\right)^{n-1}\left(1 - \frac{r_B/x}{1 + r_B/x}\right)$$

EQUATIONS 6-58

and the weight fraction of A and B sequences (Equations 6-59):

$$w_A(n) = n\left(\frac{r_A x}{1 + r_A x}\right)^{n-1}\left(1 - \frac{r_A x}{1 + r_A x}\right)^2$$

$$w_B(n) = n\left(\frac{r_B/x}{1 + r_B/x}\right)^{n-1}\left(1 - \frac{r_B/x}{1 + r_B/x}\right)^2$$

EQUATIONS 6-59

Composition as a Function of Conversion

Having expressed some useful parameters in terms of reactivity ratios through the use of conditional probabilities, we are now going to consider a few examples, to give you a feel for what happens in a copolymerization. Actually, we are going to consider just one example and you are going to do the rest.

Before getting to that, there is something important you should realize about batch copolymerizations, by which we mean a reaction where monomer is not being continuously introduced. The copolymer composition and the parameters that measure the departure from randomness and things like run number can change considerably with conversion. In other words, the type of copolymer that you get at the beginning of the reaction can be very different to what you get near the end, when most of the monomer has been used up.

Conversion in this case is defined as the fraction of total monomer that has been used up (as opposed to just one of the reactants). Accordingly, if we start with monomer compositions of $[A]_0$ and $[B]_0$ and after the reaction has proceeded for a while the monomer composition is $[A]$ and $[B]$, the conversion is defined to be (Equation 6-60):

$$Conversion = \frac{([A]_0 + [B]_0) - ([A] + [B])}{([A]_0 + [B]_0)}$$

EQUATION 6-60

Now, if you recall the copolymer equation relating the composition of the copolymer formed at any instant of time (F_A, F_B), to the monomer feed composition (f_A, f_B) in a batch copolymerization (Equation 6-5), it should be clear that unless you have $r_A = r_B = 1$, so that $F_A = f_A$, then one of the monomers is going to be used up faster than the other (unless $r_A < 1$, $r_B < 1$ and you start with a monomer composition corresponding to the azeotrope condition). That means copolymer composition varies with conversion—we say there is *compositional drift*.

In other words, you cannot simply use the copolymer equation to calculate copolymer composition and assume $[A]$ and $[B]$, hence f_A or f_B are constant over the entire course of the copolymerization. However, it is reasonable to assume that over some small interval of conversion, say 1%, the monomer concentration in the feed remains essentially constant. Then you can use the simple procedure given in Figure 6-21. Obviously, the smaller you make the conversion interval, the more accurate your calculations will be. Unless you are a screaming masochist, this is not the type of calculation you want to make with your pocket calculator. But it readily lends itself to the construction of a nice little computer program, which can also be used to calculate the other parameters we have mentioned.

The composition of a vinylidene chloride (VDC)/vinyl chloride (VC) co-polymer, calculated as a function of conversion using such a program, is shown in Figure 6-22.

COPOLYMER COMPOSITION AS A FUNCTION OF CONVERSION

1. Choose an interval of conversion where a specified amount of total monomer is used up (e.g., 1%).

2. Assume that [A] and [B] are constant over this interval.

3. Starting with concentrations $[A]_0$ and $[B]_0$ (i.e., the beginning of the reaction), calculate how much monomer of each type is incorporated into the copolymer.

4. Now calculate how much monomer of each type is left after this interval of conversion.

5. Repeat this step for the next interval of conversion and keep going until the total conversion is 100%.

FIGURE 6-21 Simple method for determining copolymer composition drift.

We started with equal concentrations of the two monomers and used $r_{VDC} = 4.0$ and $r_{VC} = 0.2$. You can see that the copolymer produced initially is rich in VDC, but the amount of this monomer in the copolymer decreases with conversion, as it is used up at a faster rate than VC. Polymer chains containing equimolar amounts of VDC and VC are not formed until there is 63% conversion. Beyond this point the copolymer produced is rich in PVC. Obviously, if this polymerization were driven to a high degree of conversion, the average value of the copolymer composition must be 50/50 (because that's the monomer concentration you started with), but this masks the very broad distribution of chain compositions that would be present in this sample.

We can also calculate the number average length of sequences of A (VDC) and B (VC) units at any given degree of conversion as shown in Table 6-1. Initially, the number average length of VDC units is ~5, while that of VC units is ~1. These values change systematically throughout the polymerization until, at 80% conversion, there are long sequences of VC units. This is just a reflection of composition as opposed to being due to the presence of a blocky microstructure, however, as the calculated values of χ indi-

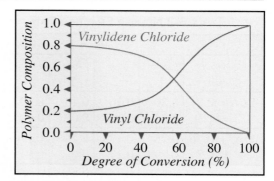

FIGURE 6-22 Polymer composition as a function of conversion for the VC/VDC system (initial concentrations $f_1 = f_2 = 0.5$).

cate that the copolymers are essentially random.

As shown in Table 6-2, the calculated number of A and B sequences at specific values of copolymer composition (corresponding to specific values of conversion) show that initially the vast majority of VC segments are present as isolated units, while there is a broad distribution of VDC sequence lengths. At high degrees of conversion, as you might expect, this situation is reversed. At 60% conversion the polymer composition is nearly equimolar and the distribution of VC and VDC units is almost the same, with 50% of the units present as monomers, 25% as

TABLE 6-1 COMPOSITION AND SEQUENCE DATA FOR THE COPOLYMERIZATION OF VDC AND VC

Monomers	*Monomer Conc.*	*Conversion (%)*	*Polymer Composition*	$\overline{l}_A , \overline{l}_B$ *Values*	χ *Value*
VDC = VC =	0.50 0.50	*Initial*	0.81 0.19	5.0 1.2	1.03
VDC = VC =	0.42 0.48	10	0.79 0.21	4.5 1.2	1.04
VDC = VC =	0.34 0.46	20	0.76 0.24	4.0 1.3	1.04
VDC = VC =	0.20 0.40	40	0.68 0.32	3.0 1.4	1.05
VDC = VC =	0.08 0.32	60	0.51 0.49	1.9 1.9	1.06
VDC = VC =	0.01 0.19	80	0.13 0.87	1.1 7.5	1.02

TABLE 6-2 NUMBER FRACTION DATA FOR THE COPOLYMERIZATION OF VDC AND VC

	Number Fraction of $(A)_n$ and $(B)_n$ Sequences									
Polymer Comp.	$n = 1$	$n = 2$	$n = 3$	$n = 4$	$n = 5$	$n = 6$	$n = 7$	$n = 8$	$n = 9$	$n = 10$
VDC = 0.81	0.20	0.16	0.13	0.10	0.08	0.07	0.05	0.04	0.03	0.03
VC = 0.19	0.83	0.14	0.02	0.00	0.00	0.00	0.00	0.00	0.00	0.00
VDC = 0.76	0.25	0.19	0.14	0.11	0.08	0.06	0.04	0.03	0.03	0.02
VC = 0.24	0.79	0.17	0.04	0.01	0.00	0.00	0.00	0.00	0.00	0.00
VDC = 0.51	0.52	0.25	0.12	0.06	0.03	0.01	0.01	0.00	0.00	0.00
VC = 0.49	0.54	0.25	0.12	0.05	0.02	0.01	0.01	0.00	0.00	0.00
VDC = 0.13	0.89	0.10	0.01	0.00	0.00	0.00	0.00	0.00	0.00	0.00
VC = 0.87	0.13	0.12	0.10	0.09	0.08	0.07	0.06	0.05	0.04	0.04

dimers (i.e., AA or BB sequences) and 12% as trimers (AAA or BBB sequences).

The corresponding weight fraction of sequences is different (Table 6-3). For example, at the beginning of the reaction, the number fraction of single or isolated VDC units is the predominant species present, but on a weight basis, sequences containing 4 and 5 VDC units predominate. This, as you might guess, is analogous to the number and weight average molecular weight distributions found in linear polycondensations.

Now it's your turn! Having taken you through one example in detail, we now leave it to you to do three more, each chosen because they illustrate a particular type of copolymerization. Using the reactivity ratios and initial molar concentrations given

in Table 6-4, write a program that gives plots of copolymer composition as a function of conversion, l_A, l_B, and χ at chosen concentrations, and also the number fraction of A and B se`quences. Compare and contrast the results obtained from all four examples by constructing a table of values of these quantities for polymer chains formed from equimolar concentrations of the monomers. This is repeated in the "Study Questions" at the end of this chapter and should keep you out of trouble for a while!

Penultimate Model

There are a number of systems where penultimate effects are known to occur (e.g., in the free radical polymerization of methyl

TABLE 6-3 WEIGHT FRACTION DATA FOR THE COPOLYMERIZATION OF VDC AND VC

	Weight Fraction of $(A)_n$ and $(B)_n$ Sequences									
Polymer Comp.	$n = 1$	$n = 2$	$n = 3$	$n = 4$	$n = 5$	$n = 6$	$n = 7$	$n = 8$	$n = 9$	$n = 10$
VDC = 0.81	0.04	0.06	0.08	0.08	0.08	0.08	0.07	0.07	0.06	0.05
VC = 0.19	0.69	0.23	0.06	0.01	0.00	0.00	0.00	0.00	0.00	0.00
VDC = 0.76	0.06	0.09	0.11	0.11	0.10	0.09	0.08	0.07	0.06	0.05
VC = 0.24	0.62	0.26	0.08	0.02	0.01	0.00	0.00	0.00	0.00	0.00
VDC = 0.51	0.27	0.26	0.19	0.12	0.07	0.04	0.02	0.01	0.01	0.00
VC = 0.49	0.29	0.27	0.19	0.11	0.07	0.04	0.02	0.01	0.01	0.00
VDC = 0.13	0.79	0.17	0.13	0.00	0.00	0.00	0.00	0.00	0.00	0.00
VC = 0.87	0.02	0.03	0.04	0.05	0.05	0.05	0.05	0.05	0.05	0.05

TABLE 6-4 SELECTED REACTIVITY RATIOS AND INITIAL CONCENTRATIONS.

	r_A	r_B	f_A	f_B
Styrene (A)/Maleic Anhydride (B)	0.04	0.015	0.75M	0.25M
Vinylidene Chloride (A)/Methyl Acrylate (B)	1.00	1.00	0.50M	0.50M
Acrylonitrile (A)/Allyl Methacrylate (B)	9.55	0.52	0.50M	0.50M

methacrylate and 4-vinyl pyridine). At first sight, it may seem a trivial task to establish whether the terminal or penultimate model applies to a given copolymerization, but as we will see, this is actually a difficult problem requiring very precise data. Let's start at the beginning with the algebra. We now have eight rate constants to worry about (Figure 6-23) and, hence, four different reactivity ratios which are defined in Equations 6-61:

$$r_A = \frac{k_{AAA}}{k_{AAB}} \qquad r_B = \frac{k_{BBB}}{k_{BBA}}$$
$$r_A' = \frac{k_{BAA}}{k_{BAB}} \qquad r_B' = \frac{k_{ABB}}{k_{ABA}}$$

EQUATIONS 6-61

Four independent conditional probabilities can now be written using these four reactivity ratios, in the same way as two conditional probabilities could be written for the terminal model in terms of its two reactivity ratios (Equations 6-62):

$$P\{B/AA\} = \frac{1}{1 + r_A x}$$
$$P\{A/BA\} = \frac{r_A' x}{r_A' x + 1}$$
$$P\{B/AB\} = \frac{r_B'/x}{1 + r_B'/x}$$
$$P\{A/BB\} = \frac{1}{1 + r_B/x}$$

EQUATIONS 6-62

The other four conditional probabilities can simply be determined by subtracting

each of these equations from unity (Equations 6-63):

$$P\{A/AA\} = 1 - P\{B/AA\}$$
$$P\{B/BA\} = 1 - P\{A/BA\}$$
$$P\{A/AB\} = 1 - P\{B/AB\}$$
$$P\{B/BB\} = 1 - P\{A/BB\}$$

EQUATIONS 6-63

Just as in the derivation of the copolymer equation for the terminal model, we start with a reversibility relationship $P_3\{AAB\} = P_3\{BAA\}$. Now we must use second-order Markovian statistics to write this in terms of conditional probabilities (Equation 6-64):

$$P_2\{AA\}P\{B/AA\} = P_2\{BA\}P\{A/BA\}$$

EQUATION 6-64

But, $P_2\{BA\} = P_2\{AB\}$ and $P_2\{AA\} = P_1\{A\}P\{A/A\}$. Substituting we get Equation 6-65:

$$P_1\{A\}P\{A/A\}P\{B/AA\} =$$
$$P_1\{A\}P\{B/A\}P\{A/BA\}$$

EQUATION 6-65

Now, as $P\{A/A\} = 1 - P\{B/A\}$ it follows that (Equation 6-66):

$$P\{B/A\} = \frac{P\{B/AA\}}{P\{B/AA\} + P\{A/BA\}}$$

EQUATION 6-66

FASCINATING POLYMERS—FLUOROELASTOMERS

Republic F-84F Thunderstreak jet fighter, circa 1952 (Courtesy: March Field Air Museum, California—www.marchfield.org).

Viton® A

$$-\left[CH_2-CF_2\right]_x-\left[CF_2-\underset{\underset{CF_3}{|}}{CF}\right]_y-$$

VF₂   *HFP*

$$-\left[CH_2-CF_2\right]_x-\left[CF_2-\underset{\underset{CF_3}{|}}{CF}\right]_y-\left[CF_2-CF_2\right]_z-$$

TFE

Viton® B

With Roy Plunkett's discovery of PTFE in 1938 (see *Polymer Milestones*—Chapter 8) and war clouds on the horizon, DuPont's scientists embarked on a frenzied search to find new synthetic fluorinated elastomers that might have some of the same inertness, solvent resistance and high-temperature stability as their other fluoropolymers. There were some interesting fluoro-containing elastomers made independently by DuPont chemists (most notably, Ford, Hanford and Schroeder) during and immediately after WWII, but they did not amount to much. There was too much else going on. But things were about to change in the 1950s. A critical need for high-temperature stable, fuel-resistant, elastomers emerged. Conventional elastomers (including the recently introduced silicones) used in gaskets, seals and hoses were simply not good enough to withstand the solvents and temperatures experienced in the rapidly advancing military jet engines of the day. So the military once again supplied the incentive to develop new materials. Hanford, who was now working for M. G. Kellogg, remembered the earlier work on vinylidene fluoride (VF₂) at DuPont by Ford and produced a very stable elastomer of VF₂ and chlorotrifluoroethylene. This polymer was manufactured under the name KEL-F®. The 3M Company subsequently bought out the rights to this fluorinated elastomer. But DuPont wasn't sitting still and under the direction of Herman Schroeder, the company developed the Viton® family of fluoroelastomers, which were also based on VF₂. The scientists at DuPont decided to focus on copolymers of fluoroethylenes, like VF₂, and perfluoroolefins, like tetrafluoroethylene (TFE) and hexafluoropropylene (HFP). It was found that superb high-temperature stable elastomers could be made from VF₂ and HFP, which DuPont called Viton® A. In fact, after examining research samples of Viton® A, the Air Force pronounced that it was a superior material to anything they had. They wanted more, and they wanted it soon! Meanwhile, Viton® B, a terpolymer of VF₂, HFP and TFE, which was even more resistant to temperature and solvents, was developed. Both these fluoroelastomers, which had service temperatures of around 200°C, were commercialized in 1958.

TERMINAL GROUP	ADDED GROUP	RATE CONSTANT	FINAL PRODUCT
- - - - AA*	A	k_{AAA}	- - - - AAA*
- - - - BA*	A	k_{BAA}	- - - - BAA*
- - - - BB*	A	k_{BBA}	- - - - BBA*
- - - - AB*	A	k_{ABA}	- - - - ABA*
- - - - AA*	B	k_{AAB}	- - - - AAB*
- - - - BA*	B	k_{BAB}	- - - - BAB*
- - - - BB*	B	k_{BBB}	- - - - BBB*
- - - - AB*	B	k_{ABB}	- - - - ABB*

FIGURE 6-23 Penultimate model.

We bet if you never see another *P{something}* again in your life you'll probably be perfectly happy, but hang in there, because we're almost done!

In our discussion of the terminal model we obtained (Equation 6-67):

$$\frac{P_1\{A\}}{P_1\{B\}} = \frac{P\{A/B\}}{P\{B/A\}}$$

EQUATION 6-67

So, upon substitution of Equation 6-66 we get Equation 6-68:

$$\frac{P_1\{A\}}{P_1\{B\}} = \frac{1 + \dfrac{P\{A/BA\}}{P\{B/AA\}}}{1 + \dfrac{P\{B/AB\}}{P\{A/BB\}}}$$

EQUATION 6-68

which, upon further substitution, finally gives us the copolymer composition equation for the penultimate model (Equation 6-69):

$$y = \frac{P_1\{A\}}{P_1\{B\}} = \frac{1 + \dfrac{r_A' x(1 + r_A x)}{1 + r_A' x}}{1 + \dfrac{r_B'/x(1 + r_B/x)}{1 + r_B'/x}}$$

EQUATION 6-69

Testing the Models

The copolymer equation can be written in a general form (Equations 6-70):

$$y = \frac{P_1\{A\}}{P_1\{B\}} = \frac{1 + (r_A)x}{1 + (r_B)/x}$$

For the Terminal Model:

$$(r_A) = r_A \qquad (r_B) = r_B$$

For the Penultimate Model:

$$(r_A) = \frac{r_A'(1 + r_A x)}{1 + r_A' x}$$

$$(r_B) = \frac{r_B'(1 + r_B/x)}{1 + r_B'/x}$$

EQUATIONS 6-70

It should then be possible to test whether or not experimental data obtained over the whole composition range can be fit to just two reactivity ratios, either by an iterative process or by examining a Fineman-Ross type of plot. Unfortunately, the greatest sensitivity to deviations from the terminal model occurs at high and low values of x, where inherent errors are the greatest. Similar limitations apply to other measurements, such as the number average runs of A and B units determined by NMR spectroscopy. Accordingly, very precise measurements are necessary if penultimate effects are to be established unambiguously.

The probability arguments we have reproduced on the preceding pages can also be applied to describe the sequence of isomers in polymers like atactic polypropylene, or the placement of *cis*-1,4 and *trans*-1,4 units in diene polymers. But if we write one more probability equation you will probably hurl this book from the highest tower you can find, so we will defer further discussion until the chapter on spectroscopy.

RECOMMENDED READING

F. A. Bovey and P. A. Mirau, *NMR of Polymers*, Academic Press, San Diego, 1996.

M. Chanda, *Introduction to Polymer Science and Chemistry*, Taylor & Francis, Boca Raton, 2006.

J. L. Koenig, *Chemical Microstructure of Polymer Chains*, J. Wiley & Sons, New York, 1982.

G. Odian, *Principles of Polymerization*, 3rd. Edition, Wiley, New York, 1991.

STUDY QUESTIONS

1. Compare and contrast the nature of the steady-state assumption used to describe the kinetics of free radical homopolymerizations and copolymerizations.

2. Using the steady-state approximation, derive the copolymer equation for the free radical synthesis of monomer M_1 with monomer M_2. Express your answer in terms of the mole fraction of monomer 1 in the copolymer (F_1) and the mole fraction of monomer 1 in the "feed" (f_1).

3. For the copolymerization of monomer CRUD (1) with monomer YUCKO (2), the reactivity ratios are $r_1 = 0.6$, $r_2 = 0.4$. At what mole fraction of crud in the feed will the composition of the copolymer exactly equal this feed composition? (Hint: does the word *azeotrope* ring a bell?)

4. On the basis of their reactivity ratios (Table 6-5 below), predict what types of microstructures (e.g., tendency to alternating, tendency to blocky) you would expect in a free radical copolymerization of the following monomer pairs:

   (A) acrylamide and acrylonitrile
   (B) ethylene and methyl acrylate
   (C) 1,3-butadiene and vinyl stearate (vinyl octadecanoate)
   (D) chlorotrifluoroethylene and ethyl vinyl ether.
   (E) styrene/butadiene

Assume equimolar concentrations of monomer and also assume you are only examining copolymers produced in the initial stages of the reaction.

5. Predict the ratio of butadiene to styrene repeating units initially produced by free radical copolymerization of an equimolar mixture of the two monomers at 50°C. (Use Table 6-5 below.)

6. A number of years ago we asked one of our students (Dr. Yun Xu) to free radically copolymerize styrene with a (protected) vinyl phenol monomer. We needed to determine the reactivity ratios and Dr. Xu obtained the following data for monomer M_1 (which happened to be the vinyl phenol monomer).

TABLE 6-5 REPRESENTATIVE REACTIVITY RATIOS FOR MONOMER PAIRS

	M_1	M_2	r_1	r_2
(A)	Acrylamide	Acrylonitrile	1.04	0.94
(B)	Ethylene	Methyl acrylate	0.2	11
(C)	1,3-Butadiene	Vinyl stearate	34.5	0.034
(D)	Chlorofluoroethylene	Ethyl vinyl ether	0.008	0
(E)	Styrene	1,3-Butadiene	0.58	1.35

MONOMER 1 IN FEED (WT %)	CONVERSION (%)	MONOMER 1 IN COPOLY- MER (WT %)
11	6	12
33	8	24
50	6	44
73	6	69
90	8	86

A. Plot copolymer composition as a function of monomer feed. What does this suggest about the nature of the copolymerization?

B. Given the molecular weight of vinyl phenol is 120 and styrene is 104, calculate the reactivity ratios describing this copolymerization using the Fineman-Ross method.

C. Comment on the nature of this plot, in the sense of where you think the greatest source of errors lies.

D. Recalculate the reactivity ratios using the Kelen-Tüdõs method (this will keep you up at night). Contrast your results to those obtained in Part C.

7. The next few questions pertain to statistical relationships. Try to keep a grip on your sanity.

Consider a copolymerization of monomers of type A with monomers of type B.

A. Write an expression for the triad probability $P_3\{ABB\}$ in terms of two tetrad probabilities.

B. Using a pentad sequence, give an example of the equality of a reversible sequence.

C. Show that $P_2\{AB\} = P_2\{BA\}$. (Start from $P_1\{A\}$.)

D. Express the pentad sequence $P_5\{BABAB\}$ in terms of first order Markovian statistics.

8. From an NMR experiment it was determined that in a copolymerization of monomer A with monomer B the number fraction of BAAAB pentad sequences was 0.04, while the number fraction of AB diad sequences was 0.32. Calculate the number fraction of sequences of three A's (AAA), $N_A(3)$.

9. Consider a copolymerization of equimolar amounts of A and B monomers where $r_A = 5$ and $r_B = 0.2$. Calculate the mole fraction of sequences with 1, 2, 3, 4, 5, and 6 A units and plot your results on a graph. Do the same for the B units. Would you expect these results to vary with the degree of conversion? If not, why not?

10. Here's the question you've been dreading. Using the reactivity ratios and initial molar concentrations given in Table 6-4, write a program to obtain plots of copolymer composition as a function of conversion, $\bar{l}_A$, $\bar{l}_B$, and χ at chosen concentrations, and also the number fraction of A and B sequences. Compare and contrast the results obtained from all four examples by constructing a table of values of these quantities for polymer chains formed from equimolar concentrations of the monomers. That should keep you out of trouble for a while!

There's an alternative. If you've got a truly evil polymer science professor, he or she will assign different monomer pairs (in Table 6-4) to different groups of students and have you compare your results in class. Do we know how to make your life difficult, or what?

11. This final question asks you to try and determine if a set of experimental data fits the terminal or penultimate model. For a copolymerization of A and B monomers (at low degrees of conversion) the following data were obtained.

MOLE RATIO x = A/B	$\bar{l}_A$	MOLE RATIO x = A/B	$\bar{l}_A$
2.8	3.80	1.0	2.40
2.2	3.34	0.6	2.00
1.8	3.05	0.4	1.76
1.4	2.74	0.2	1.44

Which model gives the best fit to the data? If applicable, estimate the values of r_A and r_A'.

7

Spectroscopy and the Characterization of Chain Microstructure

INTRODUCTION

If you have been working your way through this epic in a more or less linear fashion, then you might have started to ask yourself some fundamental questions: such as, "How do you know if a vinyl polymer is isotactic, or atactic, or whatever? How do you know the composition and sequence distribution of monomers in a copolymer? How do you know the molecular weight distribution of a sample?" This last question will have to wait until we discuss solution properties, but now is a good point to discuss the determination of chain microstructure by spectroscopic methods. The techniques we will discuss, infrared and nuclear magnetic resonance spectroscopy, can do a lot more than probe microstructure, but that would be another book and here we will focus on the basics.

We will start by discussing some general aspects of spectroscopy, which is the study of the interaction of light (in the general sense of electromagnetic radiation) with matter. When a beam of light is focused on a sample a number of things can happen. It can be reflected, or if the sample is transparent to the frequency or frequencies of the light, simply transmitted with no change in energy. Some of the light can also be scattered and some absorbed; it is this latter phenomenon that interests us here. Light is only absorbed if its energy corresponds to the energy difference between two quantum states of the sam-

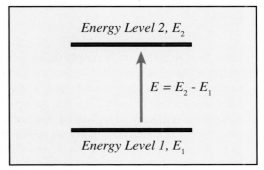

FIGURE 7-1 Schematic diagram of the transition between two energy states.

ple (Figure 7-1). The relationship between this energy difference and the frequency of the absorbed light being given by the Bohr frequency condition:

$$\Delta E = E_2 - E_1 = h\upsilon$$

EQUATION 7-1

where h is Planck's constant and υ is the frequency of the light in cycles/sec (Hertz, Hz).

In discussing light and spectroscopy we will be dealing with various parameters, the wavelength, frequency, circular frequency, and in infrared spectroscopy, the wavenumber. There's a good chance that all this stuff has melded into a formless mass in your mind, so we will spend the next few paragraphs reminding you of some basics.

If you think of light acting as a traveling

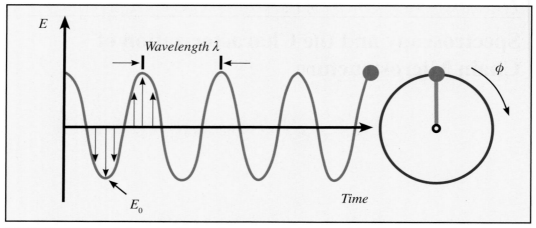

FIGURE 7-2 Schematic diagram of light acting as a traveling wave and described in terms of circular motion.

wave, then wavelength is simply the distance between the peaks and the frequency is the number of peaks that would pass a given point in a second (Figure 7-2). So, the speed of light, c (cm/sec), must equal the wavelength, λ (cm), times how many wavelengths go past a point per second, v, as expressed in the Equations 7-2:

$$c = v\lambda \qquad \lambda = \frac{c}{v}$$

EQUATIONS 7-2

Light can also be described in terms of circular motion, in that its sinusoidal oscillations are equivalent to the projection of a vector traveling in a circular path (Figure 7-2). The period, or time taken to go one full revolution (2π radians) is simply (Equation 7-3):

$$Period = \frac{2\pi}{\omega}$$

EQUATION 7-3

where ω is the angular frequency in radians/sec ($= d\phi/dt$). The period of a revolution (how long it takes the vector to make a full circle in seconds) must obviously equal the reciprocal of the frequency (sec^{-1})—Equations 7-4:

$$v = \frac{c}{\lambda} = \frac{\omega}{2\pi}$$

Hence:

$$\omega = 2\pi v$$

EQUATIONS 7-4

Infrared spectra are invariably reported in terms of wavenumbers instead of wavelength. One is just the reciprocal of the other (Equation 7-5):

$$\tilde{v} = \frac{v}{c} = \frac{1}{\lambda}$$

EQUATION 7-5

Simple Harmonic Motion

Simple harmonic motion, such as the (undamped by frictional forces) sinusoidal oscillation of a weight suspended by a spring can also be thought of in terms of the projection of a vector traveling in a circular path. This is something you should have covered in your elementary mechanics classes, of course, but we will reexamine it here, first because it is important in infrared spectroscopy, and second because it provides some illumination concerning resonance.

We will start with a very simple model, a

mass suspended by a spring. We will assume the spring obeys Hooke's law; in other words the restoring force generated by stretching the spring is proportional to the amount of extension, say x (generally a good approximation for small displacements)—see Figure 7-3.

Using one of Newton's laws of motion, the restoring force generated by the spring, f, can then be written (Equation 7-6):

$$f = -kx = m\frac{d^2x}{dt^2}$$

EQUATION 7-6

The constant of proportionality, k, is the "stiffness" of the spring, usually referred to as the *force constant* by infrared spectroscopists. Now let's solve this second-order differential equation for the frequency. There are ways of doing this directly, but the usual approach is to assume you know the answer already and "try" the following solution (Equation 7-7):

$$x = \cos\omega_0 t$$

EQUATION 7-7

You then get (Equation 7-8):

$$\frac{d^2x}{dt^2} = -\omega_0^2 \cos\omega_0 t = -\omega_0^2 x$$

EQUATION 7-8

which, when substituted into our original equation, gives (Equation 7-9):

$$\omega_0 = \sqrt{\frac{k}{m}}$$

EQUATION 7-9

Because we are dealing with an oscillatory or sinusoidal function, the quantity ω_0 is the "circular" frequency in radians/sec. The frequency in cycles/sec is (Equation 7-10):

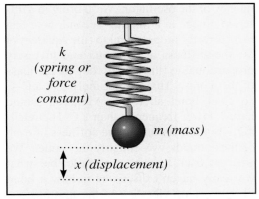

FIGURE 7-3 Schematic diagram of a spring that obeys Hooke's law.

$$v_0 = \frac{1}{2\pi}\sqrt{\frac{k}{m}}$$

EQUATION 7-10

There are a few things we want to say about this equation, but first let's make the model correspond to a simple diatomic molecule by considering the stretching vibrations of a spring connecting two balls of mass m_1 and m_2 (Figure 7-4). We won't go into the algebra, but just give you the answer, which turns out to have exactly the same form as before (Equations 7-11):

$$\omega_0 = \sqrt{\frac{k}{m_r}}$$
$$where: \ m_r = \frac{m_1 m_2}{m_1 + m_2}$$

EQUATIONS 7-11

The first thing we want you to observe is that we used a zero subscript on the omega in order to indicate that this is the "natural" fre-

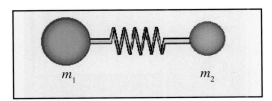

FIGURE 7-4 Schematic diagram of a simple diatomic molecule.

quency of the oscillator. We'll bring in a different omega in a minute. Second, note that ω_0 is directly proportional to (the square root of) the stiffness of the bond and inversely proportional to (the square root of) the mass of the atoms. This would imply that a C–C covalent bond stretching vibration would have a lower frequency than a C–H stretching vibration, assuming the stiffness of each of these bonds was roughly the same. We will get back to this in a while, but what we really haven't discussed is how a bond like this starts vibrating in the first place—how a molecule interacts with the light that impinges on it.

Some insight into this can be gained by first looking at forced oscillations. Let's say that there is an external force that varies with time, $F(t)$, acting on our oscillator. The equation of motion becomes (Equation 7-12):

$$m\frac{d^2x}{dt^2} = -kx + F(t)$$

EQUATION 7-12

Now let's say that this force is oscillating sinusoidally with time (Figure 7-5), just as light does (Equation 7-13):

$$F(t) = F_0 \cos \omega t$$

EQUATION 7-13

One solution to the resulting equation is Equation 7-14:

$$x = C \cos \omega t$$

EQUATION 7-14

Which gives us Equation 7-15:

$$-m\omega^2 C \cos \omega t =$$
$$-m\omega^2 C \cos \omega t + F_0 \cos \omega t$$

EQUATION 7-15

Hence, we arrive at Equation 7-16 (Note that we substituted $k = m\omega_0^2$):

$$C = \frac{F_0}{m(\omega_0^2 - \omega^2)}$$

EQUATION 7-16

In other words the function $x = C \cos \omega t$, a sinusoidally varying quantity of frequency ω and amplitude C, is a solution to the problem of the "forced" oscillator. The amplitude of the forced motion, C, is given by Equation 7-16. So, our hypothetical molecule will oscillate with the same frequency as that of the driving force, but its amplitude varies with the difference between the natural frequency of the oscillator and the frequency of the applied force. This amplitude becomes very large as ω approaches ω_0, a condition we call resonance. (In a real molecule it wouldn't become infinite, because the spring (bond)

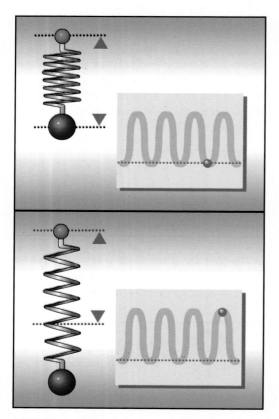

FIGURE 7-5 Schematic diagram of the periodic forced oscillations of a simple Hookean spring.

would break first). This condition holds not only for vibrations, but any sort of transition. If we let electromagnetic radiation impinge on a sample and if its frequency, hence energy ($h\upsilon$), corresponds to the energy difference between two quantum states, then resonance will be achieved, light will be absorbed, and the atom or molecule will be "excited," *providing that there is a mechanism for an interaction between the light and the material.*

Molecular Transitions

The mechanism of interaction, the type of transition that can occur and the spectroscopic technique necessary for detection varies with the energy, hence wavelength (or frequency), of the light. The electromagnetic spectrum is, of course, a continuum, but humankind has given various parts names, or associated various parts of it with certain applications (radar, radio, etc.), as illustrated in Figure 7-6. In order to provide an overview we'll work our way through this spectrum starting from the high energy end (high frequency, short wavelength) and briefly discuss the types of atomic or molecular transitions that can be excited.

At the highest energy end we have γ-rays and X-rays. γ-Rays excite transitions between energy states within the nucleus of an atom (and this is the basis for Mössbauer spectroscopy). Absorption of X-rays, on the other hand, involves the inner shell electrons. Radiation in this energy range can also beat the living hell out of a polymer, breaking bonds, knocking out atoms and generally creating molecular pandemonium. If carefully directed, however, things like γ-rays can do useful things, as in the radiation cross-linking of polyethylene. But we digress; moving down the energy scale we find that ultra-violet (uv) and visible light also excite electronic transitions, in this case the less strongly bound electrons found in the bonds that hold molecules together. This is the basis for uv-visible spectroscopy.

As we decrease the frequency of the radiation that impinges on our hypothetical sam-

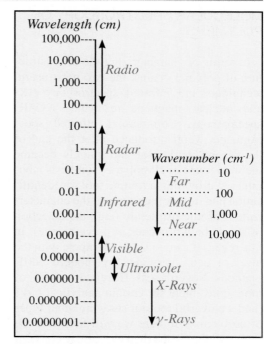

FIGURE 7-6 The electromagnetic spectrum.

ple, we find that its energy is no longer sufficient to excite electrons. Molecules have vibrational energy levels, however, and various stretching and bending modes of vibration are excited in the infrared frequency range, corresponding to wavelengths or about 0.1 to 0.00025 cm, or wavenumbers of about 10 to 4000 cm^{-1}.

Finally, in the microwave and radio frequency range, there are transitions associated with the rotational energy levels of small molecules, but not polymers. In the absence of an applied magnetic field, there are no other absorptions associated with molecular processes in this frequency range. However, if we now trundle up a bloody great magnet and apply a large magnetic field to our polymer, we would now (in principle) find a wealth of absorptions in this region of the spectrum. This is because certain nuclei (protons, deuterons, ^{13}C, ^{15}N, ^{19}F, etc., have magnetic dipole moments by virtue of their "spin" (but some, like ^{12}C, have zero spin). There is then an energy difference between the state where a dipolar nucleus is aligned with the field and the state where it is not.

MOLECULAR SPECTROSCOPY OF POLYMERS

In terms of characterizing the microstructure of polymer chains, the two most useful techniques are infrared spectroscopy (IR) and nuclear magnetic resonance (NMR) spectroscopy. Commercial infrared spectrometers were introduced after the end of the second world war and quickly became the workhorse of all polymer synthesis laboratories, providing a routine tool for identification and, to a certain degree, the characterization of microstructure (e.g., the detection of short chain branches in polyethylene). In this regard it can no longer compete with the level of detail provided by modern NMR methods. Nevertheless, IR remains useful or more convenient for certain analytical tasks (and a powerful tool for studying other types of problems). So here we will first describe both techniques and then move on to consider how they can be applied to specific problems in the determination of microstructure.

Instrumentation

The details of spectroscopic instrumentation belong in a book dedicated to the subject, but it is still important that you have a feel for the nature of the experiment. The basic elements of many types of instruments, including IR spectrometers, are illustrated in Figure 7-7. First, you need a source of radiation in the appropriate frequency range and then you need to get the light focused on the sample. In order to determine how much radiation has been absorbed and at what frequencies you then need to separate the light according to wavelength and measure its intensity (the amount of light hitting a detector at that wavelength). In the schematic picture (Figure 7-7) this is accomplished by a prism (the first IR spectrometers actually had prisms), which disperses the light according to wavelength or frequency. The light is then focused onto a narrow slit that determines the resolution of the instrument. A narrow slit allows light in just a narrow frequency range to hit the detector, permitting the separation of absorption bands that are close in frequency. However, the narrower the slit, the less energy reaches the detector and the more background noise you get from stray light and the electronics of the system. Most modern instruments therefore use interferometers that don't need slits at all. But you get the idea—shine a range of frequencies onto your sample, separate the light according to wavelength and measure how much light has been absorbed at each frequency (by comparing intensities to light from the source that hasn't passed through the sample).

This isn't the only way to do the experiment. If you could find a source that emitted light at a specific frequency and if this frequency could be changed or tuned across a broad range, then you wouldn't need a dispersive element like a prism (or gratings, or an interferometer) at all. This is essentially how NMR works in older instruments. A

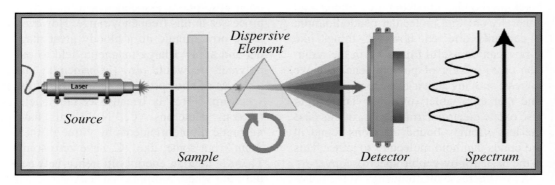

FIGURE 7-7 Schematic diagram of an infrared spectrometer.

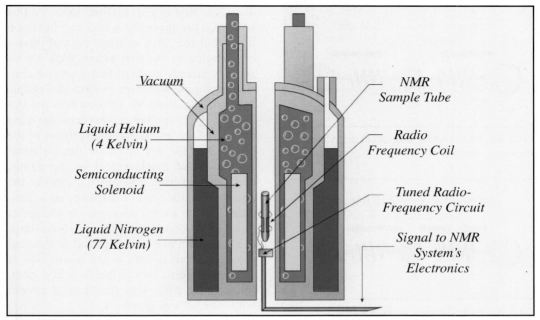

FIGURE 7-8 Schematic diagram of an NMR spectrometer (redrawn from an original figure by Bovey and Mirau in *NMR of Polymers*, Academic Press, 1996.)

sample is placed in a uniform magnetic field (for reasons we will explain shortly) whose field strength can be varied and a transmitter applies a radio-frequency field by means of an exciting coil (Figure 7-8). By varying the frequency of this latter field at fixed magnetic field strength, or by varying the magnetic field strength at constant frequency, resonance conditions can be found and detected. But a lot of time can be spent "scanning" between resonance conditions and modern instruments use something called pulse techniques.

Infrared Spectroscopy

We have mentioned that absorption of light in the infrared region of the spectrum involves the excitation of molecular vibrations and also shown that resonance occurs when the frequency of the incident light corresponds to one of the natural frequencies of vibration of a material. But, we did not explain how the light interacted with the sample to "force" this vibration. It turns out that even if the frequency of the infrared

radiation corresponds to what we call one of the fundamental normal modes of vibration of a material (like the CH_2 wagging mode of polyethylene), it will only be absorbed under certain conditions. The rules determining optical activity are known as selection rules and have their origin in quantum mechanics.

You no doubt have enough trouble in your life without having to think about quantum mechanics when you're trying to study polymers, so we will only mention the classical interpretation, which, although limited, is enough to give you a physical picture. In classical electromagnetic theory an oscillating dipole is an absorber or emitter of radiation. Accordingly, the periodic variation of the dipole moment of a vibrating molecule results in the absorption or emission of light of the same frequency. The requirement of a change in dipole moment is fundamental and not all modes of vibration result in such a change. Take, for example, CO_2, a symmetric molecule with no net dipole moment in the unperturbed state. One of its normal modes of vibration (a molecule only has a certain number of these) is the symmetric

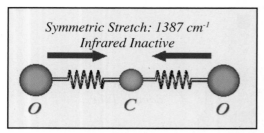

FIGURE 7-9 Schematic diagram of the symmetric stretching vibration of CO_2.

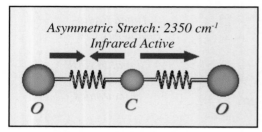

FIGURE 7-10 Schematic diagram of the asymmetric stretching vibration of CO_2.

stretch, illustrated in Figure 7-9. This motion does not change the symmetry and hence, the overall dipole molecule of the molecule, so there is no interaction with infrared radiation. (There is something called a Raman scattered line at this frequency, but unless you are a spectroscopy geek you probably don't want to know this.) Figure 7-10 illustrates the asymmetric stretch of CO_2. In this case, there is a change in the dipole of the molecule upon vibration and this mode is infrared active (and incidentally, happens to be Raman inactive).

The activity of normal modes can actually be predicted from symmetry considerations

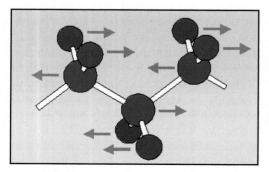

FIGURE 7-11 Schematic diagram of the CH_2 wagging mode of polyethylene near 1175 cm^{-1}.

alone, but that is a specialized topic. At this point you are probably wondering how you figure out the form of these normal modes of vibration in the first place. Well, we've actually already covered that when we considered simple harmonic motion of a simple diatomic oscillator. We showed how the frequency of vibration depended upon the mass of each of the atoms and the stiffness of the bonds. We did not calculate the (relative) amplitudes of displacement of the atoms, which would have given us the form of the motion (like the relative displacements illustrated for the stretching modes of CO_2), but this is easily done. Of course, we should have used quantum mechanics instead of classical mechanics, but as long as the assumption of simple harmonic motion holds (and it does, for the most part), you get the right answer with the classical approach.

Calculating the normal modes of vibration of large molecules is not something you can do with a pencil and paper (unless the molecule is highly symmetric, so that you can use some tricks to break up the problem into smaller bits), so our knowledge of the normal modes of vibration of complex molecules had to wait for the development and availability of computers. Even now, we can only calculate the form of the normal modes of ordered polymers (like the CH_2 wagging mode of polyethylene shown in Figure 7-11). However, thanks to the systematic observations of numerous spectroscopists over the years, it has been established that certain functional groups (e.g., methyl, ester carbonyl, etc.) absorb infrared radiation at certain characteristic frequencies or in certain narrowly defined frequency ranges (there is a sensitivity to local environment that is useful in structural studies). This has allowed the widespread use of the technique for the identification of unknown samples and quantitative analysis of things like methyl group content and copolymer composition in certain systems.

Unlike proton NMR (which we will discuss shortly), where the relative intensities of the peaks can be directly related to structural features, infrared band intensities are related to the concentration of corresponding functional groups through something called an absorption coefficient, which differs from one band to the next. In fact, the intensity of light that impinges on a sample (I_0) is attenuated in an exponential fashion according to the sample thickness, b, as illustrated in Figure 7-12. The quantities a and c are the absorption coefficient, which depends upon the nature of the chemical group, and the concentration of these groups present in the sample, respectively. In older publications it was usual to plot a spectrum as the intensity of light emerging from a sample (I) ratioed against an unattenuated beam (I_0) to give a so-called transmission spectrum (with the peaks pointing "downward"). These days it is more usual to plot the absorbance spectrum [$A = ln(I_0/I)$], which is linearly related to the concentration of the functional group that gives rise to a specific band. However, you have to know the absorption coefficient to relate A to c. But, we won't be discussing the details of quantitative analysis here, so let's move on and consider the molecular origin of NMR spectroscopy.

NMR Spectroscopy

The nuclei of certain isotopes possess what can be thought of as a mechanical spin, or *angular momentum*. In the quantum mechanical description this is characterized by the *spin number, I,* which has integral or half-integral values. The spin number is related to the mass and atomic number, as summarized in Table 7-1.

This spin gives rise to a magnetic field, so that the nucleus can be thought of as a small magnet with a magnetic moment μ. It turns out that the nucleus of the most com-

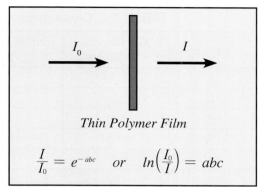

$$\frac{I}{I_0} = e^{-abc} \quad or \quad ln\left(\frac{I_0}{I}\right) = abc$$

FIGURE 7-12 Schematic diagram and equations for infrared quantitative analysis.

mon isotopes of carbon and oxygen, ^{12}C and ^{18}O, are non-magnetic ($I = 0$) and therefore do not have an NMR spectrum. In terms of characterizing (organic) polymers the most important isotopes are ^1H, ^{13}C and, to a lesser extent, ^{19}F, which is lucky for us, because all of these have a spin number of 1/2, which makes things relatively simple.

If a magnetic nucleus is introduced into a uniform external magnetic field of strength H_0, it assumes a set of $2I + 1$ discrete (quantized) orientations. Thus the nuclei of interest to us, ^1H, ^{13}C and ^{19}F, which all have $I = 1/2$, can only assume one of two possible orientations, corresponding to energy levels of $\pm \mu H_0$ (Figure 7-13). The low energy orientation ($-\mu H_0$) corresponds to the state where the nuclear magnetic moment is aligned parallel to the external magnetic field, while the high energy orientation ($+\mu H_0$) corresponds to the antiparallel (opposed) orientation. Resonance occurs and radiation is absorbed at the radio frequency v that corresponds to the energy difference between these two magnetic states (Equation 7-17):

TABLE 7-1 SPIN NUMBERS

MASS NUMBER	ATOMIC NUMBER	SPIN NUMBER
Odd	*Even or Odd*	*1/2,3/2,5/2, ---*
Even	*Even*	*0*
Even	*Odd*	*1,2,3, ---*

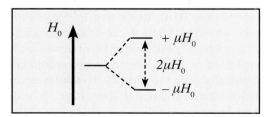

FIGURE 7-13 Schematic diagram of the two orientations for the case where $I = 1/2$.

TABLE 7-2 CHARACTERISTIC NMR RESONANCE FREQUENCIES

ISOTOPE	ABUNDANCE %	NMR FREQUENCY[a] (MHz)	RELATIVE SENSITIVITY[b]	SPIN NUMBER
^{1}H	99.98	42.6	1.000	1/2
$^{2}H\,(D)$	0.016	6.5	0.0096	1
^{13}C	1.11	10.7	0.0159	1/2
^{14}N	99.64	3.01	0.0010	1
^{15}N	0.37	4.3	0.0010	1/2
^{19}F	100	40.0	0.834	1/2
	[a]In a 10 kG field		[b]For equal numbers of nuclei at constant H_0	

$$\Delta E = h\upsilon = 2\mu H_0$$

EQUATION 7-17

Obviously, the frequency of the applied radio frequency field necessary to achieve resonance depends on both the strength of the uniform applied external field, H_0 and the size of the nuclear magnetic moment, μ. For ^{1}H in a magnetic field of 14,000 Gauss the resonance frequency is about 60 megacycles/sec (60 MHz). Some other characteristic resonance frequencies are summarized in Table 7-2. Note also the natural abundance of specific isotopes summarized in this table. ^{13}C is only present to the extent of 1.1%, and given the fact that ^{13}C nuclei only produce 1/64 the signal that ^{1}H nuclei yield, the relative sensitivity of a ^{13}C NMR experiment is about 6000 times less than that of a ^{1}H experiment. Fortunately, this is a problem that modern instrument makers have solved and high quality ^{13}C NMR spectra can now be obtained routinely.

If the resonance frequency of all nuclei of the same type, say the 6 protons in ethanol (Figure 7-14), the active ingredient of beer, were all identical, only one resonance or absorption peak would be observed and the life of an NMR spectroscopist would be dreadfully dull and tedious. It would serve the wretches right, but they have lucked out. There are subtle differences in NMR frequencies such that CH_3 protons resonate at different frequencies to CH_2 groups, and

so on. This is because the nuclei are in different molecular environments and surrounding electrons "shield" the nuclei to different extents. This shielding is a result of the motion induced in the electrons by the applied magnetic field, which in turn generates a small magnetic field at the nucleus that (generally) opposes the applied magnetic field. Accordingly, the magnetic field experienced by a nucleus is not identical to the applied field and varies with chemical structure.

In order to resolve these subtle effects you have to apply large magnetic fields. Those spectrometers that generate the largest fields actually use superconducting solenoid magnets. However, the frequency of the radiation that needs to be applied to the sample to achieve resonance also depends upon the strength of this magnetic field, which means that a sample of, say, ethanol will have resonances at different frequencies in different instruments. In order to be able to compare spectra taken by different instruments, something called the *chemical shift* has therefore

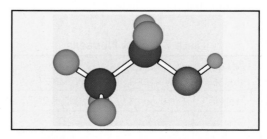

FIGURE 7-14 Chemical structure of ethanol.

been defined. This involves two things. First, the frequency scale is defined relative to an internal standard. Tetramethylsilane, TMS [Si(CH$_3$)$_4$], is commonly used, because all its protons are equivalent and therefore only one NMR peak is observed (Figure 7-15). Also, the spectra of most organic molecules are "downfield" from TMS, so the resonance due to this material only appears at the edge of the spectrum and there is no problem with overlap of resonances. The second thing you have to do is normalize for the strength of the field, because shifts relative to TMS will still depend on H_0. This is accomplished by dividing by the frequency of the spectrometer. Accordingly, the chemical shift, δ, is given by (Equation 7-18).

$$\delta(ppm) = \frac{(\upsilon_{TMS} - \upsilon_S)10^6}{\upsilon_{Spectrometer}}$$

EQUATION 7-18

Because the spectrometer frequency is in MHz, while the differences in the resonance frequencies between the sample and TMS are in Hz, the numerator is multiplied by 10^6, so as not to deal with awkwardly small numbers. The units of the chemical shift are thus referred to as parts per million (ppm). This is a bit confusing when you first encounter NMR spectroscopy, because it's also the type of unit that is bandied about by people earnestly looking to protect you from trace amounts of what might be nasty, cancer causing stuff in your food or water. Just keep in mind that ppm as used here has nothing to do with concentration. Typically, for ^{1}H, a range of 10 ppm covers most organic molecules, but for ^{13}C the corresponding range is much bigger, $\sim$ 600 ppm. We'll get to ^{13}C spectra in a while, first let's look at the proton NMR spectrum of ethanol, because this allows us to say something about spin-spin interactions or coupling.

^{1}H NMR

The first thing you should do is check out the low resolution NMR spectrum of etha-

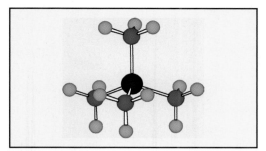

FIGURE 7-15 Chemical structure of tetramethyl-silane.

nol shown in Figure 7-16. There are three absorption peaks, corresponding to the OH, CH$_2$, and CH$_3$ groups, respectively. Note that the areas of these peaks are in the ratio of 1:2:3. In NMR, the band intensities give a direct measure of the number of nuclei they represent. (This is not true in IR spectroscopy where the *absorption coefficients* of the bands are all different.)

Now compare this spectrum to one taken at higher resolution (Figure 7-17). The peaks due to the CH$_2$ and CH$_3$ protons appear as multiplets (but the total relative area of each group maintains the 1:2:3 ratio). The methyl absorption is split into a triplet of bands (relative areas 1:2:1) while the methylene absorption is split into a quartet with relative areas of 1:3:3:1. This splitting is due to the magnetic field of the protons on one group influencing the spin arrangements of the protons on an adjacent group. The observed multiplicity of a given group of equivalent protons depends on the number of protons

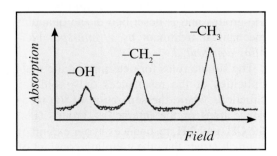

FIGURE 7-16 Low resolution ^{1}H spectrum of ethanol (redrawn from the data of L. M. Jackman and S. Sternhell, *Nuclear Magnetic Resonance Spectroscopy in Organic Chemistry*, Pergamon Press, 1969).

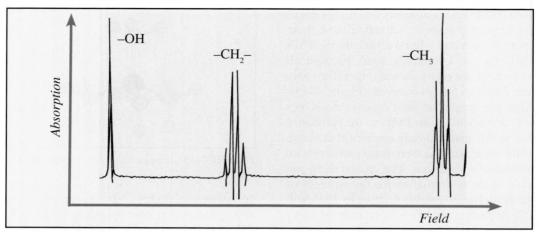

FIGURE 7-17 High resolution ^{1}H spectrum of ethanol (redrawn from the data of L. M. Jackman and S. Sternhell, *Nuclear Magnetic Resonance Spectroscopy in Organic Chemistry*, Pergamon Press, 1969).

on adjacent atoms, n, and is equal to $n + 1$. Thus the two CH_2 protons in the ethyl group of ethanol split the CH_3 resonance into a triplet, while the three protons on the CH_3 group split the CH_2 resonances into a quartet.

For chemically simple molecules such as this, the relative intensities of a multiplet are symmetric about their mid-point and are determined by statistical considerations, the intensities being (more or less) proportional to the coefficients of the binomial expansion. In other words, they go as Pascal's triangle, in a doublet 1:1, a triplet 1:2:1, a quartet 1:3:3:1, etc. The spacing between all the split resonances, the triplet for CH_3 and the quartet for CH_2 is the same and is independent of the strength of the applied field. This spacing depends on the interaction between the groups and is described in the quantum mechanical treatment by a *spin-spin coupling constant, J*.

The simple rules for determining the multiplicities of the resonances only hold for situations where the separation of the resonance lines of the interacting groups (i.e., the CH_2 and CH_3 resonances in our example) is much larger than the coupling constant of the groups. If the two are of the same order of magnitude, then these multiplicity rules don't apply—more lines appear and simple patterns of spacings and intensity are no longer found.

At this point, you may be wondering why the two protons on the CH_2 group don't interact with one another. In general, chemically equivalent nuclei don't interact through spin-spin coupling. This is a consequence of the quantum mechanical selection rules, so unless you want to get into this, take our word for it! However, there are cases, and we'll see some examples in the NMR spectra of polymers, where two magnetically nonequivalent protons on the same carbon can give rise to a maximum of four lines. We'll take as an example the CH_2 group of $(\pm)$-2-*iso*-propylmalic acid, shown in Figure 7-18. This molecule has *two non-equivalent protons* on a single carbon, H_A and H_B. The ^{1}H NMR spectrum, scale expanded in the region 2.5 to 3.1 ppm, shows four lines, with the two central lines being more intense than the two satellite lines (Figure 7-19). In this case, *the chemical shift difference is of the same order of magnitude as the coupling constant*, so the simple rules for multiplicities don't apply. This system is designated an *AB pattern*, in that the two protons involved (shown as a darker shade of green in Figure 7-18) have slightly different chemical shift. Of course, if the two methylene protons were equivalent, we would only have seen one line. We will be looking for similar differences in chemical shifts when we consider tacticity measurements in polymers.

^{13}C NMR

Unlike hydrogen, where the most abundant isotope (^{1}H) has nuclear spin (see Table 7-2), the most abundant isotope of carbon has a quantum spin number $I = 0$, and thus does not have an NMR spectrum. For the isotope that does have nuclear spin, ^{13}C, the relative sensitivity of a ^{13}C NMR experiment is about 6000 times less than that of a ^{1}H experiment, as we have already mentioned, but this is a problem that has been overcome by contemporary instrument makers. Incidentally, this was probably a good thing in the early days of NMR spectroscopy, because ^{13}C-^{1}H coupling would have seriously complicated the proton NMR spectrum.

The effect of ^{13}C-^{1}H coupling in the ^{13}C NMR spectrum can now be eliminated by a technique called proton decoupling. Furthermore, because of the low natural abundance of ^{13}C, the spin-spin coupling between these nuclei is negligible in unenriched compounds (assuming random placement of these isotopes). High quality ^{13}C NMR spectra of organic compounds can now be routinely obtained, as illustrated by the spectrum of 2-ethoxyethanol shown in Figure 7-20. The advantage of ^{13}C relative to ^{1}H NMR spectroscopy is the higher resolution that can be obtained in the latter. ^{13}C resonances of organic compounds are found over an enormous chemical shift range of 600 ppm and one can frequently identify resonances for individual carbon atoms in a molecule, as also illustrated by the spectrum (Figure 7-20). Note, however, that the lines are not of equal intensity, even though each is assigned to an individual carbon atom.

Unfortunately, in ^{13}C NMR spectroscopy one cannot simply relate the relative intensities to the number of equivalent carbon atoms in a molecule; relaxation phenomena and something called the nuclear Overhauser effect have to be taken into account. But it is not our purpose to drag you through the subtleties of this technique, you must get this from a more specialized treatment. Here we will simply give you a feel for the type of information on polymer microstructure

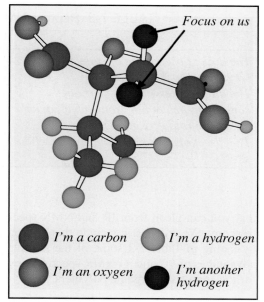

FIGURE 7-18 Molecular model of (±)-2-*iso*-propylmalic acid.

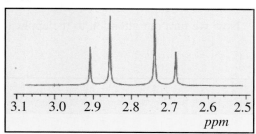

FIGURE 7-19 ^{1}H NMR spectrum of (±)-2-*iso*-propylmalic acid, scale expanded in the region from 2.5 to 3.1 ppm.

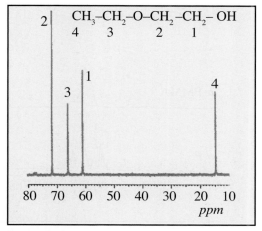

FIGURE 7-20 Proton decoupled ^{13}C NMR spectrum of 2-ethoxyethanol.

TABLE 7-3 RELATIVE NMR SPECTROSCOPIC ADVANTAGES

^{13}C NMR	^{1}H NMR
The direct observation of molecular backbones. *The direct observation of carbon containing functional groups that have no attached protons (e.g., -CN).* *The direct observation of carbon reaction sites.*	*The ease of quantitative analysis.* *The rapidity of analysis time.* *The direct observation of OH and NH groups (undetectable by ^{13}C NMR).* *The separation of olefinic and aromatic protons (olefinic and aromatic carbon resonances overlap).*

that you can glean from IR and NMR spectroscopy. We will just conclude this section by listing in Table 7-3 some of the relative advantages of ^{13}C and ^{1}H NMR spectroscopy for the analysis of organic polymers.

CHARACTERIZATION OF MICROSTRUCTURE

Now we turn our attention to the application of spectroscopic techniques to the characterization of the microstructure of polymer chains. Both NMR and IR can measure lots of other things, of course, but that level of detail does not belong here. However, we would be remiss if we did not at least mention what has been one of the most useful applications of IR and NMR spectroscopy over the years: the identification of polymers. So let's start there.

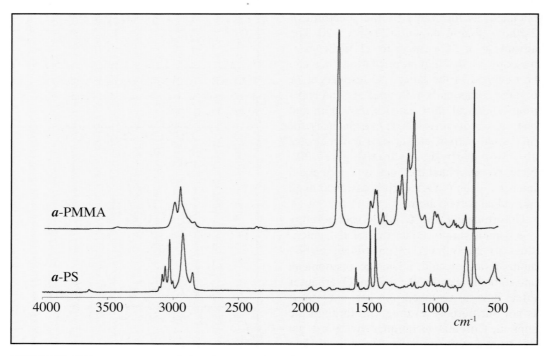

FIGURE 7-21 Room temperature infrared spectra of atactic polystyrene (*a*-PS) and atactic poly(methyl methacrylate) (*a*-PMMA) (recorded by Dr. Maria Sobkowiak).

Polymer Identification

Infrared spectroscopy can be applied to the characterization of polymeric materials at various levels of sophistication. As most commonly used, it is a rapid and easy method for the qualitative identification of major components through the use of group frequencies and distinctive patterns in the "fingerprint" region of the spectrum. Let's look at a couple of examples.

In Figure 7-21, room temperature infrared spectra of films of the amorphous polymer glasses, atactic polystyrene (*a*-PS) and atactic poly(methyl methacrylate) (*a*-PMMA) are compared. Note that the spectra are quite different and it is very easy to distinguish between the two polymers. The polymer chain of *a*-PS is composed of chemical repeat units ($-CH_2-CHC_6H_5-$). The chemical repeat unit is relatively large and, to a first approximation, we can assume that a given repeat unit does not know, in infrared terms, that the adjacent unit exists. Accordingly, the gross features of the spectrum will reflect those of a low molecular weight analog of the polymer repeat unit. What can we immediately gain from just a cursory glance of the IR spectrum of *a*-PS? First of all, from group frequency correlations we can determine that the sample contains aliphatic and aromatic groups from the bands observed in the 2800 to 3200 cm^{-1} region of the spectrum. Secondly, we can initially eliminate such groups as hydroxyls, amines, amides, nitriles, carbonyls etc., which all have distinctive group frequencies. Thirdly, the presence of a group of distinctive and relatively sharp bands (e.g., the band at about 700 cm^{-1} and the two bands near 1500 cm^{-1}) that are characteristic of monosubstituted aromatic rings readily leads one to the conclusion that the spectrum resembles that of a styrenic polymer.

When we look at the spectrum of *a*-PMMA we find bands that are associated with aliphatic CH_2 and CH_3 groups in the CH stretching and fingerprint regions, but the dominant feature is the presence of the carbonyl stretching vibration at about 1720

cm^{-1}. The presence of this band, characteristic bands attributed to the $-C-O-$ stretching vibration near 1200 cm^{-1}, and the precise pattern of the bands in the fingerprint region, lead to the conclusion that this polymer is poly(methyl methacrylate).

One can also use solution and solid state NMR spectroscopy in a similar manner to identify polymers (there are plenty of examples coming up shortly). But we do not want to give the impression that the identification of polymers, and especially copolymers, is as easy as we have implied in the simple infrared example shown in Figure 7-21. There are many subtleties and many industrial scientists have made a career specializing in polymer identification. However, this is enough for our purposes here and we now move on to the characterization of polymer microstructure.

Branching

From about 1940 into the 1960's, infrared spectroscopy was the workhorse of most polymer analytical laboratories, first as a routine tool for identifying unknown samples, then increasingly as a tool for analyzing structure and conformation. For example, it was originally thought that polyethylene produced by high pressure free radical polymerization was linear, but one of the first infrared spectroscopic studies of polymers showed that there were far more methyl groups present than could be accounted for by end groups alone. This conclusion was based largely on an analysis of the bands in the C–H stretching region of the spectrum, between 3000 and 2800 cm^{-1}, as shown in Figure 7-22. In later work, a band near 1375 cm^{-1} was used (Figure 7-23), which shows the spectrum of a branched sample. Compare this spectrum to the one beneath it, obtained from a linear sample. However, this analysis was (at least initially) difficult, because the 1375 cm^{-1} band overlaps some vibrational modes due to CH_2 groups. Furthermore, what you get is a measure of the total branch content. Although extremely useful, it would also be nice to know the length and distribution of

FASCINATING POLYMERS—VERSATILE POLYETHYLENES

Paul Hogan and Robert Banks
(Source: ConocoPhillips).

You may recall that Fawcett and Gibson discovered low density polyethylene (LDPE) in the 1930s (see *Polymer Milestones,* Chapter 3). What happened next in the polyolefin story was truly fascinating. In the late 1940s, Robert Banks and Paul Hogan commenced a research collaboration at the Phillips Petroleum Company in Bartlesville, Oklahoma. Their research concerned finding new ways to convert ethylene and propylene, products of natural oil and gas refining, into useful gasoline components. Similar research was also being performed at Standard Oil of Indiana. The third player in the field was Karl Ziegler, who we have highlighted in a another *Polymer Milestone* (Chapter 3, page 75). The metal oxide catalysts developed at Phillips and Standard Oil used different metals, but worked in roughly the same manner. Hogan and Banks performed an experiment where they modified a nickel oxide catalyst by adding a small amount of chromium oxide. This they packed in a column through which they passed a mixture of propylene and propane. They anticipated a conventional mixture of low molecular weight hydrocarbon products. But, something else, a white solid, also came out of the column. It was polypropylene! The two scientists then devoted all their attention to their chromium catalyst, using it to polymerize ethylene at relatively low pressure and producing a much higher density polyethylene (HDPE) than anything available at that time. It was marketed in 1954 under the brand name Marlex. HDPE, which is far harder, stiffer and more thermally stable than LDPE, has been used to produce everyday articles including outdoor furniture, toys, catheters, garbage cans, laundry baskets, car bumpers, milk containers etc. Ironically, in the late 1950s the majority of Marlex HDPE production was used to manufacture the hula hoop—the going craze or obsession of the day. So what was the difference between LDPE and HDPE? In characterizing polyethylenes, the most important difference is the degree of short-chain branching, which is reflected in the number of methyl groups per 1000 carbon atoms (see table below). Thus the high-pressure PE chain has many more branches and is more irregular and less crystalline than the Ziegler PE. In turn, the PE formed by the Phillips and Standard Oil metal oxide catalysts have even less branches and the chains are "straighter." This is reflected in the increased density (ability to pack efficiently) and crystalline melting points of these polymers. The degree and perfection of crystallinity has a profound effect on physical and mechanical properties.

	High Pressure	Ziegler	Metal Oxide
Molecular Weight (M_n)	20,000	15,000	15,000
Number of Me Groups/1000 Carbons	30	6	<0.15
Density (g/cc)	0.92	0.95	0.96
Crystalline Melting Point (°C)	108	130	133

Source: K. J. Saunders, *Organic Polymer Chemistry*, Chapman and Hall (1976).

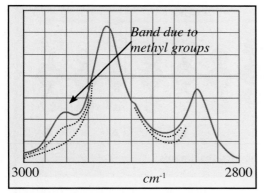

FIGURE 7-22 Infrared spectra in the C–H stretching region (Redrawn from the original spectrum reported by J. J. Fox and A. E. Martin; *Proc R. Soc. London, A,* **175**, 208 (1940)).

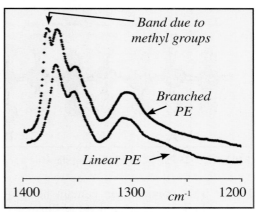

FIGURE 7-23 Infrared spectra of two polyethylenes in the region from 1200–1400 cm^{-1}.

the branches that are present.

The advent of ^{13}C NMR allowed an analysis of branching at a far greater level of detail because the ^{13}C shifts of paraffinic hydrocarbons depend strongly on their proximity to tertiary carbons (i.e., the branch points). As a result, the ^{13}C NMR spectrum of a (short chain) branched polyethylene sample is rich in detail, as can be seen in the example shown in Figure 7-24. It was originally thought that

ethyl branched predominated in low density polyethylene, but NMR demonstrated that they were, in fact, butyl branches. However, there are also significant amounts of amyl, hexyl and longer branches present.

Sequence Isomerism

NMR can also provide clear evidence of sequence isomerism in polymers. Take, for example, poly(vinylidene fluoride), or

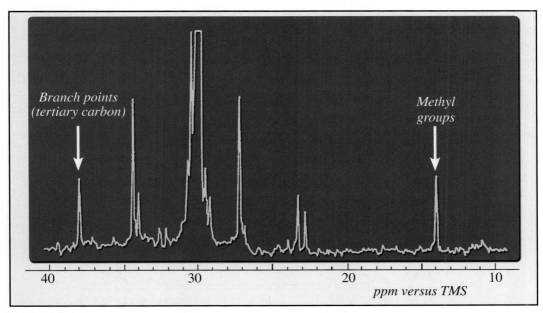

FIGURE 7-24 ^{13}C NMR spectrum of a branched polyethylene in the region from 10-40 ppm (redrawn from the original spectrum reported by F. Bovey, *Chain Structure and Conformation of Macromolecules,* Academic Press, New York, 1982).

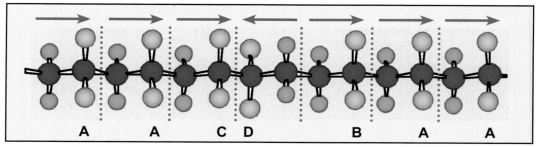

FIGURE 7-25 Part of a PVDF chain showing a segment inserted "backwards."

PVDF. This polymer can contain head-to-head and tail-to-tail sequences, as illustrated in Figure 7-25. Because the ^{19}F nucleus has a spin number of 1/2 and its natural abundance is 100%. ^{19}F NMR spectroscopy is a powerful tool for studying fluoropolymers like PVDF. The fluorine atoms are the little yellow balls and you'll notice that one of the units in Figure 7-25 has been put in backwards, giving a head-to-head and tail-to-tail sequence. The peaks marked A, B, C and D in the spectrum of PVDF (Figure 7-26) correspond to the fluorine atoms labeled in the same way in the figure. Of course, what you would really like to know is the number fraction of VDF monomers that are incorporated backwards into the chain. This information can be obtained from the relative intensities of the bands, but to extract it we have to revisit probability theory. (Oh no! the nightmare continues!)

If you recall the stuff you should have learned in Chapter 2, then we don't have to

tell you that for the general case of a vinyl monomer, $CH_2=CXY$, the head is defined as the CXY end, while the tail is the CH_2 end (Figure 7-27). We define the probability of "normal" addition (top right in the figure) as $P_1\{TH\}$ and the probability of "backward" addition as $P_1\{HT\}$. It can only be one or the other, so it follows that (Equation 7-19):

$$P_1\{TH\} + P_1\{HT\} = 1$$

EQUATION 7-19

Diads are related through Equations 7-20:

$$\begin{aligned} P_1\{TH\} &= P_2\{TH\text{-}TH\} + P_2\{TH\text{-}HT\} \\ &= P_2\{TH\text{-}TH\} + P_2\{HT\text{-}TH\} \\ P_1\{HT\} &= P_2\{HT\text{-}HT\} + P_2\{HT\text{-}TH\} \\ &= P_2\{TH\text{-}HT\} + P_2\{HT\text{-}HT\} \end{aligned}$$

EQUATIONS 7-20

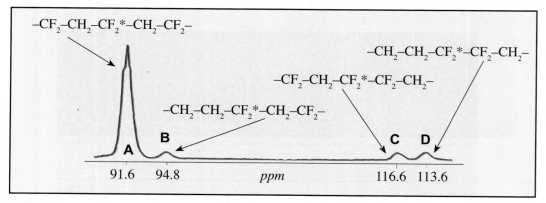

FIGURE 7-26 The ^{19}F NMR spectrum of PVDF at 56.4 MHz [redrawn from an original figure by C. W. Wilson III and E. R. Santee Jr., *J. Polymer Sci.—Part C*, **8**, 97 (1965).]

Think about it: a TH, for example, must be followed by another TH or an HT. Similarly, it must be preceded by a TH or an HT. It then follows that (Equation 7-21):

$$P_2\{HT\text{-}TH\} = P_2\{TH\text{-}HT\}$$

EQUATION 7-21

which means that the number of head-to-head diads is always equal to the number of tail-to-tail diads.

In the same way triad sequences can be calculated from diad sequences, as shown in Equations 7-22, and if you have a weekend or two to spare you can continue calculating higher order sequences from those of lower order (tetrad from triads, etc.).

$$\begin{aligned} P_2\{TH\text{-}TH\} &= P_3\{TH\text{-}TH\text{-}TH\} \\ &\quad + P_3\{TH\text{-}TH\text{-}HT\} \\ &= P_3\{TH\text{-}TH\text{-}TH\} \\ &\quad + P_3\{HT\text{-}TH\text{-}TH\} \end{aligned}$$

$$\begin{aligned} P_2\{TH\text{-}HT\} &= P_3\{TH\text{-}HT\text{-}TH\} \\ &\quad + P_3\{TH\text{-}HT\text{-}HT\} \\ &= P_3\{TH\text{-}TH\text{-}HT\} \\ &\quad + P_3\{HT\text{-}TH\text{-}HT\} \end{aligned}$$

$$\begin{aligned} P_2\{HT\text{-}TH\} &= P_3\{HT\text{-}TH\text{-}TH\} \\ &\quad + P_3\{HT\text{-}TH\text{-}HT\} \\ &= P_3\{TH\text{-}HT\text{-}TH\} \\ &\quad + P_3\{HT\text{-}HT\text{-}TH\} \end{aligned}$$

$$\begin{aligned} P_2\{HT\text{-}HT\} &= P_3\{HT\text{-}HT\text{-}TH\} \\ &\quad + P_3\{HT\text{-}HT\text{-}HT\} \\ &= P_3\{TH\text{-}HT\text{-}HT\} \\ &\quad + P_3\{HT\text{-}HT\text{-}HT\} \end{aligned}$$

EQUATIONS 7-22

You then get additional reversibility relationships, in this case (Equations 7-23):

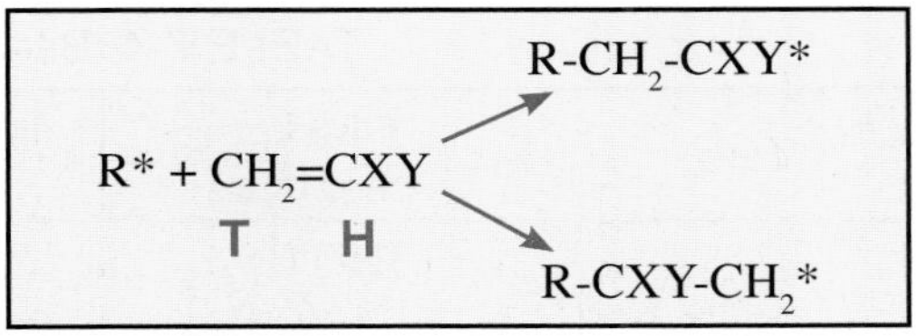

FIGURE 7-27 Sequence isomerization.

$$P_3\{HT\text{-}HT\text{-}TH\} = P_3\{TH\text{-}HT\text{-}HT\}$$
$$P_3\{TH\text{-}TH\text{-}HT\} = P_3\{HT\text{-}TH\text{-}TH\}$$

EQUATIONS 7-23

If we want to calculate the number or fraction of "backward" units in a sample, we need to determine $P_1\{HT\}$ ($= 1 - P_1\{TH\}$). This determination will depend upon what information is available in the NMR spectrum. Often, diad, triad or even higher sequence information is available, but the bands must be assigned correctly. There are 4 possible diad sequences, 8 possible triad sequences, 16 possible tetrads, and so on. However, not all of these are distinguishable. For example, NMR cannot distinguish between {TH-TH-TH} and {HT-HT-HT}—this is equivalent to reading the chain in one direction or the other. A summary of observable diad and triad sequences is given in Table 7-4. Also shown in this table are the Bernouillian probabilities (you remember—the probability of a unit being placed in a particular fashion doesn't depend on what's already there). Note that these results predict that we should see 3 NMR lines if we are observing diad information and 4 for triads (there would be 5 lines for tetrads and 6 for pentads). Even more interesting, n of the observed NMR lines (assuming they are resolved in the spectrum) will have the same intensity if n-ad information is available. This just means that if we have triad information available, 3 of the 4 observed lines will have the same intensity. For example, if $P_{TH} = 0.9$ and if we are indeed observing triad information, we would expect to see one dominant line with a normalized intensity of 0.73 and 3 other lines each with an intensity of 0.09.

TABLE 7-4 BERNOULLIAN PROBABILITIES

Type	NMR Observables	Indistinguishable Sequences	Bernoullian Probability
Diads	(TH-TH)	$P_2\{TH\text{-}TH\}$ $+ P_2\{HT\text{-}HT\}$	$P_{TH}^2 + (1 - P_{TH})^2$
	(TH-HT)	$P_2\{TH\text{-}HT\}$	$P_{TH}(1 - P_{TH})$
	(HT-TH)	$P_2\{HT\text{-}TH\}$	$P_{TH}(1 - P_{TH})$
Triads	(TH-TH-TH)	$P_3\{TH\text{-}TH\text{-}TH\}$ $+ P_3\{HT\text{-}HT\text{-}HT\}$	$P_{TH}^3 + (1 - P_{TH})^3$
	(TH-TH-HT)	$P_3\{TH\text{-}TH\text{-}HT\}$ $+ P_3\{TH\text{-}HT\text{-}HT\}$	$P_{TH}^2(1 - P_{TH}) + P_{TH}(1 - P_{TH})^2$ $= P_{TH}(1 - P_{TH})$
	(TH-HT-HT)	$P_3\{TH\text{-}HT\text{-}TH\}$ $+ P_3\{HT\text{-}TH\text{-}HT\}$	$P_{TH}(1 - P_{TH})$
	(HT-TH-TH)	$P_3\{HT\text{-}TH\text{-}TH\}$ $+ P_3\{HT\text{-}HT\text{-}TH\}$	$P_{TH}(1 - P_{TH})$

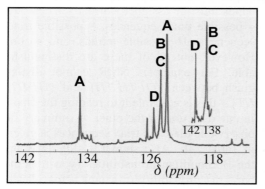

FIGURE 7-28 ^{13}C NMR spectrum of a polychloroprene in the region from 110–140 ppm.

The ^{19}F NMR spectrum shown in Figure 7-26 is actually very close to this, with 5–6% of the monomers reversed.

PVDF is just one polymer where sequence isomerism has been obseved using NMR spectroscopy. The ^{13}C NMR spectrum of polychloroprene, for example, has also been analyzed by one of your authors (in collaboration with his coworkers) and the olefinic region is shown in Figure 7-28. It proved possible to assign lines to triad sequence isomers of the *trans*-1,4 units (TH and HT) and also triad sequences containing the *cis*-1,4

TABLE 7-5 POLYCHLOROPRENE ^{13}C NMR BAND ASSIGNMENTS

$-CH_2-C^*Cl=CH–CH_2-$	$-CH_2-CCl=C^*H–CH_2-$	Designation	Assignment
134.9	124.1	A	(TH-TH-TH) + (HT-HT-HT)
134.6	124.9	B	(HT-TH-TH) + (HT-HT-TH)
133.9	124.9	C	(TH-TH-HT) + (TH-HT-HT)
133.5	125.8	D	(HT-TH-HT) + (TH-HT-TH)
134.1	126.6	E	(TH-CS-TH)
134.3	124.7	F	(TH-TH-CS)
135.1	124.1(?)	G	(CS-TH-TH)
-	127.6	H	(HT-CS-TH)

unit (CS). Assignments are summarized in Table 7-5 (for clarity, only some of these are shown in Figure 7-28).

Structural Isomerism

Both NMR and vibrational spectroscopy can be used to detect and study structuctural isomerism in diene polymers, but as we will see, there are difficulties with a number of these techniques. Take, for example, the infrared spectrum of poly(2,3-dimethylbutadiene), shown in transmission (i.e., with the bands pointing down) on the top of Figure 7-29. There is no band near 1660 cm^{-1}, where you would expect C=C double bonds to absorb. There is a beautiful, strong Raman line at 1665 cm^{-1}, however, as also shown in the figure. This is because of the symmetry in the repeat unit of this polymer. Remember that in CO_2 the symmetric stretching mode was IR inactive and only appeared in the Raman spectrum? Well the same thing happens here. Because of the center of symmetry present in the structure, there is mutual exclusion of the modes, those bands that appear in the infrared spectrum don't appear in the Raman, and vice-versa. (Look at Figure 7-30—if a molecule has a center such that a line drawn from any atom in one part of the molecule through this point encounters an equivalent atom equidistant from this point in the other half, it possesses a center of symmetry or inversion.) To obtain complete information on this sample you would have to obtain both spectra. Obtaining the Raman spectrum of industrial samples can be difficult, however, because of a fluorescence usually associated with impurities. Nevertheless, you can clean up your sample and usually get a spectrum. And if all you were trying to do was identify a sample that had essentially no structural defects this would be fine.

Often, however, you are trying to determine how many *cis*-1,4-; 1,2-; 3,4-; etc. units there are in a sample that may largely consist of *trans*-1,4- units, as in polychloroprene, for example. Consider the spectra marked A and B shown in Figure 7-31. Sample A was polymerized at −40°C, while sample B

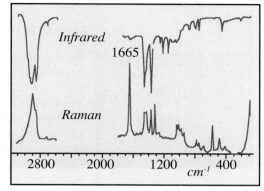

FIGURE 7-29 Infrared and Raman spectra of poly(2,3-dimethylbutadiene).

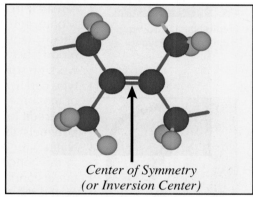

Center of Symmetry
(or Inversion Center)

FIGURE 7-30 Model of poly(2,3-dimethylbutadiene).

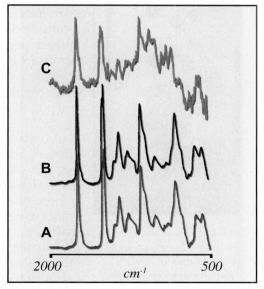

FIGURE 7-31 Infrared spectra of two different polychloroprenes (A and B). Difference spectrum (C = A − B).

POLYMER MILESTONES—GIULIO NATTA

Giulio Natta (Courtesy: Politecnico di Milano).

Giulio Natta, an Italian Professor at the Politecnico di Milano, who shared the 1963 Nobel Prize in chemistry with Karl Ziegler, made major contributions to the field of polymer science. In the middle of 1954, Natta and his coworkers, having been informed by Ziegler of the nature of his original HDPE catalyst, used it to polymerize propylene and obtained a partially "crystalline," essentially isotactic, polypropylene. Subsequently, Natta and his coworkers made changes to the catalyst, replacing $TiCl_4$ with $TiCl_3$, which increased the isotacticity of polypropylene from 50–70% to 80–90%. Natta's research in this area was funded by the Italian company Montecatini and a patent was filed jointly soon thereafter. Unfortunately, this caused considerable "bad blood" between Ziegler and Natta. In addition to his work on studying, modifying and understanding Ziegler-type catalysts and how they affect the spatial orientation of monomer insertion during polymerization, Natta gets the credit for his seminal physical characterization of the polymers that he and his coworkers produced. It was Natta who introduced the unsuspecting polymer world to the concept of stereoregular polymers—although we understand that it was his wife, Rosita, who actually suggested the terms isotactic (from the Greek *iso* meaning equal) and syndiotactic (from the Greek *syndio* meaning "every two"). Paul Flory, who was to get his Nobel Prize later, remarked in a letter to Natta in 1955, "The results disclosed in your manuscript are of extraordinary interest; perhaps one should call them revolutionary in significance." A. V. Tobolsky, another famous polymer scientist, wrote, "Natta and his coworkers (Milan) using Ziegler-type catalysts, prepared isotactic polypropylene and polystyrene, among others. It is Natta who first recognized the chemical revolution taking place." Thinking you've made something and proving it are two entirely different beasts. Organic chemists are often confident that they have performed a specific synthesis only to be bought back to earth by the results of physical characterization. Natta and his very talented team of polymer scientists were experts in the characterization of polymeric materials. Infrared spectroscopy played an important role in the characterization of polypropylene among other polymers and copolymers. There were many scientists that made significant contributions, but the fractionation studies of Piero Pino and the X-ray studies of Paolo Corradini deserve special mention.

Our Italian friend, Prof. Gus Zerbi, in Milano, attempting to straighten out Natta's original model of isotactic polypropylene.

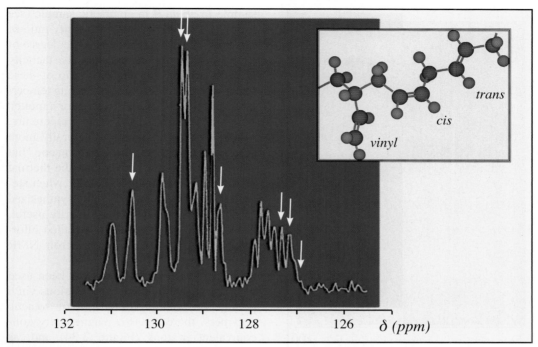

FIGURE 7-32 ^{13}C NMR spectrum in the region from 124–132 ppm of a polybutadiene containing 34% *trans*-1,4-, 24% *cis*-1,4- and 42% 1,2- placements [redrawn from an original figure by K-F. Elgert, G. Quack, and B. Stutzel, *Polymer* **16**, 154 (1975)].

was polymerized at −20°C. This results in a difference in microstructure, but it can be seen that the spectra of the two samples are very similar, simply because there is only a small difference in the content of structural irregularities. Nevertheless, with modern computer-based instruments it is possible to subtract one spectrum from the other (A − B) to give a difference spectrum (C) revealing the presence of bands that can be assigned to *cis*-1,4-, 1,2-, and 3,4- units.[6] The concentration of each of these structural irregularities is not easily quantified, however, and for that type of analysis you would have to turn to ^{13}C NMR.

There are only minimal chemical shift differences between *cis* and *trans* isomers in the ^{1}H NMR spectra of polymers like polybutadiene and polyisoprene, so this tool is less useful in characterizing structural isomers in diene polymers.

[6] M. M. Coleman, R. J. Petcavich and P. C. Painter, *Polymer*, **19**, 1243 (1978).

On the other hand, ^{13}C NMR is extremely effective in characterizing these types of polymers. These spectra not only display well-separated lines for the various structural isomers that may be present, but also provide sequence distribution information and readily allow quantitative analysis. For example, the proton decoupled ^{13}C NMR spectrum of polybutadiene shown in Figure 7-32 has bands that can be assigned to specific triad sequences. We have marked with arrows those that have a central *cis* unit, like the sequence shown at the top of the figure.

Tacticity in Vinyl Polymers

The measurement of tacticity in vinyl polymers is an area of characterization where NMR stands supreme. This is not to say that infrared spectroscopy is not sensitive to the tacticity of a polymer. In polypropylene, for example, the ratio of the intensity of the 998 cm[-1] band to that of the 973 cm[-1] band varies from about 1 in a completely isotactic

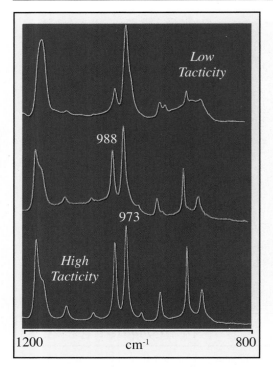

FIGURE 7-33 Infrared spectra of polypropylenes of varying tacticities.

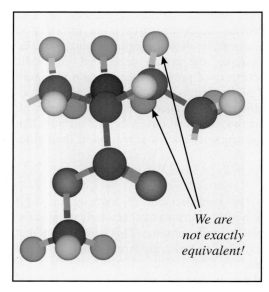

We are not exactly equivalent!

FIGURE 7-34 The repeat unit of poly(methyl methacrylate).

sample to about 0 in an atactic sample (see Figure 7-33). But what the intensity ratio of this band actually represents is the degree of chain order in the sample, not the tacticity as such, so that the intensity ratio of these bands also decreases with increasing temperature and becomes close to zero for a molten sample. Measured at the same temperature, a highly isotactic sample is naturally more ordered than one that is less isotactic, but the degree of order depends on the thermal history of the sample. Of course, when stereoregular polymers were first synthesized IR spectroscopy was extraordinarily useful, but it cannot give the sort of detailed information that can now be obtained from NMR spectroscopy.

Proton NMR spectroscopy has been used to characterize the tacticity of various vinyl polymers in solution. In the case of isotactic polymers, there are two magnetically nonequivalent protons (Figure 7-34) and, as we discussed earlier in this chapter, this can result in the appearance of four bands (the chemical shift difference is of the same order of magnitude as the coupling constant, so the simple rules for multiplicities don't apply and we get what we called an AB pattern). On the other hand, in syndiotactic polymers the two methylene protons are equivalent and we observe only one line. Let's look at this in more detail, using poly(methyl methacrylate) (PMMA), as an example, because bands due to various tactic sequences are particularly well resolved in the ^1H spectrum of this material.

Common or garden PMMA, the stuff most people refer to as Plexiglas®, is atactic (*a*-PMMA), but both the isotactic (*iso*-PMMA) and syndiotactic (*syn*-PMMA) forms can be polymerized. The arrangements of the chemical groups along the chain are shown schematically in Figure 7-35 and we will start by considering the isotactic form, where groups of the same type are (stereochemically) on the same side of the chain. Here it is easy to see that the two hydrogen atoms of the CH_2 groups (the little balls) are not magnetically equivalent. One is always in the environment of two methyl groups, while the other lies between two ester groups. In contrast, for a purely syndiotactic polymer, the CH_2 protons are magnetically eqivalent, each being flanked on one side by a methyl group and on the other by an ester. Naturally, atactic polymers have both types of CH_2 units.

As you should now anticipate, because the two protons of the methylene group of *iso*-PMMA are not magnetically equivalent there is a small difference in chemical shift, resulting in the four-line AB pattern shown schematically in Figure 7-36. On the other hand, the CH_2 protons of the syndiotactic polymer are equivalent, there is no spin-spin interaction and just one line is observed. An atactic sample should resemble a combination of these two, of course. However, given the spectrum on the bottom, we would not be able to tell if the sample was a mixture of isotactic and syndiotactic chains, or was actually atactic. This is because the methylene protons are only sensitive to what are called diad structures (the stereochemical arrangements of two adjacent units).

However, the lines due to backbone methyl groups are sensitive to triad sequences, highlighted on the schematic spectrum shown in Figure 7-35. The differences in the magnetic

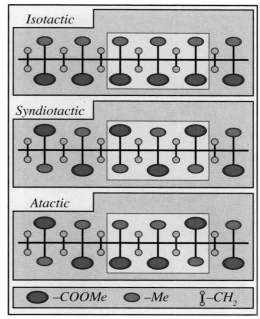

FIGURE 7-35 Schematic diagram of different poly(methyl methacrylates).

environment of the (backbone) methyl group result in subtle differences in the chemical shift of the methyl proton. The representation of the spectrum also shows this sche-

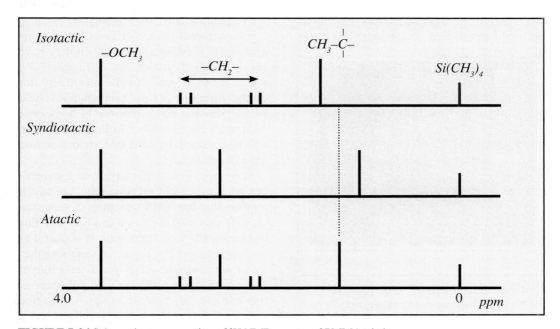

FIGURE 7-36 Schematic representation of ^{1}H NMR spectra of PMMA triads.

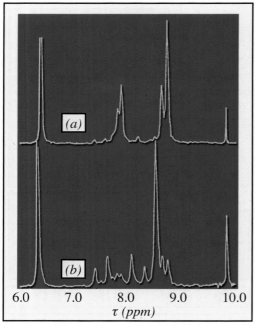

FIGURE 7-37 60 MHz ¹H NMR spectra of PMMA [redrawn from an original figure by F. A. Bovey, *High Resolution NMR of Macromolecules,* Academic Press (1972)].

matically, with a line due to a heterotactic sequence falling between lines due to isotactic and syndiotactic triad sequences.

An atactic polymer will also have isotactic and syndiotactic triad sequences, of course, but we'll get to that in a minute. First, we

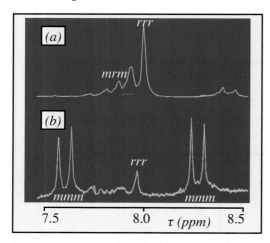

FIGURE 7-38 220 MHz ¹H NMR spectra of PMMA [redrawn from an original figure by F. A. Bovey, *High Resolution NMR of Macromolecules,* Academic Press (1972)].

wanted you to realize that triad or higher sequence data is necessary in order to adequately describe stereoregularity.

We've shown you a schematic representation of what we would expect the proton NMR spectra to look like, what about the real thing? The ¹H NMR spectra of a predominantly syndiotactic PMMA sample is shown at the top (a) in Figure 7-37, while the spectrum of a predominantly isotactic sample is shown at the bottom (b). The essential features of the experimental data indeed correspond to the schematic representation of our expectations, but it can be seen that there are signs of additional splittings. These are due to higher order sequences, which can be seen in the spectra of samples obtained on more powerful instruments, such as those shown in Figure 7-38. Here the CH_2 region of the proton NMR spectrum of predominantly syndiotactic PMMA (top) and predominantly isotactic PMMA (bottom) are again compared and now tetrad data are resolved. The tetrads are labeled *mmm*, *rrr*, *mrm*, etc. (only a few major assignments are shown—further details may be found in Bovey's book), signifying sequences of *meso* and *racemic* diads. We told you what these were in Chapter 2, but we suppose that we're going to have to remind you!

Two adjacent monomer units in the same chain are called *meso diads* when they have the same configuration, as in Figure 7-39, where it can be seen that the substituent lies on the same side of the chain when it is held in an extended conformation. In a *racemic diad*, such as that shown in Figure 7-39, the substituents are opposite one another across the extended chain.

Very nice, you might say, now I remember what meso and racemic units are and it appears that the NMR spectrum is very sensitive to this stuff, but what can I do with this information? Well for a start, it is useful to put a number on the degree of stereoregularity of your sample. Also, you would like to know how this can change with polymerization method, say anionic as opposed to free radical. But to understand the full power of this technique, we have to revisit probability theory. Try not to scream with joy!

As you might guess if you were thinking about it, the theory and methodology we will use here corresponds to what we used to describe copolymer sequence distributions. Now the "monomers" consist of the same chemical unit, but have their functional groups arranged in different steric configurations. When these units are added to a chain they will form either a *meso diad*, designated *m*, or a *racemic diad*, designated *r* when taken with the repeat unit previously at the end of the chain (Table 7-6). This nomenclature can be extended to sequences of any length, as shown in the second column of the table. For example, a *syndiotactic triad* is designated *rr*, a *heterotactic triad, mr*, and so on. If you're having trouble visualizing what, say, an *mmr* sequence looks like in terms of the arrangement of the groups, check out the third column. The fourth column describes the *Bernoullian* probability of obtaining a particular sequence.

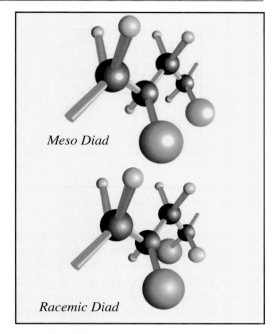

FIGURE 7-39 Meso and racemic diad models.

TABLE 7-6 CONFIGURATIONAL SEQUENCES

TYPE	DESIGNATION	PROJECTION	BERNOULLIAN PROBABILITY
Diad	*meso, m*		P_m
	racemic, r		$(1 - P_m)$
Triad	*isotactic, mm*		$P_m^{\ 2}$
	heterotactic, mr		$2P_m(1 - P_m)$
	syndiotactic, rr		$(1 - P_m)^2$
Tetrad	*mmm*		$P_m^{\ 3}$
	mmr		$2P_m^{\ 2}(1 - P_m)$
	rmr		$P_m(1 - P_m)^2$
	mrm		$P_m^{\ 2}(1 - P_m)$
	rrm		$2P_m(1 - P_m)^2$
	rrr		$(1 - P_m)^3$

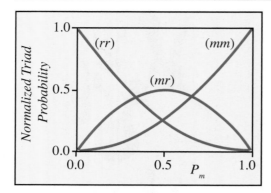

FIGURE 7-40 Bernoullian triad probabilities.

If you just walk into a classroom and yell "Bernoullian statistics" you can almost guarantee that your students will scatter in terror. But this stuff is actually pretty easy. If the addition of a "monomer" with a particular configuration is *independent* of the stereochemistry found at the end of the growing chain, then we say that propagation obeys Bernoullian statistics. The probability of generating a meso sequence can then be described by a single parameter P_m. Conceptually, this is the same as reaching into a large jar containing balls marked "m" and "r" and withdrawing a ball at random. P_m will simply be the (number) fraction of balls marked m in that jar. Naturally, the probabilty of forming a racemic sequence between the monomer at the end of the chain and the one being added is just $(1 - P_m)$. The probabilities of various triad and tetrad sequences then follows directly and these are listed in the last column of Table 7-6.

One important thing to realize is that even though the stereochemical structure, m or r, formed by the addition of a monomer does not depend on the configuration of units that were at the end of the growing chain, this does not mean that $P_m = 0.5$. The probability of forming a meso diad could be much greater than this, but the addition of units would still follow random statistics, and the probability of obtaining various diad, triad and tetrad sequences would follow the relationships given in the table. A plot of triad relationships as a function of P_m is shown in Figure 7-40. As you might expect, if you

think about it, the proportion of *mr* units is a maximum when $P_m = 0.5$.

If propagation obeys Bernoullian statistics, then the relative proportions of *mm*, *rr* and *mr* sequences, calculated from the relative intensities of corresponding NMR bands, would lie on a single vertical line drawn on Figure 7-40, corresponding to a single value of P_m. Poly(methyl methacrylate) produced free radically usually follows (within error) Bernoullian statistics, but those produced anionically do not. However, conformity of triad data to this plot (i.e., fitting data to a value of P_m) does not prove that the propagation mechanism obeys Bernoullian statistics, we test that using some conditional probability arguments. We'll get to that shortly; first we have to plough through a lot more algebra.

In Chapter 5 we used the symbol $P_n\{ \ \}$ to describe the probability of finding a particular "n-ad," a sequence in the progression diad, triad, tetrad, pentad, etc. (e.g., $P_3\{ABA\}$). Here (m) is equivalent to $P_2\{m\}$, the fraction of m *diads*, while $P_2\{r\}$ is fraction of r *diads*. The probability of finding a heterotactic *triad*, *mr*, is designated (mr), and so on.

Different types of brackets are used to make a distinction between *distinguishable* (n-*ads*) and *indistinguishable* [n-*ads*]. If your first reaction to this sentence is "say what?", don't panic, this is straightforward stuff. Check out the schematic depictions of *mm* and *rr* triads shown at the top of Figure 7-41. It doesn't matter if you "read" the chain from left to right or right to left, the result is the same. The sequences shown on the bottom, however, read *mr* in one direction and *rm* in the other. But NMR doesn't "read" a chain in a particular fashion, to the spectrometer an *rm* triad is the same as an *mr* triad.

So, for the sequence shown at the top of Figure 7-42, we would expect to see three NMR lines, one for *rr* diads, one for *mm* diads and a line for *mr* and *rm* sequences that appears at the same chemical shift. Mathematically we can express the number fraction of distinguishable triads (mr) in terms of the number fraction of indistinguishable

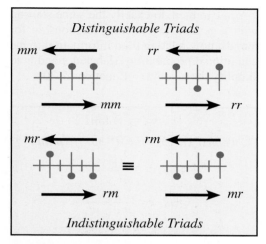

FIGURE 7-41 Schematic "ball and stick" diagram depicting triad sequences.

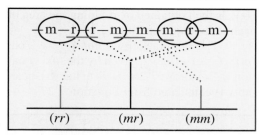

FIGURE 7-42 Schematic diagram illustrating indistinguishable sequences observed in NMR.

sequences, $[mr]$ and $[rm]$ (Equation 7-24):

$$(mr) = [mr] + [rm]$$

EQUATION 7-24

tinguishable n-ad sequences (hence what's observed in an NMR experiment) and those of higher order (Equations 7-26).

$$(m) = (mm) + (mr)/2$$
$$(r) = (rr) + (rm)/2$$
$$(mm) = (mmm) + (mmr)/2$$
$$(mr) = (mmr) + 2(rmr)$$
$$= (mrr) + 2(mrm)$$
$$(rr) = (rrr) + (mrr)/2$$

EQUATIONS 7-26

Now, if you recall that reversibility relationships are general (e.g., $P_2\{AB\} = P_2\{BA\}$—Chapter 6), then it follows that:

$$(mr) = 2[mr] = 2[rm]$$

EQUATION 7-25

There are a bunch of other relationships that are also obtained by exactly the same arguments as those used in our discussion of copolymerization, and these are summarized below in Table 7-7. And we're not done yet! There's also the relationships between dis-

At this point you're probably throwing up your hands in despair, thinking that your authors are nothing but algebra junkies. But now we can start using this stuff. Let's begin by assuming that you have a spectrum where you have established that you are observing triad data. Perhaps the intensity pattern will look something like the schematic diagram shown in Figure 7-42. You want to know the % isotacticity (m) of this sample. The first thing you do is measure the areas of the three bands, $A_{(rr)}$, $A_{(mr)}$, and $A_{(mm)}$. The number fraction of each type of triad, (mm), (mr), and

TABLE 7-7 PROBABILITY RELATIONSHIPS

GENERAL RELATIONSHIPS	REVERSIBILITY RELATIONSHIPS	LOWER/HIGHER ORDER RELATIONSHIPS
$(m) + (r) = 1$ $(mm) + (mr) + (rr) = 1$ $\Sigma(all\ tetrads) = 1$ *etc.*	$[mr] = [rm]$ $[mmr] = [rmm]$ $[mrr] = [rrm]$ $[mmmr] = [rmmm]$ *etc.*	$[mr] = [mmr] + [rmr]$ $[rmr] = [rrmr] + [mrmr]$ *etc.*

(rr), is then obtained by dividing the area of each individual band by $A_{(T)}$ $(= A_{(rr)} + A_{(mr)} + A_{(mm)})$. The % isotacticity is **not** $A_{(mm)}/A_{(T)}$, but $(m) = (mm) + (mr)/2$, given by the first equation in Equations 7-26. Hence, the % isotacticity is calculated from Equation 7-27.

$$\% \ Isotacticity = \frac{A_{(mm)} + A_{(mr)}/2}{A_{(T)}}$$

EQUATION 7-27

Propagation Mechanisms

Now let's say you wanted to determine the propagation mechanism in a particular polymerization. How exactly do you do this? We have to start by considering some conditional probabilities. We only need two, the first being the probability that the next unit added to a chain forms an r configuration, given that the preceding two units formed an m, $P(m/r)$. The second probability is the other way around, $P(r/m)$. These are related to the remaining two conditional probabilites, $P(r/r)$ and $P(m/m)$ by Equations 7-28.

$$P(r/m) = 1 - P(m/m) = u$$
$$P(m/r) = 1 - P(r/r) = w$$

EQUATIONS 7-28

Then starting from reversibility relationships for triads you can relate diad information to these conditional probabilities, as shown in Equations 7-29.

$$[mr] = [rm]$$
$$(m)P(r/m) = (r)P(m/r)$$
$$(r) = 1 - (m)$$
$$(m)P(r/m) = (1 - (m))P(m/r)$$
$$(m) = \frac{P(m/r)}{P(r/m) + P(m/r)} = \frac{w}{u + w}$$
$$(r) = \frac{u}{u + w}$$

EQUATIONS 7-29

You can then do exactly the same starting with a general reversibility relationship for tetrads, now relating triad information to the conditional probabilities (labeled u and w to simplify the algebra)—Equations 7-30.

$$[mmr] = [rmm]$$
$$(mm)P(r/m) = (r)P(m/r)P(m/m)$$
$$(mm) = \frac{w(1 - u)}{u + w}$$
$$(mr) = \frac{2uw}{u + w}$$
$$(rr) = \frac{u(1 - w)}{u + w}$$

EQUATIONS 7-30

We'll be merciful and stop after just giving you the relationships for tetrads obtained starting from reversible pentad sequences (Equations 7-31).

$$(mrr) = \frac{2uw(1 - w)}{u + w} = \frac{(mr)(rr)}{(r)}$$
$$(rmr) = \frac{u^2 w}{u + w} = \frac{(mr)^2}{4(m)}$$
$$(rrr) = \frac{u(1 - w)^2}{u + w} = \frac{(rr)^2}{(r)}$$

EQUATIONS 7-31

(But if our students put us in a bad mood, we make them derive this stuff as a homework problem, just to make them suffer.)

You can now test the models. So far we have only talked about Bernoullian statistics, where the stereochemistry of the addition of a monomer does not depend upon the configuration at the end of the chain. Just to make sure you understand, from NMR triad information—i.e., (mm), (mr) and (rr) data—we can readily calculate the conditional probabilities $P(r/m)$, $P(m/m)$, $P(m/r)$ and $P(r/r)$ as shown in Equations 7-32. Bernouillian statistics are followed if Equations 7-33 are obeyed.

$$P(r/m) = u = \frac{\frac{1}{2}(mr)}{(m)}$$

$$P(m/r) = w = \frac{\frac{1}{2}(rm)}{(r)}$$

$$P(m/m) = 1 - u = \frac{(mm)}{(m)}$$

$$P(r/r) = 1 - w = \frac{(rr)}{(r)}$$

EQUATIONS 7-32

$$P(r/m) = P(r/r) = P(r) = 1 - P(m)$$

and:

$$P(m/r) = P(m/m) = P(m)$$

EQUATIONS 7-33

Think about it: what these equations are saying is that if the probability of an *r* given an *m* is the same as *r* given an *r* (and similarly, the probability of a *m* given an *m* is the same as *m* given an *r*), the configuration of the group at the end of the chain doesn't matter. Note that if $P(r/m) \neq P(r/r)$ and/or $P(m/r) \neq P(m/m)$ one can only say that Bernoullian statistics are not obeyed.

If the stereochemistry of addition does depend upon the configuration found at the end of the chain, whether it is *m* or *r*, then we have a terminal model, or first-order Markovian statistics. At minimum we need tetrad data from NMR—i.e., data for (*mmm*), (*mmr*), etc.—to test for the terminal model. Remember, we can always calculate triad data from tetrad data using the relationships previously given in Equations 7-26. Equations 7-34 relate the relevant conditional probabilities to observable tetrad and triad sequences.

$$P(m/mm) = \frac{(mmm)}{(mm)}$$

$$P(m/rm) = \frac{(mmr)}{(mr)}$$

$$P(m/mr) = \frac{2(mrm)}{(mr)}$$

$$P(m/rr) = \frac{(mrr)}{2(rr)}$$

EQUATIONS 7-34

First-order Markovian statistics (the terminal model) are followed if Equations 7-35 are obeyed.

$$P(m/mm) = P(m/rm) = P(m/m)$$

$$P(m/mr) = P(m/rr) = P(m/r)$$

$$P(r/rr) = P(r/mr) = P(r/r)$$

$$P(r/rm) = P(r/mm) = P(r/m)$$

EQUATIONS 7-35

There are higher order Markovian models and non-Markovian models. The Society for the Prevention of Cruelty to Students only permits us to consider the two models above. But this should give you an idea of how to proceed if you ever need to go further. (God forbid!)

Copolymer Analysis

Now we move on to consider the analysis of copolymers. There are usually two things we would like to know. First, the composition of the copolymer and, second, some measure of sequence distributions. Again, in the early years, before the advent of commercial NMR instruments, infrared spectroscopy was the most widely used tool. The problem with the technique is that it requires that the spectrum contain bands that can be unambiguously assigned to specific functional groups, as in the (transmission) spectrum of an acrylonitrile/methyl methacrylate copolymer shown in Figure 7-43 (you can tell this is a really old spectrum, not only because it is plotted in transmission, but also because the frequency scale is in microns).

Things become a little more difficult in materials such as ethylene/propylene copolymers, for example, because vibrational modes in different units can "couple" and are then no longer associated with a particular type of monomer. Or the bands due to specific functional groups (e.g., CH_2 or CH_3) can overlap other modes or each other (e.g., in the CH stretching region of the spectrum) and be difficult to separate. Adding to this

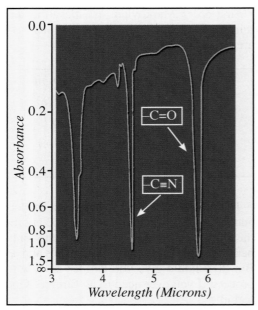

FIGURE 7-43 IR spectrum of an acrylonitrile/ methyl methacrylate copolymer [redrawn from an original figure in R. Zbinden, *Infrared Spectroscopy of High Polymers*, Academic Press, (1964)].

are problems involving baselines, etc., that before the advent of modern computerized instruments were difficult to solve and introduced considerable errors. Nevertheless, you can usually find a couple of bands on which to base your analysis (like the 1155 cm⁻¹ propylene and 720 cm⁻¹ ethylene bands shown in Figure 7-44) and a lot of good analytical work was accomplished over the years.

As in other areas of the characterization of polymer microstructure, NMR can provide more information. Nevertheless, infrared still has its uses. It is often much faster and

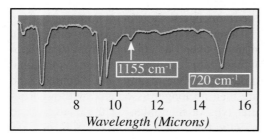

FIGURE 7-44 Infrared spectrum of an ethylene/ propylene copolymer. (Redrawn from Gardner et al, *Rubber Chem & Technology*, 1971).

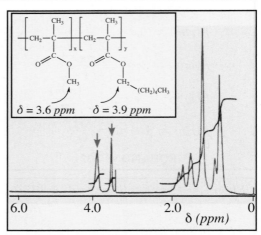

FIGURE 7-45 ¹H NMR spectrum of a methyl methacrylate/hexyl methacrylate copolymer.

easier to use for certain tasks. For example, Paxson and Randall[7] established an empirical relationship between the intensity of the 720 cm⁻¹ CH_2 rocking mode and the composition of ethylene/propylene copolymers high in propylene content by calibrating with careful NMR measurements, allowing the subsequent routine use of this band for copolymer composition measurements. But, this is getting into more detail than you need to know in this general overview. We'll simply finish up this chapter by considering how NMR can be used to measure copolymer composition and sequence distributions.

Both ¹³C and ¹H NMR are powerful tools for the analysis of copolymer composition. Consider, for example, a copolymer of methyl methacrylate and hexyl methacrylate, segments that differ only in the CH_2 units in the side chain, as shown in Figure 7-45. The rather complex collection of proton NMR lines between 0.5 and 2.5 ppm are assigned to the alkyl methylene and methyl protons of this copolymer. Here, the analytical power of NMR is evident, as two isolated NMR lines near 3.6 and 3.9 ppm, assigned to -OCH_3 and -OCH_2- groups, respectively, can be used for analysis. Because these lines are assigned to the three protons of the -OCH_3 group and the two protons of the -OCH_2- group, you

[7] J. R. Paxson and J. C. Randall, *Analytical Chemistry*, **50**, 1777 (1978).

divide the areas of the two NMR lines by 3 and 2, respectively, and the copolymer composition then follows directly from Equation 7-36.

$$\% \ MMA = \left(\frac{A_{3.6 \ ppm}/3}{A_{3.6 \ ppm}/3 + A_{3.9 \ ppm}/2} \right) \times 100$$

EQUATION 7-36

An interesting set of copolymers that we synthesized for our some of our research on polymer blends[8] are those containing styrene and 4-vinyl phenol. Because side reactions involving the phenolic hydroxyl group occur during the direct polymerization of 4-vinyl phenol (VPh), protected monomers are commonly employed. In our studies we used VPh protected with the *tert*-butyl-dimethyl-silyl group (*t*-BSOS) to prepare copolymers of styrene and *t*-BSOS. These were subsequently hydrolyzed to produce styrene-co-vinyl phenol (STVPh) copolymers, as illustrated in Figure 7-46. We then used ^{1}H NMR spectroscopy to identify the copolymers and follow the deprotection step. Figure 7-47 compares the ^{1}H NMR spectrum of a parent styrene-co-*t*-BSOS copolymer (denoted A) to that of the product after the desilylation process, STVPh (B). The NMR lines at $\delta = 0.95$ and 0.16 ppm in spectrum (A) are assigned to the methyl and *tert*-butyl substituents in the *tert*-butyldimethylsilyl monomer, and their absence in the NMR spectrum of the deprotected copolymer (B) clearly indicates the elimination of the *tert*-butyldimethylsilyl group.

The copolymer composition of the various styrene-co-*t*-BSOS copolymers synthesized was also determined from ^{1}H NMR spectra similar to that shown in Figure 7-47 (A). It is a little more complicated than the methyl methacrylate-co-hexyl methacrylate copolymers that we just discussed, but the principle is the same, and you just have to keep track of the number and type of protons that contribute to a given NMR line or group of lines.

[8] Y. Xu, P. C. Painter and M. M. Coleman, *Polymer*, **34**, 3010 (1993).

As we inferred above, the NMR line at 0.16 ppm corresponds to the six protons present in the two methyl substituents of the *t*-BSOS. Similarly, the NMR line at 0.95 ppm corresponds to the nine protons present in the three methyl substituents of the *t*-butyl group

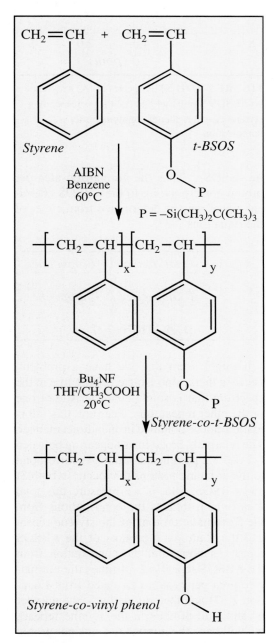

FIGURE 7-46 Synthesis of styrene/vinyl phenol copolymers.

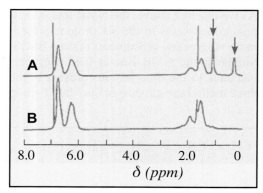

FIGURE 7-47 ^{1}H NMR spectra of (A) a styrene-co-t-BSOS copolymer and (B) the corresponding styrene-co-vinyl phenol copolymer after deprotection.

$$A_{St}^H =$$
$$\frac{(Total\ area\ from\ 6.2\ to\ 7.2) - 4A_{tBSOS}^H}{5}$$
EQUATION 7-38

Thus, the % styrene in the copolymer is simply given by Equation 7-39:

$$\% Styrene = \frac{A_{St}^H}{A_{St}^H + A_{tBSOS}^H} \times 100$$
EQUATION 7-39

in t-BSOS. Thus the normalized area per proton corresponding to the t-BSOS chemical repeat may be determined from either of the Equations 7-37:

$$A_{tBSOS}^H = \frac{Total\ area\ of\ 0.16\ ppm\ line}{6}$$
$$= \frac{Total\ area\ of\ 0.95\ ppm\ line}{9}$$
EQUATIONS 7-37

It might seem that we have a problem, because there is no isolated NMR line in the spectrum that is solely characteristic of one or the other repeat unit. Fear not, this information can be obtained in an indirect manner. The relatively broad lines appearing between 6.2 and 7.2 ppm in spectrum (A) correspond to the aromatic protons that occur in both St and t-BSOS. So, if we measure this total area (which reflects the contributions from the 5 aromatic protons of the styrene repeat and the 4 aromatic protons of the t-BSOS repeat), subtract out the contribution from the t-BSOS repeat (i.e., 4 times the normalized area per proton calculated from Equation 7-37) and then divide by 5 (the number of aromatic protons in the styrene repeat), we will have the normalized area per proton corresponding to the St chemical repeat unit. In summary (Equation 7-38):

A final example of the application of ^{13}C NMR spectroscopy is taken from our work[9] on copolymers of methacrylates and vinyl phenol which were synthesized using similar chemistry to that shown previously for the styrene-co-vinyl phenol copolymers. In Figure 7-48 are typical ^{13}C NMR spectra of an ethyl methacrylate (EMA) copolymer containing 52 mole % EMA before (top) and after (bottom) deprotection. The absence of the NMR peaks at around 0 ppm (the two methyl carbons attached to silicon), 19 ppm (the tertiary carbon of the t-butyl group), and 26 ppm (the three methyl carbons on the t-butyl group) after desilylation, clearly indicates the absence of any residual t-butyldimethylsilyl groups.

From our discussion of the statistics of copolymer sequence distributions in Chapter 6, and the raft of NMR spectra to which you have just been exposed, you might expect that the observation and assignment of sequences in the high resolution NMR spectra of copolymers would be reasonably straightforward. Indeed, if one is dealing with comonomers that do not contain asymmetric centers, such as $CH_2=CX_2$, we find the number of distinguishable sequences to be manageable. For example if the two monomers are denoted A and B, then there are three distinguishable diad sequences, AA, BB and AB ($\equiv$ BA), six distinguishable triad sequences, AAA,

[9] Y Xu, P. C. Painter, M. M. Coleman, *Polymer*, **34**, 3010 (1993).

BBB, ABA, BAB, AAB (≡ BAA) and BBA (≡ ABB), etc. In general, for sequences of length n, the number of distinguishable sequences $N(n)$ is given by:

$n =$	2	3	4	5	6
$N(n) =$	3	6	10	20	36

However, if one factors in the additional complications of stereochemistry, which occurs if one uses comonomers that contain asymmetric centers, such as $CH_2=CXY$, the problem rapidly becomes unmanageable. For example, Figure 7-49 shows a schematic representation of six distinguishable diad sequences.

In general, for the case of two $CH_2=CXY$ monomers the number of distinguishable sequences, $N(n)$, of length n is given by:

$n =$	2	3	4	5	6
$N(n) =$	6	20	72	272	1056

Now add the possibility of sequence isomerization ("backwards addition") and the resulting high resolution NMR spectrum becomes a nightmare, as the sheer number of lines, many of which are unresolved and overlap, overwhelm our ability to analyze the data. In fact, it can sometimes be an advantage to limit the amount of information by degrading resolution or recording the NMR spectra on instruments with less powerful magnetic fields.

Accordingly, we will just show one example where NMR spectroscopy has been employed successfully to study copolymer sequences. Figure 7-50 shows the 60 MHz ^{1}H NMR spectrum of a vinylidine chloride-co-isobutylene (VDC-co-IB) copolymer containing 70 mole % VDC. Also included in the figure are the spectra of pure PVDC [denoted (A)] and pure PIB (B) in the same region. Note that both VDC and IB fall under the category of monomers of the type $CH_2=CX_2$.

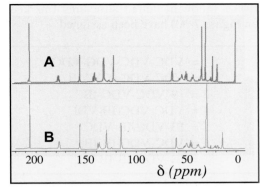

FIGURE 7-48 ^{13}C NMR spectra of (A) an ethyl methacrylate-co-*t*-BSOS copolymer and (B) the corresponding ethyl methacrylate-co-vinyl phenol copolymer after deprotection.

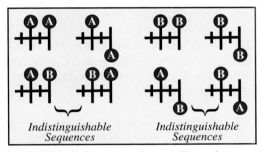

FIGURE 7-49 Schematic diagram illustrating indistinguishable diads for two different monomers of the type $CH_2=CHX$.

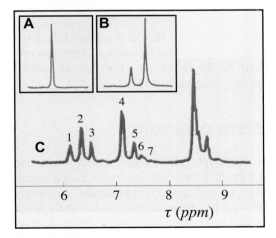

FIGURE 7-50 ^{1}H NMR spectra of (A) poly(vinylidine chloride) homopolymer (B) polyisobutylene homopolymer and (C) a vinylidine chloride-co-isobutylene copolymer [redrawn from an original figure by F. A. Bovey, *High Resolution NMR of Macromolecules,* Academic Press (1972)].

The following tetrad structures (marked on Figure 7-50) have been assigned:

```
1 = VDC-VDC-VDC-VDC
2 = VDC-VDC-VDC-IB
3 = IB-VDC-VDC-IB
4 = VDC-VDC-IB-VDC
5 = IB-VDC-IB-VDC
6 = VDC-VDC-IB-IB
7 = IB-VDC-IB-IB
```

RECOMMENDED READING

F. A. Bovey and P. A. Mirau, *NMR of Polymers*, Academic Press, New York, 1996.

D. I. Bower and W. F. Maddams, *The Vibrational Spectroscopy of Polymers*, Cambridge Solid State Science Series, Cambridge University Press, 1989.

J. L. Koenig, *Chemical Microstrucure of Polymer Chains*, Wiley, New York, 1982.

J. L. Koenig, *Spectroscopy of Polymers*, American Chemical Society, Washington, 1992.

P. C. Painter, M. M. Coleman and J. L. Koenig, *The Theory of Vibrational Spectroscopy and Its Application to Polymer Materials*, Wiley, New York, 1981.

STUDY QUESTIONS

1. Although infrared spectroscopy was one of the first tools applied to the characterization of polymer microstructure, for many tasks it has been supplanted by NMR spectroscopy. In about a page or so briefly outline why this is so.

2. Infrared spectroscopy is still widely used for identification and other structural studies. Go to the literature and write a brief essay outlining one such application.

3. Another short essay: outline some of the advantages and problems in applying ^{13}C NMR to the analysis of copolymer sequence distributions.

4. Again in a short essay, describe the physical basis for infrared and NMR spectroscopy (i.e., what energy levels in a molecule are excited, and so on).

5. Revisit the proton NMR spectrum of a methyl methacrylate-co-hexyl methacrylate copolymer shown in Figure 7-45. The areas of the lines at 3.6 and 3.9 ppm were determined to be 36.9 and 61.3, respectively. How much hexyl methacrylate was incorporated into the polymer?

6. You are given an ^{19}F NMR spectrum of poly(vinylidene fluoride) (PVDF) that is sensitive to triad information. What would you expect the normalized intensity of the most intense line in the spectrum to be if 20% of the structural units are incorporated "backwards" into the chain. (Assume Bernoullian statistics.)

7. The tacticity of poly(methyl methacrylate) (PMMA) polymers is particularly amenable to characterization by proton NMR spectroscopy.
   A. Draw and label a schematic diagram of the NMR spectrum you might expect to observe for the methylene and α-methyl protons of an atactic PMMA.
   B. Draw and label a schematic diagram of the proton NMR spectrum you might expect for the methylene and α-methyl protons for a copolymer consisting of blocks of pure syndiotactic followed by pure isotactic sequences of PMMA.
   C. Briefly, describe two methods that you could use to distinguish between a sample of atactic PMMA and a mixture of isotactic and syndiotactic PMMA.

8. NMR was used to determine triad sequences in a sample of polystyrene. It was found that $(mm) = 0.42$ and $(mr) = 0.12$. Calculate the percent isotacticity (m). The conditional probabilities $P(r/m)$ and $P(r/r)$ were also determined from this NMR data. The two values were found to be the same (within error). Is this consistent with the terminal model?

9. Methyl methacrylate was synthesized under different polymerization conditions to give two samples of PMMA (denoted I and II in Table 7-8) that were synthesized under different polymerization conditions. An NMR analysis of these samples gave the following data regarding triad and tetrad sequences.

TABLE 7-8 NMR RESULTS

	I	II
(*mm*)	0.060	0.040
(*mr*)	0.340	0.175
(*rr*)	0.600	0.785
(*rrr*)	0.455	0.710
(*mrr*)	0.280	0.155

Are these results consistent with Bernoullian or Markovian statistics?

8

Structure and Morphology

INTRODUCTION

So far in this book, we have focused on aspects of polymer synthetic chemistry and what can be considered "local structure," the arrangements of units in a chain and how these can be characterized spectroscopically. In the next few chapters our focus shifts to a more global scale and involves the physics and physical chemistry of polymer materials. We will start with the shapes or conformations available to chains in solution and the solid state, how these chains interact with one another and other molecules (e.g., solvents), and the conditions under which chains can organize and aggregate into larger scale structures, as in crystallization (or, more briefly, some of the fascinating morphologies formed by block copolymers).

Before we get to some of the details of how polymer molecules organize themselves into structures such as the spherulites shown in Figure 8-1, it is useful to review a few fundamental things. First, what are the basic states of matter, in the sense of solid/liquid/gas, found in most materials and do polymers behave the same way as smaller molecules? Second, we should review the nature of intermolecular forces between molecules, because it is the magnitude of these relative to thermal energy (kT or RT) and hence molecular motion that determines the state of a polymer at a particular temperature. Once these fundamentals have been established we will discuss structure.

FIGURE 8-1 Nylon spherulites observed under a polarized optical microscope.

STATES OF MATTER

Let's start by considering a low molecular weight material (e.g., water) at a temperature above its boiling point—in other words the forces of attraction between the molecules are not enough to overcome thermal motion and the material is a gas. As this material is cooled, the balance changes and there is a temperature at which condensation occurs, forming a liquid. This can be followed by measuring volume as a function of temperature, as illustrated schematically in Figure 8-2. However, there is a very large change in volume on going from the gas to the liquid state, and we did not have enough room to plot the volume of the gas on the graph (Figure 8-2). Upon further cooling, most materials crystallize, again with an abrupt

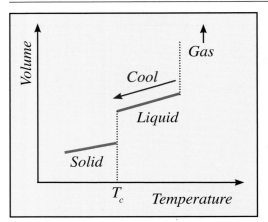

FIGURE 8-2 Schematic diagram of the volume of matter as a function of temperature

change in volume, although this change is much smaller.

However. certain materials, like window glass, have difficulty in crystallizing, sometimes because of their irregular structure and sometimes because it takes a while for the initially formed very small crystals, called nuclei, to form. If a crystallizable material is cooled at a fast enough rate, another transition is reached and this is called the *glass transition temperature* or T_g. At this temperature the material has contracted to the point that molecules become more or less frozen in position and cannot rearrange themselves to form crystals. The material still takes on the properties of a solid, but the arrangement of the molecules remains disordered, as in the liquid state. It is thought that all materials can in principle form glassy solids, but many crystallize so quickly (e.g., metals) that obtaining them as glasses takes extraordinary effort.

These states of matter are summarized in Figure 8-3. It is not meant to be exclusive, in that there are other molecular arrangements, such as those found in the liquid crystalline state, that are both interesting and of great technological importance. But that is beyond the scope of what we want to cover here. (However, you may want to read about the discovery of Kevlar® which involves liq-

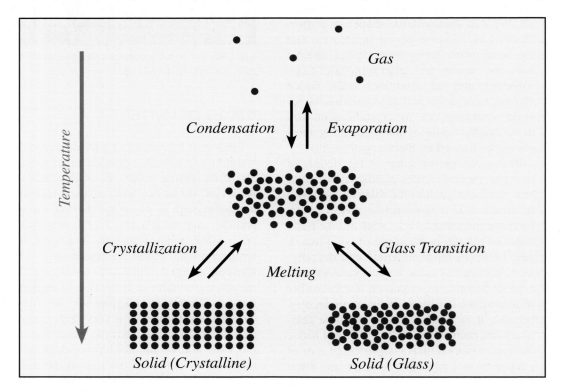

FIGURE 8-3 Schematic diagram of the states of matter.

uid crystals and is mentioned as a *Polymer Milestone* later in this chapter.)

Well then, do polymers crystallize and form glassy solids? Of course they do! But they cannot be found in the gaseous state, at least in the sense that you can heat a polymer liquid and have molecules evaporate, as illustrated schematically in Figure 8-4. Because they are so large, the forces of cohesion are such that a polymer will degrade before the melt boils. There are other differences. For example, polymers only crystallize in a partial, chain-folded fashion (more on this later).

Moreover, the polymer liquid state has certain characteristics that we normally associate with solids. This is because of chain entanglements. (Think of the difference in properties between those horrible little spaghettios that you get out of a can and a good dish of real spaghetti!) If all this wasn't enough, certain polymers (e.g., atactic polystyrene) can never crystallize, because of the constraints imposed by their microstructure. Furthermore, there are states of matter such as gels and liquid crystalline polymers that are again somewhat outside the usual categories. Before you get to the details of some of this fascinating stuff you should have a good grasp of the nature of bonding and intermolecular interactions, however, so if you've forgotten this stuff carefully read the next section; otherwise jump to conformations (page 211)!

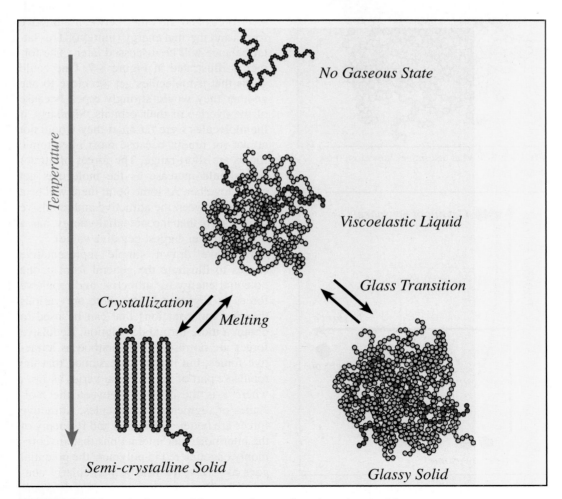

FIGURE 8-4 Schematic diagram of the states of matter for polymeric materials.

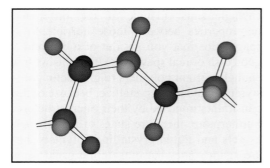

FIGURE 8-5 Model of PE chain.

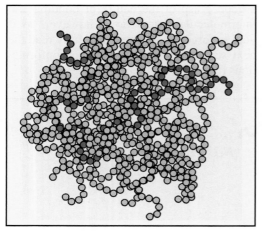

FIGURE 8-6 What determines how close these chains pack?

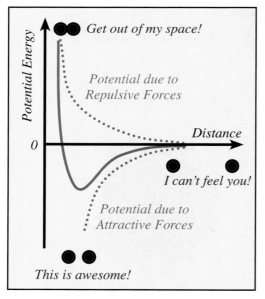

FIGURE 8-7 Schematic diagram of potential energy versus distance.

Bonding and Intermolecular Interactions

The first thing we have to do is distinguish between the chemical bonds that connect the atoms in a typical polymer, such as the covalent bonds shown in the segment of a polyethylene chain in Figure 8-5, and the physical forces that act between the chains (Figure 8-6). We will assume you did not just take Drivers Ed. and Self Esteem in high school and know what a covalent bond is. What concerns us more are the forces that provide cohesion in the solid state and the melt and it is those interactions that we will review here.

In considering the state of a material (solid, liquid or gas) we need to consider two sets of balances, one between attractive and repulsive forces and the other between attractive forces and thermal energy (motion). This latter balance will be discussed later. The former is illustrated in Figure 8-7. One could guess that if molecules get too close to one another they would strongly repel, because of the overlap of their orbitals. Similarly, if the molecules were far apart they would not attract (or repel), because most interatomic forces are short range. The forces of attraction should increase as the molecules get closer together. At some point there will be a balance between the attractive and repulsive forces such that the potential energy has a minimum (i.e., largest negative value).

We have drawn simple representative curves to illustrate the general form of the potential energy for attractive and repulsive forces in Figure 8-7, but there are various theories and equations that can be used to obtain a more formal description. Repulsive forces are not as well understood as attractive forces, but it is often assumed that the repulsive part of the potential varies as $1/r^{12}$, where r is the distance between the molecules or segments of molecules. Attractive forces are better understood and for many of the intermolecular interactions that are commonly encountered in polymers the potential goes as $-1/r^6$. We will be particularly concerned with *dispersion forces*, *dipole/dipole interactions*, *strong polar forces* and *hydro-*

gen bonds; and to a lesser extent, *coulombic interactions*.

The most frequently encountered forces that act between molecules or polymer segments are listed in Figure 8-8 in order of increasing "interaction strength." In general, interactions between non-polar molecules are considered to be "weak" relative to those between molecules or segments that are more polar. If you haven't a clue about the origin of these interactions—press on!

Dispersion interactions have their origin in fluctuating charge distributions within molecules (i.e., the shape of the electron probability clouds)—see Figure 8-9. The time average of these fluctuations in non-polar molecules is zero, but at some moment in time an instantaneous dipole can induce a corresponding dipole in a neighboring molecule or segment, so that there is a net attractive force between them. This perturbation of the electronic motion of one molecule by another can also be related to its perturbation by light as a function of energy (frequency). This in turn can be expressed in terms of the variation of refractive index with frequency, or the dispersion of light (hence, the name dispersion forces).

Polar molecules or polymers are those where the distribution of electrons in certain bonds is such that they have a permanent dipole moment. In polymer materials, this usually means there is a heteroatom such as O, N, Cl, F, etc., somewhere in the repeat unit, as in polyacrylonitrile (PAN) and poly(vinyl chloride) (PVC). There is a certain force of attraction between such dipoles that will depend upon their distance apart and relative orientation (Figure 8-10). For molecules where the dipoles are not too large, we can assume that there is a "randomizing" effect due to thermal motion, but in more strongly polar materials there is a degree of alignment of the dipoles. Furthermore, these permanent dipoles can induce dipoles in neighboring groups. Accordingly, for molecules containing polar functional groups we need to consider the sum of these effects as well as contributions from dispersion forces.

There is no simple, universally accepted

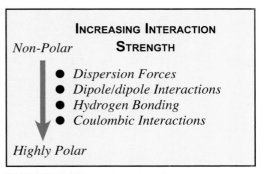

FIGURE 8-8 Interaction strengths.

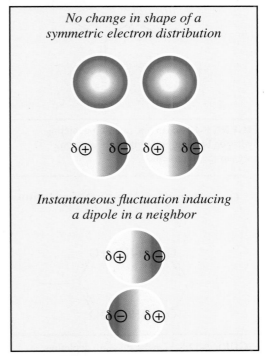

FIGURE 8-9 Schematic diagram depicting the origin of dispersion forces.

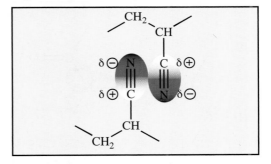

FIGURE 8-10 Schematic diagram depicting dipole/dipole interactions in segments of polyacrylonitrile chains.

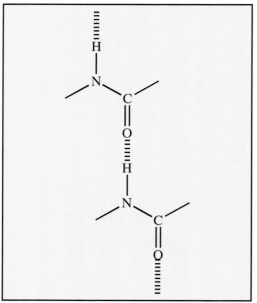

FIGURE 8-11 Schematic diagram depicting hydrogen bonding in segments of a nylon, polypeptide or protein.

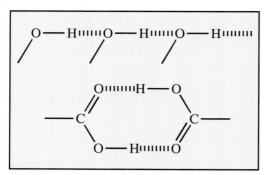

FIGURE 8-12 Schematic diagram depicting hydrogen bonding in hydroxy and carboxylic acid segments.

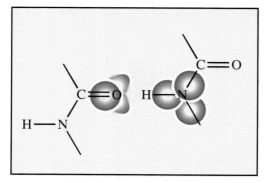

FIGURE 8-13 Schematic diagram depicting dipoles in amide groups.

definition of a hydrogen bond, but the description given by Pauling[10] comes close to capturing its essence:

> . . . *under certain circumstances an atom of hydrogen is attracted by rather strong forces to two atoms instead of only one, so that it may be considered to be acting as a bond between them. This is called a hydrogen bond.*

The N—H and C=O groups of nylon (and polypeptides and proteins) interact in this manner (Figure 8-11). Hydrogen bonds form between functional groups A—H and B such that the proton usually lies on a straight line joining A—H''''''B. The distance between the nuclei of the A and B atoms is considerably less than the sum of their van der Waals (the "hard core") radii of A and B; i.e., the formation of the hydrogen bond leads to a contraction of the A—H''''''B system. The atoms A and B are most often (but not exclusively) the most electronegative, i.e., fluorine, oxygen and nitrogen. Hydrogen bonding can occur in various patterns, the most common being chain-like arrangements (as shown for amide groups (Figure 8-11) and the hydroxyl (-OH) groups), and the cyclic structures characteristic of carboxylic acids (Figure 8-12).

The hydrogen bond is largely electrostatic in nature and charge density studies of acetamide[11] show that there is an electron deficiency at the position of the hydrogen bonding proton, resulting in a strong dipole near the end of the N—H bond of the amide group. There is an equivalent dipole at the "acceptor" oxygen atom due to the lone pair electrons. We have illustrated some of this charge density very roughly and schematically in Figure 8-13. This would suggest that dipole/dipole interactions make a considerable contribution to the overall strength of the hydrogen bond. Also, note the absence of charge density in the hydrogen bond itself

[10] L. Pauling, *The Nature of the Chemical Bond.* Third Edition. Cornell University Press, Ithaca, New York (1960).

[11] E. D. Stevens, *Acta Cryst.*, **B34**, 1864 (1978).

(i.e., between the H and O atoms in the figure). As mentioned before, the hydrogen and oxygen atoms approach closely; their separation being less than the sum of their van der Waals (hard core) radii.

Ionomers are usually copolymers where the minor component (~5%) has a functional group that can form strong ionic interactions. In ethylene-methacrylic acid copolymers, for example, the proton of the carboxylic acid can be exchanged to form a salt. In Figure 8-14 we have represented the structure as a dimer that would act to cross-link the chains. The structure of ionomers is actually far more complicated than this and the ionic domains phase separate from the non-polar parts of the chains into some form of cluster, as illustrated previously in Figure 6-4 (page 138). It is these clusters that act as cross-links, as ionic bonds are considerably stronger than hydrogen bonds (although ionic interactions can be of the order of 100 kcal/mole, the small clusters found in ionomers presumably makes the interaction strength far less than this).

Table 8-1 summarizes the approximate strengths of the major types of interactions.

CONFORMATIONS

We are going to start our discussion of

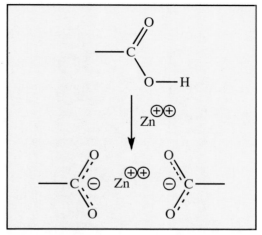

FIGURE 8-14 Schematic diagram depicting a zinc salt.

structure by considering the single chain and the types of shapes it can fold itself into. These conformations can be subdivided into two broad categories, *ordered* and *disordered*. The helical arrangements of beads shown in Figure 8-15 (the *zig-zag* shape can also be considered a *simple helix*) are obviously ordered. The so-called random coil or piece of cooked spaghetti shape shown in Figure 8-16 is disordered. As we shall see, there are a lot of such random arrangements available to a flexible chain.

TABLE 8-1 SUMMARY OF INTERACTION STRENGTHS

TYPE OF INTERACTION	CHARACTERISTICS	APPROXIMATE STRENGTH	EXAMPLES
Dispersion Forces	*Short Range Varies as* $-1/r^6$	*0.2–0.5 kcal/mole*	*Polyethylene Polystyrene*
Dipole/dipole Interactions (Freely Rotating)	*Short Range Varies as* $-1/r^6$	*0.5–2 kcal/mole*	*Polyacrylonitrile Poly(vinyl chloride)*
Strong Polar Interactions and Hydrogen Bonds	*Complex Form, but also Short Range*	*1–10 kcal/mole*	*Nylons Polyurethanes*
Coulombic Interactions (Ionomers)	*Long Range Varies as 1/r*	*10–20 kcal/mole*	*Surlyn®*

Increasing Interaction Strength

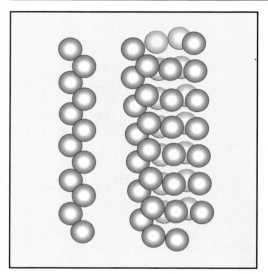

FIGURE 8-15 Schematic diagram depicting ordered chain conformations.

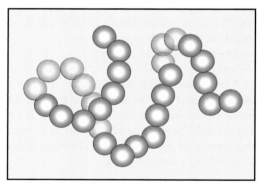

FIGURE 8-16 Schematic diagram depicting a disordered chain conformation.

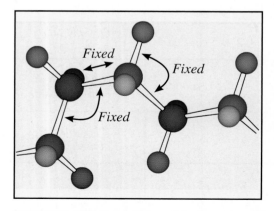

FIGURE 8-17 Ball and stick model of a section of a polyethylene chain.

It's easy to see how a piece of spaghetti or a collection of beads on a string can be folded into different shapes, but how about molecules? The covalent bond is stiff and cannot be deformed very much. Neither can certain covalent bond angles, like the CCC and CCH bonds illustrated in the section of a polyethylene chain shown in Figure 8-17. But, for reasons to do with the symmetry of the structure of their molecular orbitals, free rotations around single bonds (but not double or triple bonds) are theoretically allowed. This would make a long chain of such bonds very flexible indeed. But, there is a problem. The hydrogen atoms on adjacent carbons along the chain (or any other chemical groups attached to these atoms) can get in one another's way or, to be a bit more precise, their orbitals overlap and repel. This is perhaps not easily appreciated from looking at ball and stick type models. Space-filled models, which give a much better feel for the spatial extent of molecular orbitals, do a much better job, as illustrated for a part of a vinyl chain in Figure 8-18. So, first we need to consider how much this overlap, or steric hindrance to bond rotation, affects the flexibility, or number of conformations available to a chain as a whole.

An easy way of doing this is by looking at a very simple saturated hydrocarbon molecule (i.e., one with single bonds only) such as ethane, C_2H_6. Figure 8-19 shows two of the many possible arrangements of the hydrogen atoms relative to one another. In one arrangement, the *eclipsed* conformation, the hydrogens on adjacent carbon atoms line up and are as close as they can get to one another. In the *staggered* conformation the hydrogens line up with the gaps between those on the adjacent carbon atom and are as far apart as they can get. Which of these arrangements do you think is the most favorable, or as we should say, has the minimum (potential) energy? The correct answer is, of course, staggered! In the staggered position, the steric hindrance and hence repulsion between hydrogens on adjacent carbon atoms is at a minimum and hence so is the potential energy, as illustrated in Figure 8-20. As the

C—C bond rotates, the potential energy goes from a maximum when two specific hydrogens are aligned, to a minimum as bond rotations take them into the gap or staggered position, and then back to a maximum as the next hydrogen on the adjacent carbon comes into the eclipsed position. If you are having trouble visualizing this, we suggest that you might want to look at the animation in Chapter 7 (page 6) of our *Polymer Science and Engineering* CD.[12]

So, how does this apply to a polymer chain such as polyethylene, the simplest possible example? Here again the minimum energy conformation is found when the hydrogens are as far away from one another as possible, in a conformation we call *trans*. There is also a second shallower minimum for the *gauche* conformation. Actually, there are two gauche positions for the bond that is rotating (Figure 8-21). Again, check out the movie in Chapter 7 (page 9) of our CD to get a better view of these conformations.[12]

So why don't polymers like polyethylene (or molecules of ethane) just sit in their minimum energy conformation? In fact, in the crystalline state they usually do. Staudinger originally thought that they also would in the melt, behaving something like rigid rods rather than random coils. What we really have to consider is the height of the potential energy barrier relative to thermal energy, which we usually express in terms of kT (on a molecular basis) or RT (when we're talking about moles). If you've forgotten all this stuff and the fact that temperature is simply a measure of motion (proportional to the mean kinetic energy of the molecules in a system), it would be a good idea to read our brief review of thermodynamics and its statistical basis (Chapter 10). Essentially, at any temperature an isolated chain will have a distribution of conformations, so many bonds will be trans and many will be gauche.

Not only that, but if the chain is in a melt of other chains, or in solution (i.e., has not

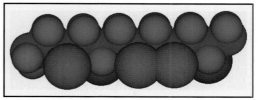

FIGURE 8-18 Space-filling model of a section of a vinyl polymer chain.

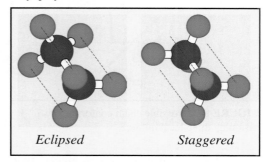

Eclipsed   *Staggered*

FIGURE 8-19 Two conformations of ethane.

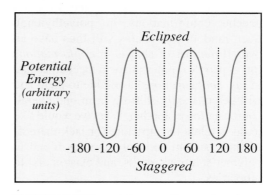

Eclipsed

Potential Energy (arbitrary units)

-180 -120 -60   0   60 120 180

Staggered

FIGURE 8-20 Potential energy diagram for ethane.

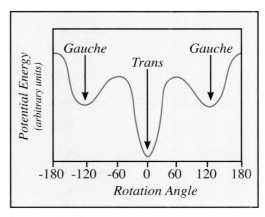

Gauche   *Trans*   *Gauche*

Potential Energy (arbitrary units)

-180 -120 -60   0   60   120   180

Rotation Angle

FIGURE 8-21 Potential energy diagram for polyethylene.

[12] P. C. Painter and M. M. Coleman, *Polymer Science and Engineering—Disk 1*, CD-ROM, DEStech Publications, Lancaster, PA (2004).

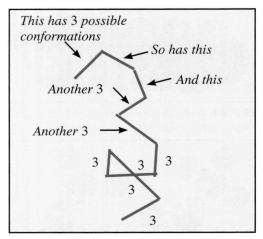

This has 3 possible conformations

So has this

And this

Another 3

Another 3

3   3   3

3

3

FIGURE 8-22 Possible chain conformations assuming three local minima.

crystallized or frozen into a glass), then the bonds are in constant motion, flipping between energy minima (like the trans and gauche conformations in polyethylene). Such random coil chains will thus have an overall number of shapes or *conformations* (or, as people doing statistical mechanics like to say, *configurations*) available to them and will be constantly shifting from one overall conformation to another. What we would like to do is relate the structure (or lack thereof) of such random coil polymers (as found in polymer solutions, melts, and elastomers) to properties.

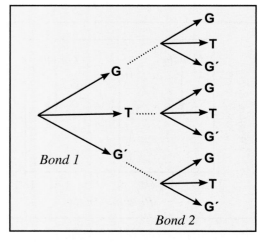

G
T
G′
G
T
G′
G
T
G′

G

T

G′

Bond 1

Bond 2

FIGURE 8-23 Pascal's triangle.

The Number of Chain Conformations

This, at first, may seem to be an impossible task. How can we relate properties to an unruly and disorganized bunch of chains that are constantly squirming about like a bunch of snakes on a rampage? Furthermore, even an approximate estimate of the number of conformations available to a chain makes the task seem so daunting that you want to go and lie down until the problem goes away. For example, let's assume that we have a chain with three local minima, like trans, gauche and the other gauche in polyethylene, but these local minima have the same energy (and each bond is therefore equally likely to be in any one of these states). Let's also assume that a bond can only be found in one of these three conformations (i.e., when it clicks from one to another it does so instantaneously). This assumption edges us towards something called the rotational isomeric states model, but that is an advanced topic that we will do our best to avoid. The first bond can therefore be found in any one of three conformations, as can the second, the third, and so on (Figure 8-22).

So, how many configurations are available to the first two bonds taken together (ignoring redundancies—more on this in a minute)? If you said 9, you've got it! You're actually constructing a form of something called Pascal's triangle, which for this problem would look something like that shown in Figure 8-23. For each of the 3 conformations of bond 1 (G, T or G′) there are 3 possible arrangements of the second bond (G, T or G′), giving a total of 9. What about a chain with 10,000 bonds? The answer would seem to be simple; $3^{10,000}$ or $10^{4,771}$. This is a bloody enormous number! But you have to be careful in doing these types of calculations. For example, it does not matter which end of the chain you start from when you begin counting. In other words, a sequence such as TTGTG′TG, etc., imposed from one end would give the same overall chain shape or conformation (or configuration) as the same sequence imposed from the other. The conformations would be indistinguishable. To take out this particular redundancy we would have to divide the above number by some factor (~2). But this does not matter for the purposes of this discussion, all we care about is that even for a chain of modest length there are an enormous number of possible conformations, even when we take redundancies into account. How are we supposed to obtain any sort of relationship between molecular and macroscopic properties from such a mess?

The Chain End-to-End Distance

It is this enormous number that saves us, because it allows a statistical approach that gives tremendous insight into the nature of these materials and their properties. (Those of you that are familiar with statistical mechanics knew this, of course!) What we need, however, is a parameter that tells us something about the shape of the chain. One that is extremely useful is the distance between the ends of the chain (Figure 8-24), which we will call *R*. The distance between the ends will be equal to the chain length if the chain is fully stretched out;

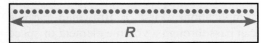

but may approach zero if the chain is squished in on itself, forming a compact ball. Intuitively, one would expect the conformations of most chains to lie somewhere between these extremes. For example, in his classic book on *The Physics of Rubber Elasticity* Treloar presented a picture of the general form of a random coil for a polyethylene chain by fixing the backbone CCC bond angles but allowing bond rotations to take on any one of six equally spaced values (60°, 120°, etc.), The value assigned to each successive bond in the chain was then chosen by throwing a dice. A more recent depiction of a chain of 200 segments, where the choice of bond rotational angles were limited to trans and gauche and the occurrence of these was weighted according to their potential energy, is shown in Figure 8-25.

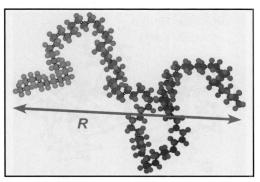

FIGURE 8-24 The end-to-end distance.

Random Walks and Random Flights

We will want to know two things about the conformations of disordered polymer chains: the average distance apart of the chain ends or, as we say, the *end-to-end distance*, and the distribution of end-to-end distances. Obtaining the average is easier for most students to understand, so we will do that first. This problem is similar to a number of others in physics that can be related to a *random walk* or *random flight*.

The classic example is Brownian motion, where the path of a (relatively) large particle

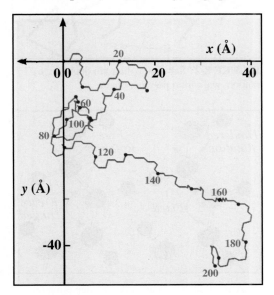

FIGURE 8-25 Conformation of a chain of 200 segments [redrawn from the figure by J. Mark and B. Erman, *Rubberlike Elasticity: A Molecular Primer.* John Wiley and Sons, New York (1988)].

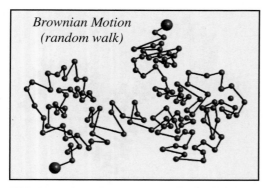

FIGURE 8-26 Schematic diagram depicting Brownian motion (redrawn from an original figure in J. Perrin, *Atoms*, English translation by D. L. Hammick, Constable and Company, London, 1916).

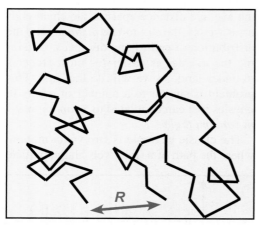

FIGURE 8-27 Schematic diagram depicting a random walk for a polymer chain.

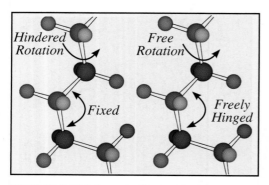

FIGURE 8-28 Schematic diagram depicting on the left a "real" chain with fixed bond angles and hindered bond rotation and on the right a freely jointed chain with unrestricted bond angles and bond rotations.

immersed in a liquid (e.g., a grain of pollen in water, as originally observed by the English botanist, Brown), is traced by noting its position at regular intervals. The particles actually take a random path between the observation points, rather than move in a straight line, a result of thermal collisions with the molecules of the solvent—see Figure 8-26. But the problem remains the same: what is the distance between the starting point (first observation) and the finishing point (last observation) after a walk of N steps? (When all this was being figured out at the beginning of the last century, it was important stuff. Einstein's theory of Brownian Motion allowed macroscopic observations of particle diffusion to be related to fundamental molecular parameters. The Frenchman, Jean Perrin won a Nobel Prize for his work in this area and they named a square after him in Paris. Your intrepid authors braved the rigors of French cuisine just to check this out.)

For a polymer chain, we will be considering steps of equal length, defined by the chemical bonds, rather than observations at given time intervals (Figure 8-27). But it is easy to see that the problem is essentially the same. Except that now we have some complications. First, a Brownian particle can pass through a volume element of space that it has previously traversed. A polymer chain is sterically excluded from an element of volume occupied by other bits of itself. Also, the particle could randomly take steps in any direction in 3-dimensional space. The "steps" taken by a chain are constrained by the nature of the covalent bond and the steric limitations placed on bond rotations.

To begin with, however, we are simply going to ignore all the problems. Instead of dealing with fixed bond angles and hindered rotation, we will assume the bonds are freely hinged and there are no barriers to rotation (Figure 8-28). We will also neglect excluded volume,[13] the fact that the chain cannot pass

[13] The term excluded volume is used here to describe so-called intramolecular excluded volume. There is also something called intermolecular excluded volume that we will mention when we consider dilute polymer solutions.

through itself. Nevertheless, we will still get the right answer, less a fudge factor, for a polymer chain in the melt and concentrated solutions! (Dilute solutions are more complicated.) Also, to begin with we are only going to consider a 1-dimensional walk. In fact, this is all we really need. You can imagine a 3-dimensional walk projected onto say the *x*-axis of a Cartesian system, then the steps will vary from 0 to *l*, the bond length, but there will be some average value, denoted <*l*>, that we can use.[14] It is then a simple matter to appropriately sum the contributions of projections in all three spatial directions (remember Pythagoras and his theorem?) to get the end-to-end distance *R*.

Let's consider the problem in what has become a politically incorrect manner. Imagine you were sampling some beer in an English pub (Figure 8-29) and you have a couple too many. (Something we never did when young—we were too busy memorizing tables of integrals and pondering Shakespeare's sonnets!) You stagger out onto the fortunately quiet road that runs east/west.

You are so inebriated that after each step you fall down and forget which way you're going, so you take random steps (of some average length <*l*>) in the east or west direction. After *N* steps you don't bother to get up any more and fall asleep. How far from the pub will you be?

Next morning, realizing that this is an important scientific question, you decide to repeat the experiment. Also, vaguely recalling things you learned in a statistics class, you decide that this process needs to be repeated a large number of times. Not wishing to totally destroy your liver, even in the service of science, you get all your friends to help and each morning you all note down where you fell asleep (after *N* steps). You

[14] This is actually calculated using the same arguments as we are going to use for the random walk!

FIGURE 8-29 This pub is actually in Savannah, Georgia, but they serve a great fish and chips!

arbitrarily decide that distances from the pub in an Easterly direction are positive, while those towards the West are negative, so you can figure out in which direction everybody ended up.

The experimental values you obtain will look something like those listed in Table 8-2. To make things easy let's just assume you all took little steps, averaging just 1 ft each, so the values of *R* are simply the number of steps (in ft). What, intuitively, do you think will be the average distance that you all end up from the pub? Yeah, even if you are not a rocket scientist, you will guess that all the distances in the plus direction will cancel with those in the minus direction and the average value of *R*, <*R*>, will be zero. If you and your friends make a graph of the number of walks that ended up a distance *R* from the pub against *R* it would look some-

TABLE 8-2 RESULTS OF A 1-DIMENSIONAL DRUNKEN WALK

WALK	R
1	+30
2	+50
3	-40
4	-50
5	+40
6	-30
ETC.	ETC.

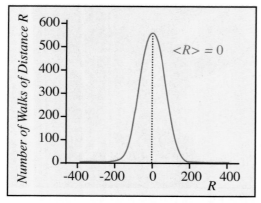

FIGURE 8-30 Average end-to-end distance.

FIGURE 8-32 A one-dimensional walk!

thing like the graph shown in Figure 8-30, if you did the experiment a very large number of times!

Another way to graph this is to plot the fraction of walks that end up a distance R from the pub (i.e., you simply divide by the total number of walks). Each of these values then also represents the probability that a walk of N steps will have an end-to-end distance $R*$. What you are actually plotting is a probability distribution, $P(R)$, and if you know some statistics you may guess that the shape of the curve will be Gaussian (for large N). (See equation on y-axis of Figure 8-31.)

You may notice that although the average distance everybody ended up from their starting point was zero, only a small fraction of the total number of walks ended back in the pub again (just as well!). Let's now say you wanted to know the distance between the

pub and where on average people ended up regardless of direction. This is the quantity we actually want to determine for our much less interesting problem of the end-to-end distance of a chain.

The first thing we do in order to work out the average distance between the chain ends, regardless of direction, is square all the values of R, thus making them all positive. Then, having calculated R^2 for all the walks, we determine the average value, $<R^2>$. Now, we want an average number, not its square, so we take the square root of this to get the root mean square (rms) end-to-end distance, $<R^2>^{0.5}$. What we would now like to know is how $<R^2>^{0.5}$ or R_{rms} varies with N. There are various ways of doing this, but we will use the simple yet ingenious method described by Feynman.[15]

Keep in mind we are just considering a 1-dimensional walk (Figure 8-32). We start with the seemingly trivial problem of calculating the average of the squares of walks of a single step, $<R_1^2>$. After one step R^2 is always 1, again assuming the length of each step is just one unit, hence $<R_1^2> = 1$.

If after $N-1$ steps have been taken in a particular walk the distance between the ends is R_{N-1}, then R_N is given by (Equations 8-1):

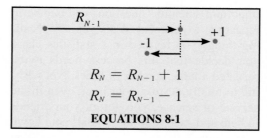

$$R_N = R_{N-1} + 1$$
$$R_N = R_{N-1} - 1$$

EQUATIONS 8-1

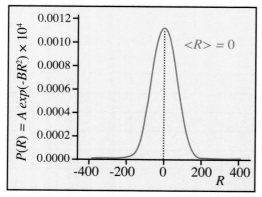

FIGURE 8-31 Average end-to-end distance in terms of $P(R)$.

[15] R. P. Feynman, R. B. Leighton and M. Sands, *The Feynman Lectures on Physics*, Addison-Wesley, 1971.

We would then expect to get as the average of a whole bunch of walks of N steps one of each of these outcomes 50% of the time. Averaging the two possible values of R_N^2 (Equations 8-2),

$$R_N^2 = R_{N-1}^2 + 2R_{N-1} + 1$$
$$R_N^2 = R_{N-1}^2 - 2R_{N-1} + 1$$

EQUATIONS 8-2

we obtain Equation 8-3. This is simply accomplished by adding them together and dividing by 2:

$$\langle R_N^2 \rangle = \langle R_{N-1}^2 \rangle + 1$$

EQUATION 8-3

It is now easy to obtain R_N by a process of successive substitution (Equations 8-4).

$$\langle R_2^2 \rangle = \langle R_1^2 \rangle + 1 = 2$$
$$\langle R_3^2 \rangle = \langle R_2^2 \rangle + 1 = 3$$
$$\vdots$$
$$\langle R_N^2 \rangle = N$$

EQUATIONS 8-4

If we now generalize by letting each step (or bond) have a length l, we obtain Equations 8-5.

$$\langle R^2 \rangle = Nl^2$$
$$or$$
$$\langle R^2 \rangle^{0.5} = N^{0.5} l$$

EQUATIONS 8-5

This is a really interesting result, because if we assume a chain of 10,000 (freely hinged and rotating) bonds each of length 1 unit, $<R^2>^{0.5} = 100$!!! If we could then grab the ends of such an "average" chain with some magic molecular tweezers and pull, it could follow the applied force by simply rearranging its conformations through bond rotations. Only when the chain becomes fully stretched out, at 100 times its original end-

to-end distance, would the bonds experience a stress. Moreover, the fully extended chain is in a less probable state (there is only one fully stretched-out conformation out of the enormous number available to the chain), so if we let the ends go it would, through thermal motion, rearrange its shape to one that is more probable. This entropic driving force is the basis for rubber elasticity. Clearly, this is a fundamental property of disordered polymer chains.

We will consider rubber elasticity in more detail later, but before you jump to that topic (if that's what you want to do next), it is important to consider probability distribution functions in a little more detail. The distribution of 1-dimensional drunken walks we obtained was shown earlier in Figure 8-31 and for large N the curve takes a Gaussian shape, by which we mean the probability distribution has the mathematical form given in Equations 8-6.

$$P(R) = A \, exp(-BR^2)$$

where for this problem:

$$A = \left[\frac{2\pi}{3}\right]^{-3/2} \langle R^2 \rangle^{-3/2}$$

$$B = \left[\frac{3}{2}\right] \langle R^2 \rangle^{-1}$$

EQUATIONS 8-6

A distribution function that describes the end-to-end distance regardless of direction can be obtained by fixing one end of the chain at the origin of a coordinate system and then finding the probability that the other end lies in an element of volume $4\pi R^2 dR$ (Figure 8-33). This is called the radial distribution function and is simply given by Equation 8-7:

$$W(R) = P(R) \times 4\pi R^2 \, dR$$

EQUATION 8-7

The two distribution functions are plotted in Figure 8-34. The second moment (don't ask) of the radial distribution function, $W(R)$,

FASCINATING POLYMERS—NATURAL RUBBER

Tapping rubber (Source: The Malaysian Rubber Board).

Just think for a moment, if there was no such thing as rubber or rubber-like materials (collectively referred to as elastomers), what would our world be like? How about going back to the days of wooden wheels? (A ride in your author's Jaguar certainly wouldn't be so fast or smooth!) How would you keep your pants up, your knickers on, your breasts from sagging? How would you make running shoes, panty hose or form-fitting swimming suits? (Your authors, sexist curs that they are, would certainly miss the latter!) How would you seal or caulk anything? Just look around you, elastomers are an essential part of modern life and it is the macromolecular nature of elastomers that is responsible for their unique properties. Your authors still have vivid and unsettling memories of certain high school teachers, who took obvious delight in teaching incredibly boring history classes, where one was expected to memorize the names and dates of long deceased kings, queens, Popes, military leaders, explorers and other scoundrels, who were deemed to have done something significant (more often than not, something quite brutal and despicable). Being cursed with a mildly sadistic streak, we thought, "Why not do the same to our readers?" Poor old Christopher Columbus, a man who has suffered his fair share of revisionist attacks in recent years, actually did do something significant! Columbus observed the indigenous people of Haiti playing with balls that were made from the gum of a tree. After his second voyage to the New World he brought back some of these little rubber balls, which unsurprisingly bounced a lot more than the stuffed leather balls then available in Europe. (There are also later reports from missionaries of people wearing crude shoes fashioned out of this elastic gum.) Natural rubber is an unsaturated hydrocarbon polymer that is found in Hevea trees and has the chemical name *cis*-1,4-poly-isoprene. It is composed of just carbon and hydrogen and has a repeating unit with the molecular formula, C_5H_8. The number of repeating units, n, in a typical natural rubber varies, but often exceeds 15,000 (i.e., a molecular weight of greater than one million g/mole). If a rubber tree is deliberately scored, or even damaged accidentally, a white latex slowly oozes out, similar to that shown in the picture (incidentally, this is not the sap of the tree). This latex, which has the consistency of heavy milk, is mainly composed of microscopic particles of rubber, a polymer that is hydrophobic (water hating) and insoluble in water. Thus, the latex is not a solution, but actually an emulsion of suspended rubber particles in water that is stabilized by natural proteins and surfactants. Milk, not a bad analogy, is also an aqueous emulsion, but in this case, the polymers are proteins and other natural polymers. Emulsion paints are similarly constituted. If the natural rubber latex is permitted to dry, or an acid (vinegar, for example) is added to the latex, a highly elastic natural rubber gum coagulates out. This is the stuff that Columbus saw the Haitians playing with.

gives the value of $<R^2>^{0.5}$, while the position of the peak is actually the most probable value, equal to $(2/3)^{0.5}<R^2>^{0.5}$. [We told you not to ask—now you realize you need to understand more about distribution functions. This would be good for you (like cod liver oil), and might even result in you understanding number average and weight average molecular weight.]

Real Chains

Real chains are not freely hinged and have steric restrictions on bond rotations, of course. But these problems can be handled.[16] Fixing the bond angle introduces a correlation between one bond direction and the next, such that $<R^2>$ is now given by Equation 8-8:

$$\langle R^2 \rangle = Nl^2 \left(\frac{1 + cos\,\theta}{1 - cos\,\theta} \right)$$

EQUATION 8-8

where θ is the CCC backbone bond angle. A similar factor accounts for the effect of hindered rotation (Equation 8-9):

$$\langle R^2 \rangle = Nl^2 \left(\frac{1 + cos\theta}{1 - cos\,\theta} \right) \left(\frac{1 + \eta}{1 - \eta} \right)$$

EQUATION 8-9

where η can take on various forms, depending on the nature of the rotational potential function. We do not want to go into the details of developing and discussing these equations, but there are two important points we wish to make. First, even sterically allowed bond rotation angles will not be independent of one another, as the equations above assume, in that certain sequences of, say, trans and gauche conformations could lead to contact or overlap of units that are otherwise separated along the chain (Figure 8-35). Nevertheless, using simplifications and approximations (as in the Rotational Isomeric States

[16] P. J. Flory, *Statistical Mechanics of Chain Molecules*, Hanser Publishers, 1969.

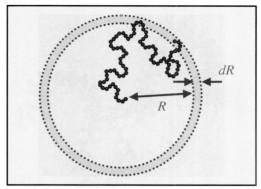

FIGURE 8-33 Schematic for the radial distribution function.

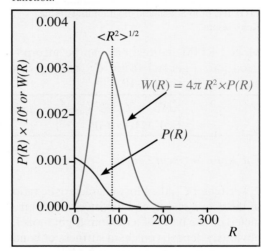

FIGURE 8-34 The radial distribution function.

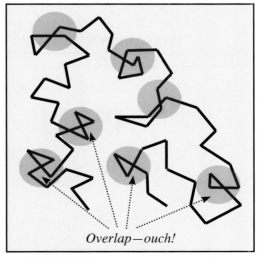

Overlap—ouch!

FIGURE 8-35 Schematic diagram illustrating the overlap of units.

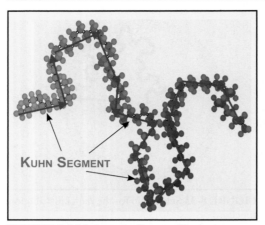

FIGURE 8-36 Schematic diagram illustrating Kuhn segments.

Model, RISM) corrections can be incorporated into a general factor C_∞, and we can now write Equation 8-10.

$$\langle R^2 \rangle = C_\infty N l^2$$
EQUATION 8-10

The Kuhn Segment Length

The factor C_∞, the Flory characteristic ratio of the actual end-to end distance to that predicted on the basis of a random flight model, obviously depends on chain stiffness or bond rotational freedom (Equation 8-11).

$$C_\infty = \frac{\langle R^2 \rangle}{N l^2}$$
EQUATION 8-11

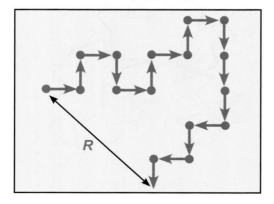

FIGURE 8-37 Schematic diagram depicting a 2-dimensional random walk.

It does not affect the exponents in the equation (i.e., the dependence of $\langle R^2 \rangle$ on N), however, but simply introduces a prefactor. This suggests a different approach, where we consider the number of adjacent bonds whose combinations of allowed rotations essentially behave like a freely jointed unit when taken collectively. We would then have N_K effective segments each of length l_K, known as the Kuhn segment length (Figure 8-36), defined in Equation 8-12:

$$N_K l_K^2 = \langle R^2 \rangle = C_\infty N l^2$$
EQUATION 8-12

Self-Avoiding Walks

There is one final element in our discussion of real chains, the fact that in the course of its random flight a chain cannot pass through itself, unlike the Brownian motion of a particle suspended in a solvent. It would then seem more appropriate to construct a self-avoiding walk model where the chain cannot pass through other bits of space already occupied by itself. This is a more difficult problem and we will only give you the results. We'll start with the 1-dimensional self-avoiding walk which is trivial, because this only goes in one direction. The end-to-end distance is then simply given by $R = Nl$. The problem of a self-avoiding walk in two or three dimensions is more difficult,[17] but in two dimensions (Figure 8-37) <R> goes as (Equation 8-13):

$$\langle R^2 \rangle^{0.5} = N^{0.75} l$$
EQUATION 8-13

While in three dimensions Equation 8-14 is appropriate:

$$\langle R^2 \rangle^{0.5} = N^{0.6} l$$
EQUATION 8-14

[17] For a very nice discussion of this problem see R. Zallen, *The Physics of Amorphous Materials*, Wiley (1983).

In general, (Equation 8-15):

$$\langle R^2 \rangle_{SAW}^{0.5} = Constant \times N^\nu$$

EQUATION 8-15

where $\nu = 3/(d + 2)$ and d is the dimensionality of space you're working in (i.e., $d = 3$ is the world as we perceive it; physicists just love this dimensionality stuff).

So now do you think $<R>$ should go as $N^{0.6}$ in the polymer solid state? It doesn't! Neutron scattering experiments show that $<R>$ actually varies as $N^{0.5}$, the random flight result. Flory saw this clearly, before the advent of such experiments. It is only in dilute solution in a good solvent that $<R>$ $\sim N^{0.6}$. To understand this properly you have to consider a statistical mechanical model where intermolecular interactions are also included. We ignored these in the preceding discussion. Crudely stated, in the solid state a polymer segment is surrounded by other segments of its own type and it does not "know" if these are from its own chain or a different chain (see Figure 8-38). It therefore gains nothing from spreading out. In a dilute solution of a good solvent, the chain segments would prefer to be next to solvent molecules. To accomplish this the chain does spread out, giving an end-to-end distance that is close to the Flory result ($<R> \sim N^{0.6}$). We will discuss this in a little more detail in the sections on polymer solutions.

In summary, from our brief look at chain conformations we have figured out a way to describe a collection of random chains and realized that among other things, a qualitative understanding of rubber elasticity immediately follows. Moreover, we hope you see how a pathway to more rigorous and quantitative work opens up.

ORDER AND MORPHOLOGY

So far we have focused on disordered polymer chains. Now we turn to the question of order and morphology. Morphology is a word that was stolen from the world of biology and botany, where it means the study

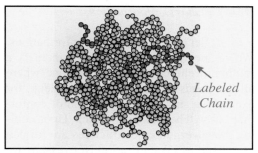

FIGURE 8-38 Schematic diagram of a labeled chain (dark red) where hydrogens have been replaced with deuterons so that they can be "seen" in neutron diffraction experiments.

of form and structure. It has a corresponding meaning when applied to polymers and has historically been taken to mean the study of order in polymers that crystallize. (Remember, not all do, as the picture shown earlier in Figure 8-4 illustrates.) However, block copolymers in which the chain conformation in each block is disordered are also capable of self-assembly to form some beautiful and fascinating morphologies (Figure 8-39). We will also briefly consider some of these structures and qualitatively discuss the factors that lead to the microstructures that are formed.

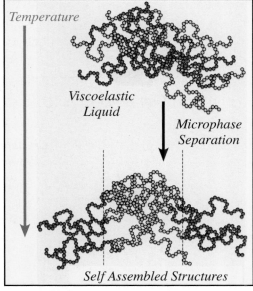

FIGURE 8-39 Schematic diagram of microphase separation.

Ordered Chain Conformations

First, let's revisit chain conformations. We've mentioned that although there is a minimum energy conformation, one where all the bonds are trans in polyethylene, for example, a statistical distribution of conformations will be found in the melt. Upon cooling, however, ordered structures are formed as a result of crystallization (for reasons we consider later). So, the initial questions we want to answer are first, what is the shape or conformation of the chains in the crystal and second, how are they arranged relative to one another?

Crystal structures are determined by X-ray diffraction—you know, Bragg's Law and all that reciprocal space stuff. If you've yet to study this technique, then all you really need to know for now is that a monochromatic (i.e., one wavelength or frequency) X-ray beam, when passed through a perfect single crystal, will give a pattern of spots which (in the old days) were recorded on a photographic plate (Figure 8-40). From this pattern of spots one can, with some effort, reconstruct the crystal structure.

It is not always easy to grow single crystals that are large enough for the experiment, but if you have a powder consisting of a bunch of small crystals, you can still use the technique. In this case all the spots from the randomly arranged different crystals now merge to give rings. The position of these rings still provides a lot of information about the crystal structure, but less than what you would get from a perfect single crystal.

The intensity of scattering measured radially from the center gives a pattern (diffractogram) similar to that shown in Figure 8-41. Large single crystals give very sharp peaks —plot (*a*), while smaller less perfect crystals result in broader diffraction rings, hence broader lines or bands—plot (*b*). Finally, at temperatures above the melting point, a very broad peak is observed—plot (*c*), sometimes accompanied by a weak secondary band. Although the sample is now disordered, there is still a close packing of molecules and hence a reasonably well-defined number of nearest neighbors, scattering from which gives this broad line.

If we now look at the diffractogram of a sample of a low molecular weight analog of

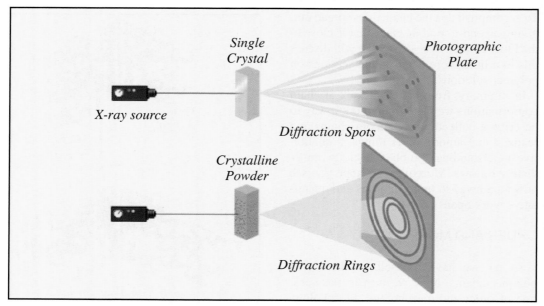

FIGURE 8-40 Schematic diagram of the scattering of X-rays by single crystals (top) and a crystalline powder (bottom).

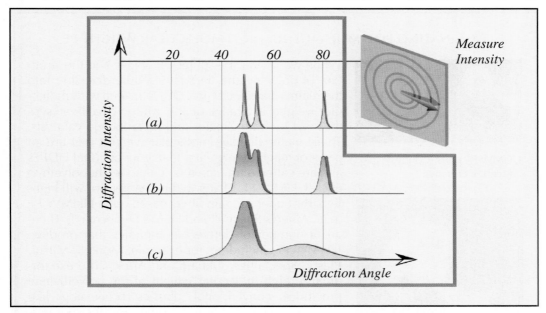

FIGURE 8-41 Schematic diagrams of X-ray diffractograms.

polyethylene, an *n*-alkane, where the chains are relatively short but all the same length (e.g., $C_{44}H_{90}$), then in the crystalline state we would see relatively sharp bands (Figure 8-42). These would disappear above the melting point and be replaced by the single broad line that is characteristic of amorphous material. In polymers such as polyethylene, both of these bands are observed at the same time at temperatures below the melting point (Figure 8-43). Also, polymers melt over a range of temperatures, rather than at a single well-defined point, a topic we will discuss more completely in a separate section.

According to the phase rule, this should not be. A single component material should be either crystalline or amorphous at a given temperature, not both at the same time. Now all crystalline materials have defects, but we're not talking about grain boundaries here. Some polymers are barely 50% crystalline. Thus "polymers had laid upon them the curse of not obeying thermodynamics," as Hoffman[18] put it. Of course, this invocation of the phase rule assumes the material can

achieve equilibrium, which for various reasons polymers do not. But now we can add to our list of questions, which have essentially become:

• What is the conformation of the chains

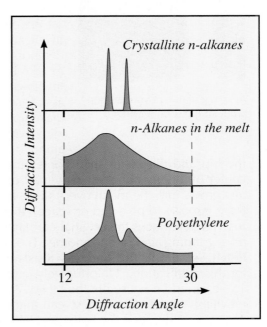

FIGURE 8-42 Schematic diagrams of various X-ray diffractograms.

[18] J. D. Hoffman, G. T. Davis, J. I. Lauritzen in *Treatise on Solid State Chemistry*, N. B. Hannay, ed., Vol. 3, Ch 7, Plenum Press, New York (1976).

FASCINATING POLYMERS—ULTRA HIGH MOLECULAR WEIGHT PE

UHMW-PE parts (Courtesy: Okulen, Germany).

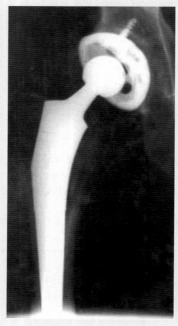

An X-ray of one of your authors' artificial hip (Courtesy: Dr. Wayne Sebastianelli).

While we are on the subject of X-rays and the structure of polyethylene, let's take a little diversion into the realm of artificial hips! Polyethylene with molecular weights in a range of 3–6 million have been synthesized by both Ziegler and metallocene catalysts. These materials have molecular weights that are an order of magnitude greater than conventional HDPEs and are called ultra high molecular weight polyethylenes (UHMW-PE). Later in this chapter, we will highlight the super strength fibers based upon UHMW-PE (e.g., Allied's Spectra® and DSM's Dyneema®). There are a number of different companies that produce UHMW-PE, including, for example, Montell, Mitsui, Plastic Specialties, Okulen and Artek. The extraordinarily high molecular weight of the polyethylene precludes conventional plastics processing like injection molding, blow molding, and thermoforming (their melt viscosities are too high). Thus parts are usually made by compression molding or sintering and machining methods akin to those used in powder metallurgy. UHMW-PE is inert, very tough and incredibly resistant to abrasion. It is self-lubricating and finds application where high wear resistance is essential. This leads us to the artificial hip and the reason why one of your authors can still beat the pants off the other at golf! In 1997, one of your authors, in an act of pure desperation—his golf game was deteriorating faster than his arthritic joints!—decided to take the plunge and have his right hip replaced. In the operation—which, let's not kid ourselves, is akin to sophisticated butchery—the ball and cup that make up the hip joint are replaced by titanium metal alloy parts. As you can see from the X-ray of your author's upper leg and hip, the cup is screwed into the hip and the metal shaft containing the ball is literally "hammered" into the center of the leg bone (cleverly, part of the shaft is porous and bone matter actually grows into the metal over time). But, in the X-ray, between the ball and cup, there appears to be a gap or space. It is here that we find a machined UHMW-PE insert that replaces the original cartilage material and prevents the two metal parts from grinding against one another. The properties of UHMW-PE, that it is inert, very tough, self-lubricating and very resistant to abrasion make it the material of choice for this application. Unfortunately, artificial hips currently have lifetimes of about 15 years, and it is usually the UHMW-PE insert that fails. Very small particles are inevitably worn off and these can initiate an immune response and inflammation. And now a question for you. As UHMW-PE is a highly crystalline material, why don't we see it in the X-ray? We'll let you figure this out!

POLYMER MILESTONES—ROY PLUNKETT

Like polyethylene, polytetrafluoroethylene (PTFE) has a very simple chemical repeat unit, $(-CF_2-CF_2-)_n$, and its discovery is a great story that in large part depended on Lady Luck, as opposed to design. Dr. Roy Plunkett, who obtained his Ph.D. from The Ohio State University, was hired by DuPont in 1936 and assigned to work on a new Freon®-type refrigerant, $CClF_2CHF_2$. In order to synthesize this compound, he needed to prepare about 100 pounds of tetrafluoroethylene gas (TFE or $CF_2=CF_2$) as an intermediate. This he did with the aid of his technician, Jack Rebok. He then stored the purified TFE in 1- to 2-pound quantities in small steel cylinders. On April 6, 1938, Jack Rebok took out one of the steel cylinders for an intended experiment and observed that there was no TFE gas emerging when he opened the valve to the cylinder. After poking the valve with a wire, the two men soon established there was indeed no TFE gas in the cylinder. The dilemma, however, was that the weight of the cylinder indicated that the gas had not escaped. They unscrewed the valve and carefully tipped the cylinder upside down. A little white powder fell out! But not enough to account for the total weight of the cylinder. Mainly out of curiosity, they decided to cut the cylinder open. (We wonder just how many chemists would have decided to go this far and not consider it simply a mistake and throw it out.) More powder was discovered packed in the bottom of the cylinder. They had discovered Teflon®. Plunkett, in his article *"The History of Polytetrafluoroethylene: Discovery and Development"* (referenced at the end of this chapter), suggests it was perhaps a bit of all three. Major discoveries usually are. However, Plunkett's first reaction, and that of his technician, was one of frustration, because it would mean that they would have to start their experiment all over again. They didn't quite realize what they had! But Plunkett was aware that the TFE had somehow polymerized. With 20/20 hindsight, given that Plunkett worked for a company that was at the forefront of polymer synthesis, he should have perhaps predicted that the polymerization of TFE was possible, if not likely. Frankly, Plunkett and Rebok were bloody lucky that they were not blown to bits! Within a few weeks of laboratory testing, PTFE was found to be inert to all the solvents, acids and bases available. It was then quickly established by DuPont's polymer chemists that TFE could be free radically polymerized in water. Plunkett's discovery is legendary and for the 50th anniversary of the discovery of Teflon®, DuPont established the Plunkett Award in his honor, for innovative applications of this unique material.

Roy Plunkett. (Courtesy: DuPont).

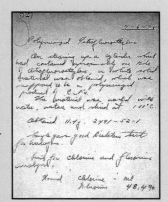

Page of Plunkett's laboratory notebook recording the discovery of PTFE (Courtesy: Dr. Peter Plimmer).

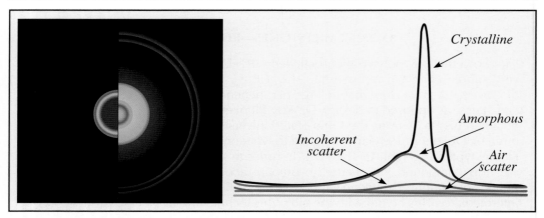

FIGURE 8-43 X-ray diffractogram of polyethylene.

in the crystalline domains and how are they stacked relative to one another?

• What is the overall shape and form of the crystals?

• What are the relative arrangements of the crystalline and amorphous parts?

X-ray diffraction data from good single crystals allows the determination of structure in the form of a so-called unit cell, a simple example of which is shown in Figure 8-44. The rest of the crystal structure is then obtained by packing identical cells along the crystallographic axes, in this case the x,y,z directions of a Cartesian co-ordinate system.

Unfortunately, apart from one or two polymers that can be polymerized in the solid state from single crystals of monomers (e.g., polydiacetylenes), polymers cannot be obtained as single crystals. Most often, patterns are obtained from oriented fibers and

structure determination depends on constructing a model and comparing the diffraction pattern predicted from the model to that observed experimentally. The model is then refined until agreement is obtained. The crystal structure of polyethylene obtained in such a fashion is shown in Figures 8-45 and 8-46.

Note that individual chains pass through many unit cells. Note also that the polyethylene chains are in their minimum energy all-trans (zig-zag) conformation and are close packed to maximize intermolecular interactions. Branches could not be accommodated in this structure and such "defects" are generally excluded from crystalline domains.

This suggests the following questions. Can a random ethylene/propylene (EP) copoly-

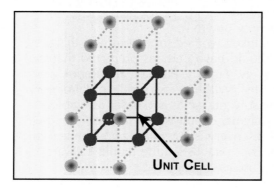

FIGURE 8-44 Schematic diagram of a unit cell.

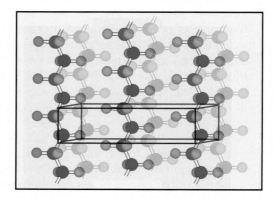

FIGURE 8-45 Side view of the unit cell of PE determined by C. W. Bunn, *Fibers from Synthetic Polymers*, R. Hill, Ed., Elsevier Publishing Co., Amsterdam (1953).

mer crystallize (one with 50% of each type of unit present)? And what about atactic polypropylene *a*-PP?—Figure 8-47. These two questions are just too easy; of course the EP copolymer and *a*-PP cannot crystallize, they are too irregular. But isotactic polypropylene does. This polymer also sits in its minimum energy conformation, in this case a 3_1 helix (three residues or chemical repeat units per turn of the helix), constructed from a trans-gauche sequence of bond conformations (Figure 8-48). In this figure, we show just the carbon atoms (for clarity) and the view from on top of the chain demonstrates how this conformation minimizes steric overlap of the pendant methyl groups (i.e., the CH_3 units sticking out from the side of the backbone). Figure 8-49 shows how the chains stack in the crystal.

Although most polymers sit in their minimum energy conformation in the crystalline state, the molecules are actually trying to minimize their total free energy, the sum of both intramolecular and intermolecular interactions. Accordingly, certain polymers that can fold into a number of different conformations of almost equal energy can exist in a variety of crystal forms. This is called *polymorphism*. For example, in nylon 6,6 the chains align so as to allow the formation of hydrogen bonds between the amide groups (Figure 8-50). Interactions are maximized and free energy minimized if these hydrogen bonds are essentially linear (i.e., the O⋮⋮⋮⋮⋮H–N atoms are aligned). However, as a result of maximizing the hydrogen bond interactions, the CH_2 groups do not sit in their preferred all-trans conformation and they do not stack as efficiently. Accordingly, there is a second crystalline form of almost equal energy where the hydrogen bonds are less linear in their alignment while the CH_2 groups are more comfortably arranged. Which form you get depends upon how you crystallize the sample.

Let's summarize what we have just learned. Polymer crystallization is incomplete (i.e., polymers are "semi-crystalline"). X-ray diffraction experiments tell us the chain conformation and the mode of packing.

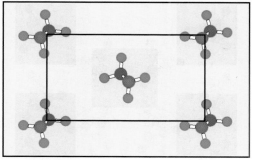

FIGURE 8-46 Top view of PE unit cell [C. W. Bunn, *Fibers from Synthetic Polymers*, R. Hill, ed., Elsevier Publishing Co., Amsterdam (1953)].

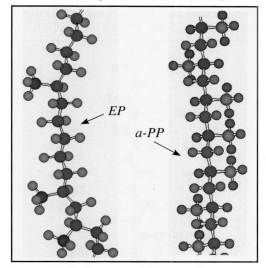

FIGURE 8-47 Ball-and-stick models of a part of a chain of an EP copolymer (left) and *a*-PP (right).

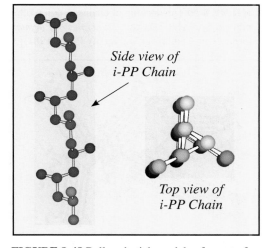

Side view of i-PP Chain

Top view of i-PP Chain

FIGURE 8-48 Ball-and-stick models of a part of a chain of an *i*-PP in a 3_1 helix conformation.

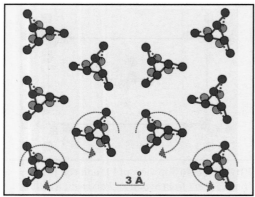

FIGURE 8-49 The packing of chains in *i*-PP [redrawn from data of G. Natta and P. Corradini, *Nuovo Cimento*. Suppl. to Vol. 15, **1**, 40 (1960)].

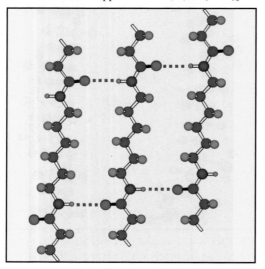

FIGURE 8-50 The packing of chains in nylon 6,6 [redrawn from D. C. Bassett, *Principles of Polymer Morphology*, Cambridge University Press (1981)].

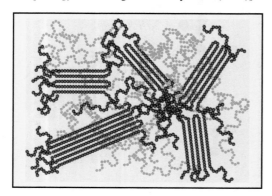

FIGURE 8-51 Schematic diagram depicting the fringed micelle model.

Chain Arrangements and Morphology

We have learned that polymer crystalline domains co-exist with (most often) extensive amounts of amorphous or unordered material. Now we wish to explore the nature of this co-existence. Do some chains lie entirely within the ordered parts while others do not participate in crystallization at all, or do chains traverse both regions of order and disorder? If you think about it a bit, the fact that polymers have a distribution of chain lengths makes the first hypothesis unlikely. Also, such a structure would have little cohesive strength and it would be easy to separate or pull apart the material. This leads us to the first really useful model of polymer structure, the so-called *fringed micelle* model (Figure 8-51). It made a lot of sense, not only in terms of explaining certain mechanical and other properties, but also in terms of the semi-crystalline nature of polymers. One could imagine crystallization starting at various points, incorporating different bits of various chains, but at some stage the remaining amorphous parts would become so entangled and "knotted-up" that crystallization would cease.

Polymer Single Crystals

Ideas about polymer morphology started to change in 1957, when three different groups separately and independently succeeded in growing so-called single crystal lamellae (they are not really single crystals in the crystallographic sense). The central idea was beautiful in its simplicity: grow polymer crystals from dilute solution thereby minimizing the effect of chain entanglements and maximizing order. The crystals grown in this fashion are too small to be seen by the naked eye, but in the electron microscope appear like flat plates, being very thin in one direction and long in the others (think of a sheet of paper), as in the micrograph of polyethylene shown in Figure 8-52. There are also some areas of spiral growth and multilayers can be seen in the bottom right-hand corner, but we will neglect these details and simply focus on the basic features of lamellar structure.

The shape and form of the crystals, or their habits depend on a lot of things, including the crystallography of chain packing, the concentration of the solution from which the crystals are grown and, most crucially, the temperature of crystallization and the nature of the solvent. These two are related variables, in that the crucial parameter is actually the degree of undercooling—the difference in temperature between the dissolution or melting temperature (which varies with solvent) and the crystallization temperature. The basic habit of polyethylene single crystals is a lozenge-shaped crystal, which in its original form is more like a hollow pyramid that collapses upon drying. (Hence the "gash" in the crystals shown in Figure 8-53.)

At higher crystallization temperatures (>80°C for crystals grown from xylene) the lozenges tend to become truncated at their ends, while at lower temperatures something called dendritic growth is observed (Figure 8-54). Many books have been written about polymer morphology alone and it is not our purpose to cover the details of this subject. Instead, we will focus on a crucial general feature. Electron diffraction experiments show that the chains are oriented perpendicular to the (apparently) flat surface of the crystal. The crystals are only about 100–150 angstroms thick, whereas a typical chain may be more than 500 angstroms long, leading to

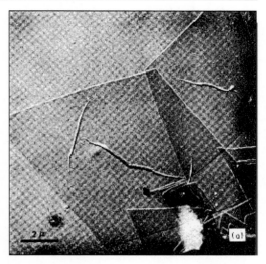

FIGURE 8-53 Single crystals of PE [reproduced with permission from D. Renicker and P. H. Geil, *J. Appl. Phys.*, **31**, 1916 (1960)].

the suggestion that the chains must be folded back and forth in some manner.

Chain Folding

It was initially thought that chain folding in polymers like polyethylene would take place in some sort of regular, ordered fashion, as observed in certain cyclic paraffins (Figure 8-55), with some specific sequence

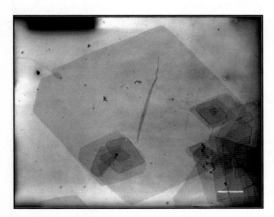

FIGURE 8-52 Single crystals of PE (Courtesy: Ian Harrison, Penn State University).

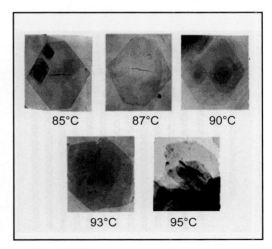

FIGURE 8-54 Single crystals of PE grown at different temperatures (Courtesy: Prof. Ian Harrison, Penn State University).

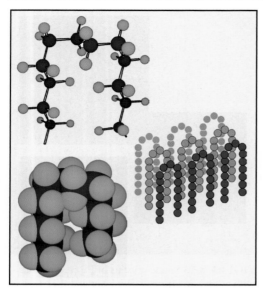

FIGURE 8-55 Three different representations of chain folding.

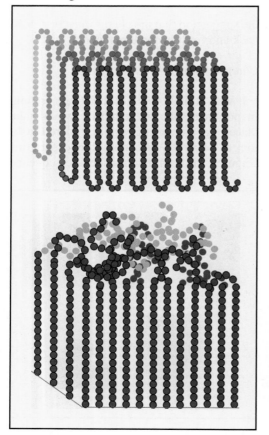

FIGURE 8-56 Chain folding models: (top) adjacent reentry and (bottom) switchboard.

of (perhaps somewhat distorted) gauche and trans bond conformations providing regular adjacent reentry of the chains into the crystal lamellae (top of Figure 8-56). This idea was anathema to Flory, who considered that the fold surface should be largely disordered, with chains entering and leaving the crystal at random. This was called the switchboard model (bottom of Figure 8-56), and is a great analogy, at least if you remember what an old telephone switchboard looks like (Figure 8-57).

The controversy concerning the relative merits (or lack thereof) of these two models occasionally reached polemical proportions and culminated in a famous meeting of the Faraday Society in 1979. The arguments essentially revolved around the requirements of steric packing at the interface and the interpretation of neutron scattering data. From all reports, Flory felt he had been "ambushed" at this meeting and returned home seriously displeased. But it is an ill-wind that blows nobody any good, and Flory proceeded to develop a statistical mechanical model of crystal surfaces and interphases that (among other things) allows the calculation of the number adjacent re-entry sites in single crystals, which Flory initially determined to be less than 40%. The precise value depends upon a number of variables (e.g., chain flex-

FIGURE 8-57 A classic old telephone switchboard. (Believe it or not, these were still being used when your authors were young lads!)

ibility), however, and has subsequently been revised upwards to values as high as 80% for very flexible chains. It seems safe to conclude that for single crystal lamellae, most chains re-enter within about three lattice sites of the surface, which for polyethylene would correspond to a surface region of about 15Å, in good agreement with experiment.

Crystals grown from the melt are less ordered, which brings us to the subject of melt crystallization.

Melt Crystallization

The most striking feature of crystallization from the melt is the appearance of spherulites. These can be observed under crossed polarizers[19] in the optical microscope during crystallization and are in a sense transitory, in that they eventually impinge on one another to form straight or hyperbolic boundaries. Figure 8-58 shows a picture (snap-shot) of growing spherulites that was taken from a movie that is on our CD—you may want to play this movie as it is fascinating to see how spherulites grow and impinge upon one another.[12] The bright contrast is due to something called birefringence and is characteristic of a crystalline entity. The "Maltese cross" pattern indicates the presence of radial order.

The initial development of spherulites depends on how they are nucleated. (We will consider things like homogeneous and heterogeneous nucleation in the chapter on crystallization.) Some appear to develop from single crystal lamellar-type structures, while electron micrographs of others show a sheaf-like stage that begins with a fiber-like structure, as illustrated in Figure 8-59. These can also be seen using atomic forces microscopy, as shown in Figure 8-60.

[19]The incident light is polarized in one direction and after passing through a thin film of the sample is viewed through a second polarizer placed at 90° to the first. In this position no light is transmitted unless there is some interaction with the sample, which in turn usually means there must be some degree of order.

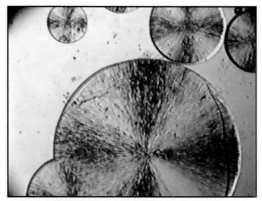

FIGURE 8-58 Growing spherulites (Courtesy: Prof. Jim Runt, Penn State University).

As with single crystal lamellae, there are a range of complex morphologies that can be observed, but we wish to focus on just a few key features of the ordinary common or garden spherulites. Figure 8-58 showed an actual micrograph of spherulites in the pro-

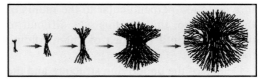

FIGURE 8-59 The initial development of spherulites (redrawn from D. C. Bassett, *Principles of Polymer Morphology*, Cambridge University Press, 1981).

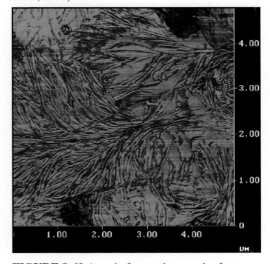

FIGURE 8-60 Atomic force micrograph of spherulites (Courtesy: Prof. Jim Runt, Penn State University).

FIGURE 8-61 Schematic diagram of the final stage of the growth of spherulites.

cess of growing, while the schematic figure (Figure 8-61) illustrates how they appear once the spherulites impinge on one another. (Spherulites aren't blue; one of the guys working with us indulged in some artistic license!) Microscopy and X-ray diffraction studies indicated that the chains are oriented in a direction perpendicular to the radius of the spherulite. Initially this was difficult to understand, until the idea of chain folding took hold.

Details of the underlying structure and the relative arrangement of the crystalline and amorphous parts become more apparent if we

examine the results of an ingenious experiment by Keith and Padden, who crystallized a 10% mixture of isotactic polypropylene in 90% atactic polypropylene. The amorphous atactic polymer was then removed by washing with solvent (which does not dissolve the crystalline component at ordinary temperatures), revealing a pattern of branched lamellar-like structures (Figure 8-62). The lamellar arms of the spherulite consist of chain-folded material, but this folding is probably less regular than that found in solution-grown single crystals. Furthermore, there are so-called tie-molecules, chains that emerge from one lamellar arm, coil around in the amorphous inter-lamellar region for a while, and then enter another lamellar arm, as in the old fringed micelle model (see Figure 8-63).

Size and Properties

Although we have shown a number of micrographs, it is beyond the scope of this treatment to discuss the details of experimental methodology and the tremendous art and craft that sometimes goes into sample preparation. Nevertheless, before finishing this section with a description of the morphology of fibers and block copolymers, it is useful to summarize the relative size of

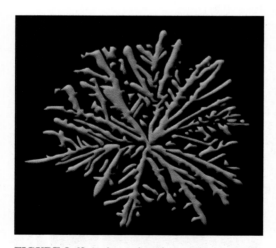

FIGURE 8-62 An isotactic polypropylene spherulite grown from a mixture of 10% isotactic polypropylene and 90% atactic polypropylene [redrawn from a micrograph of H. D. Keith and F. J. Padden, *J. Appl. Phys.*, **35**, 1270 (1964)].

FIGURE 8-63 Schematic diagram illustrating chain folding in a spherulite.

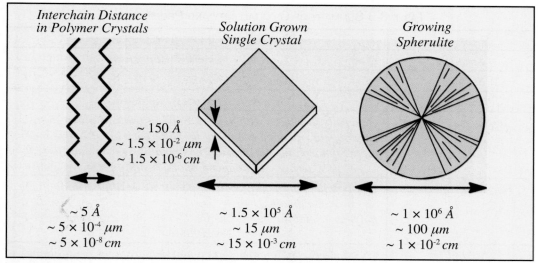

FIGURE 8-64 Summary of the relative size of different structures.

the structures we have discussed and make some brief comments on the relationship of morphology to properties, to give you a feel for the importance of the subject. (See Figure 8-64, which also serves to remind you of the relationships between the units we have used!)

Crystallization and Properties

We discuss the details of things like mechanical properties in a separate chapter, but one might intuitively expect strength and stiffness to increase with crystallinity, if for no other reason than the material is more dense and there are therefore more chains per unit volume. Furthermore, intermolecular interactions are optimized in the crystalline state and this should also give rise to a greater cohesive strength. The relationship is a lot more subtle than that, however. For example, a mat of agglomerated solution-grown single crystals would not have much strength, because there is nothing tying the lamella together. The generally tough nature of certain polymers and their ability to undergo plastic deformation depends upon the presence of *tie molecules*. Also, because density increases during crystallization (the chains are better packed), the sample shrinks

towards each nucleus, sometimes leading to weak points and even voids at the spherulite boundaries, if they are too large. Accordingly, it is often better to nucleate a large number of spherulites (that are then on average smaller) when processing plastics.

Properties will also depend upon the nature of the amorphous domains, whether they are glassy or rubbery, and so on. So, structure/property relationships is a big, complex subject, but certain "rules of thumb" can be summarized, as shown in Table 8-3.

Fibers

Because of chain folding, melt-crystallized polymers are not as strong as they could be. One can envisage that under a load a sample will at some point "yield," with chains in the amorphous domains becoming oriented in the draw direction while the lamellar arms of the spherulite undergo shear and whole sections are pulled out. This process is illustrated in Figure 8-65.

If, instead of the imperfect alignment you get from this type of drawing process, we could align the chains in a fully stretched out very regular manner, then we might imagine that the fiber so formed would be very strong, with the strong covalent bonds taking

TABLE 8-3 SUMMARY OF CRYSTALLINITY AND PROPERTIES

PROPERTY	CHANGE WITH INCREASING DENSITY
Strength	Generally increases with density
Stiffness	Generally increases with density
Toughness	Generally decreases with density
Optical Clarity	Generally increases with density. Semi-crystalline polymers usually appear opaque because of the difference in refractive index of the amorphous and crystalline domains, which leads to scattering. Will depend upon crystallite size.
Barrier Properties	Small molecules usually cannot penetrate or diffuse through the crystalline domains, hence "barrier properties," which make a polymer useful for things like food wrap, increase with degree of crystallinity.
Solubility	Similarly, small molecules cannot penetrate the crystalline domains, which must be melted before the polymer will dissolve. Solvent resistance increases with degree of crystallinity.

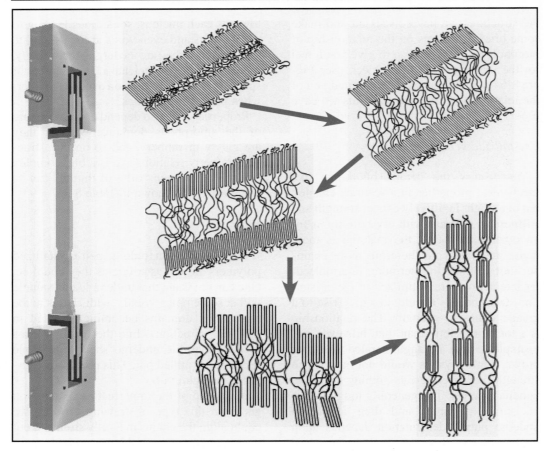

FIGURE 8-65 Expanded diagrams of the processes occurring when a semi-crystalline polymer is deformed under load.

the load, as opposed to the situation found in spherulitic samples, where weaker inter-chain forces, tie molecules and the physical entanglements found in the amorphous domains are crucial. Actually, the inter-chain forces in a fully stretched out collection of chains are equally important in preventing the chains slipping past one another under a load. However, in high molecular weight samples a chain is much longer than the fold period found in melt-crystallized material, so that their cumulative strength over the length of a chain is greater than the strength of an individual covalent bond, which then becomes the limiting factor in defining the strength of these materials.

We try to produce this fully stretched out state when we make commercial fibers, but by and large we do so imperfectly. To appreciate this you should take the plastic (polyethylene) thingies (we have no idea what they're really called) that hold together cans of soda (or pop, as we used to call them) and pull slowly. A neck will develop and the sample can be "cold drawn" to many times its original length. The chains in this drawn region will be only partly stretched out and oriented and the morphology is not well-defined, however. Better, but still imperfect results can be obtain by melt-drawing samples, the way many textile fibers are processed, or by spinning and drawing from solution. The most perfectly aligned samples are obtained using recent innovations in processing technology, where polymers are drawn from a gel, for example (see the *Fascinating Polymer* aside on the next page), or by synthesizing really rigid chains that cannot bend much to begin with (e.g., Kevlar® — the subject of a *Polymer Milestone* presented below), although these can be difficult and expensive to process.

Block Copolymers

The final topic we wish to consider in this section is block copolymers. These are fascinating materials at a number of levels. First, they are technologically important. For example, triblock copolymers of styrene and butadiene are used to make thermoplas-

tic elastomers, or impact resistant glassy polymers, depending on the length, arrangement and relative proportions of the blocks. Polyurethanes are an example of what are called segmented block copolymers and can be rubbery or rigid materials depending on the length of the hard or soft segments. (We discuss these in a little more detail in the chapters on "Synthetic Rubber" and "Polyurethanes" in our other *The Incredible World of Polymers* CD.) Here we will just consider some basic aspects of the morphogy of these materials.

The fascinating thing about block copolymers is that if you can take two monomers, say styrene and isoprene, and synthesize the corresponding homopolymers, (atactic) polystyrene and polyisoprene (containing various structural isomers), you find that they have no order, they are amorphous materials. However, if you instead synthesize a diblock copolymer out of the same monomers, you find that the resulting chains can "*self-assemble*" to form all sorts of fascinating microstructures, depending on total chain length and the relative lengths of the two blocks (assuming the chains and blocks have a narrow molecular weight distribution). For example, electron micrographs of two styrene/isoprene diblock copolymers, obtained by Richard Register of Princeton University, are shown in Figure 8-66.

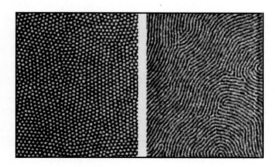

FIGURE 8-66 Electron micrographs of styrene (S)/isoprene (I) block copolymers: molecular weight of the blocks (S/I) are 68,000/12,000 g/mol (left) and 30,000/11,000 g/mol (right) [images courtesy of C. K. Harrison, P. M. Chaikin, and R. A. Register, Princeton University].

FASCINATING POLYMERS—SUPER HIGH-STRENGTH PE FIBERS

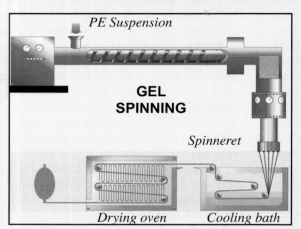

Schematic diagram depicting the gel spinning of polyethylene.

Theory tells us that the more tightly packed the polymer chains, the higher the strength and modulus. It follows, therefore, that we should be able to make super-high-strength polyethylene (PE) fibers that are stronger and stiffer than Kevlar®. It was the development of gel spinning ultra-high molecular weight PE that led to the super-high-strength PE fibers that are known commercially as Spectra® and Dyneema®. Why ultra-high molecular weight? It is now well-established that the strength (tenacity) of PE fibers increases with molecular weight or polymer chain length. In the molten state, the chains of PE resemble a bowl of spaghetti, and the PE chain is said to be in a random coil configuration. If you were to stretch the polymer chain to its maximum length before failure (breaking the chain), it would assume a planar zig-zag configuration (dictated by the angles between adjacent carbon atoms). Maximum fiber strength will be attained if you can now arrange for these extended planar zig-zag chains to pack in a crystal. Tenacity increases with chain length, because the greater the chain length (molecular weight), the smaller the number of packing defects per unit volume in the fiber. In a conventionally drawn fiber there are blocks of lamellae linked by tie molecules but with chain fold "defects." In a highly aligned fiber, like that of Kevlar®, there are substantially less defects. But, what if we could approach the level of perfection that would be close to a "continuous crystal." Gel spinning comes close to achieving this goal. The major problem with polyethylene (PE), and all conventional fibers for that matter, is that even though the polymer chains are aligned and extended during extrusion through a spinneret, the polymer inevitably relaxes and loses some alignment before it is quenched. Thinking about this dilemma, a Dutch scientist, A. J. Pennings, came up with an interesting idea in the early 1970s. He spun-oriented fibers from a flowing dilute solution of PE. Pennings was able to avoid chain folding and grow oriented fibrillar crystal fibers that had tenacities of 44 gpd! Building on this idea, scientists at the Dutch State Mines (DSM) in Holland developed a gel spinning method for the commercial production of super-high-strength PE fibers. A solution of an ultra-high molecular weight PE (about 5×10^6 g/mol) was prepared in a mineral oil solvent at elevated temperatures using a mixing extruder. A gel-like structure is formed and this is the key to the successful formation of super-high-strength material. It has fewer entanglements, but enough to maintain a gel. The gel-like fibers that emerge from the spinneret are then drawn some 300 times their original length. The resulting fibers are very highly oriented (>95%) and are highly crystalline (about 85%). This, in a nutshell, is how DSM's Dyneema® and Honeywell's Spectra® fibers are produced.

POLYMER MILESTONES—STEPHANIE KWOLEK

Here's a great story involving high strength fibers. In the 1960s, DuPont, and similar large chemical companies, were staffed overwhelmingly with male scientists and engineers. Female scientists and engineers were rare, and to gain acceptance and respect they had to overcome the formidable prejudices of many of their male counterparts. Stephanie Kwolek delicately stepped into this environment, she was a woman slight in physical stature, but not intellect. Her seminal contributions to the quest of super-high-strength fibers was to lead to the discovery of Kevlar®. Kwolek was an outstanding experimentalist—she repeatedly would have to show others (usually men!) how to do things after they had failed to duplicate her results—and she had the tenacity (no pun intended!) of a bulldog. In 1964, Stephanie Kwolek was assigned the task of preparing fibers from completely para-substituted aromatic polyamides. She succeeded in producing

DuPont advert for Kevlar® featuring Stephanie Kwolek (Source: Hagley Museum).

the pure amino acid and polymerized it in tetramethylene urea to make a polymer designated 1,4B, which is analogous to nylon 6. Finding a solvent for 1,4B was a problem. Frustrated, she turned to concentrated sulfuric acid, made a 5% solution of 1,4B, and using a syringe, produced a fiber by squirting the solution into a water bath. After some crude stretching and annealing at 500°C (this is bloody hot!), she produced a fiber with reasonable tenacity (2.5 gpd) and an extraordinarily high modulus (400 gpd × 100). A definite reason for celebration! But, concentrated sulfuric acid is miserable stuff and hardly a benign solvent. Stephanie Kwolek turned her attention to finding a more acceptable solvent. She found that after temperature cycling with a combination of tetramethylurea and LiCl, a solution of sorts could be obtained; but it was turbid. This suggested that there were fine particles present. So she filtered the turbid "solution" through a very fine filter, expecting to obtain a clear true solution. But, try as she might, the resultant liquor was turbid. Moreover, the "solution" was opalescent when stirred and had a textured appearance. It also was not as viscous as might be expected and, in fact, an experienced technician refused to try to make fibers from the "solution" because it had bits floating in it (which would clog the spinneret) and it flowed like water. Not easily discouraged, Kwolek persuaded the technician to give it a go. The spinning worked like a charm! After heat treatment at 500°C the fibers had tenacities of 7 gpd and moduli of 900 (gpd × 100). This was four times the modulus of glass fibers—something that had never before been achieved with an organic fiber! Kwolek was intrigued with the strange properties of her 1,4B solution in tetramethylurea and what she had succeeded in making was a liquid crystalline solution of 1,4B, from which she produced her high modulus fiber. The rest, as they say, is history. Stephanie won many honors, including one for Creative Invention from the American Chemical Society. She was inducted into the National Inventors Hall of Fame in 1995 and she is in the select group of scientists that are in DuPont's Lavoisier Academy.

The micrograph on the left consists of a polygrain body-centered cubic structure (see Figure 8-66). In other words, if you look carefully you can see different regions of order that meet at so-called grain boundaries, just like in the crystal stuctures of things like metals. The micrograph on the right of Figure 8-66 consists of cylinders lying in a plane.

Because the blocks in these (and other) diblock copolymers have a uniform size, they can arrange themselves in a regular manner, the form of which depends upon the relative lengths of the blocks. If the molecular weight or block length of one of the components of the diblock, say A, is much shorter than the other, B, then you get spheres of A forming a body-centered cubic lattice in a matrix of B [Figure 8-67(*i*)]. At larger A block sizes (but with the A block length still shorter than B) you get a cylinders of A arranged in a hexagonal lattice, again in a matrix of B blocks [Figure 8-68(*ii*)]. When the blocks are of approximately equal size layered structures or lamellae are formed [Figure 8-67(*iii*)].

These are the principal morphologies found in diblock copolymers, but under certain conditions other periodic structures are formed. One such morphology, where both phases are continuous (connected to all other phases of the same type), is illustrated in Figure 8-67(*iv*). It was originally called an ordered bicontinuous double diamond structure, but has more recently been related to minimal surface structures known as gyroids.

As you might guess, if you go on to consider things like triblock copolymers, the number of possible structures becomes very large, but we won't go there. Instead, we will conclude this section by briefly considering why block copolymers self-organize in the first place.

As you will see when you get to Chapter 11 and study polymer solutions and blends, most homopolymers do not want to mix with one another, they phase-separate. This is because unlike low molecular weight materials their entropy of mixing is very small, although positive and still favorable to mixing. But, if the interactions between the components are just governed by dispersion or weak polar forces, then the components "like" molecules of their own type more than those that are different, or, more accurately, they interact more strongly with similar molecules. As a result, the energy change upon mixing is positive or unfavorable to forming a single phase. The interaction term in the free energy then usually overwhelms the small entropy term and phase separation into large domains of essentially pure homopolymers occurs. However, in block copolymers this phase separation can only go so far, because chains of different chemical structure are covalently linked together, as in the triblock copolymer illustrated previously in Figure 8-39.

As we will see in Chapter 11, it is usual to describe interactions in polymers using the Flory interaction parameter χ, which varies as the inverse of the temperature. Thus if χ is relatively small it is possible to form a single phase at temperatures below the degradation point of the polymer. Then, as you cool and χ gets larger, the system phase-separates. The temperature or value of χ at which this occurs, the order-disorder transition, varies with composition and it is possible to

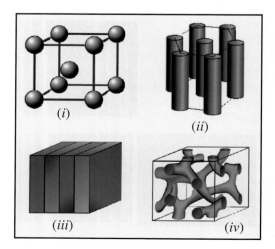

FIGURE 8-67 Schematic diagrams of the principal morphologies occurring in block copolymers. (*i*) Spheres in a body centered cube, (*ii*) cylinders, (*iii*) lamellea and (*iv*) bicontinuous structures.

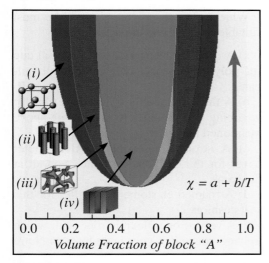

$\chi = a + b/T$

Volume Fraction of block "A"

FIGURE 8-68 Schematic diagrams of the principal morphologies occurring in block copolymers. (*i*) spheres in a body centered cube, (*ii*) cylinders, (*iii*) lamellea and (*iv*) bicontinuous structures.

construct a phase diagram, such as the one displayed schematically in Figure 8-68 for a diblock copolymer (real phase diagrams are more complex and not as symmetrical, because the size of the monomer units in each block are not the same, amongst other things). But, what's with all the colors, you might ask? These designate regions where different morphologies are found (note that at low volume fractions of A blocks you will get spheres of A embedded in a matrix of B, but at high volume fractions it will be the other way round). These morphologies are a result of a competition between various entropic and enthalpic factors. At large values of χ (something called the strong segregation limit), the blocks would like to minimize their area of contact, but there is an entropic price to pay for this in terms of chain extension and other factors. But that's an advanced topic that we won't cover.

RECOMMENDED READING

D. C. Bassett, *Principles of Polymer Morphology*, Cambridge University Press, Cambridge, 1981.

P. G. de Gennes, *Scaling Concepts in Polymer Physics*, Cornell University Press, Ithaca, New York, 1979.

P. J. Flory, *Principles of Polymer Chemistry*, Cornell University Press, Ithaca, New York, 1953.

R. J. Plunkett, *The History of Polytetrafluoroethylene: Discovery and Development,* in: *High Performance Polymers: Their Origin and Development,* Proceed. Symp. Hist. High Perf. Polymers at the ACS Meeting in New York, April 1986, (R. B. Seymour and G. S. Kirshenbaum, eds.), Elsevier, New York, 1987.

L. H. Sperling, *Physical Polymer Science*, 3rd Edition, Wiley, New York, 2001.

G. Strobl, *The Physics of Polymers*, Springer-Verlag, Berlin, 1996.

STUDY QUESTIONS

1. Discuss briefly the effect of increasing the degree of crystallinity on the following properties (include a short explanation).

   a)   Density
   b)   Modulus (stiffness)
   c)   Transparency

2. What is the relationship of the root mean square end-to-distance of a polymer chain in the melt to the number of segments in that chain? Include in your answer the following points:

   A. The relationship obtained for a chain with freely hinged and rotating bonds.
   B. A brief discussion of how this relationship is modified for real chains.
   C. The Kuhn segment length.
   D. The difference between a random walk and a self-avoiding walk and how this pertains to polymer chain conformations.

3. Describe the mathematical form of the Gaussian distribution and briefly discuss how this applies to polymer chains. How does the position of the peak on this distribution relate to the root mean square end-to-end distance?

4. Two copolymers of ethylene ($CH_2=CH_2$) and propylene ($CH_2=CHCH_3$) have the same ratio of monomers. However, one is rubbery at room temperature and does not stiffen until the temperature is lowered to about $-70°C$, while the other is stiffer, but still flexible, tough, and opaque at room temperature. Explain the difference.

5. Depending on temperature and thermal history, polyethylene terephthalate (PET), can exist in a number of different states (semi-crystalline with the amorphous regions glassy, semi-crystalline with the amorphous regions rubbery, amorphous glass, amorphous melt). Consider a sample of PET cooled rapidly from 300°C (state 1) to room temperature. The resulting material is rigid and perfectly transparent (state 2). The sample is then heated to 100°C and maintained at that temperature, during which time it gradually becomes translucent (state 3). It is then cooled down to room temperature and is again found to be rigid, and still translucent (state 4). For this polymer, $T_m = 267°C$, and $T_g = 69°C$.

A. Identify each of the states (as described in parentheses above).
B. In state 4 PET has a higher modulus and is less flexible than in state 3. Why?
C. In which state would you use PET for

(i) plastic pop bottles
(ii) engineering thermoplastic reinforced with glass fibers for making a car body

6. Consider the following polymers:

> Isotactic polypropylene
> Atactic polystyrene
> Low density polyethylene
> High density polyethylene
> A "soft" polyurethane
> Nylon 6

Which of these polymers would be most suitable in the following applications:

a) A cheap, transparent, stiff disposal cup for polymer science, college football tailgates.
b) A thin, reasonably transparent film for wrapping your hoagies to take to the above mentioned tailgate.
c) A strong, reasonably flexible, outer cover for the cushions on which you sit at the game.
d) A foamed elastomer as a filler for this seat cushion.
e) A very strong, flexible rope with which to hang yourself if your team loses to Notre Dame.

Note: Pick one polymer for each application. Give a brief one-sentence reason for your choice. There may be more than one correct answer for certain applications. [Hint: first ask yourself if the polymer crystallizes or not. If not, drawing on your general knowledge of the world around you, ask yourself if the polymer is glassy (relatively rigid and stiff) or rubbery at room temperature.]

7. If you take a sample of linear polyethylene and dissolve it in a large excess of xylene at 130°C and then slowly cool this dilute solution, you find that after a couple of days a fine white suspension is formed. What does this suspension consist of?

If this suspension is subsequently filtered and dried, then heated to 170°C and slowly cooled to room temperature, a different type of structure is obtained. Compare and contrast the two different types of structures.

8. Discuss the types of morphologies that can be formed when a diblock copolymer is cooled from the melt to a temperature where self-assembly occurs. Include a consideration of how structure changes with the length of each block.

9. Discuss the limitations of the freely jointed and rotating bond model for a disordered polymer chain.

9

Natural Polymers

INTRODUCTION

The portrait on the right (Figure 9-1) is that of George Emory Goodfellow, also known as "The Gunfighter's Surgeon." He seems to have had an interesting life,[20] which included treating the survivors of the OK Corral gunfight, interviewing Geronimo and playing a key role in brokering the Cuban peace settlement in the Spanish-American War. Living in what was then the "wild-west" town of Tombstone, Arizona, he observed a number of gunfights and noted on no less than three occasions incidents when thin silk handkerchiefs or scarves had "stopped" bullets. In one incident, a bullet had inflicted a fatal chest wound on its unfortunate victim, but Dr. Goodfellow observed no blood emanating from the body and noticed the edges of a silk handkerchief protruded from the entrance wound. He was able to pull on its ends and bring it out, bullet and all! The bullet had ripped apart the man's other clothing, not to mention his flesh and bones, but not two layers of silk (the handkerchief had been folded).

The silk used in nineteenth century handkerchiefs no doubt came from silkworms (*Bombyx mori*). Silks from other creatures, particularly spiders, have equally fascinating properties. Although we all recognize that Spider-Man's power to shoot silk ropes

FIGURE 9-1 George E. Goodfellow (Courtesy: Arizona Historical Society).

and swing from one building to another is fiction, real spider silk has an impressive and extraordinary range of properties. Actually, a spider makes various types of silk. "Sticky" silk, for example, is used to make webs or capture prey. Dragline silk is the strongest, as it is used to attach the spider to the web and support its weight. (This is presumably the type of silk used by Spider-Man to swing from building to building.) The toughest silk is thought to be the dragline silk from the Golden Orb-Weaving spider, which is five times stronger than steel (on a unit weight basis). Although spider silk is tougher and more elastic than silkworm silk, spiders can't

[20] See http://www.ahsl.arizona.edu/about/exhibits/goodfellow/

243

FASCINATING POLYMERS—SPIDER'S SILK

Now here's a fine specimen, something that you might wish to collect and put in your mother-in-law's bed (just kidding)! One of your authors came across this magnificent spider, which was actually about 2 inches long, in the backyard of his condo in Hilton Head, South Carolina. Incredibly the spider's web occupied an area of some 10 square feet. It was a superbly constructed intricate structure composed of an array of interlocking ultra-thin fibers. Looking at the web moving back and forth in the breeze it is difficult to imagine that a web made of such fine and delicate fibers can support the weight of the spider, and what's more, capture a gourmet dinner "on the fly"! Spiders have evolved their own sophisticated extruders and spinnerets (fiber-forming dies) which we can see if we examine them under the scanning electron microscope or SEM. Let's look at some photomicrographs kindly supplied by Dr. Kenn Gardner, an old friend (and former drinking buddy) from our graduate student days at Case Western Reserve University. Kenn is an expert in the characterization of polymers by X-ray diffraction and various microscopic methods. He also has a long-standing research interest in the incredible natural fibers formed by spiders. Spiders produce a variety of fibers that have the unique chemical, physical and mechanical properties required to produce an effective web and much can be learned from their study. In the top micrograph you see the "rear end" of a spider, showing several orifices that are analogous to extruder dies or "spinnerets." The spider extrudes the fibers used to spin her web through these orifices, as seen in the scanning electron micrograph shown at the bottom. When you next look at a fragile-looking spider's web, take a moment to ponder just what an incredible feat of engineering this is—it has to withstand high winds and trap projectiles (flies etc.) without breaking.

be "farmed" to make the stuff (they have this nasty habit of eating one another). Accordingly, there is presently a significant effort aimed at using recombinant DNA techniques to make this marvelous material in significant quantities.

Materials made from many types of natural polymers (silk, cotton, wool, hemp, leather, etc.) have been with us throughout recorded history and have played crucial roles in the rise of civilizations and the economies of tribes and nations. The famous "silk road," for example, stretching across Asia and some of the most inhospitable and hostile environments on the planet, came into being in part because of the demand for silk created by the ancient Romans. Cotton is a wonderful material, with a "feel" that is difficult to match using synthetic polymers. Yet its history in the United States is inextricably linked to the horrors of slavery. Hemp also has an interesting and controversial history. The word hemp (or industrial hemp as it is called in some sources) now usually refers to a subspecies of the Cannabis plant that was selected for traits that enhance seed or fiber production (*Cannabis sativa*). The other subspecies, *indica*, was apparently selected for its high concentration of the psychoactive drug, *delta*-9-tetrahydrocannabinol (THC) —hence the controversy! The naughty (and illegal) stuff usually comes in the form of dried flowers (marijuana), resin (hashish), or various extracts collectively referred to as hash oil. The fibers obtained from these plants have a much less controversial history of useful service to mankind, however. The Chinese used hemp for fabric, rope and fishing nets as long ago as 8000 BC. By the middle ages, sails, ropes (Figure 9-2) and rigging on sea-going ships were all made from hemp, which was also used to make canvas and paper. In the 1800s, hemp canvas was used to cover the American pioneer's wagons and 80% of all textiles for clothes, linens, rugs, quilts, sheets and towels in America were made from hemp. In spite of its many uses, the growing of hemp in the United States was banned in the 1930s. The cultivation of low THC hemp remains banned in the United

FIGURE 9-2 Modern abaca or manila rope, which is derived from a plant in the banana family, was the major line used on sailing ships because of its superior saltwater resistance (Courtesy: Phoenix Rope and Cordage Company).

States today, although it is being successfully cultivated without incident in Europe, Asia, and Canada.

Leather is also a natural material whose modern day use ranges from the mundane to, shall we say, the exotic. (Googling leather, not to mention nylon, quickly gets you to the far corners of the web.) Prehistoric peoples used the skins of animals to protect their bodies from the elements, but after a relatively short period of time the skins would have decayed and rotted away. At some unknown point in history, it was discovered that the bark of certain trees (containing "tannin" or tannic acid) could be used to convert raw skins into what we recognize today as leather. The "Iceman" discovered in the Italian Alps several years ago, dating from at least 5,000 BC, was clothed in very durable leather.

There are a staggering variety of molecules present in living materials. The size of these molecules varies from the relatively small to the very large (i.e., high molecular weight), but our focus will be on the latter. These biological macromolecules also have an astonishing range of properties and functions. Some, such as the proteins found in skin, hair, nails and claws, are structural. The polysaccharides that make up the chitin that forms the outer, protective layer of insects have a similar function. Other proteins, the enzymes, act as catalysts, promoting the complex chemical processes that are essential to life. The polymers that we call DNA (deoxyribonucleic acid) are equally remarkable, carrying and propagating information—the instructions to build a living organism like you and me.

Our goal here is to have you compare and contrast some synthetic and natural polymers, such as nylon and silk, cotton and polyesters, so we will focus largely on biopolymers that are fibrous in nature and which are used as structural materials in plants and animals. We also aim to illuminate the way nature has married structure to function in such a marvelous fashion. Hopefully you will emerge with an appreciation of the beauty and complexity of natural materials and how far we have to go in order to match the intricate and subtle ways nature has put these molecules together (e.g., what are a butterfly's wings made of, and how do they get their color?). We will approach these materials in the same way that we tackled synthetic polymers, systematically exploring microstructure, conformation and morphology. However, in the preceding chapters we have discussed these subjects almost as separate topics. Here we will try and put all these things together, starting with proteins.

POLYMER MILESTONES—OTTO RÖHM & OTTO HAAS

Otto Röhm (top) and Otto Haas (bottom) (Courtesy: Rohm and Haas Company).

You're not going to believe this, but it was the odor of dog "do-do" (feces) that was instrumental in the formation of the Rohm and Haas Chemical Company. At the beginning of the 20th century, the process of turning animal hides into leather required that they first be dehaired (using a lime solution) and then be "bated" or softened by placing the hides in fermented dog dung! (We are not making this up!) The smell emanating from a tannery of the day was appalling! Dr. Otto Röhm was familiar with this smell, as a tannery was close by the gas works where he was employed. As your authors can attest from childhood memories, gas works had their own distinctive set of unsavory "pongs," and Röhm thought that certain gas-works and tannery odors had something in common. Röhm, a quintessential German scientist-entrepreneur, experimented with concentrated gas-water mixtures and came up with a bating substitute for dog poop! He invited his German friend, Otto Haas, who was then in the United States, to join him and form a partnership to market the new bating material. After some inevitable "ups and downs," during which time Röhm came to recognize the importance of enzymes in the bating process, a partnership agreement was signed in 1907. The business grew in Germany and then in 1909 Haas returned to the United States to form an American branch of Rohm and Haas. Suffice it to say, the US company continued to grow and was poised to take advantage of the new synthetic polymers developed in the 1930s.

POLYPEPTIDES AND PROTEINS

Much to your horror, no doubt, you will find that all the elements of microstructure that we found in synthetic polymers—isomerism, branching, cross-linking, etc.—are also found in natural polymers. Of course, this sentence should probably be written the other way around, in that through synthetic chemistry we manage to duplicate some of the richness of the natural world. Whatever! Here we will start by looking at proteins, as these introduce most of the elements of microstructure we discussed in Chapter 2. First, though, let's make a simple comparison. Figure 9-3 compares the structure of nylon 6 to what can be called nylon 2. (Do you recall where the number comes from in naming nylons? Just count the carbon atoms in each repeat unit!) These are both polyamides, consisting of CH_2 groups linked by amide groups (–CO–NH–). (You can only see one of the hydrogen atoms on each CH_2 group, because the other one is hidden behind the first.) But nylon 2 has a much more commonly used name, polyglycine. Also, instead of being called a polyamide, it is much more commonly referred to as a polypeptide. There is a difference in names because we have now crossed over from the discipline of polymer science to that of molecular biology. Of course, the laws of physics and chemistry stay the same, but we find that there are different names for things because the subjects have a different history.

Proteins as Copolymers

Polyglycine is a polymer of the amino acid, glycine. It is not found in nature, but has to be synthesized in a laboratory. When it comes to proteins, nature doesn't bother with homopolymers, but goes straight to copolymers, using up to twenty different amino acids to make the chains. Each amino acid has the same backbone structure, illustrated in the segment of a chain shown in Figure 9-4, but they differ from one another in the nature of the pendant or side group (labeled

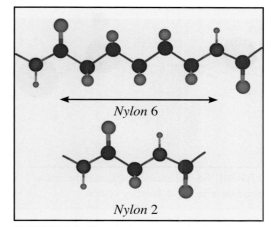

FIGURE 9-3 Chemical structure of nylon 6 and nylon 2.

R) that is attached to what is called the alpha carbon atom, C^α.

You may have noticed that there is a lot more to Figure 9-4. In addition to labeling the alpha carbon atom and pendant "R" group, you can see that there is something called the amide plane and we have used Greek letters, ϕ and ψ (pronounced fie and sie to rhyme with pie), to indicate that there are bond rotations around the C^α–N and C^α–C bonds. What about the amide or peptide bond CO–NH? To understand what's going on, we need to talk about the nature of the amide bond in a little more depth.

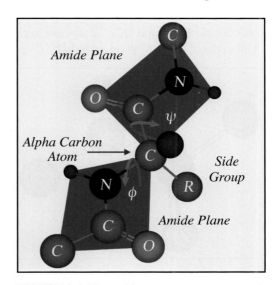

FIGURE 9-4 The amide group.

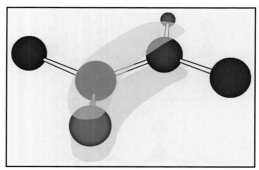

FIGURE 9-5 Schematic diagram of the amide bond showing the π-bonding system.

The Amide Bond

The way we have depicted it so far, the CO–NH linkage looks like it is a single bond. You may recall that rotation around single bonds is allowed (but the extent of such rotations may be limited by steric interference and so on). However, the carbonyl group, –C=O, is a double bond, which means there are π-bonds between electrons in the *p* orbitals of the carbon and oxygen atoms. It turns out that nitrogen atoms have a lone pair of electrons in an unhybridized *p* orbital. As a result, electron density becomes spread over all three atoms in an extended π-bonding system, as illustrated in Figure 9-5. If this talk of orbitals is giving you a headache, however, just think of it in a different way. Instead of having a single C–N bond and a C=O double bond, the OCN atoms are each liked to the other by a "bond and a half" (Fig-

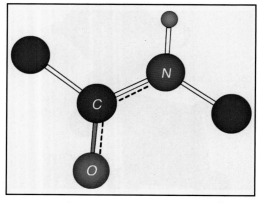

FIGURE 9-6 Schematic diagram of the amide bond showing the "bond and a half".

ure 9-6) "So what," you might say. This has consequences for the shape or conformations of polypeptide chains and hence the marvelous way they function in living systems.

Because of this partial double bond character, the amide group is essentially planar. In addition, it is stiff and rotation around the amide C–N bond is highly restricted. Accordingly, rotations can only occur around two of the three bonds in each amino acid residue in a polypeptide chain and, as we will see, these rotations are limited by steric factors. But we are getting ahead of ourselves here. What is the nature of the different groups, R, that determine the properties of each amino acid? Besides, what is an amino acid? So far we have only showed you amino acid segments that have been incorporated into chains.

Amino Acids

Amino acids have the general structure HOOC–CHR–NH$_2$. There is a carboxylic acid on one end and an amine on the other. In-between there is a carbon atom that is attached to a hydrogen atom and a group we have simply labeled R. As we have already mentioned, it is the nature of this group that varies from one amino acid to another.

There is a subtle aspect to the structure of amino acids. Any carbon atom that has four different groups bonded tetrahedrally to it, like the C$^\alpha$ atom of amino acids, can have them arranged in two different ways, one a mirror image of the other. They are isomers, the type of thing we encountered in vinyl polymers (remember isotactic polypropylene?). These two structures are optically active and, as a result, the two forms came to be called D and L optical isomers, after the direction they rotated the plane of polarized light (*dextro*, right and *levulo*, left). Only the L-amino acids are found in the proteins made in nature (although D-amino acids are found in some hormones and toxins). No one knows why. (Perhaps God is left-handed!) In any event, if you travel down a polypeptide chain going in the CO to NH direction, the L-form has the pendant R group on the left, as illustrated in Figure 9-7. (Note that in gly-

FASCINATING POLYMERS— NATURAL ADHESIVES

To your authors, who prefer to feel the earth beneath their feet and chase a little white ball around a manicured golf course, it seems preposterous to travel out to sea and sit in a boat all day fishing. This must rate as one of the most boring pastimes known to man and up there with watching paint dry! On top of that, those who own seafaring boats pay for this dubious privilege by having to battle the elements and spend an inordinate amount of time cleaning and removing foreign critters from their boats. Anyone who has tried to remove barnacles from the hull of a ship, a rock or part of the superstructure of a harbor, knows that it is not easy. These marine crustaceans (or cirripedes) tenaciously "cement" themselves to almost anything and you have to use brute force in the form of a high-speed rotating stainless steel brush with protruding metal flanges to literally rip and tear them off. Barnacles cause drag and reduce sailing efficiency. Accordingly, an enormous amount of time, effort and money is spent world-wide each year to remove barnacles, mussels and the like from fouled boats and other structures. Why do they stick so well? The incredible ability of barnacles, mussels, and the like to adhere strongly to a variety of surfaces in hostile marine environments, which can vary widely in salinity and temperature, has fascinated marine biologists and material scientists for centuries. It turns out that a barnacle starts out swimming around as a cyprid (larvae) and looks for a hard surface upon which to spend its life. After finding such a place, the cyprid attaches itself by landing head-first and releasing an adhesive cement through its antennae. This cement, a polypeptide, contains the amino acid, 3,4-dihydroxyphenyl-L-alanine (DOPA), which appears to be

Australian barnacles (Courtesy: Keith Davy—www.mesa.edu.au).

the active adhesive ingredient. Researchers, most notably, J. H. Waite, T. J. Deming, and H. Yamamoto, have studied these natural adhesives and shown that the DOPA residue can chelate to metals, interact strongly with wood and other polar organic materials and undergo a variety of chemical and oxidation reactions that lead to the rapid formation of a cross-linked polymer network. Dentists are particularly intrigued by the facile "setting" (curing) of this adhesive under water and hope that similar adhesives can be developed for dental applications.

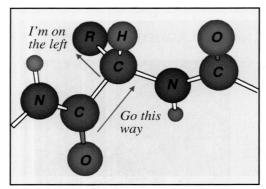

FIGURE 9-7 Schematic diagram of an L-amino acid group in a polypeptide chain.

cine, where R = H, you can't have D and L isomers, because the two hydrogen atoms are indistinguishable from one another. The mirror images are identical.)

So what is it about the pendant groups R, why are they important? It is the variety of these groups that make proteins such versatile molecules. For example, if we want to make a chain of just 100 amino acids using just the 20 utilized by mother nature, there are 20^{100} or about 10^{130} different ways of doing this! These amino acids can be classified in various ways. Some are called "essential," meaning they must be provided by the things you eat, while others are called "non-essential," because they don't have to be part of your diet. (Vegetarians have to be careful that they eat stuff that (taken as a whole) includes all the essential amino acids.)

For our purposes, it is most useful to consider amino acids in terms of three categories, non-polar, polar and neutral, as the arrangements of polar and non-polar groups along a polypeptide chain is one of the major factors that determines structure. In part, this is because most proteins operate in an aqueous environment, so a protein will fold so as to get the "water-loving" (hydrophilic) polar groups as near as possible to the surface, while the "water-hating" (hydrophobic) groups are buried in a non-polar interior environment. Rather than go through the amino acids with a discussion of their characteristics, one-by-one and bore you to death, however, we have put together a simple collection (see next

two pages) that we have called, *Fascinating Monomers*. The amino acids are grouped by character and each represented by a three-letter symbol. Glance at these quickly now, then come back and use them when you need to.

Some Final Observations on Polypeptide Microstructure

So far we have looked at two elements of microstructure that synthetic polymers and polypeptides have in common. One is isomerism and the other involves copolymers. As we will see shortly, certain fibrous proteins have repeating sequences of amino acids and this results in the formation of ordered structures. Others appear to be random copolymers of amino acids, but, in effect, they cannot be. They have evolved such that a precise number and sequence of amino acids fold to give a structure with a precise function. Mutations that result in a change in just one or two amino acids (as in sickle cell anemia) can dramatically change the conformation of the chain, such that the protein can no longer "do its thing." But our focus here is not the genetic basis of certain diseases, but structure. Accordingly, we will finish this section by mentioning one other aspect of microstructure, together with an example of an interaction that has a crucial role in conformation. These are illustrated in Figure 9-8. First, there is an intra-molecular cross-link formed between two cysteine (cys) amino acids in the form of a disulfide bridge (–S–S–). Second, there are hydrogen bonding interactions between the amide groups.

Before getting to a discussion of how these things play into structure, we just want to make two final points in our comparison of the microstructure of polypeptides and synthetic polymers. First, protein chains are linear and branching is not observed. (Some polysaccharides are branched, however.) Second, proteins (as opposed to synthetic polypeptides) are monodisperse, as the DNA of the cell's genes defines the number of amino acids and their order in a chain. Even

FASCINATING MONOMERS—NON-POLAR AMINO ACIDS

Let's start with the non-polar groups. First there are those with aliphatic side chains, shown in the first two rows below. There is also one non-polar, aliphatic amino acid that is a little strange, but very important, proline. The aliphatic side chain is in the form of a five-membered ring that links the α-carbon and amide nitrogen atoms. As a result, the latter only has one hydrogen (–NH) instead of two (–NH$_2$). Accordingly, when it is polymerized, this lone hydrogen is removed to allow the formation of a covalent bond to the next amino acid in the chain. Then it doesn't have a hydrogen atom at all and cannot take part in hydrogen bonding. In addition, the linking of the α-carbon and amide nitrogen atoms severely restricts bond rotations and hence flexibility. As a result, proline plays an important role in determining the structure and folding of proteins. The three remaining non-polar amino acids are shown at the bottom. Two are aromatic and the third, methionine has an –S– atom.

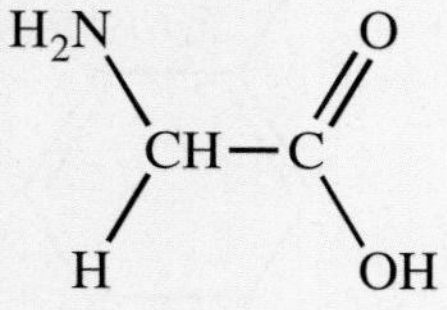

Glycine (gly)

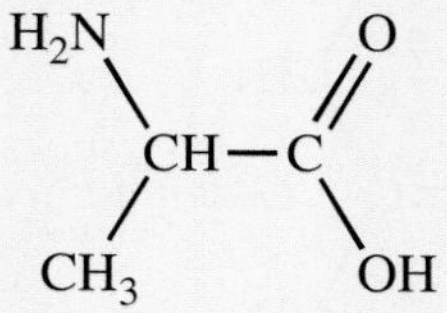

Alanine (ala)

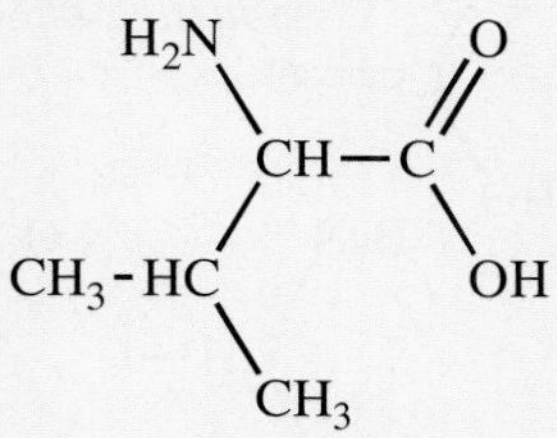

Valine (val)

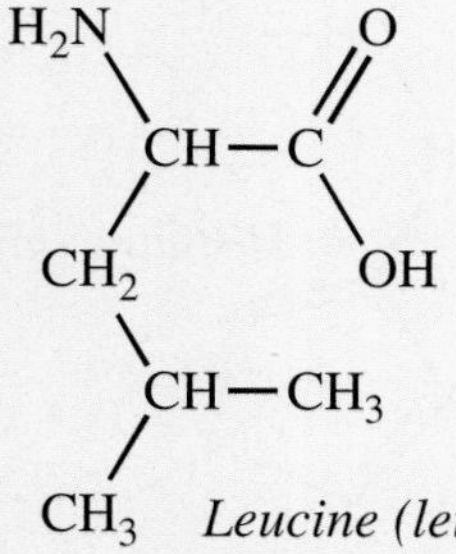

Leucine (leu)

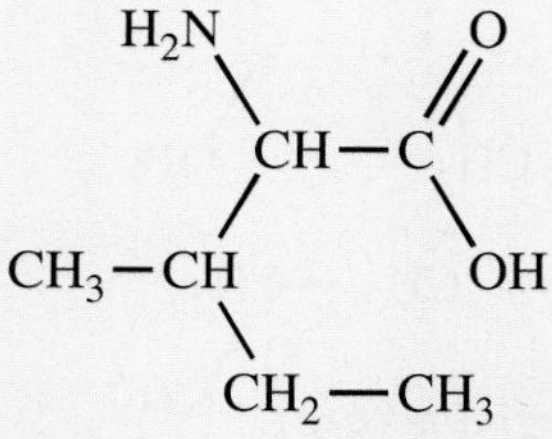

Isolucine (ilu)

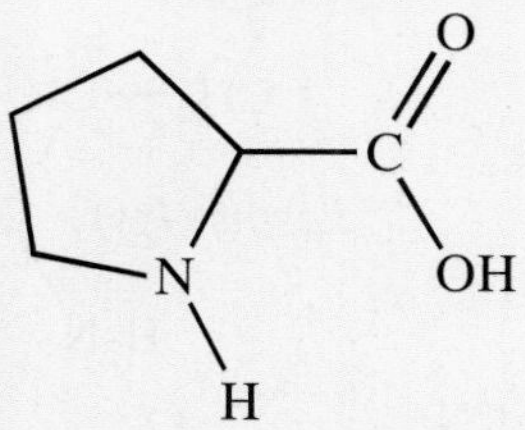

Proline (pro)

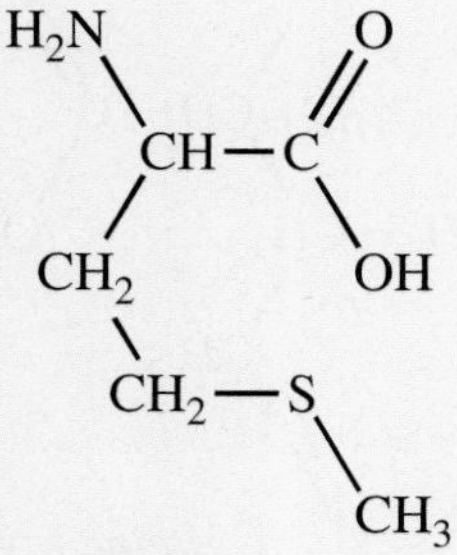

Methionine (met)

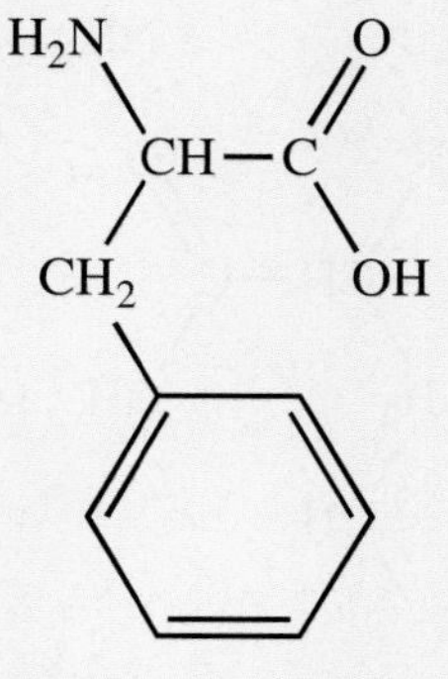

Phenylalanine (phe)

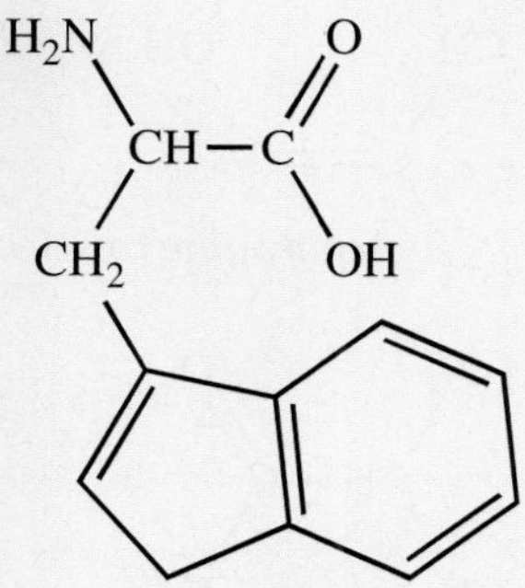

Tryptophan (trp)

FASCINATING MONOMERS—POLAR AMINO ACIDS

Polar amino acids can be acidic, basic or neutral. The structures of the three acidic polar groups are illustrated at the top of the box. Each of these can give up a proton (H^+) leaving a negatively charged group behind, $-COO^-$. The basic polar amino acids, on the other hand, have amine groups (NH_2 or NH) that can accept a proton (H^+) and become positively charged (e.g., $-NH_3^+$). Finally, there are neutral amino acids that have heteroatoms (e.g., O, N) that make them polar. Note that cysteine, has an $-SH$ group. This is important as it allows for the formation of cross-links between parts of the same chain or between different chains.

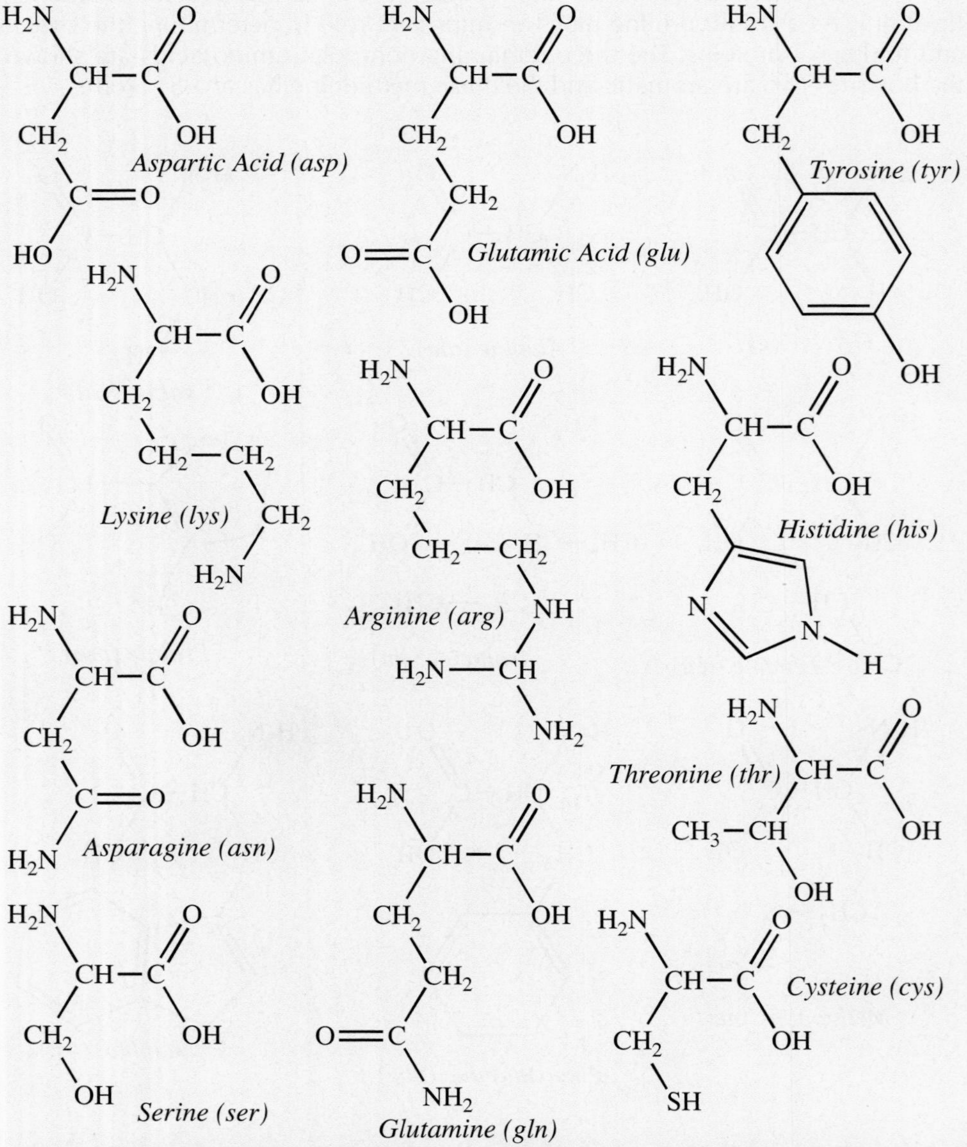

Aspartic Acid (asp)

Glutamic Acid (glu)

Tyrosine (tyr)

Lysine (lys)

Arginine (arg)

Histidine (his)

Asparagine (asn)

Threonine (thr)

Serine (ser)

Glutamine (gln)

Cysteine (cys)

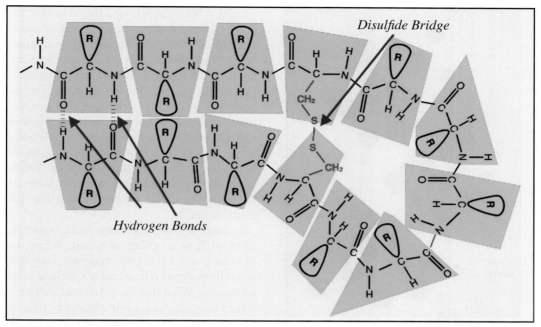

FIGURE 9-8 Schematic diagram of a polypeptide chain showing hydrogen bonds and a disulfide bridge.

the most stringently prepared synthetic polymers have a degree of polydispersity.

THE CONFORMATION OF POLYPEPTIDES AND PROTEINS

One of the great triumphs of 20th century science was the determination of the structure of many proteins and DNA (deoxyribonecleic acid) by X-ray diffraction. The excitement of the times and a flavor of the work was captured in James Watson's wonderful book, *The Double Helix*. Log on to amazon.com and order a copy immediately! We now turn our attention to some of the fruits of this work.

In describing protein structure it is usual to consider four levels of organization, termed primary, secondary, tertiary and quaternary structure. Primary structure refers to the sequence of amino acids that makes up the chain of a particular protein (or synthetic polypeptide). Secondary structure is the ordered conformation that the chain (or usually parts of chains) can twist itself into. An example, a section of an α-helical chain is shown in Figure 9.9. More on this shortly.

Tertiary structure refers to how a single chain can be folded in on itself (globular proteins are usually tightly folded and look like a knotted up piece of string). Finally, quaternary structure refers to how different molecules can pack to form an organized unit.

We have already discussed primary structure in terms of the general character of amino acids and some specific examples of amino acid sequences in certain proteins will be discussed later. Our attention now is focused on secondary structure, or conformation as we called it when we discussed synthetic polymers. There are a number of factors that affect the conformation of a polypeptide chain and a lot can be learned initially by just focusing on two of these: steric restrictions on bond rotations and the strong driving force for amide groups to hydrogen bond to one another.

In 1951, Pauling and Corey published a landmark paper on the structure of polypeptide chains. Earlier X-ray diffraction work in the 1930s had revealed the characteristics of the peptide bond and studies of fibrous proteins suggested that helical and extended chain structures were probably present. Using

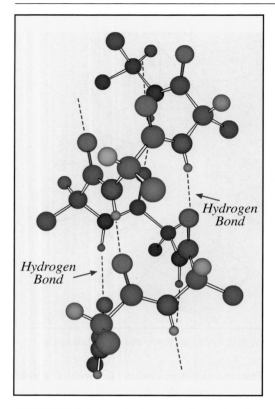

FIGURE 9-9 Schematic diagram of a section of an α-helical chain.

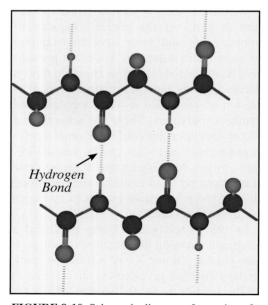

FIGURE 9-10 Schematic diagram of a section of a β-sheet.

carefully built models and the idea that the chains would fold so as to minimize steric overlap of atoms and maximize the number of hydrogen bonds that could form between amide groups, two principal structures were proposed, the α-helix and β-sheet. The α-helix has 3.6 residues per turn and is illustrated in Figure 9-9, while the β-sheet has an extended chain structure, as shown in Figure 9-10. The structure of the α-helix is stabilized by intra-molecular hydrogen bonds along the helix direction, while in β-sheets the hydrogen bonds are perpendicular to the chain direction.

In case you don't get this helix business, think of it as a spring or a coil. When we discussed synthetic polymers, we saw that isotactic polypropylene had a 3_1 helical conformation. What this means is that there are three chemical repeat units in every turn of the coil, as shown in the middle drawing of Figure 9-11. The atoms in the backbone can be viewed as tracing out the pattern of a helix. Each dot represents a chain segment or chemical repeat unit. Although the backbone atoms themselves may not lie on the path of the helix, taken together they trace out the characteristic spring-like pattern. Note that an extended chain structure can be simply viewed as a 2_1 helix. Having a fractional number of residues per turn, as in the α-helix, may seem unlikely at first glance, but think of it in a different way. It has 18 complete chemical repeat units (amino acids) in every five turns of the helix! Helices also give characteristic X-ray diffraction patterns and that is why scientists were eventually able to sort out these structures.

But why are the α-helix and β-sheet particularly favored structures? To understand this we have to say a little more about steric hindrance. First, if the ϕ and ψ (fie and sie) angles at every C^{α} atom along the chain are identical, then the conformation of the chain falls naturally into a helix. However, some combinations of rotational angles are not allowed, because they bring neighboring unbonded atoms too close together and the electrons in their orbitals strongly repel one another. Figure 9-12 illustrates a combination of ϕ and ψ angles that bring the amide carbonyls too close together. On the other hand, certain values of ϕ and ψ minimize steric repulsion, because non-covalently bonded atoms, like the carbonyl oxygen and the pendant R group, get as far away from one another as possible.

The Ramachandran Plot

Using considerations such as these, G. N. Ramachandran and his group constructed the famous Ramachandran plot, shown in Figure 9-13. The most favored regions (in terms of values of ϕ and ψ) are those shown in white on the plot. Slightly less favored regions are in light blue, while conformations that involve large repulsive forces between non-bonded atoms are in dark blue. Also shown on this plot are some red lines. The solid red lines represent right-handed helices, while the dotted red lines represent left-handed helices. (A right-handed helix has a clockwise twist as you go along it.)

As mentioned previously, if the values of ϕ and ψ on each residue in a chain are identical, the conformation falls naturally into a helix. However, many different values of ϕ and ψ result in a helix of a given type (e.g., 3_1 or 3.6_1), as indicated by the lines on the plot. Each helix will have a different pitch, however (see Figure 9-14). Certain helical structures are highly favored, not only because they minimize steric repulsions, but also because the pitch of the helix is particularly suited to the formation of hydrogen bonds. The right-handed α-helix, illustrated above in Figure 9-9, is notable in this regard and is represented by the symbol α_R on the plot. Note also that there are a couple of close-lying values of ϕ and ψ that give an α_R helix. Also, as might be expected, the β-sheet conformation is in one of the regions of the plot where repulsive forces are at a minimum. There are other helical structures indicated on this plot, such as C, the collagen coiled coil. You're probably bored with this account of conformation by now, however, so rather than tackle these systematically, let's jump to a consideration of natural materials where

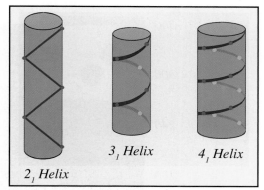

FIGURE 9-11 Schematic diagram illustrating different helices.

these conformations play a key role in structure and properties, starting with silk.

The Structure of Silk

There are all sorts of silk that are found in nature. The stuff that is usually found in textiles comes from silkworms (*Bombyx mori*). They are not really worms, but the larvae of moths. They emerge from very small eggs with an incredible lust for mulberry leaves, which they consume until they are ready to pupate and weave a cocoon around themselves. Unlike spiders, which spin silk from their rear end, silkworm silk is actually hardened saliva, which comes out of the mouth. The larva has a small spinneret on its lip, through which the silk emerges. The cocoon is formed from a single strand of silk that

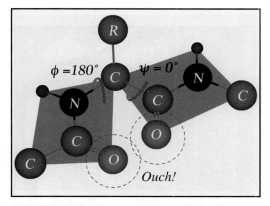

FIGURE 9-12 Schematic diagram illustrating the case when two amide carbonyl groups are too close.

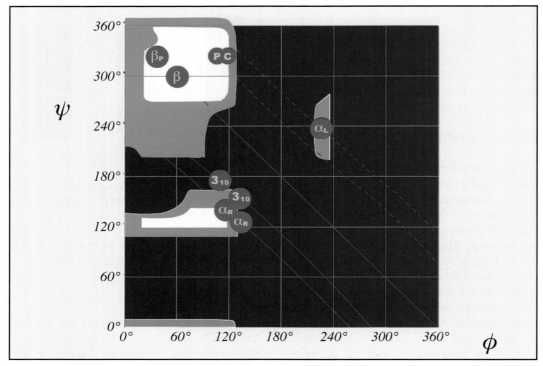

FIGURE 9-13 The Ramachandran (ϕ,ψ) plot.

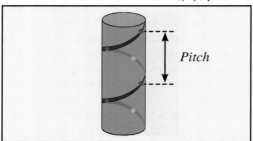

FIGURE 9-14 Schematic diagram illustrating the pitch of a helix.

FIGURE 9-15 Unraveling silk fibers from cocoons (the white things floating in the baths).

is about one mile long! (Figure 9-15 shows silk fibers being obtained from cocoons at the Silk Museum in China.) The silkworm moves its head in figure eight like patterns as it spins the cocoon. This action serves to orient the fibers. The silk inside the silkworm actually has a different structure (Silk I) than the oriented fibers produced after "spinning" (Silk II). As we will see, spider silk is similar, in that it also has a different structure before and after spinning.

Bombyx mori silk has both crystalline and amorphous domains. In the crystalline fraction, there is a high proportion of glycine (Gly), alanine (Ala) and serine (Ser) amino acid residues, present in the ratio of 3:2:1. This reflects the presence of hexapeptide sequences, Gly-Ala-Gly-Ala-Gly-Ser. You've probably forgot the structure of these amino acids, which we have therefore reproduced in Figure 9-16.

Notice that if the chain is arranged in a zig-zag or extended chain conformation, as found in the crystal structures of polyethyl-

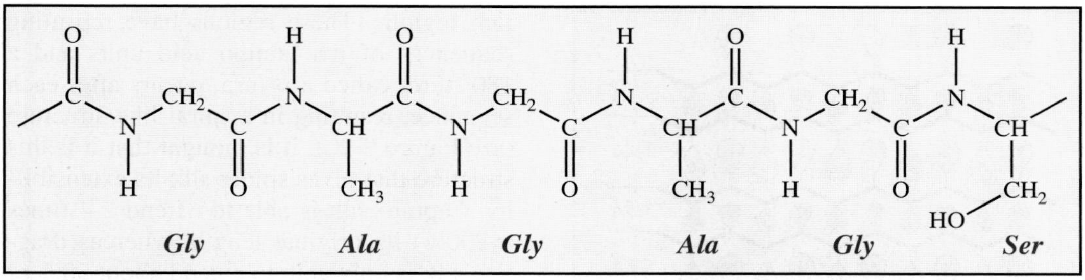

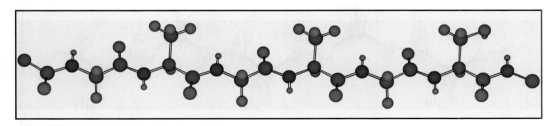

FIGURE 9-16 The predominant hexapeptide sequence of the crystalline domain of *Bombyx Mori* silk.

ene and the nylons, then all the glycine residues, which have just a hydrogen as their R group in the generic structure –HN–CHR–CO–, are all on one side of the chain, while the –CH$_3$ and –CH$_2$OH groups of alanine and serine are all on the other side of the chain. We'll come back to this in a minute. Also, as mentioned above, this extended chain conformation is one that minimizes steric repulsion and allows all the amide groups to form hydrogen bonds to one another in the form of β-sheet structures (see Figure 9-10). Unlike the hydrogen bonds in the α-helix, which are formed between amide groups in the same chain, those in the β-sheet structures of silks are usually formed between different chains. (β-Sheet structures can also form between segments of the same chain, if parts of it fold and bend back on itself.) Furthermore, the amino acid side groups are perpendicular to the plane of the sheets, so that if we take one of the chains and rotate it 90°, it appears as in Figure 9-17.

Studies using X-ray diffraction have shown that the spacing between the sheets alternates between 3.5 Å and 5.7 Å (Figure 9-18). This means that the small glycine residues must pack next to one another, while the larger alanine (and serine) units pack next to other alanines (and serines). Note that if the chains are aligned along the fiber direction, the resistance to tension is borne by the covalent bonds, making the fiber strong. Much weaker dispersion forces between the side chains hold the sheets themselves together, so that the fiber bends easily and is quite flexible. There are some subtleties to the crystal structure that we won't go into in depth, but in order to maximize the interactions and minimize the overall energy, the chains actually form a rippled or pleated structure, as illustrated in Figure 9-19.

If you've looked at the structure of some of the amino acids, it may have occured to you that glycine and alanine are the two smallest amino acids. It is presumably no accident that nature built silks using these residues, as it allows a tight packing of chains in the crystal. Silks also contain amino acids with much more bulky side groups, however, and these would obviously not fit into this structure. The chain segments that include these residues are disordered, giving silk its extensibility. In silk from *Bombyx mori* silkworms, the fibers are about 60% crystalline and extensibility is thus fairly limited.

FIGURE 9-17 A single β-sheet chain rotated 90°.

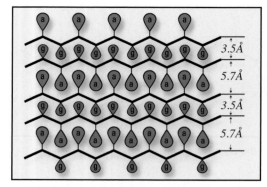

FIGURE 9-18 A single β-sheet chain rotated 90°.

Spider Silk

Spider silks are a little different and much more extensible. They are made up of two proteins, spidroin I and spidroin II, which vary somewhat with species and diet. Like silkworm silk, these proteins contain high proportions of glycine and alanine (about 42% and 25%, respectively), but differ in the proportions of the other, more bulky amino acids present. In addition, these proteins have 4–9 alanine residues strung together in a block, which, in turn, are linked by glycine

rich regions. These regions have repeating sequences of five amino acid units and a 180° turn, called a β-turn, occurs after each sequence, resulting in a spiral-like structure (see Figure 9-20). It is thought that it is this structure that gives spider silk its extensibility. Capture silk is able to extend 2-4 times (>200%) its original length, whereas dragline silk is only able to extend about 30% of its original length.

The β-turn, has an important role in the folding of many proteins, not just silks. These 180° turns are stabilized by intramolecular hydrogen bonds, as illustrated in Figure 9-21. It is thought that both spiders and silkworms incorporate structures like this in their silks. When these molecules pass through the narrow tubes to the spinnerets possessed by these creatures (silkworms in the front, spiders in the rear), the protein molecules align and crystallization of regions of the chain with a regular repeating sequence of amino acids occurs parallel to the fiber axis. The hydrogen bonds in the β-turn are reorganized to form pleated β-sheets It is these β-sheets that give silks their high tensile strength.

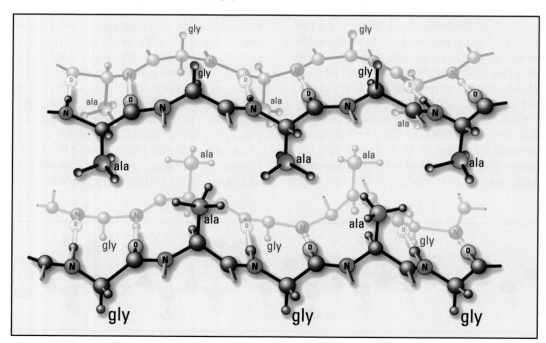

FIGURE 9-19 The rippled or pleated structure of *Bombyx mori* silk.

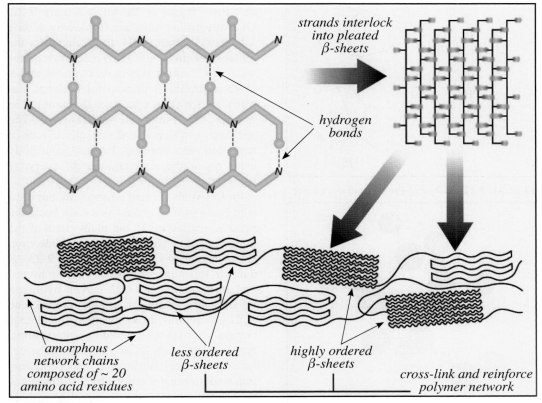

FIGURE 9-20 The structure of spider silk.

The Structure of Collagen

Collagen, another structural, fibrous protein, makes up about one-quarter of the protein in the human body and forms the rigid, inextensible fibers that are found in connective tissue–tendons, ligaments and skin. Adding mineral crystals to collagen makes bones and teeth. Collagen thus provides structure to our bodies, supporting the softer tissues and connecting them with the skeleton. As you might guess, it is also a crucial component of animal hides, from which we obtain materials like leather.

There are various types of collagen, but they all have similar structures. A crucial component of this structure is repeating sequences of three amino acids, Gly–X–Pro or Gly–X–Hypro, where Gly is glycine and Pro and Hypro are proline and hydroxyproline. X is any other residue. A –Gly–

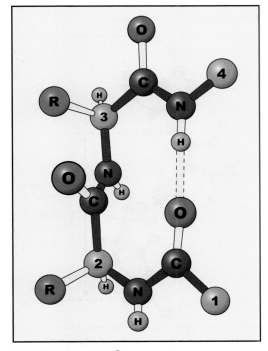

FIGURE 9-21 The β-turn structure.

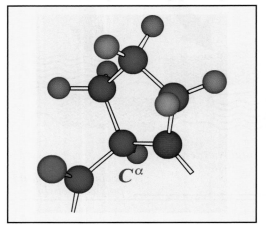

FIGURE 9-22 A Gly-Pro-Pro tripeptide sequence.

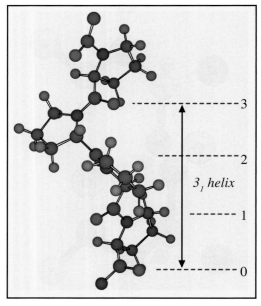

FIGURE 9-23 A pyrrolidine ring.

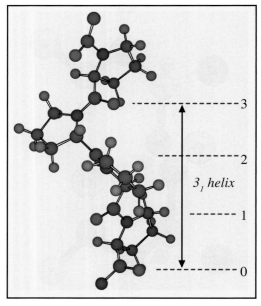

FIGURE 9-24 The 3_1 helical structure of poly-L-proline.

Pro–Pro– repeat is shown in Figure 9-22. Hydroxyproline units are formed by adding a hydroxyl group, –OH, to proline after the collagen chain is built. This reaction requires vitamin C, which our bodies cannot make. Hydroxyproline is critical for collagen stability, so if we don't get enough vitamin C in our diet, the results can be disastrous. (You get scurvy, which initially results in loss of teeth and easy bruising, because your body can't repair the wear-and-tear of everyday life.)

Proline is an unusual amino acid, having a side chain of CH_2 groups that loop back and become reattached to the main chain at the amide nitrogen (forming a pyrrolidine ring, if you wanted to know) — see Figure 9-23. As a result, this nitrogen does not have a hydrogen atom and cannot take part in hydrogen bonding interactions to carbonyl groups. The –OH group on hydroxyproline can and does engage in hydrogen bonding interactions, however, and this helps stabilize the structure of collagen. Because of the constraints imposed on bond rotations by this linking of the C^α and nitrogen atoms, a proline residue will force a bend on the conformation of the main chain.

A polypeptide chain made up of just proline units, poly-L-proline, will naturally bend into a 3_1 helical conformation, as shown in Figure 9-24. Polyglycine chains can also form a 3_1 helix, so –Gly–X–Pro– sequences also fold into such conformations. Note that the glycine units in every third position will then be directly in line (along the chain) with one another. Note also that the amide carbonyl groups extend perpendicular to the chain, so that helices consisting of –Gly–X–Pro– residues can line up, allowing the Gly and X units to form intermolecular hydrogen bonds. (Remember that the Pro unit does not have an N–H group to form hydrogen bonds.)

Collagen fibrils form various patterns in the tissues of animals. In tendons, for example, they are all oriented parallel to the length of the tendon. In cowhide (one source of leather), they form fibers that branch into all directions, forming a complex structure capable of holding a ton of cow together. All of these fibrous structures appear to be made up of tropocollagen molecules, which are about 15 Å in diameter and 2800 Å long. These tropocollagen molecules are triple helices, three helices wrapped around one another, as illustrated In Figure 9-25. Each helix consists of a slightly twisted 3_1 helix characteristic of –Gly–X–Pro– sequences. The proline units are too big and bulky to fit on the inside of this tightly coiled structure, which is responsible for the strength and inextensibility of collagen, hence the importance of having glycine every third residue. It is these small amino acids that are found on the inside of the coiled coil.

Tropocollagen molecules then pack with a displacement of one-quarter of their length, to maximize intermolecular interactions (hydrogen bonds and electrostatic interactions), as illustrated in Figure 9-26, forming collagen fibers. The tropocollagen molecules also form covalent cross-links in reactions involving lysine units. As you age, the degree of cross-linking increases, your tissues become stiffer and more brittle and your body generally goes to hell in a hand-basket!

The Structure of Keratin

There are different types of keratin and here we will focus on just one, α-keratin. This type of keratin is found in materials such as wool and is also the most important protein in hair and finger nails. As in silk and collagen, the amino acid sequence in α-keratin favors a particular secondary structure, the α-helix, which we described earlier in this chapter. (There is also β-keratin, found in bird feathers, for example, and which as the name suggests, forms β-sheet like structures.) However, unlike the other fibrous proteins we have discussed so far, in keratins there does not appear to be a repeating motif of specific amino acids, but instead the pattern involves the placement of hydrophobic (water-hating) and hydrophilic (water-loving) residues. There also appears to be a pattern in the placement of acidic and basic groups among the latter residues. X-ray diffraction studies have shown that the α-helices of α-keratin form double-stranded coiled coils. The driving force for forming these double helices appears to be the sequence of amino acid types along the chain. The geometry of the α-helix is such that every amino

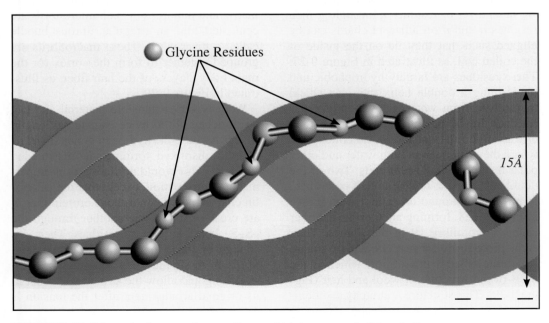

FIGURE 9-25 Schematic diagram of tropocollagen which is composed of three helices.

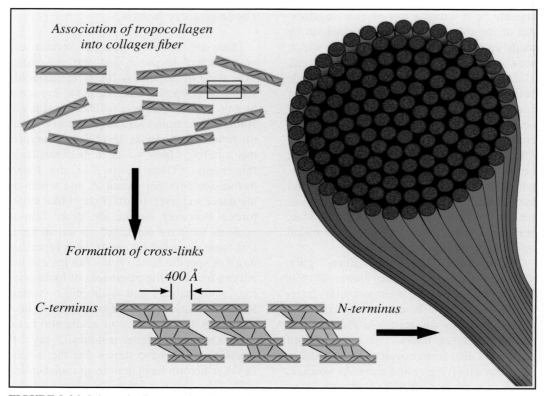

Association of tropocollagen into collagen fiber

Formation of cross-links

400 Å

C-terminus

N-terminus

FIGURE 9-26 Schematic diagram of a collagen fiber.

acid residue has the same environment as the seventh one following (and preceding) it. If we label a set of residues a through g, then residues *a* and *d* on adjacent chains can be aligned such that they lie on the inside of the coiled coil, as illustrated in Figure 9-27. These residues are largely hydrophobic and by forming a double helix they can shield themselves from water. Equivalently, polar residues lie in positions where their side chains are on the outside of the coiled coil, where they can interact with water and other polar residues on adjacent coils. Two of the double-stranded α-helical coiled coils then, in turn, wrap around one another, as shown in Figure 9-28, forming another helical fiber called a protofibril (P in the figure). Then eight protofibrils form a circular or square structure, called a microfibril, which is the basis of the structure of wool and hair (Figure 9-29). This structure is stretchy and flexible and can be compared to a rope containing various threads that are twined together.

In the structure of a single hair, microfibrils are embedded in an amphorous protein matrix and hundreds of such microfibrils are cemented into an irregular fibrous bundle called a macrofibril. These macrofibrils are grouped together to form the cortex (or the main body) layers of the hair fiber, as illustrated in Figure 9-29.

Wool is very extensible material and can be stretched out to twice its original length. This is because the hydrogen bonds in the α-helix are disrupted, forming a more extended structure. This would normally result in permanent deformation, except that the α-keratin chains and the amorphous protein matrix are cross-linked to one another through the –S–S– bonds of cystine residues. There are enough of these to prevent the chains and fibrils from slipping past one another upon stretching and allow the structure to relax to its original arrangement after the tension is removed.

Hair can be curled or straightened by rear-

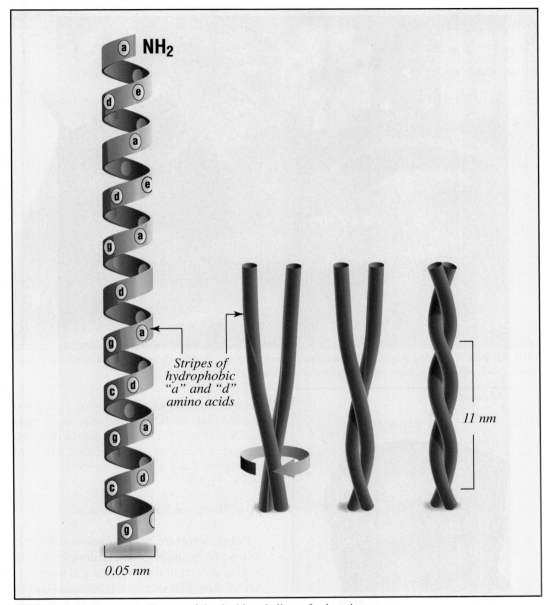

FIGURE 9-27 Schematic diagram of the double α-helices of α-keratin.

ranging these cross-links. The basis of the permanent-wave process is to chemically break these bonds, shape the hair according to desire and whim, then reform the cross-links so that the new shape holds. Of course, only that part of the hair that was processed holds this "permanent" shape. Because hair grows at the rate of about six inches per year (meaning 9–10 turns of a-helix are spun off per second), your hair eventually returns to its unruly proclivities.

As we have mentioned, there are various types of keratins, and this brings us to one final aspect of sulfur cross-link bridges. The α-keratins can be classified as "soft" or "hard" by their sulfur content. The low sulfur content keratins of wool and hair are much more flexible than the hard, high sulfur

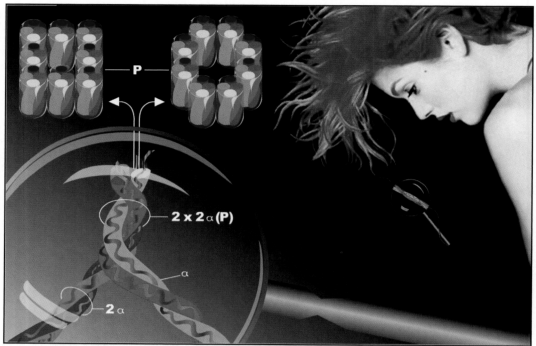

FIGURE 9-28 Schematic diagram of a protofibril (P in the figure). Eight protofibrils form a structure, called a microfibril, which is the basis of the structure of wool and hair.

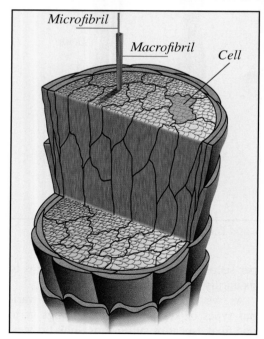

FIGURE 9-29 Schematic diagram of the structure of a single hair [redrawn from an original figure in R. E. Dickerson and I. Geis, *The Structure and Actions of Proteins*, Harper and Row (1969)].

content keratins of nails, horns, claws and hooves. Just as rubber becomes more rigid and less extensible when the degree of cross-linking is increased, so do keratins. The hard α-keratins have a higher concentration of –S–S– bridges.

Final Words on Protein Structure

Protein structure is an enormous subject in its own right and all we have done here is to outline aspects of the structure and morphology of three fibrous proteins, chosen so as to illustrate the role of the three main secondary structures, the α-helix, β-sheet and polyproline helix. Before closing this section, however, we want to briefly mention globular proteins. At first glance, these would seem to have little structure and be analogous to the random coils found in elastomers. Not so. First they are much more compact. In addition, portions of the chain often take on an ordered conformation, as in the α-helical segments of the chain of myoglobin, illustrated in Figure 9-30. This molecule has

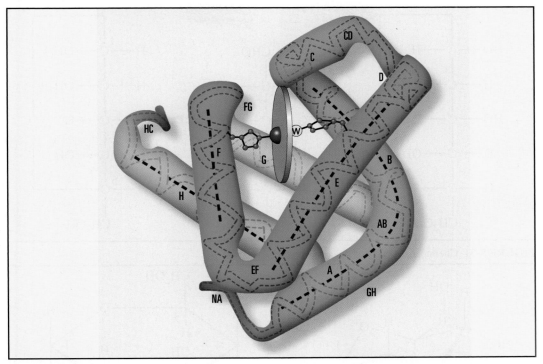

FIGURE 9-30 Schematic diagram of the structure of myoglobin.

evolved such that its structure exquisitely matches function. The eight stretches of α-helix form a box containing a heme group, which allows the molecule to store oxygen until it is required by the muscles of creatures such as whales and seals.

POLYSACCHARIDES

Polysaccharides are polymers made up of molecules of sugar strung together to make chains. (Keep in mind that the sugar you put in your tea or coffee is just one of a large family of molecules.) Sounds simple, right? Sugars consist of H–C–OH units most often linked together to form five- or six-membered rings (Figure 9-31). (The other two valences on the carbon atoms link them to other units in the rings.) However, there are layers of complexity in these structures. First, sugars are optically active (they can rotate the plane of polarized light); second, there are a number of structural isomers; third, the units can be linked to one another in a number of

different ways! (And you thought tacticity in vinyl polymers was hard!) A lot of polysaccharides are linear homopolymers, but some are branched and others form more complex combinations, like proteoglycans, which we won't mention again so as not to mess up your brain.

With all these layers of microstructure comes different ways of representing the structures. First they can be represented as an open arrangement of atoms, as shown for the structure of glucose in the center of Figure 9-31. (The stereochemistry of the molecule is then determined by the position of the –OH group on carbon atom 5. This structure is D-glucose, because the –OH group is on the right in this structural representation. (But we don't care about no stinking stereochemistry, so we'll ignore this.) The glucose ring can then be formed by linking carbon atoms 1 and 5 with an ether group (–O–) and moving the –OH group from atom 5 to atom 1. Note that this –OH group can be placed on the right- or on the left-hand side of this

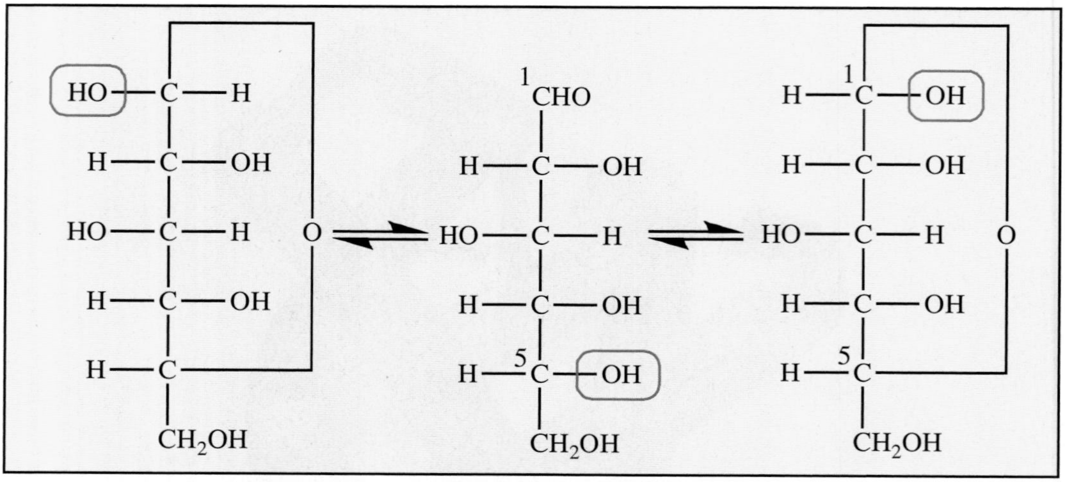

FIGURE 9-31 Chemical structure of glucose.

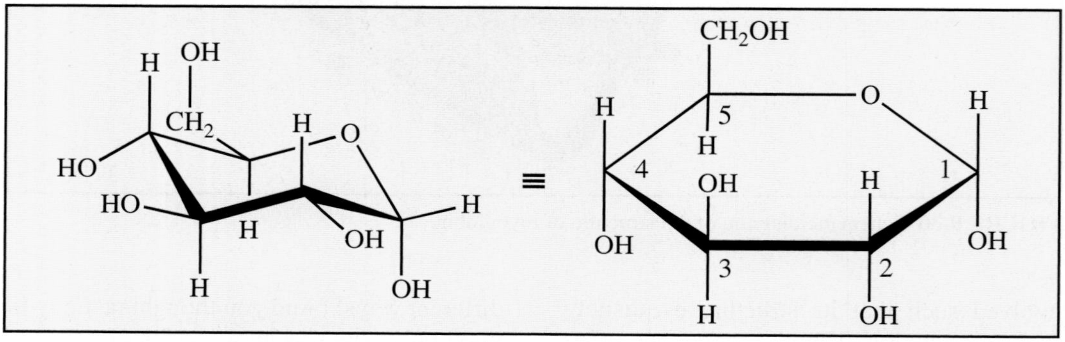

FIGURE 9-32 Chemical structure of α-D-glucose.

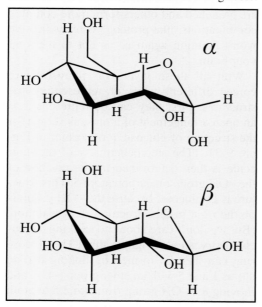

FIGURE 9-33 Chair form of α- and β-D-glucose.

structural representation of the model. We'll come back to this in a moment.

The ring structure of sugars (or monosaccharides) are more usually depicted in one of the two ways shown in Figures 9-32 and 9-33. In the structural representation shown on the right of Figure 9-32, the –H, –OH and –CH$_2$OH groups are depicted as being either above or below the plane of the ring, equivalent to being on the right or left of the representation shown on Figure 9-31. (Remember that unlabeled corners of structural representations such as this are carbon atoms.) A more accurate representation of the 3-dimensional structure of the six membered sugar ring is given by the structure shown on the left of Figure 9-32 and also in Figure 9-33. The α-D-glucose exists in a stable "chair" conformation, with the largest substituents (–CH$_2$OH or –OH) located in the equatorial position (perpendicular to the ring axis),

FASCINATING POLYMERS—WOOD AND REEDS

Wood is a natural composite, composed primarily of cellulose fibers encased in a matrix of lignin and reinforced by polysaccharides. All three are natural polymers (cellulose is a linear polymer of glucose and lignin is a phenylpropanoid network polymer). Wood, of course, has been used as a building material for houses, furniture etc., from time immemorial and it is blessed with a rare, universal appeal. Reeds are another rather unusual building material that are composed primarily of cellulose. American visitors to Essex, Suffolk and Norfolk, in a region of England called East Anglia, generally fall in love with the quaint Victorian cottages that dot the landscape. But, with their irregular dimensions, drafts, and openings that are designed to cause bodily harm, they don't want to live in them! The residents of this part of the world, however, are fiercely proud of their heritage and traditions. Cottages with thatched roofs are much sought after and it is amazing that these roofs efficiently protect the cottage from the elements and can last as long as 50 years. A thatched roof is nothing more than a collection of tightly strapped bundles of dried reeds (akin to laying bundles of drinking straws on top of one another). Laying a thatched roof is a time-consuming art which requires much skill and is labor intensive. Accordingly, building a new one in this day and age is very expensive.

Reeds growing along a bank in Norfolk, England.

A thatched cottage in Suffolk, England.

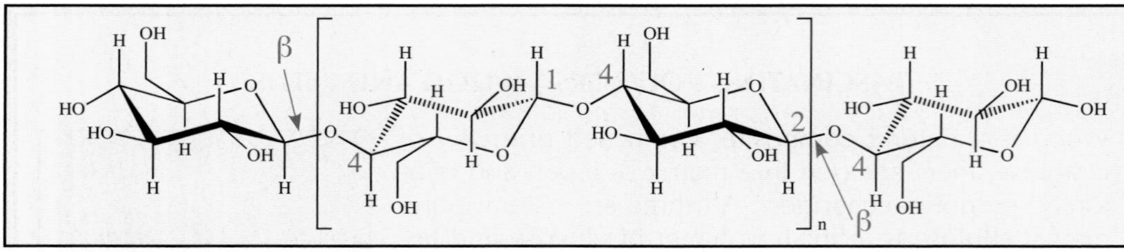

FIGURE 9-34 Chemical structure of cellulose.

except for the –OH group on C_1. The difference between α-D-glucose and β-D-glucose is simply the orientation of this –OH group (Figure 9-33).

If the relative arrangement of the other –H, –OH and –CH_2OH groups on the ring is changed, then you get different sugars. There are also other types of sugars, such as those that are in the form of five-membered rings, as we will see when we get to the polynucleotides. However, life is too short to spend a lot of time trying to memorize the structure of sugars. (Actually, it's not that bad, in that the natural polysaccharides are mostly made up of D-glucose, D-galactose, D-mannose, D-xylose, L-arabinose and D-ribose, together with some derivatives of these.)

The Structure of Cellulose

So let's not worry about the structure of sugars anymore and jump straight to a consideration of some important polymers made by stringing these monosaccharides together. We'll try to show you that the arrangement of certain –OH groups, the ones that are involved in forming the linkages between sugars, has an important affect on properties. Let's start by considering cellulose, as this is the major polymer present in fibrous materials such as cotton or hemp. Cellulose is a linear polymer of D-glucose in which the sugar units are bonded to each other by taking the β form of the molecule and linking carbon atoms 1 and 4, as illustrated in Figure 9-34. Because of this 1,4-β linkage, every glucose residue is rotated 180° with respect to its neighbor (check out the orientation of the –CH_2OH group on each neighboring unit). The 1,4-β linkage has limited confor-

mational freedom and as a result, cellulose chains are relatively stiff.

In addition to having relatively stiff chains, cellulose forms an extensive network of hydrogen bonds (some of which involve water molecules), resulting in the formation of microfibrils from extended chain sheets (Figure 9-35). There are also some amorphous polysaccharides present in a fiber such as cotton and the aggregated microfibrils have extensive pores capable of holding relatively large amounts of water (which makes them useful components of diapers!).

Energy Storage Polysaccharides: Amylose, Amylopectin and Glycogen

Starch, in foods like potatoes, contains the polysaccharides amylose and amylopectin. Like cellulose, amylose, is a linear polymer of D-glucose. However, the linkages are now 1,4-α instead of 1,4-β, as illustrated schematically in Figure 9-36. This seemingly minor change results in a significant difference in properties, however. Amylose is a much more flexible molecule than cellulose and can form a regular helix with six residues per turn.

The second major component of starch, amylopectin, is a branched polysaccharide, containing some 1,6-α linkages in addition to the 1,4-α bonding that occurs in amylose, as illustrated in Figure 9-37. This branching is in the form of small side chains (10 to 20 glucose units) that branch out from the C_6 atom. The branched structure of amylopectin hinders the formation of regular helices and amylopectin is amorphous.

Glucose is a source of energy in living things and is stored in plants in the form of

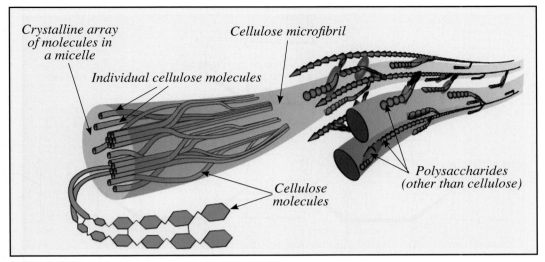

FIGURE 9-35 Schematic diagram of cellulose microfibrils.

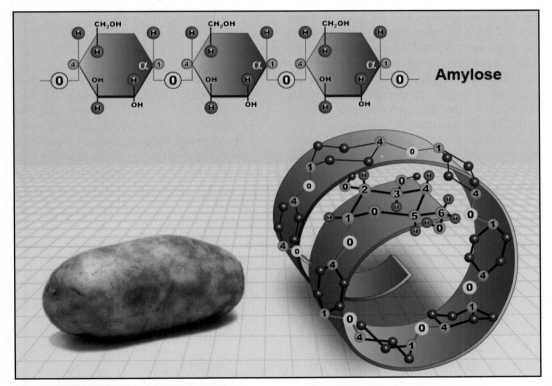

FIGURE 9-36 Schematic diagram of amylose.

amylose and amylopectin. When a cell needs glucose, it uses enzymes to split off glucose units one at a time. These enzymes cannot split the $1,4$-β linkages found in cellulose, however. Cows can digest high cellulose content plant material like grass because the bacteria that live in their stomachs have the right enzymes.

Glucose is stored in animal muscles and the liver in the form of glycogen. This polysaccharide has more branched chains than amylopectin and they are also shorter (about

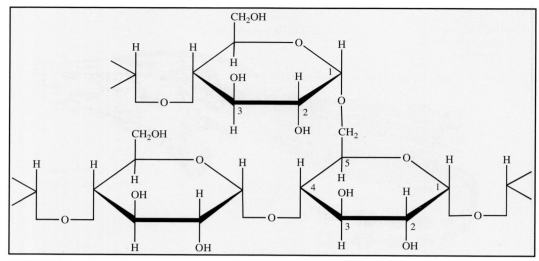

FIGURE 9-37 Chemical structure of branched amylopectin.

eight glucose units). As your body digests french fries and other starchy foods, blood glucose levels rise, and the pancreas secretes insulin. Insulin stimulates the action of several enzymes, which act to add glucose molecules to glycogen chains. After a meal has been digested and glucose levels begin to fall, insulin secretion is reduced, and glycogen synthesis stops. About 4 hours after a meal, the body again starts to break glycogen down, converted it to glucose to be used for fuel by the rest of the body.

Chitin

Finally, nature also takes polysaccharides and chemically modifies them to make useful structures. Chitin is a classic example and can be thought of as cellulose with one hydroxyl group on each monomer replaced by an acetylamino group, as illustrated in Figure 9-38. This allows for increased hydrogen

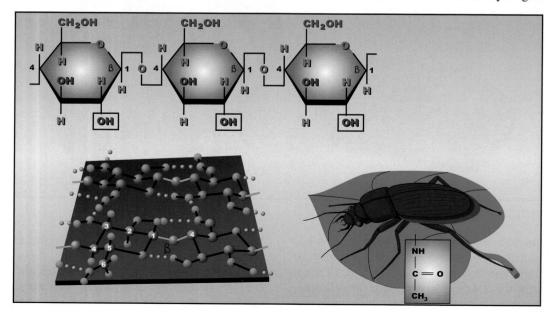

FIGURE 9-38 Schematic diagram of chitin.

bonding between adjacent polymers, giving the material increased strength.

Chitin is one of the main components in the cell walls of fungi and the exoskeletons of insects and other arthropods (you know, crabs and stuff). It is strong and relatively rigid and helps protect an insect against harm and pressure. The rigidity of a biological structure made from chitin depends on its thickness. Often, insect coats contain thick, stiff layers of chitin, while the areas around the legs and face contain very thin, pliable layers.

Chitin also appears to be finding a variety of uses in medicine. For example, it accelerates the skin healing process when applied to human wounds and surgical cloths. An acidic mixture of chitin, when applied to burns, also accelerates the healing process. Left on for a few days, it can heal a third-degree burn completely. It has been shown to support the immune system during certain kinds of illness-blocking procedures. Chitosan, chitin's spinoff, has been known to aid plants' immune systems while they are growing. When placed on a seed, it induces protective measures within the growing plant. This is cool stuff!

POLY(NUCLEIC ACIDS): DNA AND RNA

Introduction

These days, it seems that everyone has heard of DNA, probably as a result of the proliferation of TV shows that feature improbably attractive scientists investigating crime scenes. (We can't wait for the ultimate spinoff, CSI Altoona.) DNA testing is capable of astonishing specificity of identification and has become the primary tool for identifying bodies after a disaster (like 9/11), establishing things like paternity and, of course, attempting to link traces of bodily fluids or tissue left at a crime scene to a perpetrator.

Poly(nucleic acids) are fascinating molecules. They provide the mechanism by which plants and animals pass on their characteristics to the next generation. One of the great discoveries of 20th century science is the elucidation of the structure of DNA (deoxyribonucleic acid) and the essentially immediate recognition of the process by which these molecules replicate themselves, thus providing an understanding of the molecular machinery underlying the process of heredity. It is also a great story, wonderfully well told in James Watson's book *The Double Helix*. Watson, in collaboration with Francis Crick, won the race to identify the structure of DNA, thanks to some remarkable X-ray diffraction work by Rosalind Franklin, whose contributions many regard as being shamefully neglected at the time. But it's all in the book, a flavor of which can be gleaned from Watson's famous opening sentence: "I have never seen Francis Crick in a modest mood."

That the offspring of plants and animals resemble their parents has been known by humankind throughout recorded history and even ancient societies managed to manipulate this through selective breeding to obtain useful traits, such as increased crop yields. The mechanism by which the physical features of living organisms are passed on was first given form in 1865 by Gregor Mendel, who proposed that this must involve discrete units that we now call genes. Each gene codes for a particular trait, like eye color. With the advent of microscopes and staining techniques, it became recognized that living things are made up of cells and these contain a nucleus surrounded by something called cytoplasm and contained by a membrane. It is the nucleus that contains genetic information in the form of chromosomes (so called because they could be observed in living color upon treating with dyes). Genes are strung together like beads on these chromosomes. Observations showed that in the process of cell division, the chromosomes, which could be observed as long strands of material, make copies of themselves and separate, with a complete set of genetic information now residing in both cells. The question that interests us as polymer scientists is how does a chain contain information that can be replicated?

Actually, we are all quite familiar with information stored in the form of chains. The sentence you have just read is an example, with information expressed in the form of letters that are assembled to form words, which in turn are strung together to form sentences. So are there molecular letters that are strung together to form genes that are, in turn, assembled into the structures we call chromosomes? What is the nature of the "alphabet" that these molecules use and how do these molecules divide and reproduce themselves?

In the early years of the 20th century, it was thought that proteins carried genetic information, in part because there was a lot of them in the cell, but also because they have an "alphabet" of twenty different amino acids, which seemed a prerequisite for carrying the complex set of instructions necessary to make a living thing. The other main constituent of chromosomes, deoxyribonucleic acid (DNA), had been identified, but it had only four kinds of repeating units and was regarded as some sort of binding agent for proteins. In the 1940s, however, Oswald Avery discovered that when DNA was transferred from a dead strain of pneumococcus

(the bacteria that causes pneumonia) to a living strain, the hereditary traits of the dead bacteria were also transferred. DNA, not proteins, must carry the genetic code.

So, what is the structure of DNA, what are the four letters in its "alphabet," how can complex information be stored by such seemingly simple molecules (compared to proteins) and how does it reproduce itself?

First, that a code describing complex living creatures can be constructed from four "letters" should be no surprise today, when we are used to the binary code (1s and 0s) used to store information and instructions (programs) on computers. It's the way you combine these that counts. (Keep in mind that this would not have been so obvious 60 or 70 years ago.)

Structure of DNA and RNA

Turning now to the question of structure, both DNA and RNA (ribonucleic acid) can be thought of in a deceptively simple manner. They are chains of sugar molecules linked by phosphate groups. A base, consisting of nitrogen containing aromatic rings, is attached to each of these sugars. A section of a chain is illustrated in Figure 9-39. We will consider these elements one at a time.

First, there are two types of sugars used to build poly(nucleic acids), β-ribose, which has the structure shown in the center of Figure 9-40, and β-2-deoxyribose, shown at the top of this figure. They differ only in the presence of an –OH group in β-ribose that is absent in β-2-deoxyribose. This apparently minor difference has a profound effect on the conformation of the nucleic acids from which they are made and their ability to form certain structures, as we will discuss shortly.

If you've had a cup of coffee and your brain is functioning normally, you may have already guessed that β-ribose is the sugar in the backbone of ribonucleic acids (RNA), while β-2-deoxyribose forms the skeletal structure of deoxyribonucleic acid (DNA). In both types of polymers, these sugars are strung together by condensation reactions (remember those?) involving phosphoric

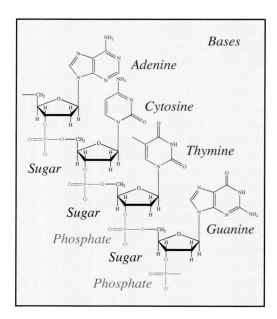

FIGURE 9-39 Section of a DNA chain.

acid (shown at the bottom of Figure 9-40) and the –OH groups at the 3 and 5 positions (illustrated in Figure 9-39).

(Incidentally, ribose is taken as a dietary supplement as it is claimed to increase muscular energy, boost endurance and promote muscle recovery after strenuous exercise. Perhaps your authors should be taking some—not that we ever exercise strenuously!)

As mentioned before, each sugar is also attached to a base, which sticks out from the side of the chain. DNA uses four different bases to make up the chain, adenine, thymine, guanine and cytosine, illustrated in Figure 9-41. RNA also has four bases, but uracil (red color) replaces thymine (blue) in these polymers (Figure 9-41). The crucial parts of these bases are the NH, N and C=O groups, which can engage in hydrogen bonding (NHııııııııN and NHıııııııO=C).

A part of a DNA chain will therefore look like Figure 9-39. To summarize, the repeat unit, called a nucleotide, consists of alternating sugar and phosphate groups. These form the polymer backbone. (The sugar units are

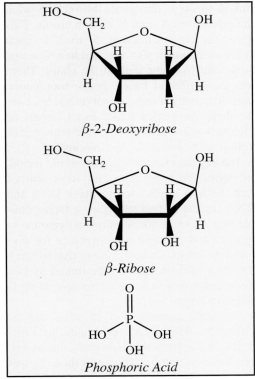

β-2-Deoxyribose

β-Ribose

Phosphoric Acid

FIGURE 9-40 Chemical structure of two sugars and phosphoric acid found in DNA and RNA.

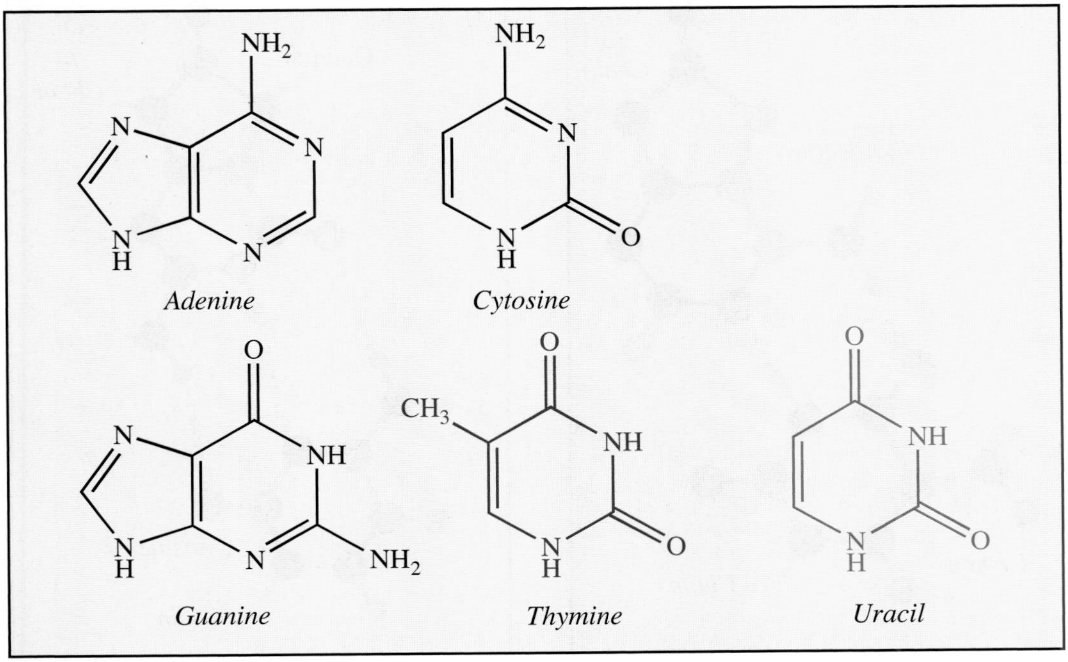

Adenine

Cytosine

Guanine

Thymine

Uracil

FIGURE 9-41 DNA and RNA bases.

linked through phosphate bridges at what are labeled the 3′ and 5′ carbon atoms. This gives the chain direction, 5′ to 3′.) Each sugar molecule is also attached to a base that sticks out from the side of the chain. There are four different bases, hence monomers, used to construct these copolymers. So, how do these molecules make exact copies of themselves and also code for the sequence of amino acids that make up proteins?

The key to replication lies in our old friend, the hydrogen bond. You may have noticed that the four bases used to make DNA and RNA consist of two subtypes: a larger double ring structure for adenine and guanine (a purine) and a single ring structure for thymine (or uracil) and cytosine (a pyrimidine). Thymine and adenine can be linked by two hydrogen bonds to form a base pair, as illustrated in Figure 9-42.

Similarly, guanine and cytosine can be linked by three hydrogen bonds, as illustrated in Figure 9-43. However, the hydrogen bonding patterns shown in these figures are just two among many that can be formed between various combinations of these bases. The crucial aspect of structure is how these patterns are coordinated with the conformation of the DNA and RNA chains. This is what Watson and Crick determined from model building and had confirmed by the superb X-ray diffraction patterns of Franklin.

Given this restriction, Watson and Crick found that only specific pairs of bases could hydrogen bond in an optimum manner (illustrated in Figures 9-42 and 9-43), an adenine with a thymine (A–T) and a guanine with a cytosine (G–C). This means that if a sequence of bases in one chain is given, then the sequence of bases on the other is automatically determined. Watson and Crick then made the following understated observation: "It has not escaped our notice that the specific pairing that we have postulated immediately suggests a possible copying mechanism for the genetic material."

Because of the complementary nature of base pairing in the two strands of the DNA double helix, one chain essentially serves as

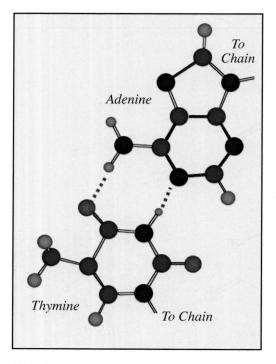

FIGURE 9-42 Thymine/Adenine base pair.

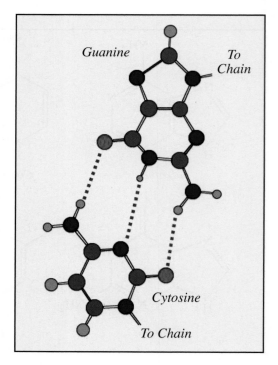

FIGURE 9-43 Guanine/Cytosine base pair.

a template for the other. Accordingly, in replication the double helix "unzips" and new chains are built on each template, as illustrated in Figure 9-44, so that two identical copies of the original DNA molecule are produced. Exactly how this is accomplished is the realm of molecular biology and we won't go there. Our interest is the structure of these polymers and how this relates to the storage and expression of information.

DNA codes for the sequence of amino acids in the proteins that make up living organisms. How does it do this? How does a message expressed in an alphabet that has

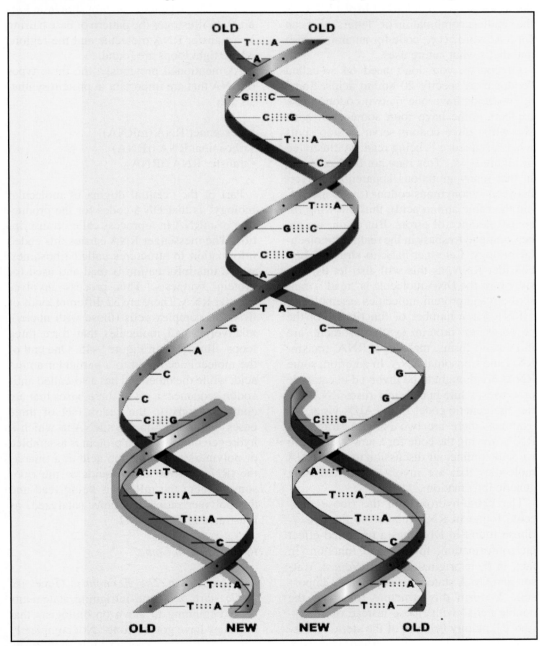

FIGURE 9-44 Schematic illustrating replication of DNA.

four letters (A, T, G and C in DNA), get translated into one with 20 (the amino acids of proteins)? Obviously, there cannot be a one-to-one mapping. Even if two bases are used to code for each amino acid, we still come up short, because there are only 16 different combinations of pairs of bases. But there are 64 combinations of triplets. This is the smallest combination of "letters" that can form a "word" (i.e., code for an amino acid) and that's what nature uses.

Of course, you don't need 64 so-called "codons" to specify 20 amino acids. Some amino acids have one or two codons (base triplets), some have four, some even have six, while three codons serve as stop signs when a sequence is being read, as illustrated in Figure 9-45. This may not be accidental, in that many mutations apparently convert between synonymous codons (those specifying the same amino acid), thus allowing for some tolerance of errors. But again, we are beginning to trespass in the realm of molecular biology. Let's get back to structure and consider RNA, as this will also let us consider how the DNA molecule is "read" (transcribed) and protein molecules assembled.

RNA has a number of functions. For the purposes of protein synthesis, there are three main forms, messenger RNA, transfer RNA and ribosomal RNA. In addition, some RNAs are thought to be involved in catalytic processes, while in certain viruses RNA carries the genetic code. In the AIDs virus, for example, there are two (single) strands of RNA carrying the code for a new virus. Here we will confine our discussion to those RNA molecules that are involved in expressing genetic information.

The extra hydroxyl on the ribose molecule found in RNA relative to the deoxyribose found in DNA has a profound effect on conformation, folding and function. In fact, in their original paper in *Nature*, Watson and Crick stated: "It is probably impossible to form this structure [meaning the double helix] with ribose instead of deoxyribose," mainly because of the steric restrictions imposed by the –OH group. However, it was subsequently found that RNA could form a double helix, although its structure is somewhat different to the DNA double helix. For the most part, however, RNA is found as single strands, but because it has complementary base pairs (adenine/uracil and guanine/cytosine), the molecules can fold back on themselves, forming hairpins, loops and bulges. The schematic picture shown in Figure 9-46 illustrates the pattern of base pairing in a transfer RNA molecule and the regions where tight loops are found.

As mentioned previously, the three types of RNA that are important in protein synthesis are:

- messenger RNA (mRNA)
- ribosomal RNA (rRNA)
- transfer RNA (tRNA)

Part of the "central dogma of molecular biology" is that DNA codes for the production of mRNA in a process called transcription. The messenger RNA carries this coded information to structures called ribosomes, where this information is read and used for protein synthesis. This process involves transfer RNA. There are 32 different kinds of tRNA in complex cells (those with nuclei), relatively small molecules that form three loops, illustrated in Figure 9-45. One end of the molecule can bind to a particular amino acid, while the other end has a so-called anticodon sequence, a set of three bases that are complementary to the codon (set of three bases) being read on the mRNA to which it hydrogen bonds. Thus a protein is assembled or polymerized one amino acid at a time as the tRNA carries amino acids to the ribosomes where the mRNA is being read and the polymerization reactions catalyzed, as illustrated in Figure 9-47.

RNA—More to Come?

An article in *The Economist* (June 16, 2007) starts with the intriguing statements "It is beginning to dawn on biologists that they may have got it wrong. Not completely wrong, but wrong enough to be embarrassing." The article goes on to explore recent

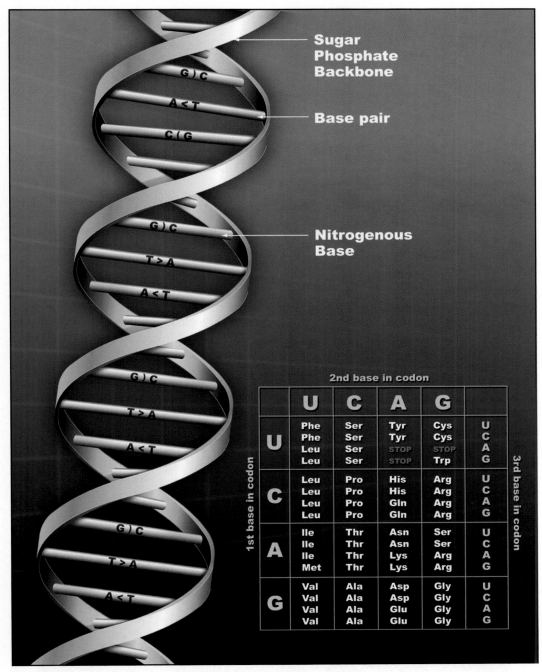

FIGURE 9-45 Example of a DNA code.

research that shows that RNA is not just a messenger and gatherer. It is now clear that some of these molecules are abundant in developing sex cells (God bless them), others can turn off entire chromosomes and so on. There is apparently a universe of RNAs that have be doing their thing unnoticed until very recently. Hopefully, this chapter has given you a feel for how structure begets function in these natural polymers.

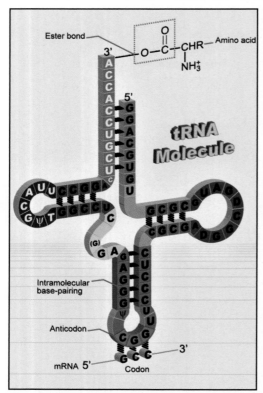

FIGURE 9-46 Schematic diagram of the transfer RNA molecule.

RECOMMENDED READING

R. E. Dickerson and I. Geis, *The Structure and Action of Proteins*, Harper and Row, Publishers, New York, 1969.

A. E. Tonelli, *Polymers from the Inside Out*, Wiley Interscience, New York, 2001.

STUDY QUESTIONS

1. Discuss the nature of the amide bond and its influence on conformation in polypeptides and proteins. (In answering this and most of the other questions listed below, a picture is worth a thousand words!)

2. Compare the chain conformations and types of crystal structures found in nylon 6 and silk. Include in your answer a discussion of the role of hydrogen bonding and other interactions.

3. What makes proline different from other amino acids? What role does it play in the structure of collagen? Describe how glycine

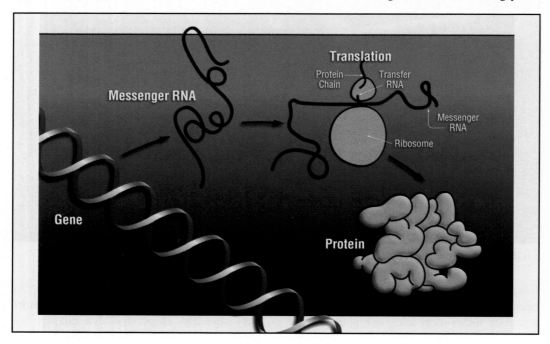

FIGURE 9-47 Schematic diagram of the gene-to-protein process.

also plays an important role in this structure.

4. Compare and contrast the structures of cellulose, amylose and amylopectin.

5. How does the chemical structure of chitin differ from that of cellulose and how does this affect structure and properties?

6. Consider the following fibrous proteins.

    A. Silk (*Bombyx mori*)
    B. Keratin
    C. Collagen

Compare and contrast the type of repeating amino acid sequences found in the ordered parts of their structures and how this affects conformation and higher order structure.

7. Explain why cellulose has a relatively high melting point (above the degradation temperature).

8. Describe the role of hydrogen bonding in maintaining the double helical structure of DNA.

9. How is the genetic information in DNA stored (i.e., what is the nature of the genetic code) and how is it replicated?

10. Go to the internet and find out in what form the genetic information of the AIDS virus is stored. Briefly summarize what you find, mentioning the nature of the molecules that store this information and how they fold.

11. Again using the internet as a source, write a brief essay on the principle fibrous protein found in hemp. Describe some of the uses that this material has been put to by humankind throughout history. Why is this material controversial and banned in the United States? Should all types of hemp be banned?

12. What is the polysaccharide that the human body uses to store energy? Describe the structure of this polymer. Does its microstructure remind you of anything you studied in Chapter 2?

10

Crystallization, Melting and the Glass Transition

INTRODUCTION

We will start this chapter on thermal transitions by assuming that you have studied thermodynamics and perhaps even aspects of statistical mechanics before you started learning about polymers. However, it has been our experience that most students digest this knowledge imperfectly, often returning from summer vacations with partially, if not totally, erased memory banks. What they remember seems dispersed in their brains like those unidentifiable and indigestible pieces of stuff you see in a Christmas fruit-cake. If such is the case with you, dear reader, then you might want to first remind yourself of some aspects of thermodynamics and its statistical basis, which we briefly review in the following section. To decide whether or not you need this review, ask yourself two simple questions:

1. What is temperature (or, more precisely, what are the molecules doing in a hot body that is different from a cold body)?

> *"Thermodynamics—A generally gloomy subject that tells us that the universe is running down, everything is getting more disordered and generally going to hell in a hand basket."*
>
> James Trefil
> N.Y. Times Book Review Section
> June 23, 1991

2. What is free energy?

If you can only answer these questions in a muddled, confused manner, read the review! Otherwise, skip to the first section on thermal transitions—crystallization.

A SUPERFICIAL AND THOROUGHLY INCOMPLETE REVIEW OF THERMODYNAMICS

We love the quote shown in the box, because in our experience, it summarizes many students' attitudes towards the subject of thermodynamics. It is also an amusing take on the second law, the part of the subject most people find confusing, because it involves the concept of entropy, which is not as readily grasped as other thermodynamic parameters, like temperature, pressure and energy. But, it is a piece of fundamental knowledge that should be understood by every scientist and engineer. We would go further. Back in the 1950s, C. P. Snow wrote an essay, *The Two Cultures*, where he noted the growing tendency of educated people to have a knowledge of only one of two broad areas: the arts and humanities, or science and engineering. He later made the comment that a person should not consider himself or herself educated if they did not know the second law of thermodynamics. So as you read what follows, just tell yourself that you're doing this for the development of culture.

The Laws of Thermodynamics

We will start with the laws of thermodynamics, whose confusing chronology was beautifully described by Atkins and is summarized in another box below.

- *The third of them, the second law, was recognized first.*
- *The first, the zeroth law, was formulated last.*
- *The first law was second.*
- *The third law is not really a law.*

P. Atkins
The Second Law
Scientific American Library
W. H. Freeman, New York, 1984

Three of these laws usually do not give students much trouble. The first law, which was actually formulated after the second law, is the conservation of energy, which makes sense to those people who believe you don't get something for nothing. The zeroth and third law deal with temperature, which most students think they understand, but sometimes don't. The big problem is usually the second law.

The purpose of this review is to remind you of a few key concepts. As such, it will not be rigorous, missing out essential caveats and details (isothermal this, adiabatic that) and so on. At the end of it, however, we hope you will have grasped or reminded yourself of the answers to the following questions:

1. What is the subject of thermodynamics?

2. What are the laws of thermodynamics?

3. What is the molecular machinery underpinning these laws?

Furthermore, you will need to have a working knowledge of entropy and free energy, as these are quantities that are essential prerequisites for the proper understanding of the physics and physical chemistry of polymers (or any other material, for that matter).

What Is Thermodynamics?

Essentially, thermodynamics expresses relationships between the macroscopic properties of a system without regard to the underlying physical (i.e., molecular) structure. A classical example of such a relationship is the equation of state of an ideal gas, which you surely remember and which is shown in Equation 10-1.

$$PV = nRT$$
EQUATION 10-1

The four variables (P, V, n, T — we assume you know what these symbols represent) specify the state of the system, hence, the name equation of state. They are not independent, in the sense that once three are specified the other is fixed by this relationship. Equations such as this are not derived from the laws of thermodynamics, but from experimental observation or physical theory. Unfortunately, obtaining equations of state for things like liquids, which will interest us when we consider polymer solutions, is a difficult problem and none of the present theories gives an entirely satisfactory fit to all the available data.

Although the main purpose of this review is to get you to an understanding of entropy and free energy, quantities we will use extensively in discussing the thermodynamic properties of polymers (melting, solutions, measurements of molecular weight, rubber elasticity, etc.), we think it is useful to first re-examine the origin of the gas laws, partly because they appear intrinsically simple, but also because they provide a concrete base on which to explore the laws of thermodynamics. We will then consider heat and work and the molecular machinery underlying the laws. This is a thing of beauty — trust us! The observed experimental relationships suddenly make complete sense and the quantity we call entropy stands revealed in all its complex glory. The development of this subject is as much an act of imagination and ingenuity as any painting by Leonardo or symphony by Beethoven. (One of your two authors, the pretentious one, prefers Mahler —there is no accounting for taste.) We will start by examining definitions of basic quantities—pressure, volume, temperature etc.— then work our way into energy and the laws of thermodynamics.

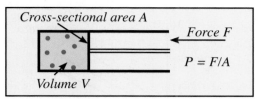

FIGURE 10-1 Schematic diagram of a gas confined in a cylinder.

The Ideal Gas Laws

Volume and pressure are easy. Volume is simply the amount of space occupied by something, in this example a gas, whose volume is actually defined by its container, in this case a cylinder with a piston in one end (Figure 10-1). (You get sick of pictures of cylinders and pistons after studying thermodynamics for a while!) Pressure is simply the force exerted per unit area on the walls of the container, including the piston ($P = F/A$).

Thanks to the newly invented air pump and with the aid of a young laboratory assistant named Robert Hooke, James Boyle (in the 1600s) determined a relationship between the pressure and the volume of a gas (he was actually studying a mixture of gases—air). He found that for a given amount of gas held at a constant temperature the pressure exerted by a gas is inversely proportional to the volume, $P \sim 1/V$, as illustrated in Figure 10-2. It was subsequently found that there are deviations at high pressures, but we'll ignore these for now.

Like pressure and volume, temperature is a thing of everyday experience and most people think they understand it, at least to the extent of being able to say that it is a measure of how hot or cold a body is relative to our senses. Giving a precise definition is more difficult, however. Also, the scales most commonly used to measure this quantity, Celsius and (less frequently these days) Fahrenheit, are arbitrary. The thermodynamic definition, ensconced in the zeroth law, essentially states that if there is no heat flow between two bodies, they are at the same temperature. This is not very enlightening and we will have to wait until we discuss what the molecules are doing to get more insight. Nevertheless, the idea that tempera-

ture determines the direction in which heat will flow (i.e., from a "hotter to a "colder"), although seemingly obvious, is crucial to the development of thermodynamics.

A relationship between temperature and the volume of a gas was discovered by Jacques Charles, who was an enthusiastic hot-air balloonist and thus interested in things that might prevent him from plunging to a nasty death. If the volume of a gas is plotted against temperature then, as with Boyle's law, there is a range of linear behavior and (in this case) deviations at low temperatures (Figure 10-3).

If we just consider the straight parts of the data, however, and make measurements at different pressures, then all the lines extrapolate to zero volume at about $-273\degree C$. Because volumes cannot be negative, this suggests that $-273\degree C$ (actually, $-273.16\degree C$) is the lowest attainable temperature.

William Thomson, Lord Kelvin, had the brilliant idea of redefining this lowest attainable temperature as zero, or absolute zero, but leaving the size of the "steps" the same as on the Celsius scale. This led to a considerable simplification of the laws of thermodynamics and also defines a point of real

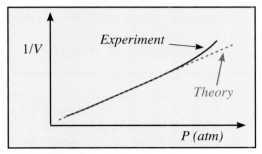

FIGURE 10-2 Schematic plot of the reciprocal of the volume of a gas versus pressure (at constant temperature).

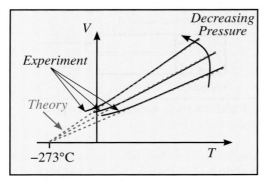

FIGURE 10-3 Schematic plot of the volume of a gas versus temperature (at constant pressure).

physical significance, as we will see later. There is also a law of thermodynamics (the third law) that deals with absolute zero—the "you can't get there from here" law—which states that you cannot get to the absolute zero of temperature in a finite number of steps.

We have considered relationships between the pressure, volume and temperature of a gas, but obviously properties also depend on how much stuff is present. To account for this properly, we have to step out of the purely thermodynamic context we have been using and introduce the atoms. Amedeo Avogadro proposed that at a given temperature and pressure, equal volumes of gas contain the same number of molecules, or $V \sim n$. There is an enormous number of molecules in (say) 1 liter of a gas, so instead of using numbers of molecules, we use moles (the word comes from the Latin, "massive heap"). The modern definition is given in the box.

---

1 mole (mol.) of particles
= the number of atoms in 12 grams of ^{12}C

1 mol. = 6.022 × 10²³ molecules

The relationships discovered by Boyle, Charles and Avogadro are limiting laws. Boyle's law, for example, does not describe the properties of any gas precisely, but is exact in the limit of zero pressure. Another way of looking at these laws is that they precisely describe the properties of a hypotheti-

cal ideal gas. (We can give some physical meaning to this later.) The combined equation is the ideal or perfect gas law and is an equation of state for an ideal (perfect) gas (Equations 10-2).

Boyle's Law:
$$P \sim 1/V \ (at \ constant \ T, n)$$

Charles' Law:
$$V \sim T \ (at \ constant \ P, n)$$

Avogadro's Principle:
$$T \sim n \ (at \ constant \ P, T)$$

Combining these relationships:
$$PV \sim nT$$

Defining R as a constant of proportionality:
$$PV = nRT$$

EQUATIONS 10-2

Note that, like all thermodynamic expressions, it is simply a relationship between macroscopic quantities—it tells us nothing about its molecular origin.

Heat, Work and Energy

With this foundation, we can now move on to a review of thermodynamics proper, in its origins the transformation of heat into work (and vice-versa), but more broadly taken to mean transformations between forms of energy. So, what is energy and how did the concept originate?

About two hundred or so years ago, heat and work were considered to be different things and had different units (something that continues to bedevil students to this day). Following Newton, work was defined as force times distance moved. Heat has a more complex and interesting history, starting from something called phlogiston and passing through Lavoisier's (Figure 10-4) concept of caloric (in which heat was considered to be a weightless form of matter that flowed in and out of materials).

Lavoisier's caloric theory was demolished by Benjamin Thomson, Count Mumford, who fled the American Colonies following the revolution and ended up a sort of mercenary in Bavaria. Observing the boring of a cannon, he realized that if caloric was the origin of the heat it should run out at some point. (Logically, if the cannon had too much caloric to begin with, it would melt itself!). Continued boring would therefore no longer heat the metal. He presciently observed that heat must be some sort of motion. After thus doing a number on Lavoisier's theory, he married Lavoisier's widow (after Lavoisier was guillotined on trumped-up charges of corruption during the French Revolution), and inherited all his money.[21] And you thought thermodynamics was boring!

The idea of energy and its conservation —the first law of thermodynamics—was directly or indirectly proposed independently by a number of people during the first half of the 19th century. Three names stick out. The first, a German physician, Robert Mayer, as a result of the observations he made as a ship's doctor on a voyage to what is now Indonesia. He extrapolated Lavoisier's idea that "burning" food creates body heat, but also provides its ability to perform mechanical work.[22] He also realized that body motion produces heat through friction and this mingles with the heat from "combustion." In other words, the bodies "combustion processes" produce motion, heat and work and these are all inter-convertible. Also recalling Lavoisier's law of the conservation of matter, he thought something similar could be said about heat. He called the capacity to do work, "force"; we now call it energy. This doesn't tell us what energy is, but then perhaps nothing can (see quote in box).

FIGURE 10-4 Lavoisier with his wife (reproduced with the kind permission of the London Science Museum).

> *"It is important to realize that in physics today we have no knowledge of what energy is. . . It is an abstract thing. . ."*
>
> R. Feynman

Mayer also realized that he needed to quantify what we now call the mechanical equivalent of heat, but relied on philosophical speculation rather than experiment. This latter field was the province of James Prescott Joule (Figure 10-5), a Manchester brewer who had skilled instrument-makers working for him, as a result of his noble and humanitarian pursuit of making better beer through control of temperature. He became a master of thermometry. With no knowledge of Mayer's work, he independently discovered the first law through painstaking measurement. He is even reported to have taken a thermometer on his honeymoon to Mont Blanc, where he attempted to measure the rise in temperature of water that should result from its passage over a waterfall. His-

[21] If you are interested in the history of this subject, one cannot do better than read *Maxwell's Demon*, H. C. von Baeyer, Random House, 1998, where we got most of the good stuff in what follows.

[22] Actually, converting "chemical" energy stored as bonds to other forms during the metabolism of food.

FIGURE 10-5 James Prescott Joule (reproduced with the kind permission of the London Science Museum).

tory does not record what his new wife made of all this. Today we call the units we use to measure energy, work and heat, joules (1 joule = 1 kg m^2 s^{-2}).

Finally, Helmholtz put the first law in mathematical form. In illustrating this law, many thermodynamic textbooks often use a picture of another damned piston and cylinder (Figure 10-6), where the change in energy of a system is equal to the heat put into that system minus the work done by that system (Equations 10-3).

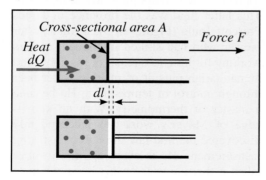

Cross-sectional area A

Force F

Heat dQ

dl

FIGURE 10-6 Schematic diagram depicting the heating of a gas in a cylinder which moves the piston.

$$dE = dQ - PdV$$

Note that:
$$Fdl = [F/A]Adl$$
$$Fdl = Pdv$$
EQUATIONS 10-3

(If your mind has wandered from the always riveting subject of thermodynamics and you wonder where the PdV came from, just recall your classical mechanics, work = force times distance moved, and use the substitution also shown in Equations 10-3.)

We do not need to take the first law any further, but do need to remind you of a couple of additional definitions. When a system takes in energy from the surroundings (or releases energy to the surroundings) we say there is a change in its internal energy. This is just the sum of the kinetic and potential energies of the atoms, molecules, or ions in the system. We also often consider changes under one of two conditions: at constant volume or at constant pressure. The relationships are summarized in Equations 10-4.

Recalling that:
$$dE = dQ - PdV$$
then at constant volume:
$$dE = [dQ]_V$$
while at constant pressure:
$$dE = [dQ]_V - PdV$$
EQUATIONS 10-4

You should remember that we call the heat absorbed from (or emitted to) the surroundings at constant pressure, dQ_P, the enthalpy, dH and it follows that (Equations 10-5):

$$dH = dE + PdV$$
If the processes are not
infinitesimally small:
$$\Delta H = \Delta E + P\Delta V$$
EQUATIONS 10-5

Entropy and the Second Law

In 1823, some 20 or so years before Helmholtz presented his mathematical formulation of the first law, Sadi Carnot, a French engineer, published the results of his studies of steam engines. His goal was " . . . to consider in the most general way the principle of the production of motion by heat." Although he based his thinking partly upon Lavoisier's caloric theory, he nevertheless came up with a form of the second law of thermodynamics and got close to formulating the first.

His major insight was to realize that steam engines (or any other mechanism for obtaining work from heat) are not only inefficient because of the vagaries of human design, poor materials, frictional losses, etc., but because this is part of the natural order of things. In other words, one cannot obtain work from heat without throwing some of that heat away (Figure 10-7). Carnot then came up with the idea of running the machine backwards, thus extracting heat from a cold body and discarding it into one that is warmer by doing work (you know, a refrigerator). He imagined a hypothetical perfect reversible engine, one that would have no frictional or other losses, and this concept and method of abstract analysis played a central role in the future development of thermodynamics. Finally, he developed the theorem that the efficiency of his reversible engine depends only on the difference in temperature of the two reservoirs, a result of enormous practical importance.

There were some contradictions in Carnot's work—a result of his reliance on the caloric theory—that were subsequently cleared up by Clausius. Clausius accepted Carnot's proposition that some heat must be thrown away when converting heat to work as a law of nature, something that cannot be proved or derived from something else, but as far as we have ever been able to tell describes the way the world works. He called it the second law of thermodynamics and then sought to recast it in a different, more general, form that did not apply to heat engines alone. He showed that an equivalent statement of the

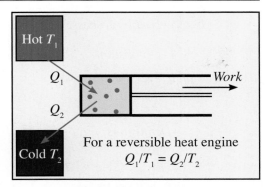

FIGURE 10-7 Getting work from heat.

second law is that heat flows naturally from a hot to a cold body, but not the other way around. In other words, just as the first law says that it is impossible to build a perpetual motion machine, the second law tells you that it would be impossible to obtain energy from the vast heat reservoir that is the ocean (for example) without having a cooler reservoir in which to discard waste heat.

Clausius, try as he might, found that he could not recast the second law mathematically in terms of energy alone. There was a missing quantity. He returned to the idea of a perfect reversible heat engine and tried to determine how much heat from a hot reservoir can be converted to work and how much must be discarded at a lower temperature. As part of this analysis, he found a simple relationship. The ratio of the heat taken from the hot reservoir to the temperature of that reservoir was equal to the ratio of heat discarded to the cold reservoir to its temperature (see Figure 10-7). Also, this relationship only held if what was then the new absolute temperature scale was used. After exhaustively analyzing many systems, he discovered that the quantity of heat over (absolute) temperature is a constant in a reversible process and increases in irreversible ones (see Equations 10-6). He arbitrarily gave this ratio the symbol S, but carefully chose the name entropy, from the Greek word for a transformation.

Accordingly, we now have a mathematical definition of entropy and the idea that it determines the direction of irreversible processes (heat flows from "a hotter to a colder"). But

In a reversible process:

$$\frac{Q}{T} = constant$$

For an irreversible process:

$$\frac{Q}{T} > 0$$

Definitions:

$$\Delta S_{rev} = \frac{Q}{T} \qquad \Delta S_{irrev} > \frac{Q}{T}$$

EQUATIONS 10-6

entropy remains an abstract quantity, one that is difficult to have the same type of "feel" for as temperature and energy. Nevertheless, it is profound in its consequences (see quote in box).

"The law that entropy always increases—the second law of thermodynamics—holds, I think, the supreme position among the laws of Nature. If someone points out to you that your pet theory of the universe is in disagreement with Maxwell's equations—then so much the worse for Maxwell's equations. If it is found to be contradicted by observation—well, these experimentalists do bungle things sometimes. But if your theory is found to be against the second law of thermodynamics I can give you no hope; there is nothing for it but to collapse in deepest humiliation."

Sir Arthur Eddington

Finally, we need to remind you about free energy. In many of our discussions we will be trying to establish whether or not a process will occur spontaneously—"will this polymer dissolve in that solvent?", for example. Conceptually, this is easily done, because once a system reaches equilibrium, its entropy is a maximum, so all we need to do is calculate if the entropy change for that process is positive. This is not so easily done for real systems, however, because they are not isolated from their surroundings. Dis-

solving a polymer in a solvent placed in a flask essentially occurs at constant temperature and pressure (atmospheric). If heat is given up during dissolution, then it is quickly absorbed into the surroundings, changing the entropy not only of the system, but the rest of the universe as well. So a new variable is needed, one that is related to the properties of the system alone and reaches a maximum or minimum when the entropy is at maximum. This is the free energy, which has two forms, the Gibbs free energy, which applies at constant pressure, and the Helmholtz free energy, which applies to constant volume processes. These are defined in Equations 10-7.

Gibbs free energy:

$$\Delta G = \Delta H - T\Delta S$$

Helmholtz free energy:

$$\Delta F = \Delta E - T\Delta S$$

EQUATIONS 10-7

The form of the equation, an energy term minus an entropy-dependent term, suggests that the free energy should be related to the maximum work that can be done by a process. And it is—the maximum work actually being equal to the negative of the free energy change during that process. Thus, from this definition, we get back to Carnot and Clausius. To many students, this is not very enlightening, however, because all we have done is substitute one abstract quantity for another. Furthermore, this is still a thermodynamic relationship. Unless we can relate quantities such as H and S to molecular quantities, we do not have a way to predict if a process will occur spontaneously. So now we must introduce the molecules, and by so doing, reveal the fundamental character of these (so far) abstract concepts.

Enter the Molecules

In his wonderful *Lectures on Physics*, Richard Feynman wrote: "If in a cataclysm all human knowledge was destroyed except one sentence that could be passed on to future generations . . . what would contain the most information in the fewest words?"

> *"All things are made of atoms—little particles that move around in perpetual motion, attracting each other when they are a little distance apart, but repelling upon being squeezed into one another."*
>
> R. Feynman

The gas laws and the laws of thermodynamics make no reference to the atomic and molecular basis of matter. The idea of atoms had been around for more than 2000 years, of course, having been introduced by the ancient Greeks. John Dalton also utilized the concept, but, as used by the chemists of that time, atoms were generally regarded as abstract units of matter that allowed the proportions of elements to be tracked during reactions and were not considered by most to be real particles.[23] (Although Dalton actually had a physical picture in mind; he regarded atoms as solid, indivisible particles surrounded by an atmosphere of heat—to accommodate Lavoisier's caloric theory.)

It is therefore remarkable that 100 years or so before the laws of thermodynamics were formulated, Daniel Bernoulli developed a "billiard ball" model of a gas that gave a molecular interpretation to pressure and was later extended to give an understanding of temperature. This is truly a wonderful thing, because all it starts with is the assumption that the atoms or molecules of a gas can be treated as if they behave like perfectly elastic hard spheres—minute and perfect billiard balls. Then Newton's laws of motion are applied and all the gas laws follow, together with a molecular interpretation of temperature and absolute zero. You have no doubt

[23] Berthelot, while Minister of Education in Paris, went so far as to decree (in 1886) that atomic theory was a hypothesis that should not be taught in schools. (Doesn't this make you think of current controversies surrounding the theory of evolution?)

studied this as the kinetic theory of gases, and if you remember the molecular meaning of temperature, just jump over the following section.

The Kinetic Theory of Gases

We will start by imagining that we have a cylinder with a piston at one end that can move without any frictional losses. (We told you would get sick of cylinders and pistons!) Then each collision of the perfectly elastic particles of the gas would move that piston a little bit. Of course, a real gas has an enormous number of atoms or particles, so the net effect of all the collisions is felt like a continuous force, rather than individual impulses. We want to calculate the force necessary to keep the piston stationary. The pressure is then this force divided by the cross-sectional area of the piston. To calculate this, we will need to sum up all the impulse forces delivered to the piston. Recalling our classical mechanics, we can write Equation 10-8.

$$F = m\ddot{x} = \frac{d}{dt}(m\dot{x})$$

EQUATION 10-8

Or, the force (F) delivered by the impact of a single particle is equal to the rate of change of momentum of that particle as a result of the collision. What we need to calculate is how much momentum is delivered as a result of all the collisions that occur in a given time, say, a second.

The problem is that the molecules are moving chaotically, in different directions with different speeds. This problem can be overcome by dealing with average values. Let's first assume that a particle strikes the piston with a component of velocity in the x-direction (normal to the piston surface area) of v_x. Equation 10-9 then gives the change in momentum.

$$Change = mv_x - (-mv_x) = 2mv_x$$

EQUATION 10-9

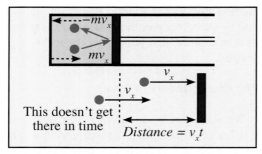

FIGURE 10-8 The kinetic theory of gases.

If all the particles happen to be moving directly towards the piston with a velocity v_x, then only those particles that are within a distance $v_x t$ of the piston, or within a volume $A v_x t$, will actually hit the piston in a time t (Figure 10-8). If there are $N = n/V$ particles per unit volume (n = total number of particles), then the number hitting the piston in time t is $NA v_x t$ and the number of collisions per second (n_c) is simply $NA v_x$. The force (F) on the piston and hence the pressure exerted by the gas are then simply calculated as shown in Equations 10-10.

$$F = (n_c)(2mv_x) = (Nv_x A)(2mv_x)$$

$$P = \frac{F}{A} = 2Nmv_x^2$$

EQUATIONS 10-10

But this is not quite right and this is where the averages come in. The molecules will not all have the same component of velocity v_x in the x-direction. Not only that, the molecules are equally likely to be moving away from the piston as towards it, so we replace v_x^2 with a value averaged over all the molecules, $<v_x^2>$ and divide by 2, to get Equation 10-11.

$$P = Nm\langle v_x^2 \rangle$$

EQUATION 10-11

This is still not right, because if the particles are moving randomly, they are equally likely to be moving with a given velocity in any direction, which leads to Equations 10-12.

The particles are equally likely to be moving in any direction, hence:

$$\langle v_x \rangle = \langle v_y \rangle = \langle v_z \rangle$$

The velocity of the particle, v, is then:

$$\langle v^2 \rangle = \langle v_x^2 + v_y^2 + v_z^2 \rangle$$

So that the average value in the x direction is:

$$\langle v_x^2 \rangle = \frac{1}{3} \langle v_x^2 + v_y^2 + v_z^2 \rangle = \frac{\langle v^2 \rangle}{3}$$

EQUATIONS 10-12

Substituting for $<v_x^2>$ in Equation 10-11, Boyle's law (PV = constant) immediately follows, as shown in Equations 10-13.

$$P = \frac{2}{3} N \left\langle \frac{mv^2}{2} \right\rangle$$

$$P = \frac{2}{3} \frac{n}{V} \left\langle \frac{mv^2}{2} \right\rangle$$

$$PV = \frac{2}{3} n \left\langle \frac{mv^2}{2} \right\rangle$$

$$P = n \frac{2}{3} \langle KE \rangle$$

EQUATIONS 10-13

(In making the substitutions, remember that $N = n/V$ is the number of particles per unit volume. KE is the kinetic energy and $<KE>$ is its average value.)

This assumes that at constant temperature the average velocity of the particles does not change.[24] Of course, there is no reason why it should. Furthermore, a comparison of this equation to the ideal gas law, $PV = nRT$, suggests that temperature, or more precisely, RT, is related to the average kinetic energy of the particles.

A more formal derivation of this relation-

[24] We have been using the word, particle, to indicate that we are still referring to a billiard ball model of atoms or molecules. Molecules also have internal motion, the vibrations of their bonds, etc., so in that case, the velocity refers to the motion of the center of mass of the molecule treated as a whole.

ship follows from a consideration of what happens when you have two gases separated by a moveable, frictionless barrier (Figure 10-9).

The molecules in one are initially moving fast while the others are moving slow. The piston will jiggle backwards and forwards, exchanging momentum between the molecules of the different gases until at equilibrium they are equal (Equation 10-14).

$$\frac{1}{2}\langle m_1 v_1^2 \rangle = \frac{1}{2}\langle m_2 v_2^2 \rangle$$

EQUATION 10-14

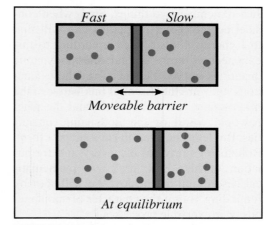

FIGURE 10-9 Schematic diagram of two gases separated by a moveable barrier.

At equilibrium the two gases are at the same temperature (the zeroth law of thermodynamics), so we could, if we wanted, define the temperature to be $T = (3/2)<KE>$. However, because the scale of temperature we use is degrees Kelvin, we need a constant of proportionality. This is k (Boltzmann's constant) when n is the number of molecules [$kT = (3/2)<KE>$) and R, the gas constant, when n is the number of moles ($RT = (3/2)<KE>$]. If you stop and think for a moment, this result immediately gives a physical meaning to the absolute temperature as the point where molecular motion ceases.

So, a simple approach based on Newton's laws of motion gives us the gas laws, a molecular interpretation of pressure and temperature and a definition of the absolute temperature. This is truly a remarkable achievement and a thing of beauty.

The Maxwell Distribution of Velocities

Maxwell, perhaps the greatest theoretical physicist between Newton and Einstein, was not happy with the use of average values for things like velocity. Furthermore, he considered the atoms and molecules of a gas to be real particles, unlike many of his contemporaries, who thought that they were merely a useful fiction that helped in the formulation of mathematical models. He had a picture of a gas consisting of innumerable numbers of these things moving about and colliding

chaotically, such that some were moving much faster than others. He turned to the use of probability theory and determined that the speed of the particles was distributed according to the familiar bell curve of a Gaussian. In other words, the fraction of molecules with a speed in the range s to $s + ds$ goes as e^{-x}, where x is proportional to s^2 (divided by kT). Alternatively, we could say that this fraction is proportional to $e^{-(\text{Kinetic Energy}/kT)}$. The same distribution applies to things like Brownian motion and the end-to-end distance of polymer chains in the melt and can be obtained from a simple consideration of coin-toss statistics. This is obviously of interest to us, so we will discuss it in a little more detail when we consider rubber elasticity.

Boltzmann generalized Maxwell's theory and found that the probability p that a molecule will be found in a state with an energy E is given by Equations 10-15,

$$p = \frac{e^{-E/kT}}{q} \quad where \quad q = \sum_i e^{-E_i/kT}$$

EQUATION 10-15

where the sum is overall the allowed energy states of the system (and is called the partition function). These equations are the heart of statistical mechanics, but we are probably already going too far in this simple review.

Just remember two things. First, when considering molecular properties, we will find that energy is distributed according to this exponential form. Second, thermodynamic quantities can be calculated from these equations; they are the essential link between the microscopic molecular world and the macroscopic world of thermodynamic quantities. But let's consider this in a simpler form, Boltzmann's original discovery and formulation of his "holy grail"—a mathematical description of entropy. We will start by reminding you of the character of spontaneous or irreversible processes.

Entropy Revisited

We have mentioned that one definition of the second law is that it defines the direction of a spontaneous process, such as heat flowing from "a hotter to a colder." Let's consider a couple more examples. If a weight is suspended above the ground (Figure 10-10), then it has a certain amount of potential energy due to the force of gravity. If somebody cuts the rope suspending this weight, then the potential energy is converted to

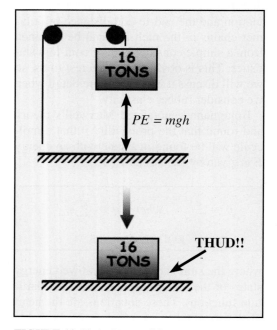

FIGURE 10-10 An irreversible process.

kinetic energy as the weight falls through the air. This, in turn, is dissipated as heat (kinetic energy of the molecules in the ground) and any work done as a result of deformation as the weight thudded to a halt.

We could put energy back into the weight by heating it up, causing its molecules to move faster, but the weight would not spontaneously jump off the ground. That is because the motions (which would now be largely associated with vibrations of the molecules, at least up until the melting point) are not coordinated throughout this macroscopic body. At any instant of time some molecules are moving this way, and some are moving another. In contrast, when falling through the air, the center of mass of the weight and hence all of its individual molecules, were moving in a coordinated direction—down! (At any instant of time the molecules would also have been vibrating around their mean position, but that does not matter.) In other words, we have quantities related to how many molecules are present and how fast they are moving (temperature, energy, etc.), but not to how things like energy and molecules are distributed in the system.

Take as another example, the situation where a gas is initially confined in one corner of a large container by a barrier. Once the barrier is removed, the gas particles disperse randomly, a classic example of the second law (that matter and energy tend to disperse chaotically) in action (Figure 10-11). You could watch these particles until the cows come home and they would never spontaneously reassemble of their own accord in the corner (that is, without doing work on the system to push them all into that corner). Clearly, how things are arranged or distributed is a crucial property, just like energy.

Boltzmann's Equation

Boltzmann perceived this link between nature's preference for disorder and entropy and, as a result of his extension of Maxwell's work, looked for the relationship through probability theory. Boltzmann knew that entropy, like volume, weight and energy, is an additive quantity. So, if you had equal volumes of a gas in a container divided by a partition, then removed the barrier, the new system would be twice the volume of each of the original "halves" and the entropy would also be twice the entropy of each of the original systems (Figure 10-12). The problem is that probabilities are multiplicative, while entropy is additive.

To see this, imagine having two separate containers, each containing just one molecule.[25] Also imagine that each of these molecules can travel at any one of only five speeds, rather than any speed in the entire distribution. There are then five ways of assigning a speed to the molecule in the first vessel and also just five ways to assign a speed to the molecule in the second vessel. Now join the containers and ask how many ways there are of assigning these five speeds to two molecules. Molecule 1 could be traveling along at, say, speed 1 at some instant of time, but at the same time molecule 2 could be traveling along at this same speed or any one of the other four that are allowed (Figure 10-13). There are therefore $5 \times 5 = 25$ ways of assigning the speeds to the molecules (not $5 + 5 = 10$).

The way an additive property can be related to a multiplicative one is through logarithms. So, by induction, Boltzmann arrived at Equation 10-16, which says that entropy is equal to the logarithm of the number of arrangements available to a system, or its probability.

$$S = k \ln \Omega$$

EQUATION 10-16

The way this works (entropies additive, probabilities multiplicative) for our simple example is shown in Equation 10-17.

$$S_1 + S_2 = k \ln \Omega_1 + k \ln \Omega_2 = k \ln \Omega_1 \Omega_2$$

EQUATION 10-17

[25] We also took this example from *Maxwell's Demon*, H. C. von Baeyer, Random House, 1998.

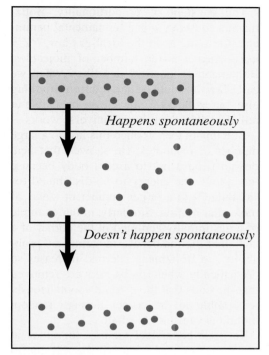

FIGURE 10-11 Another irreversible process.

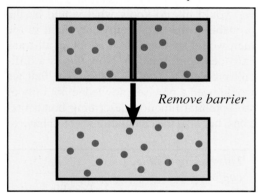

FIGURE 10-12 Mixing of two gases.

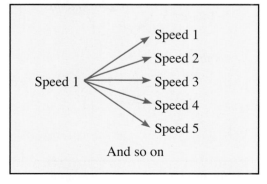

FIGURE 10-13 A simple example of distributions.

The constant of proportionality, Boltzmann's constant, k, is a fundamental parameter, the same as in the ideal gas law, $PV = nkT$, when n is the number of molecules. Boltzmann's equation describes what we call a law. It cannot be derived from anything else, but as far as we have ever been able to tell, describes the way the universe works.

Boltzmann then applied this law to a large number of examples. He showed that heat flowed from a hot to a cold body because this allows the energy to be distributed (or "assigned") to a larger number of states of molecular motion. Similarly, in our example of a gas initially confined in the corner of a container, the number of arrangements ("probability" in Boltzmann's terms) increases so dramatically when the barriers are removed that the odds that they would spontaneously reassemble in the corner through random motion is vanishingly small.

To get a better feel for this, let's look at a polymer example that will be relevant to our discussion of crystallization. Essentially, we would like to know what would be the probability that a polyethylene chain in the melt would spontaneously adopt an all-trans extended conformation. Now this is a difficult problem to tackle, in the sense that we would need a precise knowledge not only of the potential function describing bond rotations, but also the calculations would have to account for both short- and long-range steric hindrance—a certain combination of bond rotational angles would bring different parts of the chain too close together. However, considerable insight can be obtained using a simple model. We covered this in our discussion of conformations, but to save you the trouble of going backwards and forwards, we'll repeat some of the discussion here.

We assume that the chain has only three local minima, like trans, gauche and the other gauche in polyethylene (see Chapter 8!), but these local minima have the same energy (and each bond is therefore equally likely to be in any one of these states). Let's also assume that a bond can only be found in one of these three conformations (i.e., when it clicks from one to another it does so instantaneously). The first bond can therefore be found in any one of three conformations, as can the second, the third, and so on (Figure 10-14).

For each of the three conformations of bond 1 (G, T or G´) there are three possible arrangements of the second bond (G, T or G´), giving a total of 9. For a chain of 10,000 bonds there are therefore $3^{10,000} = 10^{4,771}$ conformations or shapes available to the chain (for simplicity, we're ignoring redundancies.) However, there is only one way of arranging the chain so all the bonds are trans. If all conformations or combinations of trans and gauche bonds are equally likely, then it would follow that the probability of finding a chain in the melt that had spontaneously arranged all its bonds to be trans would be vanishingly small, $\sim 1/10^{4,771}$!

This is a crude model, but hopefully you now see how the "calculus of probabilities," as Maxwell put it, explains why heat flows "downhill" (from hot to cold), why a gas expands to occupy its container and why "the world is . . . getting more disordered and generally going to hell in a handbasket"! We also hope that you now have a "feel" for entropy that cannot be obtained from the purely thermodynamic definition of heat divided by temperature. In principle, calculating the entropy of a system would now seem to be easy. Just count the num-

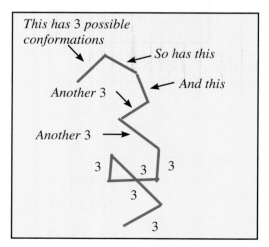

FIGURE 10-14 Possible chain conformations assuming three local minima.

ber of arrangements available to that system and take its log! Of course, in practice, this is often extraordinarily difficult and further complications arise from the constraints that you have to impose when doing the calculations.[26] Usually, you construct a (hopefully) simple model of the system and then see how closely your model gets to the reality of reproducing experimental data. This is obviously beyond the scope of what we want to do in our introduction to polymers, but we will need to use Boltzmann's equation in tackling things like rubber elasticity and polymer solutions. Before returning you to the main body of this work, however, we need to cover one final topic—free energy—as this is a quantity that we will use a lot.

Free Energy

We have previously defined the free energy in terms of the maximum work that you can get out of a process. To finish up, let's look at it in a different way, one that is more useful in understanding problems such as mixing. Let's say we are going to mix two systems. In the illustration in Figure 10-12 we used two gases, but it could be two liquids, such as oil and water. We would like to know whether or not our two systems mix. The criterion, of course, is that the entropy must increase, or the entropy change upon mixing should be positive, as required by the second law of thermodynamics ($\Delta S_{total} > 0$).

It is important to realize that this is the total entropy change, because mixing could result in heat being emitted to, or absorbed from, the surroundings, which would change its entropy. We can therefore write the total entropy change as in Equation 10-18.

$$\Delta S_{total} = \Delta S_{system} + \Delta S_{surroundings}$$
EQUATION 10-18

[26] There are various approaches, in addition to Boltzmann's equation, that can be taken to calculate thermodynamic parameters (e.g., through the use of the partition function).

However, using the thermodynamic definition of entropy we can equate the entropy change in the surroundings as the change in heat over temperature. The heat change is that given out upon mixing, of course. Assuming that our mixing will be done at constant pressure, we can also equate the heat to the enthalpy. We then get Equation 10-19.

$$\Delta S_{surroundings} = -\frac{\Delta Q_{system}}{T} = -\frac{\Delta H_{system}}{T}$$
EQUATION 10-19

Substituting and multiplying through by $-T$ we obtain Equation 10-20

$$-T\Delta S_{total} = \Delta H - T\Delta S$$
EQUATION 10-20

We can now define the free energy for a constant pressure process (i.e., the Gibbs free energy) to be as in Equation 10-21.

$$\Delta G = -T\Delta S_{total} = \Delta H - T\Delta S$$
EQUATION 10-21

Accordingly, because the entropy and free energy have opposite signs, when the entropy change is *positive*, the change in free energy is *negative* and when the entropy reaches a *maximum* (at equilibrium) the free energy is a *minimum*.

The big advantage of using the free energy is that we have now defined all our parameters in terms of changes in the system—we don't have the difficult, if not impossible, task of calculating changes in the surroundings. Of course, we still just have a thermodynamic equation relating macroscopic properties. The trick is to now relate entropy and enthalpy to molecular quantities. An example of how this is done for a simple model will be described in the chapter on polymer solutions.

THERMAL TRANSITIONS IN POLYMERS

In this section we will briefly review the character of the principal thermal transitions that occur in polymers, crystallization, melting and the glass transition, then treat these individually in more detail. Some background material has been covered in our discussion of "States of Matter," which you should also review if you've got a memory like a sieve.

Let us start with a polymer in the melt and cool it slowly while measuring its volume. If the chain microstructure is regular, as in linear polyethylene or isotactic polypropylene, then at some temperature below that at which the polymer originally melted, it will crystallize and this crystallization will be accompanied by an abrupt change in volume (Figure 10-15). If, on the other hand, we are cooling a structurally irregular polymer, like atactic polystyrene, then crystallization cannot occur, but instead, in a certain temperature range, there is a change in slope of the volume/temperature curve. The mid-point of this range is called the glass transition temperature, or T_g. Well above this temperature, the material appears to be a liquid melt, while well below it the material is glassy in its properties (relatively rigid and brittle). Around the transition it is sort of in-between, a soft glass or a stiff, permanently deformable rubber.

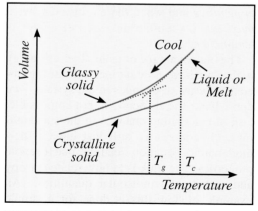

FIGURE 10-15 Schematic plots of the volume changes on cooling a polymer.

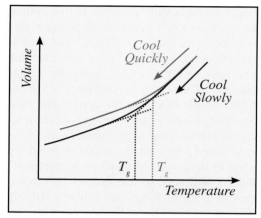

FIGURE 10-16 Schematic plots showing the dependence of the T_g on the rate of cooling.

As we will see, crystallization is governed by kinetics and if you cool a polymer quickly enough it will get to the glass transition temperature and "freeze" before it has a chance to crystallize. This is true of all materials, but some, like metals, crystallize so quickly that it is extraordinarily difficult to prepare an amorphous material by quenching from the melt. However, certain materials crystallize very slowly and preparing glassy materials is easy (think of window glass). It is possible to prepare all crystallizable polymers in the glassy state by quenching from the melt, although you have to make heroic efforts to prepare a glass from linear polyethylene.

Although crystallization is controlled by kinetic factors, it is a phenomenon that is thermodynamic in origin (a first-order phase transition). On the other hand, the glass transition may be purely kinetic. We'll say more about that shortly, but one manifestation of this is the dependence of the T_g on the rate of cooling. It is observed at a somewhat lower temperature if a sample is cooled slowly than if it is cooled quickly (Figure 10-16).

If you are unfamiliar with materials and their properties and have a basic knowledge of thermodynamics, you may by now be thinking: "To hell with kinetics, why should a polymer, or any other material, crystallize at all?" You may have recalled that in the disordered state characteristic of the melt, the system has a higher entropy than in the ordered or partially ordered state. Superficially, then, crystallization would seem to violate the second law of thermodynamics. However, when a material crystallizes it packs its molecules more efficiently, intermolecular interactions are maximized and energy is released in the form of heat (Figure 10-17). This increases the entropy of the surroundings and the total entropy change is still positive.

Hopefully, you will recall that we can express the overall change in entropy in terms of parameters describing changes within the system alone, using (for a fixed pressure and temperature) the Gibbs free energy, Equation 10-21 in the preceding section ($\Delta G = -T\Delta S_{total} = \Delta H - T\Delta S$).

What we now have is a balance of two terms. At high temperature, the $-T\Delta S$ term is positive and much larger than the negative ΔH term. The overall free energy change for crystallization would then be positive and the material would not crystallize. (Remember that the free energy has the opposite sign to the overall entropy change.) As T becomes smaller, this balance changes and at some temperature ΔG becomes negative and crystallization occurs.

The reason why there are abrupt changes in properties like volume at the melting or crystallization temperature can be appreciated by looking at hypothetical free energy curves, such as those sketched in Figure 10-18. One shows the free energy of the crystalline phase as a function of temperature, while the other shows the free energy of the melt. At high temperatures, a melt has a lower free energy than a crystal, but as the temperature is decreased the curves cross and at low temperatures it is the crystalline phase that has the lower free energy. Obviously, if we started with a crystalline polymer and increased the temperature, it would have a free energy described by the blue curve, but would "jump" to the red curve at the melting point. Note that the slopes of the two curves (i.e., $\partial G/\partial T$), are different at this point. Do you remember what this differential is? (entropy? enthalpy? volume?)

In fact, all thermodynamic parameters that can be related to the first derivative of the

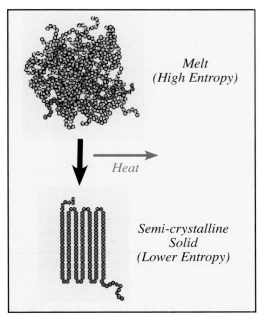

FIGURE 10-17 Entropy changes upon crystallization.

free energy (entropy, enthalpy, volume) with respect to an appropriate thermodynamic variable (here, P and T) show an abrupt discontinuity at the melting (and crystallization) temperature; hence, the latent heats that are associated with these transitions and the abrupt change in volume upon crystalli-

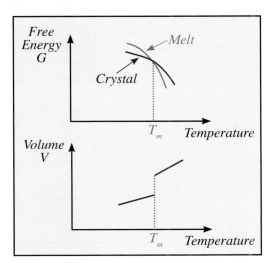

FIGURE 10-18 Free energy curves for a melt and a crystal (top) and volume as a function of temperature for a crystalline material (bottom).

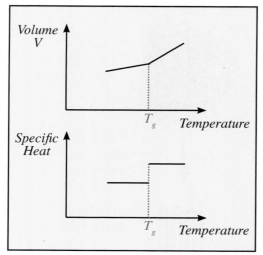

FIGURE 10-19 Changes in volume and specific heat at the T_g.

DSC is the most widely used tool in the study of polymer thermal transitions. It essentially measures the specific heat by determining the heat supplied to a sample relative to a known standard as a function of heating rate. The resulting thermograms, as they are called (Figure 10-20), show a baseline jump at the T_g (change in specific heat), but this occurs over a temperature range, not at a well-defined point. (The T_g is defined to be in the middle of the observed range.) At the melting point, there is a peak associated with the latent heat of fusion. Real data can be a lot more complicated than what we show here, but you will find that out soon enough, when you go into the lab and try and do experiments!

Having given you a general introduction to thermal transitions, you are now ready to go into crystallization, melting, or the glass transition, in more detail. We will start by considering crystallization.

zation or melting (see Figure 10-18). That is also why they are sometimes referred to as first-order phase transitions.

Similarly, if a quantity such as the volume exhibits an abrupt change in slope, which occurs at the T_g, then there is a discontinuity in quantities associated with first derivatives of this parameter, or second derivatives of the free energy (with respect to appropriate thermodynamic variables), such as the specific heat (Figure 10-19). Accordingly, the T_g may be related to a second-order phase transition, but this remains in dispute. The experimentally observed transition is clearly governed by kinetics and the standard method of measuring this transition is by differential scanning calorimetry (DSC), which measures the specific heat.

CRYSTALLIZATION

In Chapter 8, in the section on morphology, we have seen that polymers are only partially crystalline and when crystallized from solution or the melt they form chain-folded lamellar structures, as illustrated schematically in Figure 10-17 earlier. This is a consequence of the kinetics of crystallization, which we will explore here.

Polymer crystallization only occurs at significant "undercoolings" or "supercoolings." If you take a sample of a crystalline solid into the melt and then cool it, crystallization does not occur at the melting point, but just below it. This is because the initially formed crystal entity, or nucleus, has to be stable to further growth (i.e. must not re-melt!) The degree of undercooling for low molecular weight materials is usually small, but for polymers is relatively large, usually in the range of 15 –50°C (Figure 10-21). Obviously the initial "nucleation" step is more difficult in polymers and requires a bigger thermodynamic driving force ($T_m - T_c$).

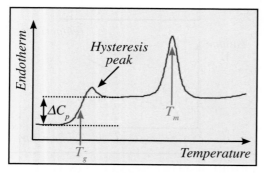

FIGURE 10-20 Schematic representation of a thermogram showing a T_g and a T_m.

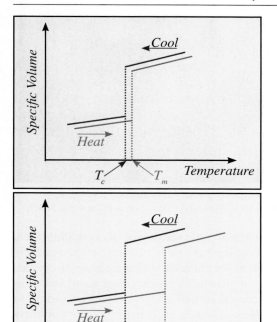

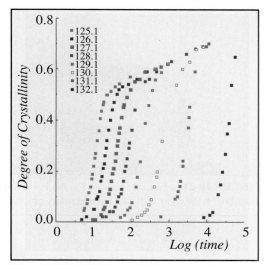

FIGURE 10-22 Graph of the degree of crystallinity versus time [redrawn from the data of E. Ergoz, J. G. Fatou, L. Mandelkern, *Macromolecules,* **5**, 147 (1972)].

FIGURE 10-21 Schematic graphs of temperature versus specific volume for (top) small molecules and (bottom) polymer molecules.

The Rate of Crystallization

If the degree of crystallinity of a polymer is measured as a function of time as it is crystallizing from the melt at constant temperature, T_c, then sigmoidal shaped curves are obtained. Curves obtained at different temperatures all have the same shape, but the rate of crystallization increases rapidly with decreasing temperature, or greater undercooling, as long as T_c remains well above the T_g (Figure 10-22).

The crystallization curves can be broken down into three parts. There is an initial *induction period* during which the primary nuclei are formed. These primary nuclei are the smallest crystalline entities that are stable enough to allow further growth at that temperature (i.e., do not re-melt). This induction period is followed by a period of fast spherulite growth called *primary crystallization* (Figure 10-23). (If you haven't observed

spherulites growing there is a great movie in our *Polymer Science* CD. Some images are reproduced in Figure 10-24). Once the spherulites fill the volume of the sample and impinge on one another crystallization does not stop, but continues at a much slower rate. This is called *secondary crystallization*.

The Avrami Equation

The time dependence of crystallization, particularly primary crystallization, has often been interpreted or modeled in terms of

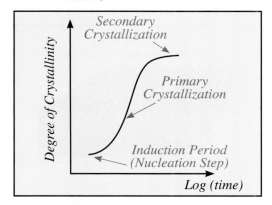

FIGURE 10-23 Schematic graph of the degree of crystallinity versus time showing the induction period, and primary and secondary crystallization.

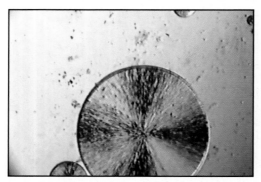

FIGURE 10-24 Spherulitic growth. On the left is a snapshot during primary crystallization, while the snapshot on the right is during the secondary crystallization stage (Courtesy: Prof. James Runt, Penn State).

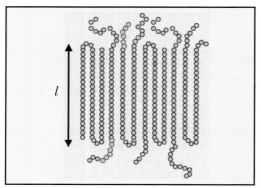

FIGURE 10-25 Schematic diagram showing the thickness of a lamellar.

the Avrami equation, derived from statistical geometrical arguments (Equation 10-22):

$$\phi_c(t) = 1 - \left(exp - (kt)^n\right)$$

EQUATION 10-22

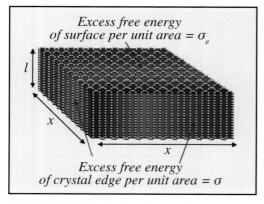

Excess free energy of surface per unit area $= \sigma_e$

Excess free energy of crystal edge per unit area $= \sigma$

FIGURE 10-26 Schematic diagram showing the parameters used in describing a polymer crystal.

where $\Phi_c(t)$ is the degree of crystallinity, k is a constant at a particular temperature and n is an exponent related to the nature of crystalline growth. This exponent should take on integral or half integral values according to the conditions of growth, but for polymers it seldom does. Nevertheless, the Avrami equation is still regarded as a useful empirical way of representing the data.

Thermodynamic Considerations

Most theories of lamellar thickness and crystal growth rate are kinetic theories and we will start our discussion of these by considering primary nucleation (not to be confused with primary crystallization). There is also something called secondary nucleation, but we will get to that later. The thermodynamic driving force for the formation of a primary nucleus can be explored using a simple lamellar model of a primary nucleus, one whose sides are of equal length, x, and whose thickness is l (Figures 10-25 and 10-26). More complex geometries only alter the constants of proportionality in the equations that follow.

The free energy of this crystal consists of contributions from the bulk, the inside of the lamellar where each segment of the chain sits in an ordered array, minimizing its free energy, and the surface's. Although the thin sides of the crystal also consist of ordered chain segments, they have a face that is exposed to solvent or the melt and thus have a higher free energy than those in the heart of the crystal. The top and bottom surfaces have an even higher free energy, because in addition to contact with the "outside", the bond rotational angles necessary for folding have a higher conformational energy than those found in the lattice. Let the bulk free energy of the crystal be Δg per unit volume (the difference in free energy between the crystal and the melt). At the equilibrium melting point (more on this shortly) Δg would be zero and just below this temperature it would be negative, driving crystallization. (Revisit Figure 10-18.)

Let's also define free energy terms for the surfaces (σ and σ_e, per unit area, because now we're considering a surface). These are actually excess terms, related to the amount by which the free energy of a segment at the surface exceeds that of a segment in the lattice. We can then write an expression for the free energy of this primary nucleus in terms of the difference between the surface and bulk terms (Equation 10-23):

$$\Delta G_{cryst} = (4xl)\sigma + (2x^2)\sigma_e - (x^2 l)\Delta g$$

EQUATION 10-23

(The last term represents the free energy that we would obtain if all the segments were in the bulk. But they're not, and the first two terms are the excess free energy that must be "added in" to account for those segments at the surface.) Clearly, crystallization will only occur when the bulk term is larger than the surface terms, so that the free energy change is negative (i.e., there must be some minimum critical size). Also, the crystal can minimize its free energy by reducing its surface area and maximizing the number of segments within the bulk. This minimum should then occur for extended chain crystals, where there would be no fold surface. However, crystallization is a kinetic phenomenon and the thermodynamically most stable form is not necessarily the lamellar thickness that grows the fastest and therefore the structure that is actually formed during crystallization.

Nevertheless, that extended chain crystals

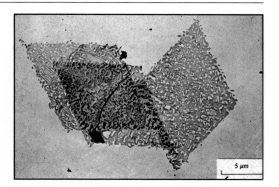

FIGURE 10-27 Electron micrograph of annealed PE single crystals (Courtesy: Professor Ian R. Harrison, Penn State).

are indeed the most stable form is demonstrated by annealing experiments (Figure 10-27). If a polymer is heated above its crystallization temperature, but below the melting point, the crystal irreversibly thickens, or, as we say, its fold period increases. For single crystal lamellae this is accompanied by the appearance of holes—the material used to make the crystal thicker has to come from somewhere!

All this begs the question: "If extended chain crystals are the thermodynamically most stable form, then why don't polymers simply crystallize in this fashion to begin with?" To find an answer let's consider the critical nucleus size, or the most probable fold period that can be obtained at a particular crystallization temperature.

Critical Nucleus Size

The critical nucleus size is given by the values of l and x that minimize ΔG_{cryst} (Equation 10-24):

$$\frac{\partial \Delta G}{\partial l} = \frac{\partial \Delta G}{\partial x} = 0$$

EQUATION 10-24

We leave the algebra to you as an exercise. Solving the two simultaneous equations you will find that the most probable thickness of the critical nucleus is (Equation 10-25):

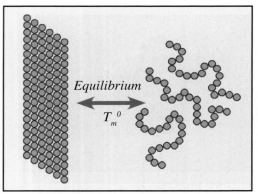

FIGURE 10-28 Schematic diagram depicting equilibrium melting.

$$l^* = \frac{4\sigma_e}{\Delta g}$$

EQUATION 10-25

It is illuminating to rewrite this equation in terms of the degree of undercooling by substituting for Δg. To do that, however, we have to digress a little and define the equilibrium melting temperature. This will also be useful in our discussions of nucleation rates and the melting point, so bear with us!

Equilibrium Melting Temperature

Consider a hypothetical, perfect, extended chain crystal in equilibrium with its melt[27] (Figure 10-28). The situation would be dynamic, with chains crystallizing and melting continuously, but at equal rates. Because the system is at equilibrium there is no net change in the total entropy or free energy of this system and we can write (Equation 10-26):

$$\Delta g = \Delta h_f - T_m^0 \Delta s_f = 0$$

EQUATION 10-26

where T_m^0 is the equilibrium melting temperature, the temperature at which a perfect extended chain crystal would melt (Figure

[27] Things either melt or crystallize at a certain temperature, they don't actually hang about in equilibrium.

10-28). The heat (enthalpy) of fusion can be measured experimentally, and so can T_m^0 (by extrapolation methods), so the entropy of fusion can be expressed in terms of these known quantities (Equation 10-27):

$$\Delta s_f = \frac{\Delta h_f}{T_m^0}$$

EQUATION 10-27

Now consider some lower temperature T_c where the system is not in equilibrium and crystallization is occurring (Figure 10-29). In this case we have (Equation 10-28):

$$\Delta g = \Delta h_f - T \Delta s_f \neq 0$$

EQUATION 10-28

And substituting for Δs_f we obtain Equation 10-29:

$$\Delta g = \Delta h_f - T_c \Delta s_f = \Delta h_f \left[\frac{T_c}{T_m^0} \right]$$

EQUATION 10-29

We are assuming Δs_f and Δh_f don't vary much with temperature. Remember that this is just the bulk free energy (per unit volume) of the crystal, it does not include any additional free energy from those segments at the surfaces. We can now return to our expression for the critical nucleus size (thickness), which was given previously in Equation 10-25, and substituting for Δg, we obtain Equation 10-30:

$$l^* = \frac{4\sigma_e T_m^0}{\Delta h_f (T_m^0 - T)} = \frac{4\sigma_e T_m^0}{\Delta h_f (\Delta T)}$$

EQUATION 10-30

where ΔT is the undercooling (Equation 10-31):

$$\Delta T = T_m^0 - T_c$$

EQUATION 10-31

Note the inverse dependence of fold period on ΔT. The larger the undercooling the smaller the fold period.

Primary Nucleation

In essence, there is a free energy barrier that has to be crossed in order to form the primary nucleus. This free energy barrier is associated with the chains disentangling and straightening themselves out. The larger the fold period, the greater the barrier and the slower the rate of nucleation. (What is the probability of finding an extended chain in the melt—go back to our review of thermodynamics if you've forgotten!). Like any other process that depends upon an energy of activation, the rate of nucleation will be given by the Arrhenius equation (Equation 10-32):

$$v_{nuc} \sim exp - \frac{\Delta G^*}{kt}$$

EQUATION 10-32

where ΔG^* is the free energy of a nucleus that has the critical size. This can be obtained by substituting l^* into the expression for ΔG_{cryst}, giving Equation 10-33:

$$v_{nuc} \sim exp - \frac{\Delta G^*}{kT} \sim exp - \frac{Const\sigma^2\sigma_e}{\Delta g^2 kT}$$

EQUATION 10-33

Again substituting for Δg we obtain Equation 10-34:

$$v_{nuc} \sim exp - \frac{Const\sigma^2\sigma_e[T_m^0]^2}{\Delta h_f^2 kT_c[\Delta T]^2}$$

EQUATION 10-34

Thus the rate of nucleation varies strongly with ΔT and for the small undercoolings necessary to get extended chain crystals the rate of nucleation would approach infinitely long time periods (i.e., $v_{nuc} \sim$ very small)— see Figure 10-30. The free energy barrier,

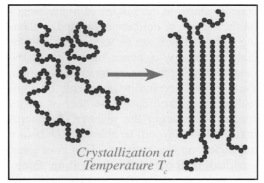

FIGURE 10-29 Schematic diagram depicting crystallization and the formation of the critical nucleus.

hence l^*, has to be small enough (hence undercoolings large enough) that crystallization can occur in ordinary time spans. So it is for kinetic reasons that polymers, at least initially, form chain-folded crystals (nuclei). What about subsequent growth? This brings us to the next part of crystallization kinetics, *primary crystallization*. But first, some final words about *primary nucleation*.

Homogeneous and Heterogeneous Nucleation

The relationship for primary nucleation (Equation 10-34) only applies to pure polymers. In most practical cases, crystallization

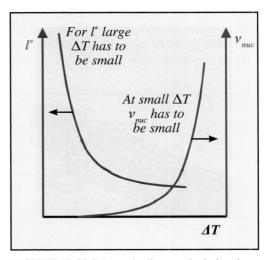

FIGURE 10-30 Schematic diagram depicting the effect of ΔT on l^* and v_{nuc}.

actually starts on the surface of impurities or on the surface of deliberately added nucleating agents. These act to reduce the interfacial free energy, σ, hence lower the degree of supercooling necessary for crystallization. Furthermore, crystallite size can be controlled by controlling the number of nuclei formed—the larger the number of spherulites the less they will be able to grow before impinging on one another. This is of great practical importance in controlling things like mold shrinkage during processing, as well as other properties.

Finally, in classical nucleation theory there are additional terms in the equation for v_{nuc}, notably a second exponential term for the activation energy of transporting a molecule across the phase boundary. It is the inverse dependence of ΔG^* on ΔT^2 that is the most strongly varying quantity and determines the nucleation rate, however, so in the interests of simplicity we ignore the other terms (for now).

Secondary Nucleation

Once primary nuclei are formed the ensuing spherulites grow radially at a constant rate. Primary crystallization, which occurs initially on the surface of the primary nucleus and then on the surface of the growing lamellar, also involves a nucleation step, secondary nucleation. It is this step that largely governs the ultimate crystal thickness and which forms the focus of most kinetic theories of polymer crystallization.

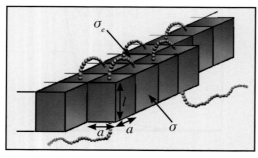

FIGURE 10-31 Schematic diagram showing the parameters used in describing secondary nucleation.

Polymer crystal growth is predominantly in the lateral direction, because folds and surface entanglements inhibit crystallization in the "thickness" direction. Nevertheless, there is a considerable increase in the fold period behind the lamellar front during crystallization from the melt and, as we have seen, polymers annealed above their crystallization temperature but below T_m also irreversibly thicken. Nevertheless, in most theories of secondary nucleation, the most widely used being the theory of Lauritzen and Hoffman,[28] it is assumed that once a part of a chain is added to the growing crystal, its fold period remains unchanged.

Lauritzen-Hoffman Theory

In this theory it is assumed that a chain stem, one fold period long, is laid down on the lateral growth face of the crystal. This is the slowest step because the stem has only one surface face on which to sit. Once this stem is in place, however, an adjacent stem is more easily laid down (i.e., has a lower free energy barrier to cross), because it can now contact two surfaces, the crystal substrate and the side face of the first stem. The row therefore quickly fills up once the first stem is deposited. The growth rate of the crystal (primary crystallization) is thus largely determined by secondary nucleation (Figure 10-31).

As in our treatment of primary nucleation, we start by considering the thermodynamic driving force and write an expression for the free energy of the nucleus in terms of a difference between the (excess) free energy of the exposed surfaces and the free energy that would be obtained if all the segments were in the bulk. The free energy change for laying down n stems is given by (Equation 10-35):

$$\Delta G_{stem} = (2al)\sigma + n(2a^2)\sigma_e - n(a^2 l)\Delta g$$

EQUATION 10-35

[28] See J. D. Hoffman, G. T. Davis, J. I. Lauritzen, Jr., in *Treatise on Solid State Chemistry*, N. B. Hannay, editor, Pergamon Press, Oxford (1989).

(Note that only two new "side" surfaces are created—the ones at the front simply replace the ones in the layer behind them.)

Obviously, the same type of equation can be written for the free energy change involved in laying down $n + 1$ stems. It is then assumed that the free energy of laying down a single stem, including the first one, is just the difference, $\Delta G(n + 1) - \Delta G(n)$. Subtracting and rearranging, left to you as an elementary homework problem, gives the following expression for the minimum fold period (Equation 10-36):

$$l^*_{min} = \frac{2\sigma_e}{\Delta h_f}\left[\frac{T^0_m}{\Delta T}\right]$$

EQUATION 10-36

This is half the value determined for primary nucleation. But, as in primary nucleation, the fold period goes as $1/\Delta T$. In other words, it increases with decreasing undercooling (smaller ΔT). The temperature dependence of the rate of secondary nucleation has a different dependence on temperature, however.

An expression for the temperature dependence of the rate of secondary nucleation can be obtained in a similar manner to the procedure used in the treatment of primary nucleation. The expression for l^*_{min} is substituted into the expression for the free energy, in this case the free energy of laying down a stem, ΔG_{stem}, to obtain ΔG^*, the free energy of a secondary nucleus that has the critical size. The rate of secondary nucleation can then be obtained from Equation 10-37:

$$\nu_{nuc} \sim exp - \frac{\Delta G^*_{stem}}{kT}$$

EQUATION 10-37

and we find that (Equation 10-38):

$$\nu_{nuc} \sim exp - \frac{Const.}{T_c \Delta T}$$

EQUATION 10-38

Of course, the thickness of the secondary

nucleus has to be larger than l^*_{min} if it is to be stable to further growth (i.e., not melt!) Nevertheless, these simple thermodynamic considerations tell us that secondary crystallization goes as an exponential in $-1/\Delta T$, whereas primary nucleation goes as an exponential in $-1/\Delta T^2$ (Equations 10-39):

$$\nu_{sec.nuc} \sim exp - \frac{Const.}{T_c \Delta T}$$

$$\nu_{prim.nuc} \sim exp - \frac{Const.}{T_c [\Delta T]^2}$$

EQUATIONS 10-39

We'll come back to this in a while. First let's look at the overall rate of crystallization, the quantity that can be experimentally determined.

Fold Period and Crystal Growth Rate

We've mentioned that the thickness, l^*, of the secondary nucleus must be bigger than l^*_{min} if it is to be stable to further growth say by an amount δl, therefore using Equation 10-36 we can write (Equation 10-40):

$$l^* = \frac{2\sigma_e}{\Delta h_f}\left[\frac{T^0_m}{\Delta T}\right] + \delta l$$

EQUATION 10-40

The theory of Lauritzen and Hoffman, perhaps still the most commonly used model for the analysis of polymer crystallization data, then seeks to evaluate δl (usually of the order of 40 angstroms) by considering the rates at which stems and folds are successively laid down. We will not go into the details of this derivation, but the expressions that have been derived for the growth rate have the following form at low undercoolings (Equation 10-41):

Growth Rate

$$\sim exp - \frac{Const\sigma\sigma_e [T^0_m]^2}{\Delta h_f kT_c [\Delta T]} \times Other\ Terms$$

EQUATION 10-41

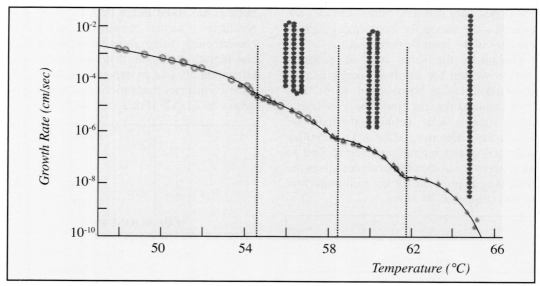

FIGURE 10-32 The rate of crystallization versus temperature for a 6000 g/mol poly(ethylene oxide) [redrawn from the data of Kovacs et al., *J. Polym. Sci., Polym. Symp. Ed.*, **59**, 31 (1977)].

So, to summarize, the theory predicts that the fold period (thickness) of the crystals increases with decreasing undercooling (i.e., higher temperature), while the rate of primary crystallization increases with decreasing temperature, at least at low undercoolings (more on this in a bit).

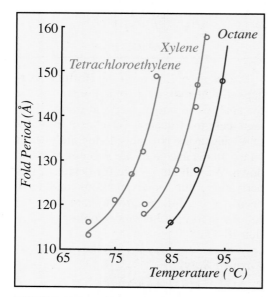

FIGURE 10-33 The fold period of PE versus temperature (redrawn from the data of Hoffman et al. cited previously).

A set of experiments demonstrating both of these relationships were performed by Kovacs et al.,[29] who crystallized a low molecular weight sample (M ~ 6000) of poly(ethylene oxide)—see Figure 10-32. At the smallest undercoolings, these chains were short enough that extended chain crystals actually formed. The crystallization rate increased with decreasing temperature, but breaks in the curves occurred as a result of a change in crystal structure, first to lamellae with once-folded chains and then to ones that have two folds. The last part of the crystallization curve corresponds to ordinary polymer behavior, where the fold period decreases continuously with undercooling, as in the plots shown in Figure 10-33 for polyethylene crystallized from various solvents.

The experiments of Kovacs et al. also demonstrate beautifully that although the extended chain crystal is the thermodynamically most stable form, when the undercooling is sufficient, kinetics favors folded chain lamellae. As we have seen, long chain polymers only crystallize at finite rates at high undercoolings, so only form folded chain structures.

[29] A. J. Kovacs, C. Straupe, A. Gonthier, *J. Polym. Sci. Polym. Symp.* Ed., **59**, 31, 1977.

Getting back to the expression for the growth rate (Equation 10-41), we specified that this applies to low undercoolings, because as the temperature is reduced the mobility of the chains decreases and their motion essentially freezes around the glass transition temperature, so crystallization must decrease dramatically as the T_g is approached. The expression for the growth rate can accommodate this by including an additional transport term. We then get (Equation 10-42):

$$\text{Growth Rate} \sim exp - \frac{K_g}{T\Delta T} \cdot exp - \frac{U^*}{R(T - T_\infty)}$$

EQUATION 10-42

where the various constants associated with the free energy barrier to nucleation have been folded into a constant K_g in the first exponential term, about which we will have more to say shortly. The second exponential term is essentially an empirical description of chain mobility, which depends on an activation energy U^* and a temperature T_∞, which is some 30°K below the T_g of the polymer being considered. This obviously has the right form—as T approaches T_∞ the growth rate decreases dramatically. We will consider equations such as this in more detail when we discuss the T_g.

Crystallization Regimes

For now, let's just focus on the secondary nucleation part of the equation for the growth rate (Equation 10-43):

$$\text{Growth Rate} \sim exp - \frac{K_g}{T\Delta T}$$

EQUATION 10-43

There are a couple of things to notice about this expression. First, there is a dependence on an exponential term in $-1/\Delta T$, unlike the rate of primary nucleation, which depends on $exp(-1/\Delta T^2)$. We'll get back to the tem-perature dependence of primary and secondary nucleation later. Furthermore, the factor K_g, which depends upon the usual suspects (σ, σ_e, Δh_f, etc.), varies with what is called the crystallization regime. In regime I, a stem is laid down and the rest of the row then quickly fills up before any further nucleation event occurs on that face. This is illustrated in Figure 10-34, which is a snapshot showing a row growing on a completed face. (There is a nice animation of regime I crystallization in our *Polymer Science* CD that you may wish to look at.)

In regime II, new secondary nuclei are formed before the rows are complete, as illustrated in Figure 10-35. (Again, we suggest that you might wish to view the animation of regime II crystallization in our *Polymer Science* CD.) There is also a regime III, where prolific multiple nucleation occurs, but we'll ignore that here. The figures and animation illustrating regime II crystallization were difficult enough!

If you think about it, you would expect regime I behavior to be observed at low supercoolings, as this is where the rate of nucleation is slower, giving the rows time to fill up. Regime II behavior should be

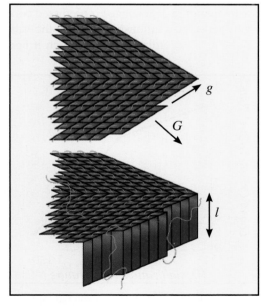

FIGURE 10-34 Schematic diagram depicting regime I crystallization.

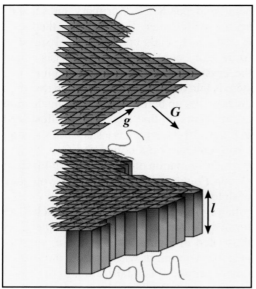

FIGURE 10-35 Schematic diagram depicting regime II crystallization.

evident at lower temperatures (i.e., higher supercoolings), and this is what is observed. Furthermore, in addition to the temperature dependence of the growth rate, the kinetic theory predicts that value of K_g for regime I growth should be twice that of regime II. Appropriate plots of experimentally determined growth rates (Figure 10-36) show that the temperature dependence is predicted correctly, as is the relationship between the values of K_g. Accordingly, even though it has

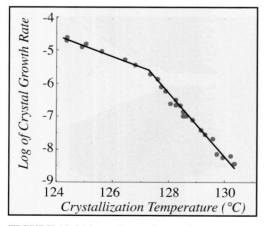

FIGURE 10-36 Log of crystal growth rate versus crystallization temperature (redrawn from the data of Hoffman et al., cited previously).

limitations (e.g., it does not treat phenomena like crystal thickening), the theory must be considered very successful.

Kinetics of Primary Crystallization

If we now consider crystallization over a wide range of temperatures, not just at low undercooling, then a plot of the crystal growth rate would look something like the plot in Figure 10-37. Crystallization at first increases rapidly with undercooling, according to the temperature dependence of the secondary nucleation rate. But as the crystallization temperature is further reduced, the mobility of the chains also decreases, slowing crystallization drastically as the T_g is approached. The shape of the curve is similar on both the high and low temperature side, because the competing factors have similar functional forms, but for entirely different physical reasons.

Of course, lower mobility affects primary as well as secondary nucleation, and both have similar functional forms (Figure 10-38). Secondary nucleation occurs more easily, by which we mean at higher temperatures (or smaller undercoolings) than primary nucleation, however, as reflected in the temperature dependencies of the nucleation rate we have developed for each. (Quick—is it primary or secondary nucleation that goes as an exponential in $-1/\Delta T^2$?) Hence, at high undercoolings (say, T_1), the rate of primary nucleation is greater than secondary nucleation and a large number of small crystals are formed. Conversely, low undercoolings (say, T_2) lead to a smaller number of large crystals. Crystallite size can affect things like mold shrinkage during processing and mechanical properties, so you should try and remember this stuff! (However, since, in practice, crystallization is usually heterogeneous, crystallite size is usually controlled by adding nucleating agents.)

Secondary Crystallization

Finally, after the growing spherulites have impinged on one another the growth rate slows considerably, but continues by a process called secondary crystallization. Various mechanisms are associated with this phenomenon, including isothermal thickening of the crystals, the formation of additional thin lamellae in the amorphous regions between the primary crystallites, or by a surface crystallization and melting process. The nature of the interphase in polymer crystals is not that well understood and is the subject of continuing research efforts and in an overview treatment such as this we have no more to say about it.

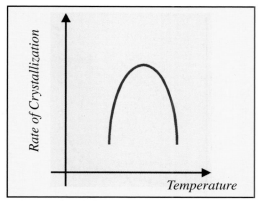

FIGURE 10-37 Schematic diagram depicting the overall rate of crystallization versus temperature.

THE CRYSTALLINE MELTING TEMPERATURE

In the sections on morphology and crystallization we have seen that long chain polymers can only form partially crystalline, folded chain structures when crystallized by cooling from solution or the melt. Because of this and the range of fold periods formed, polymers do not melt at a sharp well-defined melting point, like low molecular weight materials, but over a range of temperatures. Furthermore, the presence of defects, such as short chain branching in polyethylene, also lowers the melting temperature relative to that found in samples with more perfect linear chains. This can be illustrated by comparing the plots of volume against temperature for the relatively short (compared to polymers) paraffin $C_{44}H_{90}$ (Figure 10-39), which can crystallize without chain folding, and polyethylene. (Figure 10-40—one curve corresponds to data obtained from a sample with a broad range of molecular weight, while the other is a fractionated sample: which one's which?) The range of melting temperatures in the latter indicates the presence of crystalline lamellae of various sizes (thickness or fold periods) and various degrees of perfection. (Because of this, it is usual to take the temperature at which the last crystals disappear as the melting temperature.)

The melting point of a polymer also varies with the temperature at which it was crystallized (Figure 10-41): the higher the crystallization temperature the higher the melting

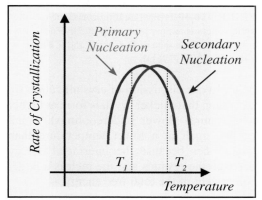

FIGURE 10-38 Schematic diagram depicting the rate of crystallization versus temperature for primary and secondary nucleation.

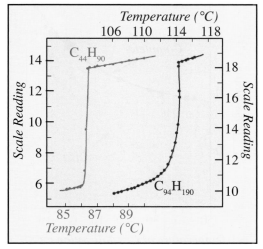

FIGURE 10-39 Melting curves of volume versus temperature for $C_{44}H_{90}$ and $C_{94}H_{190}$ (plotted from the data of L. Mandelkern, *Comprehensive Polymer Science*, Vol 2, Pergamon Press, Oxford, 1989).

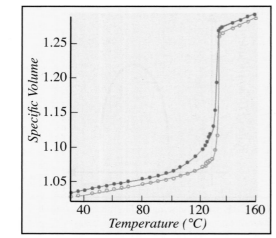

FIGURE 10-40 Typical melting curves of specific volume versus temperature for polyethylenes [plotted from the data of R. Chiang and P. Flory, *JACS*, **83**, 2857 (1961)].

point. In our discussion of crystallization we have seen that thicker crystals form at higher temperatures (lower undercoolings). Small crystals melt at a lower temperature than larger ones because the interfacial energy between the crystals and the melt is a larger component of the total free energy.

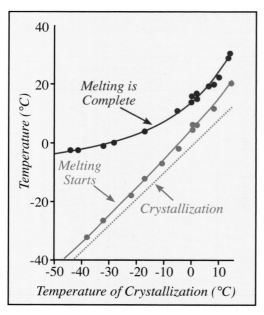

FIGURE 10-41 Schematic diagram showing the effect of crystallization temperature on the melting point.

Melting Temperature of Polymer Crystals

Recalling that the free energy of a crystal lamella is given by Equation 10-44 (we told you we'd get back to this):

$$\Delta G_{cryst} = 2x^2\sigma_e - (x^2 l)\Delta g$$

EQUATION 10-44

We can now substitute for Δg to get Equation 10-45:

$$\Delta G_{cryst} = 2x^2\sigma_e - x^2 l\Delta h_f\left[1 - \frac{T}{T_m^0}\right]$$

EQUATION 10-45

Now let's again assume that you could achieve equilibrium, this time between a folded chain crystal and its melt (see Figure 10-29 shown previously), at $T = T_m\ (< T_m^{\,0})$ we would have Equation 10-46:

$$\Delta G_{cryst} = 0$$

EQUATION 10-46

Substitution and a little bit of rearranging leads to Equation 10-47:

$$T_m = T_m^0\left[1 - \frac{2\sigma_e}{l\Delta h_f}\right]$$

EQUATION 10-47

This equation (Figure 10-42) is sometimes called "Thompson's rule" and sometimes the Thompson-Gibbs equation. It tells us that the actual melting temperature of a polymer is always less than the equilibrium melting temperature by an amount that depends on the fold period of the crystal. This is because the contribution of interfacial energy to the total free energy of the crystal—the second term in brackets in the equation for the melting temperature—is smaller for thicker crystals, therefore T_m is higher.

This also explains why polymers melt over a broad range of temperatures. Samples

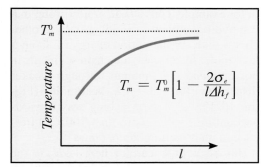

FIGURE 10-42 Schematic graph of fold period versus temperature.

cooled from the melt will experience a range of crystallization temperatures. They will then have a distribution of fold periods and, in turn, melting temperatures. Other factors that affect the fold period and hence T_m include the need to exclude defects from the lattice (e.g., short chain branches in polyethylene), molecular weight distribution (think of chain ends as defects that also tend to be excluded from the lattice), and secondary crystallization.

Factors That Influence the Melting Temperature

Having considered some fundamental aspects of polymer structure, we now wish to consider some basic questions—what is the effect of chain "defects" (co-monomers, branches etc.), molecular weight, the presence of solvent? An even more fundamental, question, and the one we will address first, is "why do polymers have the melting points they have?" Is it a matter of divine imposition that the melting point of polyethylene is about 135°C while that of Kevlar® is about 370°C? (Figure 10-43). (Note: we can only say things like "about 135°C," and use the symbol ~ to represent this, because, as we have seen, in practice polymers melt over a range of temperatures.) We're with Einstein on that one, who (in another context) said "God does not play dice with the universe.".

Effect of Chemical Structure

The reason that one polymer differs in its melting point to another is, of course, ultimately related to its chemical structure, which, in turn, determines things like the type of intermolecular interactions that occur and chain flexibility, the main factors that determine T_m. But, if you are to get a feel for this, you have to accept that, once again, thermodynamics rears its ugly head. As before, we'll start with the imaginary situation[27] of a crystal in equilibrium with its melt and write Equations 10-48:

$$\Delta G_f = \Delta H_f - T\Delta S_f$$

$$At\ Equilibrium\ \Delta G_f = 0$$

$$Hence:\ \ T_m = \frac{\Delta H_f}{\Delta S_f}$$

EQUATIONS 10-48

where the subscript f stands for fusion.

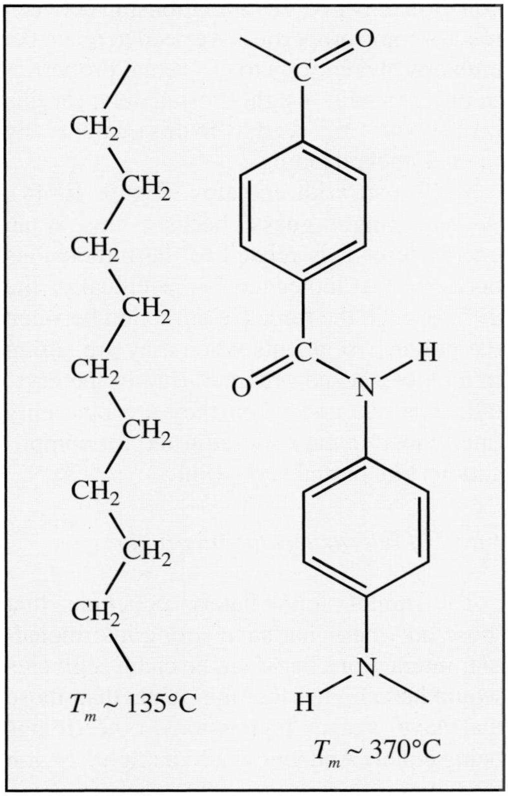

$T_m \sim 135°C$

$T_m \sim 370°C$

FIGURE 10-43 Approximate melting temperatures of PE and Kevlar®.

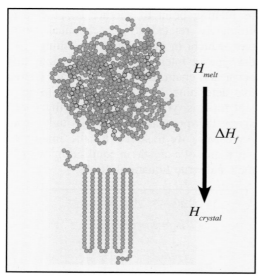

FIGURE 10-44 Schematic diagram depicting the enthalpy of fusion.

The equation, $T_m = \Delta H_f / \Delta S_f$, is all very well, but like any other thermodynamic equation only gives us a relationship between macroscopic properties. We need to relate the enthalpy and entropy to molecular properties in order to gain insight. Fortunately, for this problem we only need to do this qualitatively to gain understanding.

We'll start with enthalpy (Figure 10-44). As you might guess, because this is an energy term it is related to the interactions between the molecules[30]—specifically, the difference in the forces of attraction between the polymer segments when they are sitting in a close-packed ordered array in the crystalline lattice and when they are randomly intertwined in the melt (adjusted for complications like partial crystallinity).

Effect of Intermolecular Interactions

You might immediately conclude that those polymers that have strong intermolecular interactions between the chain segments would have higher melting points than those that have weaker interactions, other things being equal. And you would be right. Nylon

[30] We look at the details of this in our discussion of solutions.

6, for example, contains an amide group that can hydrogen bond to amide groups in other chains with an interaction energy of about 5 kcal/mole. In contrast, the chain segments in polyethylene only interact through weaker dispersion forces, ~0.2 kcal/mole (Figure 10-45). This goes a long way towards accounting for the more than 100°C difference in their melting temperature.

But be a little careful here. ΔH_f is the *change* in enthalpy on going from the crystal to the melt. In the crystal lattice the overall free energy is minimized by arranging the chains so that all the amide groups are hydrogen bonded. *Not all of these hydrogen bonds are broken in the melt.* There is actually a dynamic equilibrium, such that hydrogen bonds (or any other interaction between segments) are being broken and reformed between different partners all the time, a consequence of the continuous motion of the chains. At any instant of time there will be some groups that are hydrogen bonded and some that are "free," between partners, so to speak. The number of hydrogen bonds will depend upon the temperature, but will obviously be less than that found in the crystal-

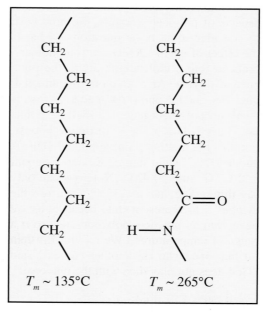

FIGURE 10-45 Approximate melting temperatures of PE and Nylon 6.

FASCINATING POLYMERS—POLYACETALS

As we are deep into the subject of crystalline polymers, let's look at a couple of useful engineering plastics. In the 1920s, one of the great polymer pioneers, Hermann Staudinger, prepared a formaldehyde polymer of reasonably high molecular weight in the absence of water at low temperatures using the initiator boron trifluoride, BF_3. The polymer had a couple of unfortunate characteristics, however, that precluded any possible

Celcon® gear wheels (Courtesy: Celanese).

commercial development. It was thermally unstable, which made processing in the melt impractical. It was also insoluble in all the common solvents of the time, so films or fibers could not be produced by spinning from a solution. After an intensive research program initiated in 1947, DuPont chemists came up with an answer. The thermal instability of polyformaldehyde was primarily caused by a depolymerization or unzipping reaction in which formaldehyde molecules were sequentially eliminated from the end of the polymer chain. By reacting the end-groups with acetic anhydride ("end-capping"), ester groups were formed which blocked this degradation mechanism. Thus was born DuPont's Delrin® family of thermoplastic acetal resins. DuPont launched their new moldable engineering resins in 1960. These materials were highly crystalline (for a plastic material), with outstanding stiffness, fatigue and creep resistance. Delrin® also had a low coefficient of friction and good wear properties. Delrin® materials were touted as replacements for metal parts in many applications such as gears, cams, springs, bearings and the like. And DuPont was in a dominant position once again with a new engineering thermoplastic. But not for long! Researchers at Celanese discovered that the DuPont patent only covered homopolymers. They quickly found that copolymers of formaldehyde (which were to be called Celcon®) not only gave materials that still had most of the favorable properties of the homopolymer, but they were easier to process and it was not necessary to perform the end-capping to prevent thermal decomposition. What Celanese research chemists found was that formaldehyde (usually in the form of trioxane) could be copolymerized with a relatively small amount of a cyclic ether, like ethylene oxide, to yield a copolymer where the majority of the chemical repeat units were still methylene oxide (CH_2O). However, every now and again there would be an ethylene oxide (CH_2CH_2O) unit incorporated in the chain. This produces an irregular chain and the copolymer does not crystallize to the same extent as the homopolymer. Thus it has a lower melting point (about 165°C). But what was more fascinating was that after a simple thermal treatment of the initially formed copolymer, some formaldehyde was eliminated and then the copolymer became thermally stable. The unstable CH_2OH end-groups "unzip" (depolymerize) until a stable ethylene oxide repeat is encountered. So unlike the homopolymer, where it was necessary to chemically modify the end-groups with acetic anhydride, this step was not needed for the copolymer.

FASCINATING POLYMERS—LIQUID CRYSTALLINE POLYMERS

Aliphatic

Swivel

Bent

"Crank Shaft"

Typical structures designed to impart tractability.

You may recall that in the early 1960s Stephanie Kwolek's studies at DuPont led to the discovery that certain solutions of aromatic polyamides were liquid crystalline. This liquid crystalline nature was a profound discovery, resulting in the development of the super-high-strength Kevlar® fibers. As you can imagine, this development reverberated around the world and caught the attention of fiber and polymer scientists in many industrial and academic research laboratories. Could similar liquid crystalline behavior be found in polymer melts? In other words, was it possible to make thermotropic polymers that could be processed from the melt? Attention was focused on the aromatic polyesters, but there were some serious problems. In the early 1970s, it was well established that polyester analogs of the aromatic polyamides, poly(p-benzoate) (PBE) and poly(phenylene terephthalate) were totally intractable, insoluble in any solvent and decomposed well before their melting points. Some clever science was needed and it came from a number of sources. From the middle of the 1970s to the early 1980s a large number of companies were involved and there was intense patent activity in the field of thermotropic polyesters. This was also a "hot" topic for academicians, who, together with scientists at Eastman, DuPont and Celanese, studied and identified the polymer chain architectural elements necessary to promote melt anisotropy at appropriate processing temperatures. This is a fine balancing act, requiring structures that reduce the melting point (by disruption of crystalline order), but at the same time preserve the melt anisotropy of aromatic polyesters. Gordon Calundann at Celanese is generally credited with the discovery of segments that impart tractability. These utilize a parallel offset or "crankshaft" geometry as illustrated in the box above. The Celanese researchers developed a family of commercial thermotropic copolyesters, called Vectra®, that are sold on the market today. The Celanese Company has gone through several metamorphoses since the 1980s and in the last decade they "spun off" their engineering polymers into a separate company called Ticona. Vectra® LCPs are one of the gems in their arsenal of high-performance engineering polymers. Liquid crystalline polymers like Ticona's Vectra®, Solvay's Xydar® and DuPont's Zenite® have excellent dimensional stability and creep resistance at very high temperatures. They are highly crystalline, thermotropic (melt-orienting) thermoplastics that have a low melt viscosity and are able to readily fill complex molds. LCPs find application in precision electrical and electronic parts, such as connectors, bobbins, relays, devices, sensors etc.

line state. The enthalpy change will depend upon the difference in these two numbers.

Other factors being equal, however, you would still expect polymers that have strong intermolecular interactions to have higher melting points, just keep in mind that if you wanted to estimate ΔH_f in a more quantitative fashion you would have to take account of the number of interactions of each type, not just their strength.

OK, now it's your turn! Supposing you could synthesize such beasts, would you expect syndiotactic polypropylene (*syn*-PP) or syndiotactic poly(vinyl chloride) (*syn*-PVC) to have the higher melting point? (If, by now, you don't know the difference between *syn*-PP and *syn*-PVC there is no hope for you—give up polymers and become a theoretical physicist—one once told us the details of chain structure are irrelevant!)

Entropy and Chain Flexibility

Now let's consider the entropy term. For a given crystal structure all the chains in the lattice are in identical conformations, the all trans, planar zig-zag conformation for polyethylene, for example. Upon melting, the chains escape the cage of the crystalline lattice and are allowed to take on a multitude of different shapes or conformations, providing there is sufficient bond rotational freedom. Some chains are far more flexible than others (Figure 10-46), so that the entropy change on going from the crystal to the melt will vary considerably with chain stiffness.

You will hopefully recall that the entropy can be related to the number of arrangements available to a chain through $S = k \ln \Omega$. It should then be immediately obvious that for a flexible polymer there is a large change in entropy on going from the crystal to the melt (Figure 10-47). If we confine our attention to the entropy associated with chain conformations, then in the crystal the number of arrangements, Ω, available to the chain is small, as large portions of the chains, those within the bulk of the crystal, are restricted to a single conformation. In the melt a flex-

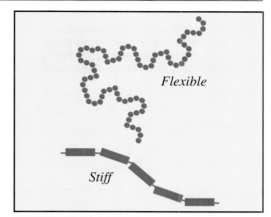

FIGURE 10-46 Schematic diagram depicting flexible and stiff chains.

ible chain has an enormous number of conformations available to it, so that (Equation 10-49):

$$\Delta S_f = k(\ln \Omega_{melt} - \ln \Omega_{crystal}) = Large$$
EQUATION 10-49

For a stiff chain, one that does not have much freedom to rotate around its backbone bonds, the opposite will be true and

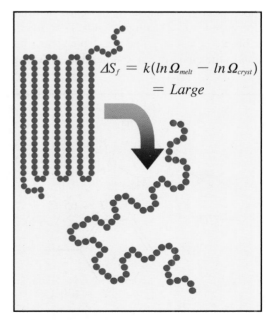

FIGURE 10-47 Schematic diagram depicting the enthalpy of fusion for a flexible chain.

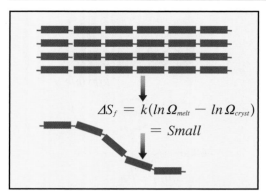

$$\Delta S_f = k(ln\,\Omega_{melt} - ln\,\Omega_{cryst})$$
$$= Small$$

FIGURE 10-48 Schematic diagram depicting the enthalpy of fusion for a stiff chain.

the entropy change will be (relatively) small (Figure 10-48). This, in turn, means that other things being equal (i.e., the ΔH_f term), polymers with stiff backbones should have higher melting points, because $T_m = \Delta H/\Delta S_f$ and if ΔS_f is small, T_m will be large.

Polymers can lack rotational freedom for various reasons. One is the nature of the functional group in the backbone. If we take polyethylene as our standard, against which we will "benchmark" other polymers (Figure 10-49), then putting an ether oxygen (or an ester) into the backbone tends to make it more flexible, because of the ease of rotation around –C–O– single bonds. In contrast, a bloody great big and rigid benzene ring makes the chain much less flexible. If ΔS_f is

larger (the chain is more flexible), T_m will be smaller, and vice-versa.

Polymers can also lack rotational freedom because of steric hindrance associated with the functional groups attached to the backbone. We know thinking is a bit of an effort, but the two missing melting temperatures are ~170°C and ~225°C (Figure 10-50). Which one of these is the melting point of isotactic polystyrene?

Effect of Solvents

We will not go into the details of the next two effects. Dealing properly with diluents such as solvents, for example, really requires that you study solution thermodynamics first. The effects are fairly easy to understand in terms of the qualitative arguments we have made so far, however, so let's start with the effect of solvent.

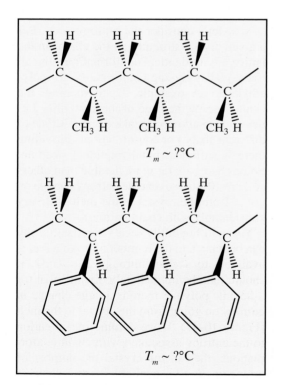

FIGURE 10-50 Isotactic polypropylene (top) and isotactic polystyrene (bottom). Which one has a T_m ~ 170°C and which has a T_m ~ 225°C?

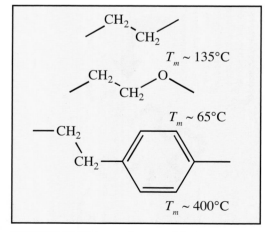

FIGURE 10-49 Approximate melting temperatures of PE, poly(ethylene oxide) and poly(*p*-xylene).

If a semi-crystalline polymer is put in a beaker full of a low molecular weight diluent such as a solvent, it usually *swells* (if the solvent is a "good" one for that polymer—see Chapter 11, page 353). The solvent diffuses into the amorphous domains, but cannot penetrate and hence dissolve the crystalline regions. The melting point of the crystal is lowered, however, for a simple reason. The entropy change on melting is now not only associated with the change in the number of conformations of the chains, but also the increase in entropy that is a result of them being able to mix with the solvent once dissolved (Figure 10-51). Remember, the larger the entropy change, the lower the melting point.

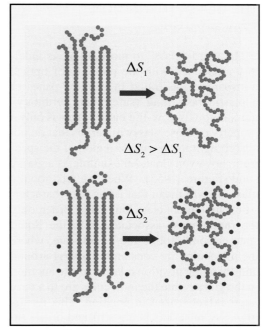

FIGURE 10-51 Schematic diagram depicting the effect of solvent on the entropy of fusion.

Effect of "Defects" and Molecular Weight

As you might intuitively expect, the random inclusion of comonomers into a chain, or the random inclusion of branches, reduces and broadens the melting point. This is because such units are different in size and/or shape to the "regular" segments and cannot be accommodated within the crystal lattice (Figure 10-52). Obviously, if the concentration of such "defects" is too high, then the polymer won't crystallize at all. In ethylene-co-propylene copolymers, for example, there is no detectable ethylene crystallinity once the propylene content reaches about 20–25%. The random presence of propylene segments shortens the length of ethylene sequences available to fit into the lattice, lowering the fold period and hence T_m. At some concentration of propylene comonomer there will not be a sufficient number of long ethylene sequences to form a nucleus that is stable to further growth upon cooling from the melt.

The effect of molecular weight can be thought about in the same way—the end groups often have a different size and/or shape to the other chain segments and are usually excluded from the lattice. Low molecular weight chains will then tend to form crystals that are thinner than their high molecular weight counterparts and hence have a lower T_m. This effect should obviously become less and less important as the chains get longer, so that the melting point does not increase continuously with molecular weight, but approach an asymptotic limit at long chain lengths.

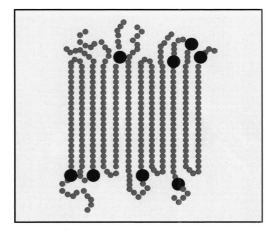

FIGURE 10-52 Schematic diagram depicting defects in the crystalline lattice.

THE GLASS TRANSITION TEMPERATURE

The word "glass" to most people is indelibly associated with the property of optical transparency, as found in window panes or a plastic beer mug (safer than "ordinary" glass in rowdy, low-life bars!). This is only a property of those glassy materials that do not absorb light in the visible region of the spectrum, however. Coal, for example, is a glassy solid (Figure 10-53). When we talk about a glass we will mean that it has two characteristics: first, there is no long range order, only the randomness associated with the liquid state; second, unlike the liquid state, where the molecules are constantly moving around and changing position relative to one another, in the glassy state the molecules are to a very large extent frozen in position. This results in glassy materials being stiff and brittle (to different degrees, compare plastic and (inorganic) glass beer mugs). We will discuss this in more detail in the sections on mechanical and rheological properties. Here our concern is thermal characteristics.

Glassy Solids and the Glass Transition

If you take a polymer like atactic polystyrene that is incapable of crystallizing and heat it, it first softens, then becomes a tacky, rubber-like material and finally takes on the liquid-like characteristics of the melt. Similarly, if you cool a piece of rubber it also changes properties, becoming rigid and brittle at very low temperatures. To emphasize, these changes in properties are not accompanied by a change in order, the polymer chains remain random coils. Also, just as the mechanical and rheological properties do not change abruptly, but over a range of temperatures, neither do other properties, as shown in these plots of volume and specific heat as a function of temperature (Figure 10-54). So, what is the nature of this transition and why should a polymer like atactic polystyrene be a glass at room temperature, but rubbers only become glassy well below room temperature? What is it about their chemical structure that results in this difference in properties?

We will start by discussing the nature of the glass transition. It all has to do with the range of motion that is allowed to a material in a particular state. First, consider a gas. Here the molecules have translational motion, bouncing around inside their container and only occasionally bumping into one another. Now compare this to a perfectly ordered crystalline solid, where all the molecules are confined to their lattice positions. There is no room for them to move relative to one another, but they can oscillate or vibrate around a mean position in the lattice. (If you are having trouble visualizing this there is a nice little movie in our *Polymer Science* CD.) Remember, temperature is a measure of the mean kinetic energy of the molecules, so above absolute zero there will always be motion of some sort.

Do you remember when you had to study a particle in a box and the harmonic oscillator in your quantum mechanics class, and unless you had a really good teacher you never exactly knew why? Well those are simple models of a perfect gas and a perfect crystalline solid. By calculating the energy levels of these systems (quantum mechanics), then calculating how a given amount of total energy is distributed among these energy levels in all the molecules in the system (statistical mechanics), you can deter-

FIGURE 10-53 Coal, a glassy solid (Source: Bayer).

mine macroscopic properties and the relationships between them (thermodynamics). This is awesome. The problem is that we don't have good simple models of the type of motion that interests us here: that found in the liquid state. Fortunately, some simple qualitative arguments serve our purpose.

Free Volume

When you heat a crystalline solid the kinetic energy and hence the amplitude of the oscillations of the molecules increases. The material expands. As it does so the attractive force between the molecules decreases (see Chapter 8, page 208). At the melting temperature this force of attraction is not enough to keep the molecules in place and the material melts. Now the motion becomes a complex coupling of vibrational oscillations and translational movement, as "holes" open up as a result of random displacements of neighbors. Clearly, there has to be enough "empty space" in the material as a whole for this to occur.

What we have just defined, in a very simplistic manner, is something called free volume. This is not the same as the empty or "unoccupied" volume found in the close-packed state (Figure 10-55). Consider the stacking of solid spheres. If we pack these in a regular or ordered fashion there will always be some gaps between the balls. If we allow these spheres to pack randomly, like peas that you throw into a colander to drain, there will be even more of this unoccupied volume, but the spheres will still be close packed, jammed up against their neighbors. However, molecules are not static, like macroscopic balls. Above the absolute zero of temperature they have motion, but in the close-packed state this will just involve vibrations around a mean position.

So, let us imagine that at very low temperatures in a material that for whatever reason has not or cannot crystallize, the molecules are more or less randomly close packed. The total volume of the system is then that of the "*hard core*" of the molecules plus the "*unoccupied*" volume between them. As the tem-

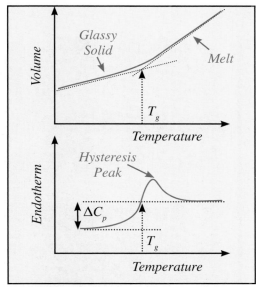

FIGURE 10-54 Schematic plots of temperature versus volume (top) and specific heat (bottom) for a glassy polymer.

perature is raised, additional or *free volume* is introduced as the amplitude of vibrational motion increases and the material as a whole expands (Figure 10-56). At some point there

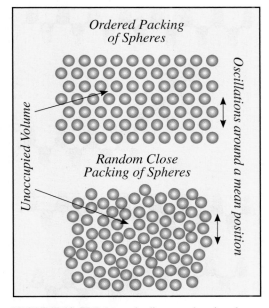

FIGURE 10-55 Schematic diagram showing the unoccupied volume and oscillations around a mean position in ordered and random close packing of spheres.

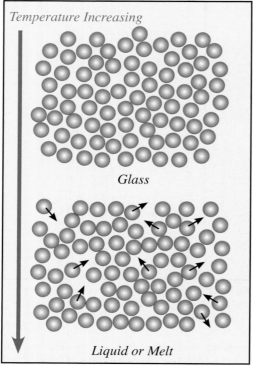

FIGURE 10-56 Schematic diagram showing the free volume as temperature is increased.

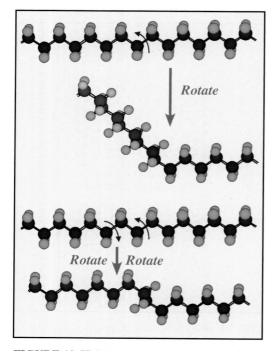

FIGURE 10-57 Schematic diagrams showing different bond rotations.

is sufficient randomly dispersed free volume that the molecules can move relative to one another and the material is then in the liquid state.

The same type of thing happens in polymers, except now the motion is more complex than in small molecules. A polymer molecule can change its shape and center of gravity as a result of bond rotations. But a rotation around a single bond, as illustrated in Figure 10-57, would require a huge amount of free volume, hence energy, to push the other chains "out of the way." The dynamics must therefore be more subtle and complex, involving coupled rotations of adjacent or nearly adjacent bonds, resulting in fairly small overall displacements of segments. But the principle is the same: the number of oscillatory motions around the bonds (coupled with intermolecular vibrations between segments of different chains) increases with temperature and at some point creates sufficient free volume that segmental motion can occur.

Free Volume, Viscosity and the Glass Transition

The free volume theory of the T_g usually defines the glass transition the other way around—the temperature where the free volume falls below a critical value as you cool a material (Figure 10-58). This would obviously result in dramatic changes in properties like viscosity. At high temperatures there is a lot of free volume and the molecules move around easily. Viscosity is a measure of the resistance to flow of a fluid and at high temperatures the viscosity is therefore low. In contrast, at low temperatures there is little free volume and practically no flow at all (we'll talk more about this when we discuss viscoelasticity). The viscosity is essentially infinite and the material is hard, rigid and brittle—what we call a glass. The viscosity should change dramatically near the transition between these states, the T_g.

The relationship between viscosity and the T_g has been described by various empirical and semi-empirical equations. Two of the better known are the Doolittle equation (Equation 10-50):

$$\eta^{-1} = A\,exp[-B/V_f]$$

EQUATION 10-50

which relates the fluidity (reciprocal of the viscosity, η) to the free volume, V_f, and the Vogel-Fulcher equation (Equation 10-51), an empirically derived relationship between the viscosity and a characteristic temperature, T_0. As the temperature T decreases and V_f or $T - T_0$ become small, the viscosity increases dramatically.

$$\eta = A'\,exp[B'/(T - T_0)]$$

EQUATION 10-51

These equations can be related to one another by assuming the free volume is proportional to the coefficient of thermal expansion, α, near T_0 [i.e., $V_f \sim \alpha(T - T_0)$]. (There is also the WLF equation, which we will get into later in Chapter 13, Mechanical and Rheological Properties.)

Thermodynamic and Kinetic Aspects of the T_g

The characteristic temperature, T_0, is usually located about 30–70°C below the experimentally measured T_g and in various thermodynamic theories it represents an equilibrium value. We will not discuss these theories, but the existence of such a thermodynamic transition remains in dispute, with some arguing that the T_g is purely a kinetic phenomenon. Certainly, the experimentally observed quantity is kinetic in character. One manifestation of this is the shift in T_g with change in cooling rate—the slower you cool, the lower the T_g (Figure 10-59). (But this also suggests that if you could cool at an infinitely slow rate you would observe the "equilibrium T_g".)

Another important manifestation of the kinetic character of the T_g is the process of

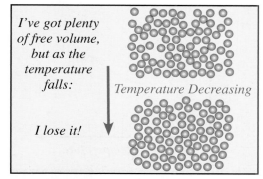

FIGURE 10-58 Schematic diagrams depicting how free volume is lost as the temperature is reduced.

"physical aging." If a polymer is cooled a few degrees below the T_g, say to point "A" on the plot (Figure 10-60), then held at this temperature, the volume of the sample decreases with time, approaching point "B" on the extrapolated melt (above T_g) volume/temperature line. This process of volume recovery clearly must involve a rearrangement or "relaxation" of the molecules to a more efficient packing arrangement. But this is a topic that we will cover in our discussion of Mechanical and Rheological Properties. Here we now turn our attention to the factors that affect T_g, things like molecular weight and chemical structure and see if we can make sense of them in terms of simple qualitative arguments involving free volume.

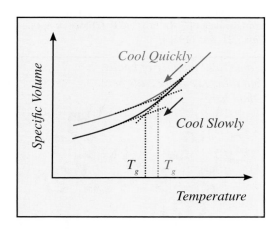

FIGURE 10-59 Schematic plots of specific volume versus temperature showing how the T_g varies with cooling rate.

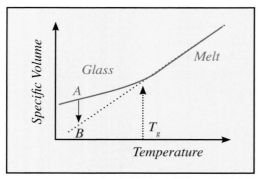

FIGURE 10-60 Schematic plots of specific volume versus temperature showing the effect of physical aging.

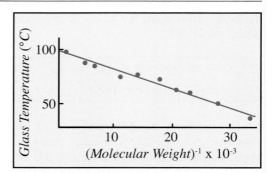

FIGURE 10-63 Schematic plot of glass transition temperature versus reciprocal molecular weight [redrawn from the data of T. G. Fox and P. J. Flory, *J. Appl. Phys.*, **21**, 581 (1950)].

Effect of Molecular Weight

We will start by considering the effect of molecular weight on T_g. The plot in Figure

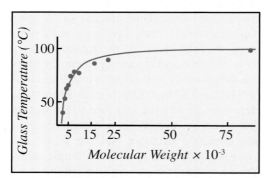

FIGURE 10-61 Schematic plot of glass transition temperature versus molecular weight [redrawn from the data of T. G. Fox and P. J. Flory, *J. Appl. Phys.*, **21**, 581 (1950)].

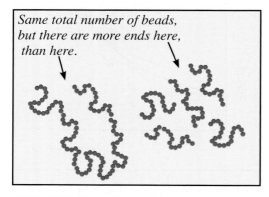

FIGURE 10-62 Schematic diagram of the number of chain ends.

10-61 shows that for low molecular weight chains the T_g increases sharply with chain length, but then levels off and approaches an asymptotic limit. This is fairly easily understood in terms of simple free volume arguments. The ends of a chain have more freedom of motion than segments in the middle (think of snapping a towel—the relatively slow motion imparted by your hand accelerates along the towel to something much faster at the end). Lower molecular weight chains have more ends per unit volume than long chains (Figure 10-62), hence a higher free volume, and therefore a lower T_g. Fox and Flory[31] used this type of argument to obtain Equation 10-52:

$$T_g = T_g^\infty - \frac{K}{M_n}$$

EQUATION 10-52

Where T_g^∞ is the glass transition temperature of a (hypothetical) infinitely long chain, $\overline{M_n}$ is the number average molecular weight and K is a constant related to parameters describing the free volume. This relationship works well for most polymers, as the plot of T_g versus the reciprocal of the molecular weight shows (Figure 10-63) for the data plotted in Figure 10-61.

[31] T. G. Fox and P. L. Flory, *J. Appl. Phys.*, **21**, 581, 1950.

FASCINATING POLYMERS—PMMA AND COCKPITS

Since we are discussing polymer glasses, this is a good place to tell you a little bit about the development of a familiar industrial glassy-like polymer that has made an impact on our everyday lives. In the early 1930s, scientists working at both Rohm and Haas A.G. in Darmstadt, Germany and ICI in England were focusing on an acrylic polymer, poly(methyl methacrylate) (PMMA). This polymer, an "organic glass" (as described by Otto Röhm—see *Polymer Milestones*, page 246), was a solid that was transparent and melt-like only

A Plexiglas PMMA cockpit canopy (Courtesy: Rohm and Haas Company).

above about 110°C. Röhm and his assistants developed techniques to make cast sheets of PMMA and form them into complex 3-dimensional shapes upon heating and stretching at elevated temperatures. But methyl methacrylate (MMA) was expensive to make and it was J. W. C. Crawford, working at ICI, who came up with a shorter and cheaper route to its synthesis. By cross-licensing the cast-sheet and monomer production patents, both companies were able to use the best technology for making "Plexiglas" (Rohm and Haas) and "Perspex" (ICI). With WWII looming on the horizon, PMMA acrylic sheet was being produced by ICI in the UK, and by Rohm and Haas in both Germany and the United States. It was all essentially being used for military aircraft. DuPont, which had cross-license patent agreements with ICI, started manufacturing cast PMMA sheet in 1939. Many of these cross-licenses conflicted with patent applications that Rohm and Haas had filed in the United States. Rather than fight a long battle in court, DuPont and Rohm and Haas signed a number of complicated agreements assigning royalty-free licenses to each other. (This is starting to look like OPEC!) And then the hammer came down! In 1942, citing violations of the Sherman Antitrust Act of 1890, the US government indicted Rohm and Haas and DuPont of criminally conspiring to control the market, production and price of acrylic sheet (with ICI, I.G. Farben and Rohm and Haas A.G. as unindicted coconspirators). The trial started on May 14, 1945, six days after hostilities ended in Europe, and lasted for some 5 weeks. Haas, who was 73 at the time and sincerely believed that he had done no wrong, was incensed that he was being tried as a common criminal and took the stand in his own defense. The jury came back with a verdict of not guilty. This was not the end, of course, and there was subsequently a civil case that took another 3 years to resolve.

FASCINATING POLYMERS—PC AND THE COMPACT DISC

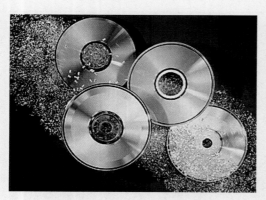

Compact discs made of Lexan® PC (Courtesy: GE Plastics).

While on a visit to the Bayer A.G. research laboratories in Leverkusen, Germany (kindly arranged by our old friend Dr. Robert Kumpf), your authors were fortunate to meet and talk with Dr. Dieter Freitag. After serving in many important research managerial positions at Bayer, Dr. Freitag became a director of the company and Head of Bayer's Plastics Research, worldwide, in 1988. He was intimately involved with and obtained key patents in the development of the polycarbonate (PC) that is used to manufacture the compact disc. As related to us by Dr. Freitag, rep-

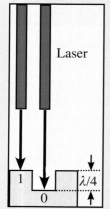

resentatives of Philips, the large European electronics company, approached Bayer in about 1978. They had with them a metal disc that contained Beethoven's 3rd symphony, digitally encoded using some 4 billion holes. The size of each hole was only about 1/10,000 mm! So how did this metal disc, encoding Beethoven's classic symphony, work? In very simple terms, a laser beam, which has a very coherent and precise frequency, is bounced off the moving disc. If there is no hole, the laser beam will be reflected and detected such that the light waves are in-phase and reinforcing—in digital terms we can call this 1. Conversely, if the beam is over a hole, which has a depth of precisely a quarter of the wavelength of the laser beam, the beam will travel an additional distance equivalent to one half a wavelength. This reflected beam is then completely "out of phase" with the incident beam, resulting in destructive interference (i.e., no light!)—we can call this 0. With 1's and 0's we have digital information, of course, and with enough holes we have the ability to encode an incredible amount of information. The question to the Bayer scientists was, "Could you use polycarbonate to accurately replicate the pattern of 4 billion tiny holes in the metal disc by injection molding?" The initial reaction: "Impossible!" And indeed the first attempts failed, because the definition of the holes could not be duplicated—rather than the nice sharp sides required for reading the digital pattern, the high melt viscosity of the polymer produced sloping sides and poor definition. Freitag and his colleagues set out to try and rectify the problem and in a superb piece of investigative science discovered that by modifying the end groups of the polymer chain with alkyl phenols, a polycarbonate could be produced that would replicate the master disc with excellent accuracy and sound quality. Not only that, but the cycle time for producing a disc was reduced to 20 secs! Phillips introduced the compact disc to the world in 1982. As Dr. Freitag likes to point out, compact (and now DVD) discs can hold information equivalent to about 300,000 pages of written text, or a stack of paper 12 meters high!

Effect of Chemical Structure

Now let's assume we're dealing with high molecular weight polymers and ask ourselves "What is the effect of chemical structure?" In other words, going back to one of the questions posed near the beginning of this section, why does polystyrene have a higher T_g than natural rubber? Just as in our discussion of T_m, this will come down to factors like chain stiffness and intermolecular interactions, although now we will couch our arguments more in terms of free volume and molecular mobility. You should have the idea by now, though, and you should be able to make an informed guess as to which of the polymers, polyethylene, atactic polypropylene, or atactic polystyrene has the highest T_g. The answer, if you have any doubt, is shown in Figure 10-64.

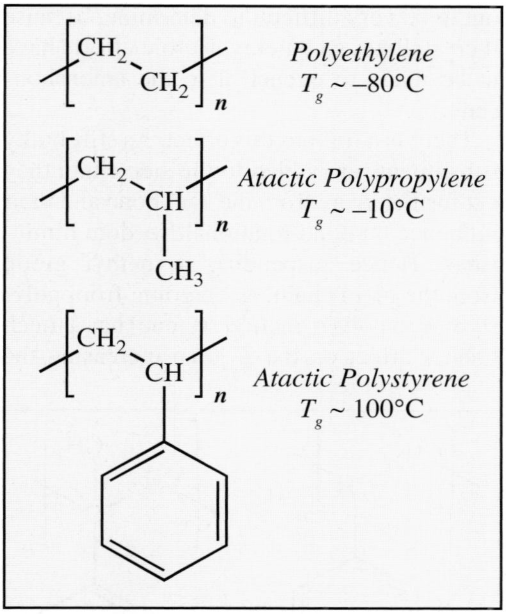

Because bond rotational freedom affects chain mobility, the effect of bulky substituent groups (like benzene rings) in the backbone is easily understood. These result in higher energy barriers to rotation, which then (on average) only occur at higher temperatures. (This can be related to free volume arguments in terms of cooperative motions—at what temperatures would these "create" enough free volume for segmental motion to occur?) The T_g of stiff chains is thus higher than flexible chains. Some examples are displayed in Figure 10-65. Note that an ether oxygen in the backbone makes the chain very flexible—there is almost free rotation around –O– bonds. Silicone rubber, poly(dimethyl siloxane), thus has a very low T_g, lower than polyethylene, our "benchmark". A bloody great big benzene ring has the opposite effect, more than canceling out the influence of the ether oxygen in poly(phenylene oxide), which is what the molecule at the bottom of Figure 10-65 is often called, even though its real chemical name is poly(2,6-dimethyl-*p*-phenylene oxide).

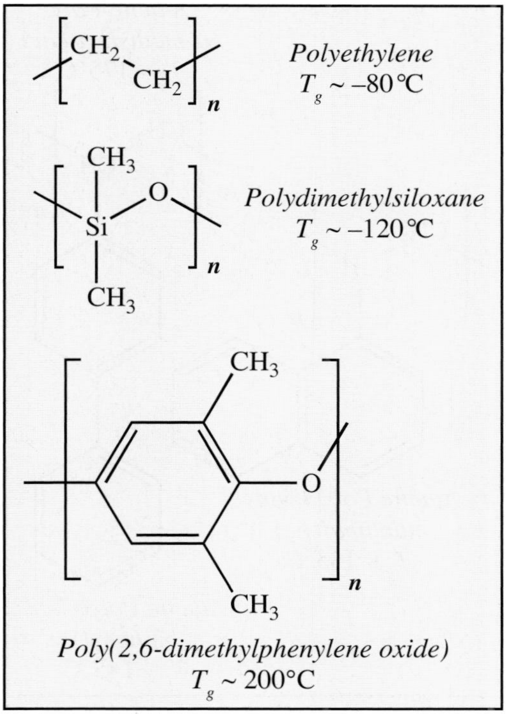

Similar arguments apply to the effect of bulky substituents, which sterically hinder bond rotations and hence raise the T_g. This can clearly be seen in the examples given in Figure 10-64. (The T_g of polyethylene turned

FIGURE 10-64 The approximate glass transition temperatures of polyethylene, atactic polypropylene and atactic polystyrene.

FIGURE 10-65 The approximate glass transition temperatures of polyethylene, polydimethylsiloxane and poly(2,6-dimethylphenylene oxide).

out to be very difficult to determine, because it crystallizes so quickly. Heroic efforts have to be made to quench it to the amorphous state.)

There is a limit to this effect. As stiff bulky substituents are added to the side chain they get further away from the backbone and their influence on bond rotational freedom diminishes. Hence suspending a methyl group from the main chain, as on going from polystyrene to poly(α-methyl styrene) has a much greater affect on the T_g than increasing the

size of the aromatic substituent (i.e., going from a benzene ring to a naphthalene ring to a biphenyl group)—see Figure 10-66.

But now it's time for you to do some work again. Instead of hanging stiff groups off the backbone, lets attach a flexible side chain and ask what would be the effect of increasing the length of this chain, as in the series of methacrylate polymers shown in Figure 10-67. Would increasing the chain length raise or lower the T_g? If you answered "lower," you're right! The T_g actually decreases with increasing side-chain length. This is because substituents closest to the chain, the methyl and ester group, provide the bulk of the steric hindrance. Because it is flexible the rest of the attached side chain can "get out of the way" of motions of the main chain through rotations around side-chain bonds. Furthermore, these side chains increase the free volume through their motions. It's like having more chain ends present and the T_g is lowered.

As you might guess, strong intermolecular attractions also act so as to raise the T_g. For example, the chlorine atom and methyl group of PVC and polypropylene, respectively, are roughly the same size and therefore have approximately the same effect on bond rotations. The polar character of the Cl atom leads to stronger forces of attraction between the chains, however, so it takes more thermal energy to initiate segmental motion in PVC. The T_g of PVC, ~87°C, is higher than that of atactic polypropylene, ~ –10°C. In terms of free volume, the interacting groups are (on average) closer together, so that there is a lower free volume (reality is probably a bit more complicated than this).

Effect of Cross-Linking and Crystallization

Cross-linking decreases the free volume, because parts of the chain are tied more closely together by covalent bonds. Hence T_g increases with the degree of cross-linking.

Crystallization works in the same way as cross-linking, raising the T_g. The amorphous parts of the sample become "tied down" in the regions between the crystalline domains

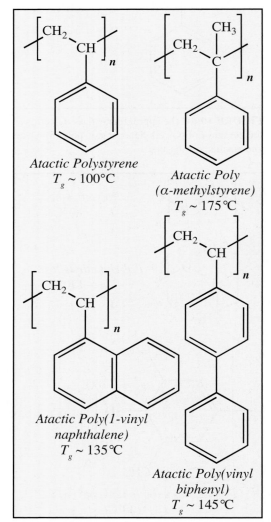

Atactic Polystyrene
T_g ~ 100°C

Atactic Poly (α-methylstyrene)
T_g ~ 175°C

Atactic Poly(1-vinyl naphthalene)
T_g ~ 135°C

Atactic Poly(vinyl biphenyl)
T_g ~ 145°C

FIGURE 10-66 The approximate glass transition temperatures of the atactic polymers polystyrene, poly(α-methyl styrene), poly(1-vinyl naphthalene) and poly(vinyl biphenyl).

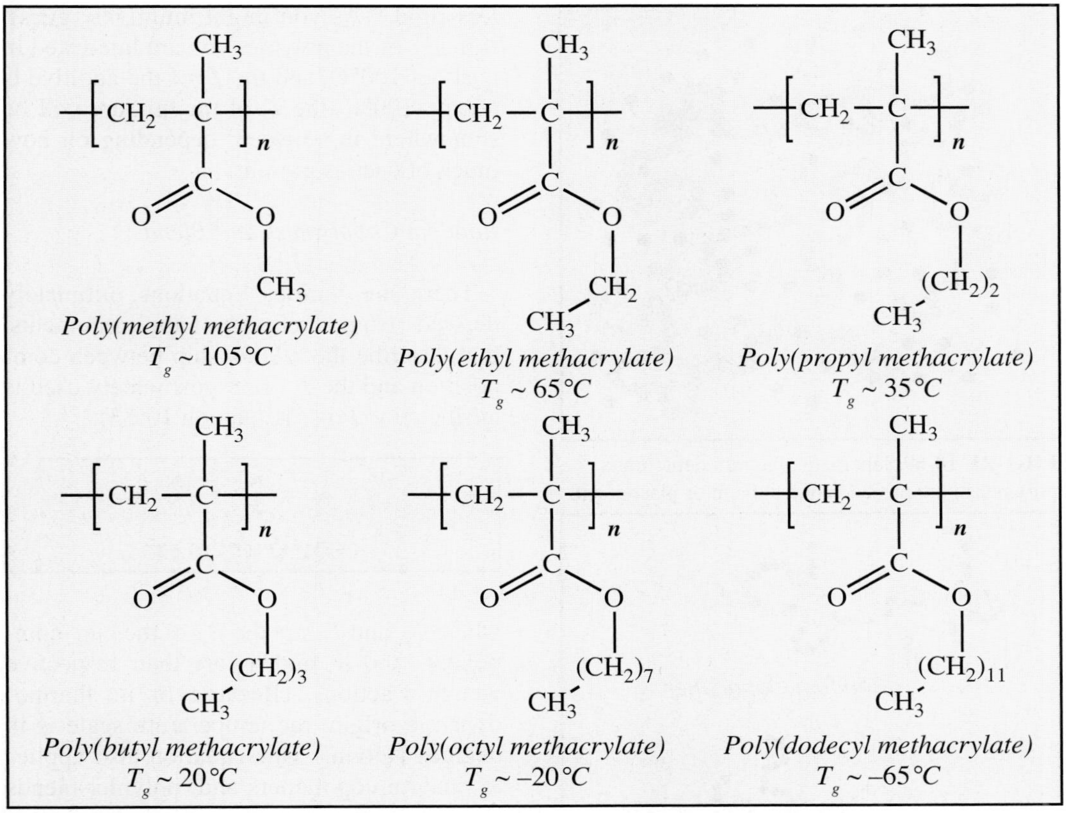

CH₃
CH₃
CH₃

$$\left[CH_2 - C \right]_n$$

Poly(methyl methacrylate)
$T_g \sim 105\,°C$

Poly(ethyl methacrylate)
$T_g \sim 65\,°C$

Poly(propyl methacrylate)
$T_g \sim 35\,°C$

Poly(butyl methacrylate)
$T_g \sim 20\,°C$

Poly(octyl methacrylate)
$T_g \sim -20\,°C$

Poly(dodecyl methacrylate)
$T_g \sim -65\,°C$

FIGURE 10-67 The approximate glass transition temperatures of various poly(*n*-alkyl methacrylates).

(the lamellar arms in melt-crystallized samples). There is less free volume, the mobility of the chains is constrained and the T_g goes up (Figure 10-68). The extent of this obviously depends on the degree of order and in some samples the T_g can be completely masked by crystallization.

Effect of Diluents

The properties of a polymer can be altered dramatically by adding low molecular weight diluents such as "plasticizers" (Figure 10-69). PVC, for example, is a rigid glassy solid at room temperature. In this form, it is used for pipes in plumbing applications. However, if you have ever owned a cheap car you are no doubt familiar with vinyl seats, which at least when the car is new are flexible. These are also made out of PVC, but in this case the polymer has been mixed with a diluent that lowers its T_g and thus makes it much

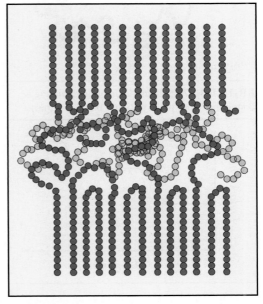

FIGURE 10-68 Schematic diagram depicting the amorphous region between lamellar molecules.

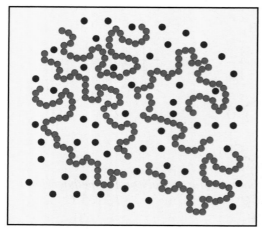

FIGURE 10-69 Schematic diagram depicting a polymer in the presence of a solvent or plasticizer.

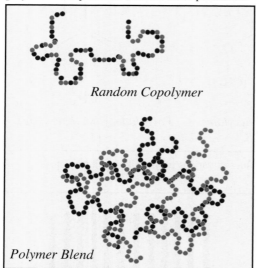

Random Copolymer

Polymer Blend

FIGURE 10-70 Schematic diagram depicting a copolymer (top) and a polymer blend (bottom).

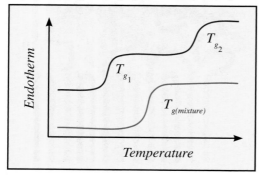

FIGURE 10-71 Schematic plots of temperature versus specific heat for a two phase system (top) and a single phase (bottom).

less rigid.[32] As you might intuitively guess, if the T_g of the polymer you are interested in is about 100°C, and the T_g of the additive is about –100°C, the T_g of the mixture will be somewhere in between, depending on how much of each is present.

Random Copolymers and Blends

There are various equations, ultimately derived from thermodynamic arguments, that describe the relationship between composition and the T_g. The most widely used is attributed to Fox[33] (Equation 10-53):

$$\frac{1}{T_g} = \frac{w_1}{T_{g_1}} + \frac{w_2}{T_{g_2}}$$

EQUATION 10-53

where T_{g_1} and T_{g_2} are the T_gs of the pure components and w_1 and w_2 are their respective weight fractions. (Because of its thermodynamic origin, the temperature scale is in degrees Kelvin.) This equation also applies to random copolymers and polymer blends (Figure 10-70), *providing the latter form a miscible mixture.* As you will see in the section on the phase behavior of polymer solutions and blends, most polymers do not want to mix with one another: they phase separate, like oil and water. In a thermogram (Figure 10-71) of such a system one would therefore expect to detect two T_gs, one for each phase. The position of the two T_gs will depend on the composition of each of the phases. In immiscible blends, each phase is almost pure homopolymer. In phase-separated polymer solutions one phase is polymer rich and the other is solvent rich, but each is a random mixture of the two components. A miscible mixture will have a single T_g, however, often at a temperature given by the Fox equation.

[32] The plasticizer also migrates out of the polymer over time, evaporating into the air and condensing on the inside of your windshield in the form of a nice yellow gunk. The T_g of your vinyl seat material thus increases, it becomes more brittle and eventually tears under the weight of your bum!

[33] T. G. Fox, *Bull. Am. Phys. Soc.,* **1**, 123 (1956).

RECOMMENDED READING

D. C. Bassett, *Principles of Polymer Morphology*, Cambridge University Press, Cambridge, 1981.

J. D. Hoffman, G. T. Davis and J. I. Lauritzen, Jr,. in *Treatise on Solid State Chemistry*, N. B. Hannay, Ed., Vol 3 Chapter 7, Plenum Press, New York, 1976.

L. Mandelkern and G. McKenna, in *Comprehensive Polymer Science*, C. Booth and C. Price, Eds., Vol 3, Chapter 7, Pergamon Press, Oxford, 1989.

L. H. Sperling, *Physical Polymer Science*, 3rd Edition, Wiley, New York, 2001.

G. Strobl, *The Physics of Polymers*, Springer-Verlag, Berlin, 1996.

STUDY QUESTIONS

1. Discuss (no more than one page) the difference in the nature of the glass transition and the crystalline melting point.

2. Explain (no more than one page) why polymers crystallize in a folded chain form. (A picture is worth a thousand words!)

3. In Figure 10-50 in the text, the chain structures of isotactic polystyrene and isotactic polypropylene were compared. We then asked you which of these polymers has the highest melting point (~225°C). Well, which one is it? Briefly justify your answer (a couple of sentences).

4. Consider the following polymers:

    A. Atactic polypropylene
    B. Polyethylene
    C. The following nylon:
        $-[(CH_2)_5CONH]-$
    D. The following polyether
        $-[CH_2-CH_2-O]-$

Which of these polymers has the high-est melting point, which has the lowest and which has no melting point at all? Briefly explain the reasons for your choices.

5. There are a class of polyesters that have the general structure:

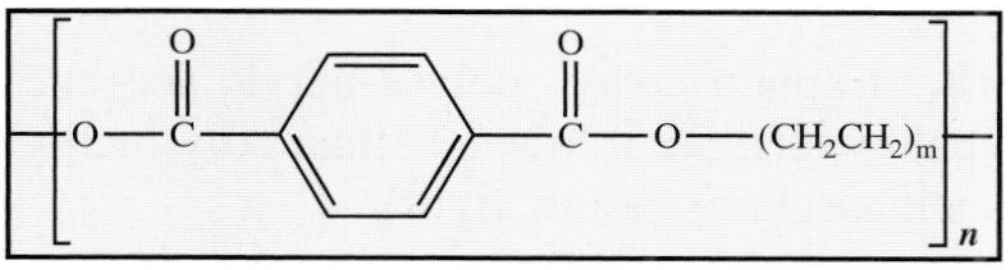

For example, in poly(ethylene terephthalate), PET, $m = 1$). How would you expect the glass transition and crystalline melting temperatures of these polymers to vary with m, given that the molecular weight of the chain is held constant? Give reasons for your answer. Also, assuming that you are now just considering one member of this class of polymers, say PET, how does the T_g vary with molecular weight (n).

6. Briefly explain why a semi-crystalline polymer in contact with a good solvent has a lower melting point than in the dry, solid state.

7. Explain why a crystallizable polymer, cooled slowly from the melt, melts over a broad range of temperature when reheated.

8. Given that the T_gs of polybutadiene and atactic polystyrene are –60°C and 100°C, respectively, calculate and plot the variation of T_g with composition for a styrene/butadiene random copolymer.

9. You are given a sample that has been identified spectroscopically as containing equal amounts of styrene and butadiene units. You figure out that there are three possible microstructures. It could be:

    A. A high molecular weight 50:50 (by weight) styrene/butadiene random copolymer.

    B. A 50:50 (by weight) block copolymer of styrene and butadiene consisting of high molecular weight blocks.

C. A 50:50 (by weight) immiscible blend of high molecular weight polystyrene and polybutadiene.

Using the results obtained from Question 8 and with suitable explanations, explain what you would expect to see in a DSC experiment for each of these possibilities.

10. Explain why poly(n-butyl acrylate) (below left) has a lower T_g than poly(methyl methacrylate) (below right).

11. You are using flexible PVC tubing in your laboratory to run water from the tap to keep your reaction vessel cool. One day, your idiot of a lab partner runs an organic solvent through this tubing. It becomes stiff and cracks. Explain why.

12. Explain why the crystallite size of a polymer varies with the degree of undercooling.

13. For reasons that are lost in the mists of time, a number of years ago your authors synthesized various random copolymers containing the comonomers, n-butyl meth-

acrylate (BMA) and 4-vinyl phenol (VPh) [*Macromolecules*, **25**, 7077 (1992)].

The glass transition temperatures of these copolymers were determined using DSC and the results are given in the following table:

VPh (wt %)	T_g (°C)	VPh (wt %)	T_g (°C)
0	20	65	138
10	44	80	151
28	81	100	170
50	118		

Compare a plot of T_g versus composition using this data to that using the Fox equation (Equation 10-53). Explain any deviations you find in the plots. (Hint: you won't find the answer in the text, you'll have to dig!)

11

Polymer Solutions and Blends

INTRODUCTION

In this chapter we are going to consider why some things mix and some things don't, like the cooking oil and malt vinegar (which we keep around to put on our chips)—see Figure 11-1. Of course, our principal concern will be polymer solutions and mixtures of different polymers, usually referred to as blends, but the principles will be quite general. As in some other sections, you will need to recall some elementary thermodynamics if you are to understand what is going on. Rather than send you back to our review, however, we will repeat a few things we said about the free energy here, because we used a mixing problem as an illustration.

Let's say we are going to mix two systems. In the snapshot shown in Figure 11-2 (you may wish to look at the nice little animation in our *Polymer Science* CD), it is two gases, but it could be two liquids, such as oil and water, or it could be a polymer and a solvent. We would like to know whether or not our two systems will mix. The criterion, of course, is that the entropy must increase, or the entropy change upon mixing should be positive, the inexorable second law of thermodynamics (Equation 11-1):

$$\Delta S_{total} > 0$$
EQUATION 11-1

It is important to recall that this is the total

FIGURE 11-1 Oil and vinegar.

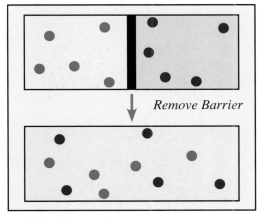

FIGURE 11-2 Two gases separated by a barrier (top) and after removal of the barrier (bottom). What you do not see is the constant random motion of the gas molecules.

entropy change, because mixing could result in heat being emitted to or absorbed from the surroundings, which would change its entropy (Equation 11-2):

$$\Delta S_{total} = \Delta S_{system} + \Delta S_{surroundings}$$
EQUATION 11-2

Calculating the entropy change in the surroundings would be hard, but because this entropy change is a consequence of an exchange of heat with the system, we can then write Equation 11-3 (assuming the process is reversible):

$$\Delta S_{surroundings} = -\frac{\Delta Q_{system}}{T} = -\frac{\Delta H_{system}}{T}$$
EQUATION 11-3

(Remember, the enthalpy change is just the heat change at constant pressure.) Substituting and multiplying through by $-T$ we get Equation 11-4:

$$-T\Delta S_{total} = \Delta H - T\Delta S$$
EQUATION 11-4

We can now *define* the free energy for a constant pressure process (i.e., the Gibbs free energy) to be Equation 11-5:

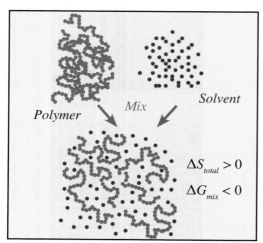

$$\Delta S_{total} > 0$$

$$\Delta G_{mix} < 0$$

FIGURE 11-3 Schematic diagram of the mixing of polymer in solvent.

$$\Delta G = -T\Delta S_{total} = \Delta H - T\Delta S$$
EQUATION 11-5

This means that when the entropy change is positive, the change in free energy is negative (Figure 11-3). The big advantage of using the free energy is that we have defined all our parameters in terms of changes in the system. Now, it would seem, our only problem is to determine whether or not the free energy change is negative for a given mixing problem. But how? The trick is to relate entropy and enthalpy to molecular quantities. Not only that, we will find that a negative free energy change is a necessary condition for mixing, but not a sufficient one. (We will see later that in addition the second derivative of the free energy with respect to composition has to be positive.)

The enthalpy and entropy of mixing in the equation for the free energy (Equation 11-6):

$$\Delta G_m = \Delta H_m - T\Delta S_m$$
EQUATION 11-6

are related to the energy of interaction between the molecules and the number of arrangements available to the system, respectively (or, more precisely, the change in these quantities upon mixing). These are to some degree interrelated. If molecules of component "A" of the mixture strongly interact with those of component "B," then you would expect that on average there would be more arrangements where A's would be next to B's than in a system where they did not attract as strongly.

It's a bit more complicated than this, in that the number of arrangements will also depend upon the strength of the interactions relative to thermal energy, RT, a measure of the average kinetic energy of the molecules (per mole). In other words, if the interaction energies between the molecules are relatively weak, say, just dispersion forces and weak polar forces (see Chapter 8, page 208), then at ambient temperatures one would expect that the disordering effect of thermal motion would result in essentially random mixing. This leads us to the approximation we will use, the *mean field assumption*, where each molecule is assumed to be acted upon by a potential that is an average taken over all the interactions in the system, rather than one determined by local composition. This, in turn, allows us to treat the entropy and enthalpy changes upon mixing as separate and additive quantities. We will start by treating the entropy.

ENTROPY OF MIXING

We start our discussion of the entropy of mixing by considering the simplest possible problem, mixing spherical molecules of equal size. If we can figure out the total number of ways all the molecules can be arranged, Ω, then we can use Boltzmann's equation to calculate the entropy (Equation 11-7):

$$S = k \ln \Omega$$
EQUATION 11-7

But where on earth do you even start on a problem like this, when you have the jumbled up, randomly arranged mess that is the liquid state? You do what you always have to do in statistical mechanics (yeah, that's what we're doing, try not to panic!): make a simple model and see how well it works. There are various approaches, but we will discuss the lattice model. This appears more like a model of the crystalline rather than the amorphous or liquid state, with each molecule sitting in a "cell" on an ordered lattice. Nevertheless, it allows you to calculate the number of arrangements or configurations in a very transparent and easy to understand manner. Amazingly enough, it also gives results that are close to those obtained by other methods (e.g. Hildebrand's approach, based on free volume arguments—see the discussion in Flory[34]).

[34] P. J. Flory, *Principles of Polymer Chemistry*, Cornell University Press, Ithaca New York, 1953.

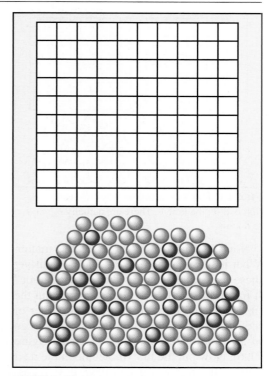

FIGURE 11-4 Schematic diagram of an empty lattice with red and blue balls representing two different spherical molecules, "A" and "B."

If we start with n_A molecules of type A and n_B molecules of type B, then also assume that there are no "holes" on the lattice,[35] then there are just $n_0 = n_A + n_B$ lattice sites. Now we take all the molecules off the lattice (Figure 11-4) and put them back on randomly, one at a time, not caring, at this point, if it's an "A" or a "B."

How many ways are there to put the first molecule on the lattice? That's right, n_0. The first molecule can go in any one of the empty lattice sites (Figure 11-5). (Again, there is a nice animation of this process on our *Polymer Science* CD that you might wish to view.)

Now let's put a second molecule on the lattice. How many ways can we do this? As there are n_0-1 empty sites for the second molecule, so there are n_0-1 ways of putting the second molecule on the lattice (Figure 11-6).

[35] This is equivalent to assuming that there is no free volume, or the fluid is incompressible.

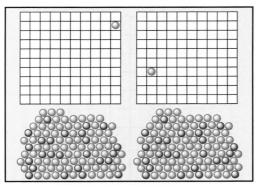

FIGURE 11-5 Two different ways of placing the first ball on the lattice. There are actually n_0 ways to do this.

Now let's get to the heart of the matter. What is the total number of ways of putting these two molecules together on the lattice? It has to be $(n_0)(n_0 - 1)$. Let's say we put the first molecule in the top left-hand corner of the lattice. Then the second molecule can go in any one of the other $n_0 - 1$ sites (Figure 11-6). Now put the first molecule in the next lattice site, and again you have $n_0 - 1$ possible ways of putting in the second molecule. And so on.

OK, how many ways can you put three molecules back on the lattice? It must be $(n_0)(n_0 - 1)(n_0 - 2)$.

So, the number of ways of putting all n_0 molecules back on the lattice is:

$$(n_0)(n_0 - 1)(n_0 - 2) \ldots (3)(2)(1)$$

or n_0 factorial, $n_0!$

You may now be thinking, "OK, that's it,

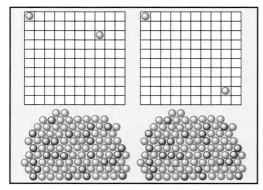

FIGURE 11-6 Two ways of placing a second ball on the lattice, given the position of the first ball. There are actually $n_0 - 1$ ways to do this.

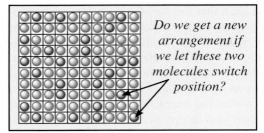

Do we get a new arrangement if we let these two molecules switch position?

FIGURE 11-7 Schematic diagram of a filled lattice. The blue balls are indistinguishable, as are the red ones.

we've got the answer. But not quite yet. What we've done is implicitly assume that all the molecules are distinguishable, as if they had little name tags: this one's Fred, that one's Doris, here comes Bert, and so on. All the "A" molecules are actually indistinguishable from one another (they're all Fred clones), and so are the "B" molecules (they're all Doris clones). So, if we take just one possible arrangement of the system, like the one shown in Figure 11-7, and switch two of the Fred's, we don't get a new arrangement. We only get a new arrangement when we switch a Fred with a Doris.

Thus we need to take out all the ways of switching n_A A's and n_B B's with one another. This is just the same as our initial problem —how many ways are there of putting n_A A molecules on a lattice of n_A sites and n_B B molecules on a lattice of n_B sites? The answer is $n_A!$ and $n_B!$, respectively. So now we need to take out $n_A!$ and $n_B!$ arrangements or configurations from our original $n_0!$ number of configurations. We don't subtract, we divide, because we're dealing with probabilities. The total number of configurations is then given by Equation 11-8:

$$\Omega = \frac{(n_A + n_B)!}{n_A! \, n_B!}$$

EQUATION 11-8

Using Boltzmann's equation (Equation 11-7) we get Equation 11-9:

$$-\Delta S_m = k(n_A \ln x_A + n_B \ln x_B)$$

EQUATION 11-9

Where x_A, x_B are the mole fractions (Equations 11-10):

$$x_A = \frac{n_A}{n_A + n_B} \qquad x_B = \frac{n_B}{n_A + n_B}$$

EQUATIONS 11-10

and we subtracted the entropy of mixing the A molecules with one another in the pure state ($n_A \ln n_A!/n_A! = 0$) and also the B's to get the entropy change on mixing:

$$\Delta S_m = S_{mixture} - S_{pure}$$

In doing the algebra, which you should work through yourself, we used Stirling's approximation (Equation 11-11):

$$\ln(n_A!) = n_A \ln n_A - n_A, \quad etc.$$

EQUATION 11-11

If you stop and think about it, Equation 11-9 is a wonderful result. We have taken a concept that many students find difficult to grasp, entropy, and through its relationship to the way energy and matter is distributed in a system ("probability," in Boltzmann's terms, or the number of arrangements or configurations as it is often written today), used a simple model to show that the entropy change on mixing is related to quantities we know or can measure: how much of each component we are mixing, i.e., the composition of the mixture.

Dealing with numbers of molecules is not always convenient, but we can use Avogadro's number to convert Equation 11-9 to a molar basis and get (Equation 11-12):

$$-\Delta S_m = R(n_A \ln x_A + n_B \ln x_B)$$

EQUATION 11-12

In what follows we will use number of molecules and moles interchangeably—you'll know what we are referring to by the constant we use (R, the gas constant, moles; k, Boltzmann's constant, molecules).

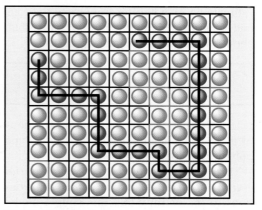

FIGURE 11-8 Schematic diagram of a filled lattice. The red balls represent low molecular weight solvent molecules and the connected blue balls represent segments of a polymer chain.

Molecules of Different Sizes

Now let's tackle a more difficult problem, mixing molecules of different size. This could be a polymer solution, where one type of molecule is far larger than the other, or it could be a polymer blend, where both molecules are large, but have different lengths. Again, some sort of simple model must be constructed in order to proceed (Figure 11-8).

The model used (independently) by Flory and Huggins (following an earlier suggestion by Fowler and Guggenheim) assumes a large molecule such as a polymer can be treated as a set of linked segments. Each of these segments is not necessarily equal in size to the chemical repeat unit, but defined to have a molar volume equal to that of the solvent.

Counting the number of ways of putting a chain on the lattice is more difficult than counting the configurations available to a set of individual small molecules, because once you place the first segment, the others in the same chain must be adjacent to one another and they must not overlap (in other words, you can only have one segment per lattice site).

Because of these difficulties, there are various levels of approximation that have been used. The simplest result, and one that Flory estimated does not differ from more rigorous and complicated treatments by more than about 5%, is given in Equation 11-13:

$$-\Delta S_m = R(n_A \ln \Phi_A + n_B \ln \Phi_B)$$

EQUATION 11-13

where Φ_A and Φ_B are the volume fractions of the A and B components (say solvent and polymer), respectively.

This is often called the *combinatorial entropy of mixing*. There are other contributions to the entropy that this simple model does not deal with. Free volume, for example, which in the lattice model approach can be handled by allowing for "holes" (empty sites) on the lattice. This is outside the scope of our discussion, however, but we will come back and qualitatively examine the effect of some of the factors we have neglected later, when we consider phase behavior.

Now let's look at our expressions for the entropy of mixing in more detail. We have two, one from so-called *regular solution theory* (Equation 11-12) and the other from the theories of Flory and Huggins (Equation 11-13). Both have the same form and tell us that the entropy change upon mixing is positive—hence favorable (remember that the log of a fraction is a negative number).

When first examined the equations don't appear to be very different, one is written in terms of *mole fractions*, while the other is in terms of *volume fractions*. The difference is significant, however, but to see that we must first remind ourselves of the nature of these quantities. To be explicit in what we are talking about we will switch subscripts, using s and p to designate solvent and polymer, respectively.

Consider the simple example shown in Figure 11-8, where we have one polymer molecule (made up of 25 segments) and 75 solvent molecules. The mole fraction of polymer is then given by Equation 11-14:

$$x_p = \frac{n_p}{n_p + n_s} = \frac{1}{76}$$

EQUATION 11-14

The volume fraction is very different. If the molar volume of a solvent molecule, hence also each of the m segments of the polymer chain, is V_s, then the volume fraction of polymer, Φ_p, is (Equation 11-15):

$$\Phi_p = \frac{n_p m V_s}{n_p m V_s + n_s V_s} = \frac{25}{100}$$

EQUATION 11-15

Note, however, that if the molecules were of equal size (i.e., $m = 1$, so that we were only mixing one "blue" molecule with 75 "reds"), then the mole fraction would equal the volume fraction. The Flory-Huggins expression for the entropy of mixing is clearly far more general and is the one we will use.

There is one more aspect of the entropy of mixing (Equation 11-13) that we need to mention. Let's say we were mixing 25 blue molecules with 75 red ones (because we're talking about the number of molecules, rather than moles, we will use k instead of R in the equation). The combinatorial entropy of mixing would be (Equation 11-16):

$$-\Delta S_m = k(75 \ln 0.75 + 25 \ln 0.25)$$

EQUATION 11-16

Now if we link the blue molecules to form a chain the entropy of mixing becomes much smaller, (Equation 11-17):

$$-\Delta S_m = k(75 \ln 0.75 + 1 \ln 0.25)$$

EQUATION 11-17

What if we also linked the red molecules to form a chain? What would then be the entropy of mixing? Of course, the entropy of mixing would be smaller still (Equation 11-18):

$$-\Delta S_m = k(1 \ln 0.75 + 1 \ln 0.25)$$

EQUATION 11-18

This trend can be more clearly seen if we

express the entropy on a per mole of lattice sites basis. This means we must divide the equation for the entropy of mixing by the number (of moles) of lattice sites, equal to the total (molar) volume, V, divided by the (molar) volume of a lattice cell, which for generality we now call V_r (number of lattice sites = V/V_r)—Equation 11-19:

$$-\frac{\Delta S_m}{R}\frac{V_r}{V} = \frac{V_r}{V}[n_A \ln \Phi_A + n_B \ln \Phi_B]$$

EQUATION 11-19

Using the definition of a volume fraction (Equations 11-20):

$$\frac{n_A V_r}{V} = \frac{n_A m_A V_r}{m_A V} = \frac{\Phi_A}{m_A}$$

$$\frac{n_B V_r}{V} = \frac{n_B m_B V_r}{m_B V} = \frac{\Phi_B}{m_B}$$

EQUATIONS 11-20

Accordingly, we obtain Equation 11-21:

$$-\frac{\Delta S_m}{R}\frac{V_r}{V} = -\frac{\Delta S_m^*}{R}$$

$$= \left[\frac{\Phi_A}{m_A} \ln \Phi_A + \frac{\Phi_B}{m_B} \ln \Phi_B\right]$$

EQUATION 11-21

In this form, the change in the entropy of mixing as molecular size increases is transparent. If the two molecules are small and of roughly equal size, then m_A and m_B are both about 1 and the entropy of mixing is large. If one of the molecules, say B, is now a polymer, then m_B, the number of segments in the polymer chain (defined now in terms of the molar volume of the solvent) becomes large and the entropy of mixing is essentially half what it would be if both molecules were "small." Finally, if two polymers are being mixed, so that both m_A, $m_B \gg 1$ (where V_r is now defined arbitrarily, usually in terms of the molar volume of one of the repeat units), the entropy of mixing, although still positive, becomes very small. Whether or not the materials form a single phase system

then depends upon the enthalpy change on mixing, which brings us to intermolecular interactions.

ENTHALPY OF MIXING

The enthalpy change, the heat or thermal energy drawn from or released to the surroundings during mixing, is a consequence of intermolecular interactions. We reviewed the types of interactions most commonly encountered in polymer systems in a separate section and if you've forgotten this stuff go back to the beginning of Chapter 8 for a review. Because we are assuming a mean field, we can only deal with those interactions that are "weak" relative to thermal energy, RT, i.e., dispersion and weak polar forces. There are ways of handling certain types of strong, specific interactions such as hydrogen bonds,[36] but we won't discuss those here.

Because most of you are probably too lazy to go and review stuff, we'll briefly mention a couple of pertinent points. First, the interaction between two molecules is described in terms of the potential energy (P.E.). When the molecules are too close they strongly repel, when they are too far apart they don't "feel" each other. There is some optimum distance apart where their interaction is a maximum, hence P.E. is a minimum, or has the largest negative or attractive value. Second, there are advanced theoretical models that deal with potential functions like this, which we will not consider, but for dispersion and weak polar forces the attractive energy varies as $1/r^6$, where r is the distance between the molecules.

One important consequence of this $1/r^6$ dependence is that it allows us to assume that only nearest neighbor interactions are important (see Figure 11-9). This is used in Flory's approach to describing the enthalpy of mixing. An alternative way of describing interactions that is also commonly used in polymer

[36] See, M. M. Coleman, J. F. Graf, P. C. Painter, *Specific Interactions and the Miscibility of Polymer Blends*, CRC Press, Boca Raton, FL (1995).

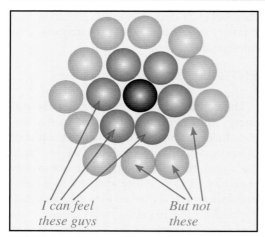

FIGURE 11-9 Schematic representation of nearest neighbor interactions.

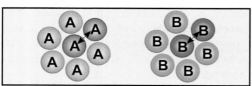

FIGURE 11-10 Schematic representation of nearest neighbor interactions in the pure state.

science involves something called *cohesive energy densities* and *solubility parameters*.[37] Because they are just alternative descriptions of the same thing and use similar assumptions they are related. Solubility parameters can be determined from pure component properties, while χ was designed to be an adjustable parameter that would be determined by a fit to experimental data. We will start by considering the Flory χ parameter.

Flory's χ Parameter

First consider interactions in the pure state (Figure 11-10). Let the attractive interaction energy between a pair of small A molecules be ε_{AA} and that between a pair of similarly sized B molecules be ε_{BB}. If we assume that we can simply add the interactions between all pairs, then we can say that the interaction of a chosen A molecule with all its nearest neighbors is $z\varepsilon_{AA}$, where z is the number of

[37] See J. H. Hildebrand, R. L. Scott, *The Solubility of Non-Electrolytes*, Third Edition, Reinhold Publishing Co. New York (1950).

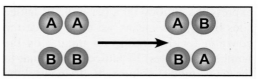

FIGURE 11-11 Schematic representation of interactions in the mixed state.

nearest neighbors. By adding all such "pairwise" interactions (being careful not to count each pair twice) we could then obtain the total energy of interaction or potential energy of the A molecules in the pure state, and in a similar fashion the B's. This total interaction energy should then be related to the heat of vaporization, the energy required to take a molecule from the pure liquid state, where it is surrounded by like molecules, to the gaseous state, where it isn't (assuming the vapor state is ideal). We will come back to this later.

What we are really interested in, however, is just a simple description of the change in energy on going from the pure to the mixed state. In order to calculate this we first consider the change in energy when we replace interactions between a pair of A molecules, AA, and pairs of B molecules, BB, with AB pairs (Figure 11-11). The energy change per AB pair is given by Equation 11-22:

$$\Delta\varepsilon = \varepsilon_{AB} - \frac{1}{2}[\varepsilon_{AA} + \varepsilon_{BB}]$$

EQUATION 11-22

(If you can't figure out where the 1/2 comes from, look at the figure and think about it a bit). Now all we need do is multiply this energy change per pair by the total number of AB pairs that are formed on mixing. (We don't worry about the AA and BB

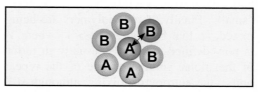

FIGURE 11-12 The probability that a B is next to an A is Φ_B.

pairs because these don't contribute to any *change on mixing*, although they must obviously contribute to the total energy of the mixture.)

If the number of AB pairs is n_{AB} we would simply have Equation 11-23:

$$\Delta H_m = n_{AB}\Delta\varepsilon_{AB}$$

EQUATION 11-23

If the mixture is random, then the probability that a B molecule is next to a chosen A is equal to the volume fraction of B's present, Φ_B (Figure 11-12). Because each A has z neighbors, the total number of B's next to a given A is then, on average, $z\Phi_B$. It follows that the total number of AB contacts is just $n_A z\Phi_B$. So the heat of mixing for a random mixture of small molecules is (Equation 11-24):

$$\Delta H_m = n_A\Phi_B[z\Delta\varepsilon_{AB}]$$

EQUATION 11-24

Flory defined the following interaction parameter, χ, which he made dimensionless by dividing by kT (Equation 11-25):

$$\chi = \frac{z\Delta\varepsilon_{AB}}{kT}$$

EQUATION 11-25

The heat of mixing is then simply (Equation 11-26):

$$\Delta H_m = kT[n_A\Phi_B\chi]$$

EQUATION 11-26

Note that if one of the components, B in this case, is a polymer, then the probabilities that unlike units are adjacent as expressed in this equation are not quite right, because each (internal) segment of the chain is covalently bonded to two others, thus reducing the probability that it can be next to a solvent molecule (Figure 11-13). Such considerations also affect the entropy term, as Flory

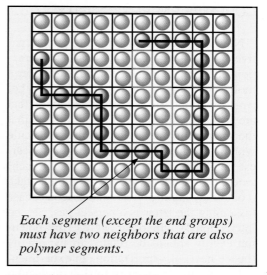

Each segment (except the end groups) must have two neighbors that are also polymer segments.

FIGURE 11-13 Schematic diagram showing the reduction in the probability that a B polymer segment is next to a solvent molecule A.

was well aware, but the approximations he made allowed the free energy of mixing to be expressed in a particularly simple form.

THE FLORY-HUGGINS EQUATION

Adding Flory's expressions for the entropy and enthalpy we arrive at Equation 11-27:

$$\Delta G_m = \Delta H_m - T\Delta S_m$$
$$= kT[n_A\Phi_B\chi + n_A\ln\Phi_A + n_B\ln\Phi_B]$$

EQUATION 11-27

Or, rearranging and converting to a molar basis we obtain Equation 11-28:

$$\frac{\Delta G_m}{RT} = n_A\ln\Phi_A + n_B\ln\Phi_B + n_A\Phi_B\chi$$

EQUATION 11-28

Huggins independently and almost simultaneously obtained a more rigorous equation for the free energy, but it is this simpler Flory expression that is most often used and which is known as the Flory-Huggins equation.

COHESIVE ENERGY DENSITY

Now let's turn our attention to an alternative but related way of describing interactions, developed by Hildebrand. Consider a molecule of type A in the pure state, in other words with a shell of nearest neighbors that are also A's. One might grasp intuitively that the energy of interaction of this molecule with its neighbors (not necessarily just nearest neighbors in the Hildebrand approach) would be equal to the energy required to take it out of the liquid and into the gaseous phase (assuming the latter is a perfect gas, i.e., no interactions!). In other words, the potential energy should be related to the energy (hence heat) of vaporization, an experimentally measurable quantity for low molecular weight materials.[38] Hildebrand defined the *cohesive energy density* of a material to be this energy of vaporization (per mole), ΔE_v (equal to $\Delta H_v - RT$, where ΔH_v is the enthalpy of vaporization), divided by the volume, V (hence energy "density"). We will let C_{AA} be the cohesive energy density of pure "A" and similarly C_{BB} be the cohesive energy density of pure "B."

What if we now have a B molecule immersed in a sea of A molecules? We can let the cohesive energy density describing the interaction of this molecule with its neighbors (actually, all the other molecules in the system) be C_{AB}. Of course, in a real mixture you don't have just one B molecule surrounded by A's, but a distribution of both A and B molecules. The simplest approach is to then assume that the interaction between a chosen B molecule and all the A molecules in the mixture depends on C_{AB} multiplied by the volume fraction of A's present, and work out the energy of interaction from there. You can be more rigorous and work from something called distribution functions, following Hildebrand, but this is something you should get into if you pursue this part of the subject in more depth. Here we will simply note that following these arguments you get Equation 11-29:

---

[38] Actually P.E. $= -\Delta E_v$.

$$\Delta H_m = (n_A + n_B)V_m\Phi_A\Phi_B(2C_{AB} - C_{AA} - C_{BB})$$

EQUATION 11-29

Hold on a minute, you might say, I may not want to work through the derivation at this point, but where did all the terms come from? Well, the cohesive energy density terms in parentheses should be obvious—it's the change in energy density per mole of mixture when a given mole fraction of A's (x_A) is mixed with a given mole fraction of B's (x_B), such that $x_A + x_B = 1$. Except that this assumes that all the AA and BB interactions in the original pure liquids are now replaced by AB interactions. (If the factor 2 worries you, go back and look at Figure 11-11.) To get the actual change in energy upon mixing we have to account for the fact that we only replace a certain fraction of AA and BB interactions with AB interactions. Hence the volume fraction terms $(\Phi_A\Phi_B)$. Also, we don't want an energy density expression, we want an energy, so we multiply by the molar volume of the mixture, V_m. Then we get an energy per mole, so we multiply by the total number of moles present to get our energy or enthalpy term.

If you're thinking about this you might see where we're going. We have an expression where we either know the parameters (composition terms) or can measure them (cohesive energy densities of the pure components). But we don't know and cannot measure C_{AB}. However, it has been shown that for *dispersion forces* and perhaps *weak polar forces* a *geometric mean assumption* is a good approximation (Equation 11-30):

$$C_{AB} = C_{AA}^{0.5}C_{BB}^{0.5}$$

EQUATION 11-30

Substituting, this gives us an expression that is a perfect square (Equation 11-31):

$$(2C_{AB} - C_{AA} - C_{BB}) = (C_{AA}^{0.5} - C_{BB}^{0.5})^2$$

EQUATION 11-31

FASCINATING POLYMERS—VARNISHES AND LACQUERS

Varnishes or lacquers are simply polymer solutions in an organic solvent. Interesting examples include shellac in alcohol and nitrocellulose in acetone, which we will mention briefly here. As a young lad, one of your authors remembers distinctly the smell of shellac varnish emanating from his father's woodworking shop. Shellac is a fascinating material. It is a natural thermoplastic resin of animal origin secreted by tiny lac beetles (the size of apple seeds) found in Southeast Asia. (Amazingly about 50,000 insects are needed to generate 1 lb of shellac!) Lac is deposited onto twigs and branches of soapberry, acacia and other trees and, following some simple chemical treatment, lac dye (reportedly used by the Romans as early as 250 AD) is separated leaving the crude shellac resin. Shellac resin is soluble in alcohol and can be applied to wood and other materials as a varnish. It was first used as a protective coating in the late 1500s and was applied routinely to furniture in the West in the early 1800s.

An old masterpiece varnished with shellac (Source: Jim Laabs Music).

Ford Model T with nitrocellulose varnish finish (Source: www.mtfca.com).

The second example involves nitrocellulose varnishes. Nitrocellulose, as you may recall, is highly inflammable and soluble in the highly volatile solvent acetone. In 1923, fast-drying varnishes were manufactured from acetone solutions of nitrocellulose and were used on car production lines in order to speed up production. It is difficult to imagine that this would get past government safety and environmental agencies today. But you can still buy nitrocellulose varnishes and they are used where coating resistance to moisture or humidity is very important. There are two serious disadvantages to these varnishes or lacquers. First, organic solvents are expensive and if simply evaporated into the atmosphere they pose significant environmental and safety concerns. Industrially, solvents must be recycled (another expense), and there is tremendous pressure on the industry to eventually convert all surface coatings to more environmentally friendly water-based systems (where a lot of progress has been made). Second, the viscosity of high molecular weight polymer solutions tends to be very high and only relatively dilute solutions (<10%) can be employed for varnishes and lacquers. At higher concentrations the polymer solution is simply too viscous to apply by brush, roller or spray gun.

FASCINATING POLYMERS—MISCIBLE PPO/PS BLENDS

Instrument panel made of GE's Noryl® resins.

In the late 1950s and early 1960s, Allan Hay and his colleagues at GE succeeded in synthesizing a new engineering thermoplastic using an unusual oxidative coupling reaction. If oxygen is simply bubbled through a reaction mixture containing 2,6-dimethylphenol and a catalyst consisting of cuprous chloride and an amine, a linear polymer, poly(2,6-dimethyl-1,4-phenylene oxide), is formed. (In order to stop reaction at the 2 or 6 positions on the aromatic ring, "blocking" groups, such as methyl groups, are necessary to produce linear polymers.) The polymer is commonly called poly(phenylene oxide) (PPO) or, less often, poly(phenylene ether) (PPE). PPO has high tensile strength, stiffness, impact and creep resistance over a temperature range from −45° to 120°C: the characteristics of a fine engineering plastic. It is remarkably stable, both chemically and dimensionally, in the presence of aqueous solutions of acids, alkalis and detergents and thus finds application in components for pumps, washing machines, and the like. But PPO has one major flaw. It is very difficult to process and this leads us to one of the more unexpected findings in polymer science. PPO has a glass transition temperature of 208°C and to obtain a melt that has a viscosity low enough for injection molding or extrusion, very high temperatures are required, well above the temperature where serious oxidative degradation occurs. This put a damper on commercial development of the polymer. But then, it was discovered, totally unexpectedly, that PPO was *miscible* with polystyrene (PS), and now the story becomes much more interesting. We have been unable to verify the following anecdote that lurks in the darker region of your authors' brains, but it makes a good story and certainly fits our preconceptions concerning scientific discovery. Folklore has it that in a GE laboratory, a Ph.D. chemist and his technician were bemoaning the fact that PPO had such a high melt viscosity. The technician said something like: "Hey, PS processes very easily and has a low melt viscosity, why don't we add some PS to PPO to reduce the overall viscosity of the melt?" "You bloody great twit! (or words to that effect)," replied the chemist, "This is why I have a Ph.D. The two polymers won't mix! Don't you know that Paul Flory, the Nobel Prize winner, has written, '. . . It is permissible to state as a principle of broad generality that two high polymers are mutually compatible with one another only if their free energy of interaction is favorable, i.e., negative. Since the mixing of a pair of polymers, like the mixing of simple liquids, in the great majority of cases is endothermic, incompatibility of chemically dissimilar polymers is observed to be the rule and compatibility the exception.'" But guess what? The technician tried it anyway and found that PPO and PS form miscible blends, leading to the family of GE's Noryl® engineering materials. The fact that PPO/PS blends are compatible (miscible) was an exception to the rule. Even today, a half a century later, one would be hard-pressed to *predict* the miscibility of PPO and PS.

So we can obtain an expression for the enthalpy of mixing in terms of known quantities.

Solubility Parameters

Hildebrand then defined solubility parameters, δ, as shown in Equations 11-32 (we don't know why; maybe he just didn't want to keep writing down expressions in the square root of a cohesive energy density):

$$\delta_A = C_{AA}^{0.5} \qquad \delta_B = C_{BB}^{0.5}$$

EQUATIONS 11-32

Thus the heat of mixing is given by Equation 11-33:

$$\Delta H_m = (n_A + n_B)V_m \Phi_A \Phi_B (\delta_A - \delta_B)^2$$

EQUATION 11-33

Now if you think about it some more, we have two expressions for the energy or enthalpy changes on mixing that are a result of interactions. If the various assumptions we have made (geometric mean, nearest neighbor interactions, pairwise additivity) are all good, then χ must be related to solubility parameter differences.

The relationship can be established by simply starting with Equation 11-25 and making a geometric mean assumption for the pairwise interaction energies (Equation 11-34):

$$\varepsilon_{AB} = \varepsilon_{AA}^{0.5} \varepsilon_{BB}^{0.5}$$

EQUATION 11-34

We therefore obtain Equation 11-35:

$$\Delta \varepsilon_{AB} = \varepsilon_{AA}^{0.5} \varepsilon_{BB}^{0.5} - \frac{1}{2}[\varepsilon_{AA} + \varepsilon_{BB}]$$
$$= \frac{1}{2}[\varepsilon_{AA}^{0.5} - \varepsilon_{BB}^{0.5}]^2$$

EQUATION 11-35

Now all you need do is substitute cohesive

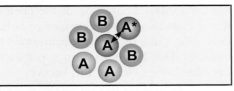

FIGURE 11-14 Interactions between two A groups.

energy (C) terms for energy terms (ε). This can be done by assuming that the cohesive energy of the pure components is obtained by summing over all the interactions between pairs of molecules for all the molecules found in the pure components. You get Equations 11-36:

$$C_{AA} = \frac{\Delta E_{vap}}{V} = \frac{n_A z \varepsilon_{AA}}{2V_A}$$
$$C_{BB} = \frac{\Delta E_{vap}}{V} = \frac{n_B z \varepsilon_{BB}}{2V_B}$$

EQUATIONS 11-36

The factor "2" comes in so you don't count the interactions twice. Think about it. If you go through the material a molecule at a time you will have situations such as those depicted in Figure 11-14 where you consider the molecule labeled A and its interactions with its neighbors, such as A*. But later in the counting process you will be considering the interaction of A* with its neighbors, including A. So you divide by 2 to eliminate the double counting.

Doing all the substitutions and assuming you are mixing molecules of equal size, so that $V_A = V_B = V_r$ (a reference volume), you get Equation 11-37:

$$\chi \approx \frac{V_r}{RT}[\delta_A - \delta_B]^2$$

EQUATION 11-37

Of course, one of the assumptions we made in treating the entropy of mixing was that the lattice cell size is defined by the size of the solvent, so in treating polymer solutions V_r is equal to the molar volume of the solvent. In treating polymer blends you need to (arbitrarily) choose one of the chemical repeat units to define your reference volume (and your value of χ will depend upon this choice). This leaves us with two things to consider before closing this section. First, how on earth do you get the solubility parameter of a polymer, because you can't measure its heat of vaporization (a polymer degrades before evaporating)? And second, how good is the equation in predicting the value of χ?

Group Contributions

The solubility parameter of a polymer can be calculated using something called group contributions. The essence of this approach

TABLE 11-1 GROUP CONTRIBUTIONS

Group	Molar Volume Constant V^* ($cm^3\ mole^{-1}$)	Molar Attraction Constant F^* [$(cal.\ cm^3)^{0.5}\ mole^{-1}$]
-CH₃	31.8	218
-CH₂-	16.5	132
>CH-	1.9	23
>C<	−14.8	−97
>C₆H₃-	41.4	562
-C₆H₄-	58.5	652
C₆H₅-	75.5	735
CH₂=	29.7	203
-CH=	13.7	113
>C=	−2.4	18
-OCO-	19.6	298
-CO-	10.7	262
-O-	5.1	95
-Cl	23.9	264
-CN	23.6	426
-NH₂	18.6	275
>NH	8.5	143
>N-	-5.0	-3

is to assume that a molecule can be "broken down" into a set of functional groups. A simple hydrocarbon, such as n-octane, is assumed to consist of CH_2 and CH_3 groups, for example. Using the energy of vaporization of a series of such paraffins (i.e., with different values of n) so-called molar attraction constants, the contribution of CH_2 and CH_3 groups to solubility parameters, can be calculated. By including branched hydrocarbons, and molecules containing other functional groups (ether oxygens, esters, nitriles etc.), a table of constants can be obtained and subsequently used to calculate the solubility parameter of a given polymer.

A compilation taken from our own work is shown in Table 11-1. To calculate the solubility parameter of a polymer one uses the simple relationship (Equation 11-38):

$$\delta = \frac{\sum_i F_i^*}{\sum_i V_i^*} (cal.\ cm^{-3})^{0.5}$$

EQUATION 11-38

where F_i^* is the molar attraction constant of the i'th group (e.g., a CH_2 group), while V_i^* is the corresponding molar volume constant of this group.

For example, say we wish to calculate the solubility parameter of poly(methyl methacrylate) (PMMA):

PMMA

then we simply need the sum of the molar attraction constants for one -CH₂-, two -CH₃, one >C< and one -OCO- (132 + 436 - 97 + 298 = 769) divided by the sum of the molar volume constants for the same set of groups (16.5 + 63.6 − 14.8 + 19.6 = 84.9). Thus the solubility parameter value for PMMA is calculated to be $769/84.9 = 9.1$ (cal. cm^{-3})$^{0.5}$. The interested reader might wish to contribute to your authors meager retirement funds and purchase the simple programs that allow you to calculate the solubility parameter of many polymers and predict whether or not two given (co)polymers are miscible that are included in our *Miscible Polymer Blends* CD.[39]

To conclude this section we will simply note that the relationship given above in Equation 11-36 does not work well for polymer solutions. A comparison of calculated values of χ to those obtained experimentally has led to the suggestion that a "fudge factor" of about 0.34 be included (Equation 11-39):

$$\chi \approx 0.34 + \frac{V_r}{RT}[\delta_A - \delta_B]^2$$

EQUATION 11-39

(This fudge factor is actually an average obtained from various studies, and so is only useful as a rough guide.)

It was initially thought that this factor had its genesis in the level of approximation used in the Flory method for counting configurations. In a series of papers published in the 1970s, however, Patterson and coworkers showed that the most likely origin of this correction term lay in so-called *free volume effects* that are neglected in the Flory-Huggins treatment (one way to account for this would be to put holes on the lattice, but that is a level of complexity you should tackle in more specialized treatments).

A more general approach is to consider the Flory χ parameter to be a free energy term that has the form (Equation 11-40):

$$\chi = a + \frac{b}{T}$$

EQUATION 11-40

where the quantity "a" can be thought of as the entropic part of χ, accounting for non-

[39] M. M. Coleman and P. C. Painter, *Miscible Polymer Blends*, CD-ROM, DEStech Publications, Lancaster, PA (2006).

combinatorial entropy changes such as those associated with free volume, while "b" would be the enthalpic part (note that "a" and "b" need not be constants, but could depend upon composition and temperature). But, enough of this, it is time to consider phase behavior!

PHASE BEHAVIOR

Now we turn our attention to the phase behavior of polymer solutions and blends and the questions we asked right at the beginning of this chapter: will a particular polymer dissolve in a given solvent or mix with another chosen polymer?

So far we have considered only one essential condition for forming a single phase system; the free energy change on mixing must be negative. There is another, as we will see shortly, but let's examine this first condition in a little more detail, using the Flory-Huggins equation as a guide.

It makes things easier to see if we consider the free energy on a per mole of lattice sites basis by dividing the Flory-Huggins expression for the free energy by the number of moles of lattice sites, V/V_r. Recall that the first two terms of Equation 11-41 are the combinatorial entropy of mixing and are always negative and hence favorable to mixing (the log of a fraction is negative).

$$\frac{\Delta G_m'}{RT} = \frac{\Delta G_m}{RT}\frac{V_r}{V}$$
$$= \frac{\Phi_A}{m_A}\ln\Phi_A + \frac{\Phi_B}{m_B}\ln\Phi_B + \Phi_A\Phi_B\chi$$

EQUATION 11-41

However the size of this term gets smaller as the size of the molecules (m_A, m_B) gets larger. On the other hand, the χ term is always positive and unfavorable to mixing in systems where the molecules interact through dispersion and weak polar forces only, as can more readily be seen if we write χ in terms of solubility parameters (neglecting any fudge factors) — see Equation 11-37 above.

This immediately tells us why recycling plastics is such a problem. It's not a con-

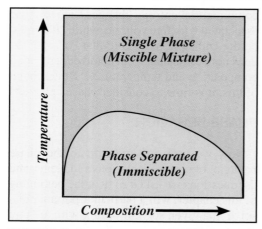

FIGURE 11-15 Schematic phase diagram.

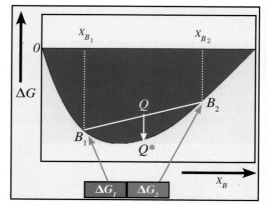

FIGURE 11-16 Free energy curve—conditions for single phase.

spiracy by the chemical companies to ruin the earth by drowning us in unwanted plastic waste. Most ordinary common or garden polymers simply do not mix. The entropy of mixing is negligibly small while the χ term is positive. Any mixture of different types of simple high molecular weight hydrocarbon polymers would be grossly phase-separated, like oil and water, by which we mean there would be large phase-separated domains and the resulting material would have little cohesive strength.[40] As a result, different types of plastics have to be separated in the waste stream.

So, it is the balance between the terms that determines whether or not the free energy will be negative. Furthermore, this balance will change with temperature. Recall that the χ term varies as $1/T$ (Equation 11-37), or, more generally as in Equation 11-40. Accordingly, as the temperature increases, χ gets smaller and conditions are more favorable for mixing.

This mirrors the experience of everyday life, where if you try to get one thing to mix with another you often heat them (Figure 11-15). However, as we will see later, polymers can also give counter-intuitive behavior, where mixtures phase-separate upon heating!

[40] Useful properties can be obtained by controlling the size of phase-separated domains, however, making the blend what some call "compatible." This is not the same as miscible or single phase.

This brings us to the next condition for forming a single phase mixture of "A" and "B" molecules. Consider a mixture where the free energy of mixing is negative across the entire composition range, going from zero in pure A (i.e., there is no free energy change on mixing something with itself) to some minimum value and back to zero for pure B. Also imagine that this curve is concave upwards across the entire composition range. Now consider taking two separate mixtures of A and B molecules, each with a different composition, corresponding to points B_1 and B_2 on the plot shown in Figure 11-16. If these two mixtures are placed in the same container, but separated by a barrier, the total free energy of mixing would simply be the composition weighted sum of the free energies of mixing of the separate mixtures, corresponding to point Q (Equation 11-42):

$$\Delta G_{total} = \Delta G_Q = x_{B_1}\Delta G_1 + x_{B_2}\Delta G_2$$
EQUATION 11-42

If the barrier is now removed and the mixtures allowed to mix (if they want to), the free energy of any new mixture would be given by the point Q^* on the free energy curve. It can be seen that (Equation 11-43):

$$\Delta G_{Q^*} < \Delta G_Q$$
EQUATION 11-43

so that the free energy change for forming the new mixture would be negative. In other words, the mixtures would mix! Clearly, any two compositions you pick would also mix, as long as the free energy of mixing curve is concave upwards, because by doing so they can get to a lower free energy. The components A and B would thus always form a single phase or miscible mixture as long as this condition held.

Now consider a situation where the free energy of mixing is still negative across the composition range, but when plotted as a function of composition the free energy curve is concave downwards in a certain range of composition, as illustrated in Figure 11-17. Again, if we take the two separate mixtures corresponding to points B_1 and B_2 and place them in a container separated by a barrier, their total free energy of mixing will be ΔG_Q, but the free energy of mixing these two mixtures with one another would be ΔG_{Q^*}, where $\Delta G_{Q^*} > \Delta G_Q$. In this case the mixtures won't mix upon removing the barrier because this would involve an increase in free energy.

If the free energy of mixing curve is examined carefully (Figure 11-17) it can be seen that any composition between the points B_1 and B_2 has a greater free energy than a mixture that is phase-separated into two mixtures of composition x_{B1} and x_{B2}. These compositions are defined by the points of contact of the line that forms a double tangent to the curve at points B_1 and B_2. Recalling simple calculus, this tangent is simply the first derivative of the free energy with respect to composition, the composition variable being mole fraction in this particular figure. This gives us Equation 11-44:

$$\left[\frac{\partial \Delta G}{\partial x_B}\right]_{x_{B1}} = \left[\frac{\partial \Delta G}{\partial x_B}\right]_{x_{B2}}$$

EQUATION 11-44

where the derivative or tangent at point B_1 is equal to that at point B_2. It is more common to express the derivatives in terms of the number of moles instead of mole fraction

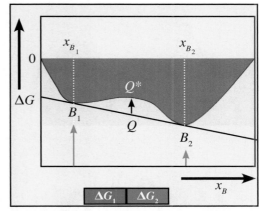

FIGURE 11-17 Free energy curve—conditions for phase separation.

(Equation 11-45):

$$\left[\frac{\partial \Delta G}{\partial n_B}\right]_{n_{B1}} = \left[\frac{\partial \Delta G}{\partial n_B}\right]_{n_{B2}}$$

EQUATION 11-45

The two results being entirely equivalent (check the definition of a mole fraction and do the differentiation if you don't believe us!). If asked, you would no doubt swear to high heaven that you immediately recognized that the derivatives given in Equation 11-45 are the *chemical potentials*. (We believe you, thousands wouldn't.) If we now use the symbols $\Delta\mu_B^1$ and $\Delta\mu_B^2$ to represent these quantities (which are also written as a difference, because they are evaluated with respect to the same reference state, usually the pure component, as that used to evaluate the change in free energy on mixing), we can write $\Delta\mu_B^1 = \Delta\mu_B^2$. Although we imagine that it is extremely unlikely that you've forgotten any of your basic physical chemistry (yeah, right), we think it useful to remind you of what this means.

Consider a phase-separated polymer solution, where one phase is rich in polymer while the other is poor in this component (Figure 11-18). [Things never phase-separate into their pure components; even in grossly phase-separated systems, like oil and water, there is always a little bit of one in the other].

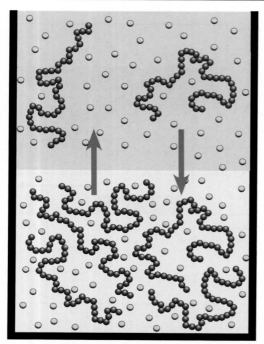

FIGURE 11-18 Schematic diagram illustrating the chemical potential.

The boundary between the phases is not an impenetrable barrier and polymer and solvent molecules move back and forth all the time. However, at equilibrium the number of polymer (or solvent) molecules moving in one direction is the same as that moving in the other. The driving force for this movement is the chemical potential (just like the driving force for water to run downhill is the potential energy due to gravity). The chemical potential is the same when there is no *net* flow between the phases (Equations 11-46):

$$\Delta \mu_B^1 = \Delta \mu_B^2 \qquad \Delta \mu_A^1 = \Delta \mu_A^2$$
EQUATIONS 11-46

It is simply a measure of how the free energy changes with composition and is the driving force for the two phases to come to equilibrium (i.e., for their compositions to adjust until Equation 11-46 is satisfied). The chemical potentials in the Flory-Huggins theory can be simply obtained by differentiating the free energy of mixing, which for

convenience we now rewrite using the subscripts s and p to designate solvent and polymer, respectively (Equation 11-47):

$$\frac{\Delta G_m}{RT} = n_s \ln \Phi_s + n_p \ln \Phi_p + n_s \Phi_p \chi$$
EQUATION 11-47

The chemical potential of the solvent is then given by Equation 11-48[41]:

$$\frac{\Delta \mu_s}{RT} =$$
$$\ln(1 - \Phi_p) + \left[1 - \frac{1}{M}\right]\Phi_p + \Phi_p^2 \chi$$
EQUATION 11-48

We will make considerable use of this equation, both in our further discussions of phase behavior and the determination of molecular weight. One immediate use is in the calculation of the phase boundary in a temperature composition plot. At a particular temperature, the chemical potential of the solvent (or the polymer) at the composition represented by B_1 is equated to that evaluated at B_2 (i.e., $\Delta \mu_s^1 = \Delta \mu_s^2$) to give two points on the phase diagram (Figure 11-19). (The equations do not have an analytical solution and have to be solved iteratively.. The process is repeated at different temperatures to give the curve. Note that as the temperature is increased the two points get closer together and finally coalesce at a single point, the *upper critical solution temperature* or *UCST*. Physicists love critical points, because weird things happen there, but that is beyond the scope of our discussion.

Returning now to our free energy curve, we have seen how the points B_1 and B_2 represent the compositions of the two phases that would be present at equilibrium, but if you examine the plot carefully you will notice that locally around these points the

[41] We leave the differentiation to you as a homework problem. Although apparently simple and straightforward, we are always amazed at how many students screw this up!

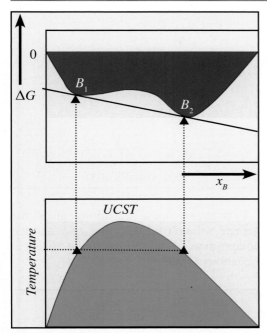

FIGURE 11-19 Schematic free energy diagram and binodal phase diagram showing the *UCST*.

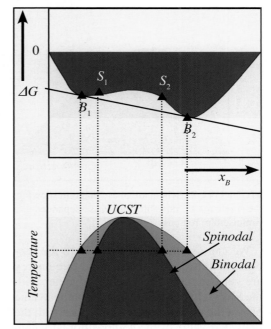

FIGURE 11-20 Schematic free energy diagram showing the inflection points S_1 and S_2 and phase diagram showing the spinodal and binodal.

free energy of mixing curve is still concave upwards until the inflection points S_1 and S_2 are reached, where the curve "turns over" (Figure 11-20). This means that mixtures that have compositions between B_1 and S_1 (and B_2 and S_2) are stable against separation into phases consisting of local compositions. There is still a driving force to phase-separate into the compositions represented by points B_1 and B_2, but there is a local free energy barrier inhibiting this process. In this metastable region phase separation proceeds by a process of nucleation and growth, like crystallization.

The points of inflection are, of course, given by the condition (Equation 11-49):

$$\frac{\partial^2 \Delta G}{\partial x_B^2} = 0$$

EQUATION 11-49

and once again the Flory-Huggins equation can be differentiated, this time to obtain an analytical solution, as we will see shortly. Again calculations can be performed at each temperature to construct what is called the

spinodal. The coexistence curve calculated by equating the chemical potentials in each phase is called the *binodal*. The metastable region lies between these curves. Note that the binodal and spinodal meet at the *UCST*, where the third derivative of the free energy must be zero (Equation 11-50):

$$\frac{\partial^3 \Delta G}{\partial x_B^3} = 0$$

EQUATION 11-50

We can now summarize our conditions for forming a single phase or miscible mixture (Figure 11-21):

a) The free energy change on mixing should be negative.

b) The second derivative of the free energy of mixing should be positive (which means a point of inflection on the free energy curve has not been reached and it is concave upwards across the composition range).

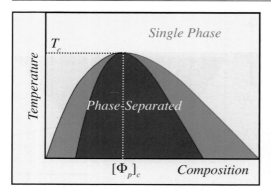

FIGURE 11-21 Schematic phase diagram showing the single phase (miscible) and phase-separated (immiscible) regions.

We can use the criteria that both the second and third derivatives of the free energy are zero at the critical point, together with the Flory-Huggins equation to obtain a value of χ at the critical point, χ_c, together with the predicted value of the composition at which the critical point occurs. In performing this calculation we can use any composition variable we like and the algebra is a little simpler if we differentiate the free energy expressed on a per mole of lattice site basis with respect to the volume fraction of one of the components. If we consider the general case of the free energy of mixing molecules of type A and B then the free energy is given by Equation 11-41 above. Having every confidence in your skills at calculus and algebra, we leave the heavy lifting to you as a homework problem. Keeping in mind that $\Phi_B = 1 - \Phi_A$ you should first get Equations 11-51:

$$\frac{\partial^2 (\Delta G_m'/RT)}{\partial \Phi_A^2}$$

$$= \frac{1}{\Phi_A m_A} + \frac{1}{\Phi_B m_B} - 2\chi = 0$$

$$\frac{\partial^3 (\Delta G_m'/RT)}{\partial \Phi_A^3}$$

$$= -\frac{1}{\Phi_A^2 m_A} + \frac{1}{\Phi_B^2 m_B} = 0$$

EQUATIONS 11-51

Rearranging these equations gives the general solution for the value of χ and the composition of the critical point given in Equations 11-52:

$$[\Phi_A]_c = \frac{m_B^{0.5}}{m_A^{0.5} + m_B^{0.5}}$$

$$\chi_c = \frac{1}{2}\left[\frac{1}{m_A^{0.5}} + \frac{1}{m_B^{0.5}}\right]^2$$

EQUATIONS 11-52

For a polymer blend, where both m_A and m_B may be ~1000, the critical value of χ will be very small. Keep in mind what this means. Any value of χ that is larger than χ_c means that the system will phase-separate at some composition. This is perhaps more easily seen if you recall that χ varies approximately as $1/T$, and a phase diagram can therefore be plotted as $1/\chi$ versus composition, if we so choose (Figure 11-22). Note also that the shape of the binodal and spinodal curves will depend upon the relative size of the molecules and will be skewed to one side or the other if m_A is not equal to m_B.

The Critical Value of χ

If we consider polymer solutions, where there is a very large mismatch in the size of the components, then we find the critical point at low polymer concentrations, as shown in some experimental data (Figure 11-23. In the Flory-Huggins theory the con-

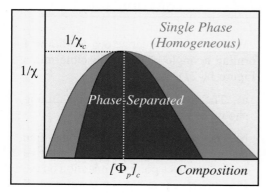

FIGURE 11-22 Schematic phase diagram of composition versus $1/\chi$.

centration at the *UCST* is given by Equation 11-53:

$$[\Phi_p]_c = \frac{1}{1 + m_p^{0.5}}$$

EQUATION 11-53

which you can find by putting $m_B = 1$ and $m_B = m_p$ in the equations derived previously.

In a similar fashion it can be shown that the critical value of χ is given by Equation 11-54:

$$\chi_c = \frac{1}{2}\left[1 + \frac{1}{m_p^{0.5}}\right]^2$$

EQUATION 11-54

or, for large m_p, $\chi_c \sim 1/2$. Again, when considering solutions we use the subscripts s and p to indicate polymer and solvent, respectively.

This can be translated into a critical value of the solubility parameter difference by noting that for *polymer solutions*, to a first approximation we can use Equation 11-55:

$$\chi_c = 0.34 + \frac{V_r}{RT}[\delta_p - \delta_s]_c^2 = 0.5$$

EQUATION 11-55

Then at room temperature RT is of the order of 600 cal/mole, while many solvents have molar volumes $V_r \sim 100$ cm³/mole, giving Equation 11-56:

$$[\delta_p - \delta_s] \approx \pm 1$$

EQUATION 11-56

A rough rule of thumb would then be that if a polymer has a solubility parameter of say, 9, it should dissolve in solvents with solubility parameters between 8 and 10. Keep in mind that such considerations, approximate as they are, *only apply to those systems that interact through dispersion and weak polar forces*.

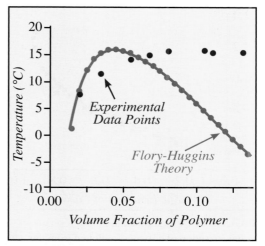

FIGURE 11-23 Experimental and theoretical plots of critical value of χ versus volume fraction of polymer [replotted from the data of Scholte, T. G., *J. Polym. Sci.*, A-2, **9**, 1553 (1971)].

So how well does the Flory-Huggins theory work? It does a good job of predicting some of the gross features of phase behavior (Figure 11-23). Measurements on various solutions have shown that χ_c is indeed close to 0.5, meaning that the theory gets the temperature of the top of the phase boundary curve, the *UCST*, nearly right. The value of the critical concentration, $(\Phi_p)_c$, predicted by the model is too small, however, and the shape of the curve deviates significantly from experiment at higher polymer concentrations.

Furthermore, the theory assumes that χ is independent of composition. Careful measurements (some experimental methods of measuring χ will be mentioned in our discussion of molecular weight) show that in most solutions χ is strongly composition-dependent, however (Figure 11-24).

If this is all that were wrong with the theory, however, it would be considered highly satisfactory and could be fixed up with some minor tinkering, such as assuming a composition dependence of χ (Equation 11-57):

$$\chi = \chi_1 + \chi_2\Phi_p + \chi_3\Phi_p^2 + \cdots$$

EQUATION 11-57

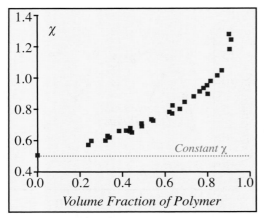

FIGURE 11-24 Experimental plot of χ versus volume fraction of polymer [replotted from the data of Krigbaum and Geymer, *J.A.C.S.*, **81**, 1859 (1959)].

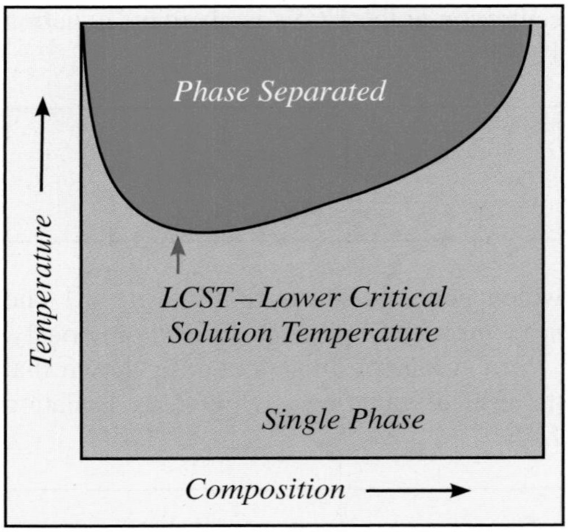

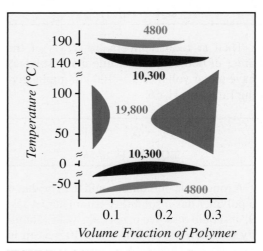

FIGURE 11-25 Schematic phase diagram showing the lower critical solution temperature.

and also making χ a free energy as opposed to an enthalpic term.

A more fundamental difficulty arose from the seminal observations of Freeman and Rowlinson,[42] that polymer solutions can phase-separate upon *heating*, giving a so-called *lower critical solution temperature*, or *LCST* (Figure 11-25). Freeman and Rowlinson didn't actually reported a phase diagram, but light scattering data obtained later by Siow, Delmas and Patterson on solutions of polystyrene of different molecular weights in acetone show beautifully both *UCST* and *LCST* behavior for low molecular weight polystyrene samples, with the curves approaching and merging to give the classic hour-glass shape curves of a grossly phase separated system for high molecular weight samples (Figure 11-26).

The appearance of *LCST* behavior, not only in polymer solutions but also in polymer blends, is due to a major factor that is neglected in the theory, i.e., free volume. This can be fixed (e.g., by putting "holes" on the lattice), but that gets us beyond the scope of an introductory treatment.

An aside is in order here. The fact that the lower critical solution temperature is at a higher temperature than the upper critical

solution temperature can be a bit confusing. The word "lower" in *LCST* is chosen to mean at the bottom (lowest temperature) of a two phase region, while the word "upper" in *UCST* designates the top.

Some Limitations of the Flory-Huggins Theory

In spite of its limitations, some of which are listed in the box below, the Flory-Hug-

FIGURE 11-26 Experimental phase diagram of polystyrene of different molecular weights in acetone [redrawn from the data of Siow, Delmas and Patterson, *Macromolecules*, **5**, 29 (1972)].

[42] *Polymer*, **1**, 20 (1959)

gins model provides what every good theory should do—insight. As the simplest possible model it is still the starting point in the analysis of solution and blend data in many studies and can be considered one of the pillars of polymer science.

DILUTE SOLUTIONS

Finally, there is one limitation of the Flory-Huggins theory that we have not discussed—the fact that it only applies to *concentrated solutions*. In treating interactions, the model assumes a random mixing of polymer chain segments with solvent molecules, as if the chain were chopped up and these segments were randomly dispersed. This is not too bad when the polymer coils in solution overlap one another considerably, but in dilute solution the picture is different (Figure 11-27). There are local regions defined by the end-to-end distance or, more accurately, the radius of gyration of the chains, where there is a certain concentration of segments, separated by regions where there are no polymer molecules at all.

Treating dilute solutions is a difficult problem, requiring some demanding mathematical tools if it is to be done rigorously. Even the approximate treatment given in Flory's

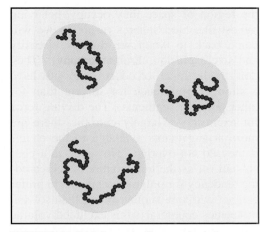

FIGURE 11-27 Schematic diagram depicting polymer chains in dilute solutions.

book is, at least to us, hard work! Here we will just qualitatively discuss three important results obtained in the treatment by Flory that have been confirmed by more complex approaches as summarized in the box below:

Taking these in order, we will first consider the intermolecular *excluded volume effect*. Essentially, this can be thought of as the chains "repelling" one another from the region of space they occupy, behaving almost like hard spheres (a notion we will come back to when we discuss viscosity and size exclusion chromatography). They don't really repel, of course, they just have a stronger preference for solvent than for other polymer segments. The driving force for exclusion is largely entropic, there are more arrangements available to the chains if they do not overlap. However, in a poor solvent (or as the temperature is decreased) the tendency for polymer segments to prefer contact with one another rather than solvent molecules counteracts this excluded volume effect and there is a temperature at which the forces are in balance. We'll come back to this shortly.

This entropic driving force also comes into play within the volume occupied by a single coil. The individual segments would like to spread out and occupy as much volume as possible, but the degree to which they can do this is constrained by the fact that they are all connected by covalent bonds (Figure 11-28). In a good solvent, where χ is small, the chain will swell, driven largely by the entropic forces favoring the dispersion of the segments over the largest volume possible. At some degree of swelling of the chain, this becomes counterbalanced by the loss of entropy associated with the chain becoming stretched out, thus having fewer conformations available to it.

Flory modeled this by considering the segments to be like a "swarm" of particles distributed about the center of gravity of the coil in a Gaussian fashion, then considered the balance between the free energy of mixing and the free energy associated with the elastic deformation of the chain. He then found an expression for a chain expansion factor, α, that minimized the free energy, which was expressed in terms of a series in the volume fraction of polymer, Φ_p. The result is shown in Equation 11-58:

$$\alpha^5 \sim m^{0.5}\left[\frac{1}{2} - \chi\right] + \frac{\Phi_p}{3} + \cdots$$

EQUATION 11-58

where m is the degree of polymerization of the chain. At low concentrations in a good solvent (Φ_p, χ, small), the amount the chain expands then varies as $m^{0.1}$. The end-to-end distance of the chain then goes as Equation 11-59:

$$\langle R^2 \rangle^{0.5} \sim \alpha m^{0.5} \sim m^{0.6}$$

EQUATION 11-59

corresponding to the result obtained for a self-avoiding walk (see Chapter 8, page 222). More rigorous treatments also give this result and neutron scattering experiments confirm that chains in dilute solution are indeed expanded (almost) to this extent. This degree of expansion is considerable for high molecular weight polymers. For example, if $m = 10,000$, then $m^{0.5} = 100$, while $m^{0.6} \sim 250$.

The degree of chain expansion also depends on χ, hence temperature (Equation 11-58). Clearly, as χ approaches values of $1/2$, α tends to zero and the chain becomes ideal. Flory reorganized his expression for the interactions in terms of a parameter, θ, and the temperature at which deviations from ideal chain dimensions occur is thus called

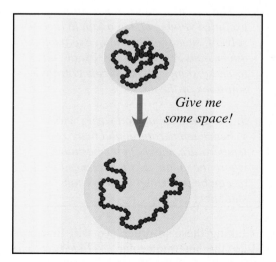

Give me some space!

FIGURE 11-28 Schematic diagram depicting expansion of polymer chains in dilute solutions.

the *theta temperature*. At temperatures just below the theta temperature, where $\chi > 1/2$, phase separation can occur, depending upon the composition of the mixture.

RECOMMENDED READING

P. J. Flory, *Principles of Polymer Chemistry*, Cornell University Press, Ithaca, New York, 1953.

L. H. Sperling, *Physical Polymer Science*, 3rd Edition, Wiley, New York, 2001.

J. H. Hildebrand and R. L. Scott, *The Solubility of Non-Electrolytes*, 3rd Edition, Reinhold Publishing Company, New York, 1950.

G. Strobl, *The Physics of Polymers*, Springer-Verlag, Berlin, 1996.

STUDY QUESTIONS

1. Why are large molecules such as polymers less likely to mix (in the sense of forming single phase mixtures) than small molecules?

2. This should be a simple question, but for some reason many students erase all knowledge of the rules of differentiation as soon as they complete their last calculus class! Simply derive the following equation for the chemical potential of the solvent in a polymer solution from the Flory-Huggins equation:

$$\frac{\Delta \mu}{RT} = \ln(1 - \Phi_p) + \left(1 - \frac{1}{M}\right)\Phi_p + \Phi_p^2 \chi$$

(Note n_p is independent of n_s, but both Φ_s and Φ_p depend on n_s.) Show every step and don't try and fudge certain lines of the derivation!

3. The values of the solubility parameter, δ_s, for certain solvents in $(cal/cm^3)^{1/2}$ are given as follows:

n-Hexane	7.2
Carbon tetrachloride	8.6
Benzene	9.2
Acetone	9.7
Methanol	14.5

Calculate the value of χ for solutions of polystyrene in these solvents (use a "fudge factor" of 0.34). Indicate which solutions are likely to be single phase and which are likely to be phase-separated. Use Table 11-1 to calculate the solubility parameter of polystyrene from group contributions.

Now do the same for poly(butadiene). Note that the solubility parameters of random copolymers can be calculated using $\delta = \Phi_A \delta_A + \Phi_B \delta_B$). Calculate the solubility parameter of a 50/50 styrene/butadiene random copolymer. Now calculate the value of χ for solutions of this copolymer in the solvents listed above. Compare the solubility of polystyrene with that of the random copolymer.

4. More differentiation, you poor soul! Obtain expressions for the critical value of χ and the concentration of polymer, $(\chi_p)_c$, at the critical point by differentiating the Flory-Huggins equation. (Hint: you may want to start from the free energy per mole of lattice sites expression and use volume fractions as your chosen concentration variable.)

5. We haven't discussed activities, vapor pressure and other aspects of the thermodynamics of liquids (optimistically assuming you've done all this in P. Chem). Nevertheless, we are sure that you recall that the vapor pressure of the solvent in a polymer solution relative to the vapor pressure of the pure solvent may, to a first approximation, be equated to the activity of the solvent, which in turn is related to the chemical potential by:

$$\ln a_s = \frac{\mu_s - \mu_s^0}{RT}$$

For the system polystyrene/cyclohexane the following values of ln a_s were obtained at 34°C [Krigbaum and Geymer, *JACS*, **81**, 1859 (1959)]:

Φ_p	ln a_s	Φ_p	ln a_s
0.343	− 0.0040	0.637	− 0.063
0.388	− 0.0042	0.690	− 0.089
0.435	− 0.013	0.768	− 0.151
0.485	− 0.018	0.818	− 0.232
0.543	− 0.024	0.902	− 0.460

A. Calculate χ for each value of Φ_p (assume m, the degree of polymerization, is large).

B. Plot a graph of χ versus Φ_p and comment on the result.

6. In an experiment you take one gram of poly(*N*-isopropylacrylamide) (PNIPA) and dissolve it in 100 grams of water at room temperature in a sealed glass vial. You notice that it is a perfectly clear (transparent) solution.

PNIPA

You then heat the vial. The polymer/solvent mixture now becomes turbid and looks like milk. You then cool the vial and the polymer/solvent mixture reverts to a perfectly clear solution. Explain your observations.

7. Discuss (~one page) some of the limitations of the Flory-Huggins theory.

8. A monodisperse polymer with a degree of polymerization of 50,000 is to be mixed with a solvent. The molar volume of a segment of the polymer is equal to that of the solvent, and their molecular weights (solvent and polymer SEGMENT) are both equal to 100. How many grams of each would have to be mixed to obtain a 50:50 mixture by VOLUME (i.e., one where the volume fraction of each is 0.5) whose total weight is 100 grams? Also, calculate the entropy of mixing, on a mole of lattice sites basis. Express your answer in terms of $\Delta S_m/R$ (i.e., a dimensionless number).

9. Discuss the Flory θ temperature by going to the literature and finding out more than what is written in the text! Include in your answer a discussion of how chain dimensions change in dilute solutions.

10. There are two types of excluded volume effects found in polymer science. Briefly describe each of them.

12

Molecular Weight and Branching

INTRODUCTION

> *"Drop the idea of large molecules
> Organic molecules with a molecular
> weight higher than 5000 do not exist."*
>
> *Advice given to Hermann Staudinger*

The fundamental characteristic of polymer materials, the ultimate source of their interesting and in some cases unique properties (like rubber elasticity), is their long chain nature. The struggle to establish the macromolecular concept was a long and difficult one, as the quote above would indicate. The ultimate test of the existence of polymers is, of course, provided by a measure of their size, but this is by no means an easy task for large molecules. In this section we will consider such methods in some detail, as they are also fundamental tools for measuring other properties, such as polymer/solvent interaction parameters, the radius of gyration of polymer coils, the degree of long chain branching and other aspects of chain microstructure.[43]

Traditional methods for measuring molecular weight date from the dawn of modern physical chemistry in the 19th century. These relied on so-called colligative property measurements (see box opposite), such as boiling point elevation and freezing

[43] When you synthesize a block copolymer do you end up with pure product or a mixture of block copolymer and homopolymers? How would you know?

point depression. [Raoult showed that one molecule of any (non-dissociating) solute dissolved in 100 molecules of solvent lowered the freezing point by a nearly constant amount.] However, even a polymer with a relatively low molecular weight of 10,000 g/mole will give very small freezing point depressions or boiling point elevations, of the order of 0.01°K. These particular methods are therefore simply not accurate enough to measure the molecular weights of very large molecules with any precision. However, there is one colligative property measurement, osmotic pressure, which we will consider in some detail.

COLLIGATIVE PROPERTIES

Colligative means "depending on the collection" and in dilute solution properties such as:

> *Vapor Pressure*
> *Freezing Point*
> *Boiling Point*
> *Osmotic Pressure*

depend upon the number of solute molecules present, or their mole fraction. From a knowledge of the weight fraction, the solute molecular weight can then be calculated. For a polymer this will give a number average molecular weight.

TABLE 12-1 MAJOR METHODS OF MEASURING MOLECULAR WEIGHT

ABSOLUTE	
End group analysis	$\overline{M}_n$
Osmotic pressure	$\overline{M}_n$
Light scattering	$\overline{M}_w$
Ultra-centrifugation	$\overline{M}_w ; \overline{M}_z$
RELATIVE	
Solvent viscosity	$\overline{M}_v \sim \overline{M}_w$
Size exclusion chromatography (SEC) (or gel permeation chromatography GPC)	Complete Distribution

But, before we get to that, you need to recall that a big problem with describing the molecular weight of synthetic polymers is that there is always a distribution of chain lengths (although certain polymerizations can give very narrow distributions) and it is necessary to define an average. We list some of the major methods for measuring the molecular weight of polymers in Table 12-1, separated into two categories. *Absolute methods* give a direct measure of a particular average, while *relative methods* require samples of known molecular weight to calibrate the instrument. We will just consider the four that are most widely used: osmotic pressure, light scattering, viscosity and size

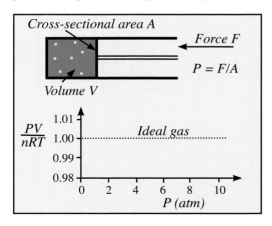

FIGURE 12-1 Plot of *PV/nRT* versus *P* for an ideal gas.

exclusion (SEC) or gel permeation chromatography (GPC).

Virial Equations

All of the methods used to determine molecular weight involve dilute solutions and all apart from SEC (or GPC) use virial equations. So, before we start discussing the details of each method you need to know what these are, in case you erased your memory banks after taking P. Chem. We will start our revision not with dilute solutions, but the ideal gas law (Equation 12-1):

$$PV = nRT$$
EQUATION 12-1

If, for example, you were to measure the volume of a gas as a function of pressure at constant temperature (Figure 12-1), then for an ideal gas you would expect a plot of *PV/nRT* against pressure to be a straight line at a value of 1 on the y-axis (Equation 12-2):

$$\frac{PV}{nRT} = 1$$
EQUATION 12-2

But this is not what you would get, of course, because the molecules interact with one another. A schematic redrawing of some real data for various gases is shown in Figure 12-2. The actual curves fit an empirical equation of state (Equation 12-3):

$$\frac{PV}{nRT} = 1 + B'P + C'P^2 + D'P^3 \ldots$$
EQUATION 12-3

This is a *virial equation*, the word virial being taken from the Latin word for force and thus indicating that forces between the molecules are having an effect. It turns out that statistical mechanical models also give equations that can be written in this form with the virial coefficients, B', C', etc., being related to various interaction parameters.

POLYMER MILESTONES—MACROMOLECULES: CLUES TO THEIR EXISTENCE

By the middle of the 19th century, it was generally accepted by the chemists of the day that atoms and molecules existed. Their understanding of inorganic compounds was better than organic compounds, which many still considered to have some supernatural "vital force" component. It was the chemist Kekulé (of benzene ring fame) who, in 1858, first introduced the concept of "structure" and the joining together of atoms by bonds. With improvements in elemental analysis, the structure of many simple molecules was elucidated. Elemental analysis, however, didn't help solve the structure of cellulose or natural rubber, materials that we now know are macromolecular in nature. In fact, it contributed to a general misconception. For example, natural rubber was found to have a composition equivalent to C_5H_8, but this only corresponds to the repeating unit of the polymer and says nothing about its long, chain-like structure.

Kekulé (Courtesy: Deutsches Museum).

It was the Scottish chemist, Thomas Graham, whose research pertaining to the diffusion of gases and liquids, not only led to Graham's law of diffusion, but also provided some of the first clues to the existence of large molecules. He observed that some natural substances, like albumin, gelatin and glue (collectively called, at that time, colloids), passed through a parchment (semi-permeable) membrane much more slowly than compounds such as sugar or salt (then referred to as crystalloids). This concept, of course, still remains the principle behind the modern dialysis machine that is all too familiar to the many people suffering from kidney disease. The tentative suggestion that the slow diffusion of a "colloid" through the membrane was due to its large molecular size was not accepted by most chemists of the time. Rather, they explained the phenomenon

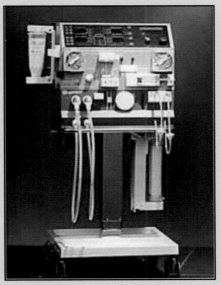

Modern dialysis equipment (Courtesy: Victrix).

in terms of micellular structures formed by the aggregation of small molecules. To be fair, this was not a foolish or "blinkered" conclusion, as there were known colloidal systems in the inorganic world (e.g., colloidal gold chloride or arsenic sulfide) that had analogous structures. As a result, it took a long time, some 60 years, before the existence of macromolecules finally gained acceptance.

POLYMER MILESTONES—MARK AND MEYER

Hermann Mark (left) with Hermann Staudinger
(Source: Deutsches Museum).

In the 1920s, it was not feasible to accurately measure the molecular weight of natural or synthetic polymers. Classical methods of molecular weight determination, those based upon colligative properties, elevation of boiling point, depression of freezing point and lowering of vapor pressure, worked very well for low-molar-mass compounds, but were essentially useless for macromolecules. Modern instrumental methods that can accurately measure high molecular weights, osmometry, light scattering and ultracentrifugation, were yet to make their mark. Staudinger (who is the subject of his own *Milestone*—see page 12) turned his attention to the measurement of solution viscosity. Differences in the viscosity of solutions of low molar mass and their analogous macromolecular counterparts are profound. (There is no perceptible increase in the viscosity of a 5% solution of ethylbenzene in benzene at room temperature over that of pure benzene. However, a 5% solution of a polystyrene that has, say, a molecular weight of 100,000 g/mol has the consistency of molasses!) Staudinger proposed a relationship between molecular weight and solution viscosity. He didn't get it quite right. In the meantime, Hermann Mark and several renowned scientific colleagues, most notably Professor K. H. Meyer, had been working on cellulose, silk, wool and rubber since the early 1920s. They were aware of and, in fact, favored, Staudinger's postulate of macromolecules, but they did not accept that his solution viscosity approach was a valid scientific proof of their existence. Actually Staudinger was quite peeved with Mark and Meyer, believing that their views ". . . do not represent anything new, but rather coincide in general with the opinions I have advocated for years and have established experimentally." But Mark and Meyer did make major contributions to the understanding of macromolecules in solution. Staudinger was wrong in his insistence that macromolecules existed as long thin rigid rods rather than intrinsically flexible chains. The disagreements between Staudinger, Mark and Meyer were relatively minor in the grand scheme of things, but at the time resulted in a number of bitter exchanges. By the late 1920s and early 1930s, the evidence accumulating as a result of these and other studies was becoming overwhelming. In particular, Staudinger's work on polyoxymethylenes; the X-ray diffraction studies of Meyer and Mark establishing the chain structure of cellulose; and, in the 1930s, the molecular weight measurement made possible by Svedberg's development of the ultracentrifuge, were decisive. One of your authors was fortunate to meet Professor Hermann Mark, who was then well into his 80s, at a polymer meeting in Köln, Germany. He still has a vivid image of this "Grand Old Master," suitcases in hand, striding purposively to catch a bus to the airport where he was to fly to yet another country to give a lecture. His energy level was quite extraordinary and when asked how he did it he responded: "I guess I have no time to get old."

Also note that at low pressures the lines are almost linear and the data could be modeled by an equation involving just the first two or three terms of the virial equation.

It was van't Hoff, winner of the very first Nobel prize in chemistry, who perceived an analogy between the properties of dilute solutions and the gas laws. We will see that many physical properties of dilute solutions, such as the amount of light scattered or the viscosity, can be written as a virial equation in the number of molecules (moles), N, or concentration of solute, c. We have written a general form of a virial equation in Equation 12-4, using the quantity P to represent some measured property of the solution and P_0 to represent the property of the pure solvent.

$$P = P_0 + BN + CN^2 + DN^3 + \dots$$
EQUATION 12-4

The first application of such equations to dilute solutions actually came from van't Hoff's measurements of the osmotic pressure of 1% solutions of cane sugar in water (relative to pure water), where the analogy to the virial equation of a gas expressed as a power series in the pressure is more direct. Accordingly, we will start our discussion of molecular weight measurements by considering osmotic pressure.

OSMOTIC PRESSURE

Osmosis, as you should have learned in high school biology, is the ability of a solvent (usually water) to pass from a dilute solution (or pure solvent) on one side of a membrane to a more concentrated solution on the other side (Figure 12-3). The osmotic pressure is the pressure that must be applied to the solution to prevent this occurring.

The nature of an osmotic pressure experiment is easily understood. Imagine a polymer solution contained in a tube that is, in turn, immersed in a reservoir of pure solvent, such that both the liquids are level at the start of the experiment (Figure 12-4). At the bottom of the tube containing the polymer solu-

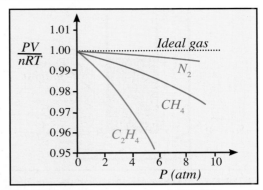
FIGURE 12-2 Schematic plot of *PV/nRT* versus *P* for non-ideal gases.

tion is a membrane and covering this, a cap. The membrane will allow the passage of small molecules, such as those of the solvent, but not large ones, like those of the polymer. Nothing happens, of course, while the cap is still on. What happens when you remove the cap? Upon removing the cap, pure solvent

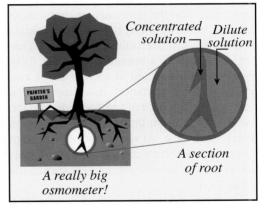
FIGURE 12-3 Schematic plot of osmosis in action.

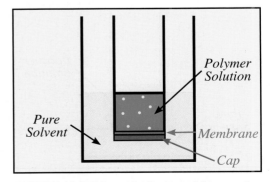
FIGURE 12-4 Schematic diagram of the osmotic pressure experiment with the cap still on.

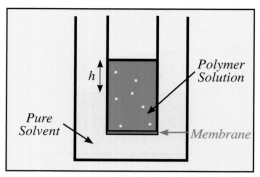

FIGURE 12-5 Schematic diagram of the osmotic pressure experiment after removal of the cap.

will flow from the solvent reservoir through the membrane and into the solution (Figure 12-5). The driving force for this is the chemical potential difference between the pure solvent and the solution, which drives solvent across the membrane, just like gravity drives water downhill. Because (in principle) the polymer cannot pass across the membrane into the pure solvent, in order to equalize concentrations the solvent attempts to make the solution infinitely dilute. It cannot achieve this, ultimately, because there is only a finite amount of solvent in the reservoir. More importantly, this flow will be opposed by the force of gravity acting on the column of solution that has risen above the reservoir surface. At some point equilibrium will be reached between the opposing forces due to gravity and the chemical potential difference.

There are two major problems with performing the experiment in this way. First the diffusion of solvent is slow, so that it would take hours or even days for equilibrium to be reached in this type of "static" experiment, all to obtain just one data point. Measurements have to be made as a function of concentration. More crucially, this type of experiment can result in large errors, because of diffusion of low molecular weight polymer, residual monomer, etc., from the solution to the solvent side of the membrane. This would alter the chemical potential difference across the membrane and hence the measured osmotic pressure.

These difficulties have been alleviated by performing the experiments in a different manner (also, membranes are a lot better than they used to be). One way is to simply apply a pressure, π, equal to the osmotic pressure, to the solution side of the membrane, as illustrated schematically in Figure 12-6. This would prevent any *net* flow. Alternatively, the deflection of the membrane could be used to measure the initial osmotic pressure directly. Molecules could still diffuse in each direction, however, but both methods provide reliable measurements of π in just a few minutes, before there has been any appreciable diffusion of solvent or solute.

As we mentioned earlier in this section, it was van't Hoff who saw an analogy between the properties of dilute solutions and the gas laws. Just as an equation of state can be written for an ideal gas, so can an equation of state be written for an ideal solution in terms of its osmotic pressure, as shown in Table 12-2. It is then a straightforward matter of remembering the definition of a mole (the weight of something divided by its molecu-

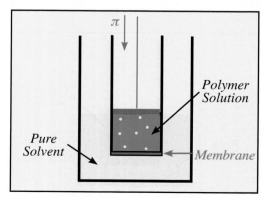

FIGURE 12-6 Schematic diagram of the osmotic pressure experiment after applying pressure, π.

TABLE 12-2 DEFINITIONS

DEFINITION	EQUATION
Ideal Gas	$PV/nRT = 1$
Ideal Solution	$\pi V/nRT = 1$
Mole	$n = w/M$
Concentration	$n/V = w/MV = c/M$
Ideal Solution	$\pi/c = RT/M$

lar weight) and using a concentration scale defined in terms of weight per unit volume (*c*) to end up with the equation of state for an ideal solution that shows an explicit relationship between the osmotic pressure of a solution (π) and the molecular weight of the solute (*M*).

For an ideal solution, then, a plot of π/c versus *c* should be a straight line at constant temperature. But, as you might expect, there is a variation with concentration (Figure 12-7). Just as with real gases, however, the data can be fit to the polynomial we call a virial equation. But, what do we do about polymers that have a distribution of molecular weights?

First reconsider the definition of concentration, *c*. This is just the weight per unit volume, which for a polymer with a distribution of molecular weights is obtained by multiplying the number of moles of chains of length *x*, N_x, by the molecular weight of each chain, M_x, and summing over all values of *x* (Equation 12-5):

$$c = \frac{w}{V} = \frac{\sum N_x M_x}{V}$$

EQUATION 12-5

Then recall the definition of the number average molecular weight and substitute to obtain an expression for *c* in terms of this quantity (Equations 12-6):

$$\overline{M}_n = \frac{\sum N_x M_x}{\sum N_x}$$

and:

$$c = \frac{\overline{M}_n \sum N_x}{V}$$

EQUATIONS 12-6

For a polymer we must write an ideal equation of state for the osmotic pressure in terms of a sum over all the moles of chains of different length, *x* (Equation 12-7).

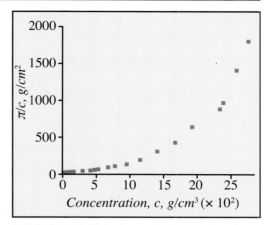

FIGURE 12-7 Plot of osmotic pressure data for solutions of polyisobutylene in chlorobenzene [replotted from J. Leonard and H. Doust, *J. Polym. Sci.*, **57**, 53 (1962)].

$$\pi V = RT \sum N_x$$

EQUATION 12-7

Substituting for $\sum N_x$ we obtain our equation of state (Equation 12-8):

$$\pi V = RT \frac{cV}{\overline{M}_n}$$

EQUATION 12-8

Then canceling *V* from each side of the equation and expanding in the virial form we get Equation 12-9:

$$\frac{\pi}{c} = \frac{RT}{\overline{M}_n} + Bc + Cc^2 + Dc^3 + \cdots$$

EQUATION 12-9

So, the osmotic pressure depends on the number average molecular weight. Now we can go back to the data and extrapolate to $\pi/c = 0$ to obtain the number average molecular weight. However, fitting a wide range of data, such as that shown in Figure 12-7, can lead to large errors. How many terms in the polynomial should you include? If you actually attempt this as an exercise you will find that you can get a large change in the inter-

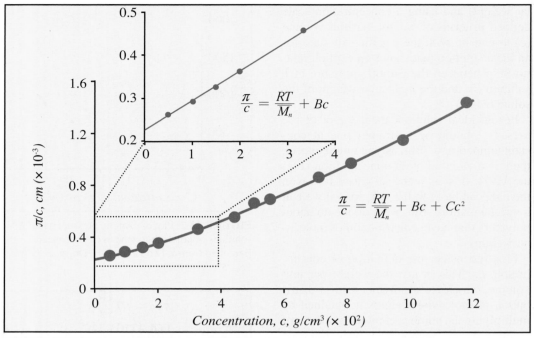

FIGURE 12-8 Plot of osmotic pressure data [replotted from the data of Leonard and Doust, *J. Polym. Sci.*, **57**, 53 (1962)].

cept as a result of just adding one additional term in the series.

However, if we just take the dilute solution data then a very good fit can be obtained by just fitting to three terms (Equation 12-10):

$$\frac{\pi}{c} = \frac{RT}{\overline{M}_n} + Bc + Cc^2$$

EQUATION 12-10

While taking even a smaller subset of the data allows a fit to a linear equation (Equation 12-11)—see insert of Figure 12-8:

$$\frac{\pi}{c} = \frac{RT}{\overline{M}_n} + Bc$$

EQUATION 12-11

The fit to the polynomial gives an intercept of 0.210, while the fit to a linear equation over a more limited concentration range gives an intercept of 0.224, a difference of

about 5%. For many purposes this may not matter too much, but if precision is important there are ways to rearrange the virial equation based on an assumption about the relationship between the second and third virial coefficients (B and C) that apparently gives more reliable results.

A final word about calculating $\overline{M}_n$. You have to be careful about the units. Some osmotic pressure data is reported in terms of the the height of a column of solvent. Think of this as the force exerted by gravity on a volume of solvent on a unit area (cm^3/cm^2 = cm). This pressure can then be expressed in terms of cgs units by first multiplying by the density, to give a mass/unit area which is how the data in Figure 12-8 were originally reported. (The units of π/c are then in cm, as in Figure 12-8.) As shown below, we multiply by the acceleration due to gravity (981 cm/sec^2) to give the pressure in terms of dyne/cm^2 (the units of π/c are then cm^{2}sec^2). The number average molecular weight is then ~ 1.13 × 10^5 g mole^{-1} as given by Equation 12-12:

$$\overline{M}_n = \frac{RT}{0.224 \times 10^3 \times 981 \, cm^2 sec^2}$$

$$= \frac{8.314 \times 10^7 \times 298}{0.224 \times 10^3 \times 981} \left[\frac{erg \, mole^{-1} \, {}^\circ K \, K^{-1}}{cm^2 sec^2} \right]$$

EQUATION 12-12

At this point you may be thinking that this is all very well, but virial equations are a bit of a fudge. However, it is relatively easy to show that they do have a basis in theory and the virial coefficients are related to interactions between the molecules. We will keep this simple by just using the Flory-Huggins theory and we will start by recalling that the driving force for solvent passing across the membrane is the chemical potential difference between the pure solvent and the solution—the solvent would like to keep passing through the membrane until the solution becomes infinitely dilute. Clearly the osmotic pressure must be related to this chemical potential difference (Figure 12-9).

Formal derivations pass through a consideration of the activity of the solvent (do you remember what that is?), but we will spare you that. Just think of it this way: the chemical potential is the potential energy (per mole) that drives solvent diffusion. This must be opposed by an equal and opposite potential energy (per mole) represented by the height of a column of solvent in the figure opposite. This is given by πV_m, where V_m is the molar volume of the solution (check the units if you don't believe us). For a dilute solution V_m is essentially equal to the molar volume of the solvent, so we can write Equation 12-13:

$$-\pi V \sim -\pi V_s = \Delta \mu_s$$

EQUATION 12-13

Now, recall the equation for the chemical potential of the solvent obtained from the Flory-Huggins theory (see Chapter 11) and substitute (Equations 12-14):

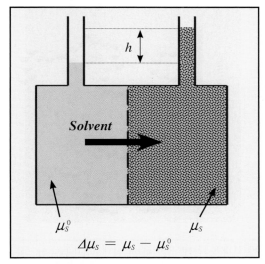

$$\Delta \mu_s = \mu_s - \mu_s^0$$

FIGURE 12-9 Schematic diagram of an osmotic pressure cell.

$$\frac{\Delta \mu_s}{RT} =$$

$$\ln(1 - \Phi_p) + \left[1 - \frac{1}{\overline{M}_n}\right]\Phi_p + \Phi_p^2 \chi$$

$$\pi = \frac{-RT}{V_s} \times$$

$$\left[\ln(1 - \Phi_p) + \left[1 - \frac{1}{\overline{M}_n}\right]\Phi_p + \Phi_p^2 \chi\right]$$

EQUATIONS 12-14

The logarithmic term can be expanded in a series (Equation 12-15):

$$\ln(1 - \Phi_p) = -\Phi_p - \frac{\Phi_p^2}{2} - \frac{\Phi_p^3}{3} - \ldots$$

EQUATION 12-15

So that Equation 12-16 written in terms of a series in the volume fraction is obtained.

$$\pi = \frac{RT}{V_s} \times$$

$$\left[\frac{\Phi_p}{\overline{M}_n} + \Phi_p^2\left(\frac{1}{2} - \chi\right) + \frac{\Phi_p^3}{3} + \ldots\right]$$

EQUATION 12-16

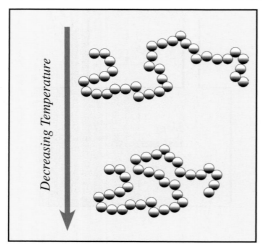

FIGURE 12-10 Schematic diagram depicting the reduction in chain expansion with decreasing temperature.

Clearly, this is a virial equation (see Equation 12-9), we just haven't bothered to convert the concentration scale from volume fraction to weight/volume. So virial equations are not just a useful mathematical fudge for fitting data, but have a basis in theory. Now if you're really good and remember some of the things we discussed in Chapter 11 you may be saying "hold your horses, Flory-Huggins does not apply to dilute solutions!" This is true; we have used it here just to illustrate how you can get equations of this form, which you still get with more appropriate and complicated models. Furthermore, the virial coefficients (B, C, D, etc.) in these models have physical meaning.[44] The second virial coefficient, for example, is related to interactions between pairs of segments, as described by the interaction parameter χ. Applied to non-dilute solutions, osmotic pressure measurements can then be used to measure χ.

Finally you should notice that at $\chi = 1/2$, the second virial coefficient vanishes (Equation 12-17:

---

[44] In more complex theories the higher order coefficients (C, D, etc.) are not constant, but are related to more complex (than pair-wise) interaction terms.

$$\frac{\pi}{\Phi_p} = \frac{RT}{V_s \overline{M}_n}$$
$$+ \Phi_p \left(\frac{1}{2} - \chi \right) + \frac{\Phi_p^2}{3} + \ldots$$

EQUATION 12-17

You may recall that the temperature where $\chi \sim 1/2$ is what Flory called the theta temperature and can now be seen to describe the situation where the second virial coefficient becomes zero (Figure 12-10). This means that at this point pair-wise interactions cancel and the chain becomes nearly ideal, as we discussed in the section on dilute solutions (Chapter 11), where we referred to the Flory excluded volume model in which the chain expansion factor is given by Equation 12-18:

$$\alpha^5 = M^{0.5} \left[\frac{1}{2} - \chi \right] + \frac{\Phi_p}{3} + \ldots$$

EQUATION 12-18

When $\chi = 1/2$, $\alpha \sim 0$.

LIGHT SCATTERING

Now let's turn our attention to the second absolute method for molecular weight measurement that we want to consider: light scattering. This provides a measure of weight average molecular weight. As with osmotic pressure measurements, the technique can also be used to measure other things, the radius of gyration of (high molecular weight) chains, for example, although we won't go into that.

A lot of students have trouble with light scattering, because at first glance the equations look fiendishly difficult. However, as with osmotic pressure we will still end up with a virial equation (Equation 12-19):

$$\frac{K(1 + \cos^2\theta)c}{R_\theta} =$$
$$\frac{1}{\overline{M}_w}[1 + 2\Gamma_2 c + \cdots]\left[1 + S\sin^2\left(\frac{\theta}{2}\right)\right]$$

EQUATION 12-19

where quantities related to the experimental measurement are on the left-hand side of the equation, we have a $1/M$ term times a virial equation, but then we have an additional intramolecular scattering term in the last set of brackets (it is the quantity S that can be related to the radius of gyration of the chain).

Unlike the equations, the light scattering experiment itself appears deceptively simple.[45] You take a solution, pass a light beam through it (one consisting of light of a single frequency or wavelength) and measure the intensity of scattered light as a function of angle of observation (Figure 12-11). You do this for the pure solvent and a bunch of different (dilute) solution concentrations, plot the data using Equation 12-19, then extrapolate to obtain the weight average molecular weight. This extrapolation, to zero concentration and angle of observation, is a bit trickier than it sounds, as you will find out.

If this is all you want to know, then we suppose you could jump straight to Zimm plots (page 375). If you want at least a feel for the origin of light scattering, however, read on! Let's start at the beginning by considering light as an oscillating electric field.

We will be considering *monochromatic light*, consisting of a single frequency or wavelength, unlike the broad range frequencies that come out of a light bulb. Also, we will initially assume the light is *plane-polarized*, meaning the oscillations of the electric field are confined to a plane, instead of being all over the place. We have probably lost a lot of you right there, because when you hear the word "field" you think of something involving grass and cows. Also, you've probably forgotten everything you ever learned about electromagnetism. Mentioning polarization merely added insult to injury. So, let's briefly review a couple of things, starting with some basic stuff about electricity and magnetism, because it is from this that our understanding of the nature of light arose.

[45] All right, the equations not only look fiendishly difficult, they are fiendishly difficult, particularly if you get into the details of fluctuations, which fortunately for you, we won't.

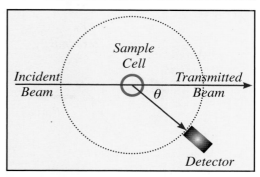

FIGURE 12-11 Schematic diagram of a light scattering experiment.

Electromagnetic Fields

You probably first learned about electricity and magnetism in high school, initially as separate subjects, which is indeed how they were treated up until about 1820. This is because they are distinct phenomena, as long as charges and currents are static. We'll allow things to move around shortly, but one of the things you surely learned in static electricity was Coulomb's law, describing the force between two charges, q_1 and q_2, as reproduced in Equations 12-20.

$$F_{cgs} = \frac{q_1 q_2}{r^2}$$

$$F_{SI} = \frac{q_1 q_2}{r^2} \frac{1}{4\pi\varepsilon_0}$$

EQUATIONS 12-20

There are two points we want to make about this law. First, note that if you go from cgs to SI units you introduce a factor $4\pi\varepsilon_0$, where ε_0 is the permittivity of vacuum. You see this term pop up now and again—now you know where it comes from. The second thing is that if there are a bunch of such charges interacting, you have to sum up all the interactions to get the force that would be experienced by a chosen charge. Things get even more complicated when the charges are moving. The most convenient way of handling these problems is by defining an abstract quantity called a *field*, which

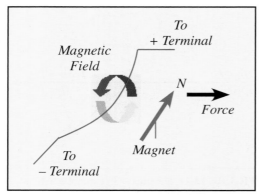

FIGURE 12-12 Schematic diagram of the link between electric and magnetic fields.

describes the forces that would be felt by a charge at a particular point in space. That's all a field is in physics, an abstract and useful construct in terms of making calculations.

The link between electric and magnetic fields was established by Hans Orsted, who observed that moving charges, like an electric current in a wire, produce magnetic fields (Figure 12-12). This observation prompted

FIGURE 12-13 James Maxwell (reproduced with the kind permission of the Deutsches Museum, Munich).

attempts to do things the other way around, get magnets to make electric fields. But, a wire carrying a *steady* electric current, which produces a magnetic field, will not induce a current in a wire placed right next to it unless, as Faraday observed, *something is changing*, such as the current in the wire. This is the part that will concern us, the observation that an *accelerating* charge produces an electric (and magnetic) field. It works the other way around as well, an oscillating field can induce an acceleration of charges! In other words, an oscillating field can induce an oscillation of charges in a material, which in turn produces an electric (and magnetic) field!

But what's all this stuff about electricity and magnetism got to do with light? The link was made by Maxwell in the 1860s (Figure 12-13). He was attempting to combine the laws of electricity and magnetism as they were understood at that time. But he found they were inconsistent until he added an extra term to complete his famous equations. This term essentially describes what we now call *electromagnetic radiation*. Furthermore, his equations contained an "electromagnetic constant," which we will label c_s^2, that turns out to be the square of the velocity of "electromagnetic influences." Maxwell calculated (from the results of static experiments) that this velocity was the same as the velocity of light and observed "we can scarcely avoid the inference that light consists of the transverse undulations of the same medium which is the cause of electric and magnetic phenomena." So light is an electromagnetic field that propagates through space. When considering scattering, however, we can just focus on the electric field. We will now turn our attention to a description of the form of this field, or, as Maxwell put it, the "transverse undulations," then consider how this field interacts with molecules.

An electric (or magnetic) field is defined by vectors that describe the forces that would be felt by a charge at some particular point in space. This field varies from point to point and can also vary with time. The electric field of light oscillates in a very simple and regular manner, oscillating sinusoidally from a maximum value of E_0, through zero to $-E_0$, then back through zero to E_0 again. This oscillation can be described in terms of circular motion and is equivalent to the projection of a vector traveling in a circular path (Figure 12-14). The period, or time taken to go one full revolution is simply $2\pi/\omega$, where ω is the angular frequency in radians/sec ($= d\theta/dt$). The oscillations of the electric field can then be written (Equation 12-21):

$$E = E_0 \cos \omega t$$

EQUATION 12-21

It probably annoys a lot of you that we call this a sinusoidal variation and then write it in terms of a cosine, but just remember your trigonometry, sines and cosines describe the same pattern, but one is simply shifted 90° or $\pi/2$ radians relative to the other. Now, if we want to describe the oscillations of the electric field in terms of the wavelength instead of ω, we note that the period is also equal to the distance traveled in one oscillation, λ, divided by the speed of light c_s.[46]

Origin of Light Scattering

Having described light as an electromagnetic field, we can now move on to consider how it interacts with matter. We've covered this; an accelerating charge produces an electric field and in the same way an oscillating field can cause charges to move. That's how the interaction occurs; the electric field of the light drives or induces oscillatory motions of the electrons in the molecule (Figure 12-15). The extent to which this occurs depends upon the details of its structure and is described by a parameter called the *polarizability*, α. This displacement of the electrons results in the formation of an instantaneous dipole moment [you remember, the product of the (induced) charges and their separation]. The magnitude of this induced dipole moment, p, is given by Equation 12-22:

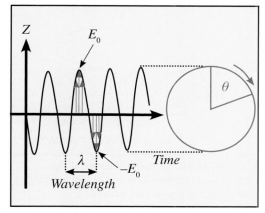

FIGURE 12-14 Schematic diagram of light as an oscillating electric field.

$$p = \alpha E = \alpha E_0 \cos\left[\frac{2\pi c_s t}{\lambda}\right]$$

EQUATION 12-22

Note that the oscillations of the dipole occur at the same frequency as the oscillations of the electric field of the light.

Now a dipole oscillating in this sinusoidal (or simple harmonic) manner is not moving with a steady velocity, but accelerating all the time (going from zero velocity at the position of maximum amplitude, where the direction of motion is changing, to a maximum velocity when the field has zero amplitude). This accelerating charge then becomes the source of a new electric field! This field oscillates with the same frequency as that of the dipole, hence the original electric field of the light, and is emitted in all directions (Figure 12-16), including back towards the source and in the forward direction, where

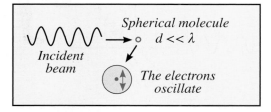

FIGURE 12-15 Schematic diagram of light impinging on a spherical molecule and inducing oscillatory motions of the electrons.

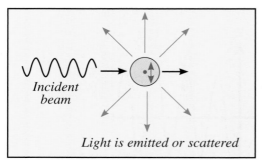

FIGURE 12-16 Schematic diagram of light impinging on a spherical molecule and inducing oscillatory motions of the electrons.

it combines with the incident beam that has passed through the sample. The scattered electric field is phase shifted (retarded) from the incident beam, however, and this is the origin of what we call the refractive index, *n*. We will come back to the refractive index in a while.

The intensity of the scattered light is not equal in all directions and, in fact, for polarized light the electric field is zero when observed at a certain angle to the incident beam. Let's look at this in a little more detail. First, we consider an isolated small, spherical molecule at the origin of a Cartesian system of co-ordinates and let the electric field of the incident beam be *polarized* or confined to the *zx* plane (Figure 12-17). In electromagnetic theory, the electric field, due to an accelerating charge, depends upon the product of the acceleration and the charge, which makes intuitive sense if you think of Coulomb's law. For our induced dipole moment this product is given by d^2p/dt^2. You

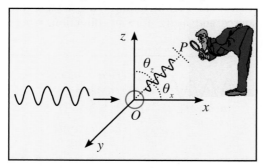

FIGURE 12-17 Schematic diagram of an incident beam of light polarized in the *zx* plane and observation of the scattered light at a point *P*.

also have to divide by c_s^2, which makes the units come out right.

Unlike Coulomb's law, however, the field does not vary with the inverse of a distance squared, but simply goes as $1/r$, where *r* is the distance from the origin. So far then, electromagnetic theory gives us for the scattered field (Equation 12-23):

$$E_{sc} \sim \frac{1}{c_s^2}\frac{d^2p}{dt^2}\frac{1}{r}$$

EQUATION 12-23

If we are observing the light at point *P*, we find that the scattered light is also polarized, now in a direction perpendicular to *OP* in the *P,z* plane. We therefore have to consider the projection of the dipole moment onto *OP*, which is obtained by multiplying by $sin\ q_z$. We then obtain the equation for the electric field given in Equation 12-24:

$$E_{sc} = \frac{1}{c_s^2}\frac{d^2p}{dt^2}\frac{1}{r}sin\ \theta_z$$

EQUATION 12-24

Recalling the expression for the induced dipole moment, defined in terms of the polarizability of the molecule (Equation 12-22), and then differentiating, we obtain our final expression for the electric field arising from the interaction of a small, single molecule with polarized light (Equation 12-25):

$$E_{sc} = \left[\frac{\alpha E_0}{r}\right]\left[\frac{2\pi}{\lambda}\right]^2 sin\ \theta_z$$

EQUATION 12-25

The intensity of scattered light, I_z', is given by the square of the amplitude of this oscillating field,[47] yielding our final expression given in Equation 12-26 (I_0 is the intensity of the incident light).

[47] If you've forgotten why this is so, think of it this way; the velocity of an oscillator is proportional to the field acting on a charge, so the (kinetic) energy or intensity developed by the field must be proportional to the square of this field.

$$I_z' = I_{0,z}\alpha^2 \frac{16\pi^4}{r^2\lambda^4}\sin^2\theta_z$$

EQUATION 12-26

Because of the dependence on $sin^2\theta$, the intensity of the light scattered at 90° to the direction of the incident beam is zero. A plot of the intensity of scattered light as a function of direction is called the *scattering envelope*. The scattering envelope in the zx-plane is shown in Figure 12-18. The full 3-dimensional figure would be obtained by rotating around the z-axis.

Having obtained an expression for the scattering of polarized light from a single oscillator we can now consider unpolarized light, then consider a dilute collection of such oscillators, as in a gas, and finally get to where we're going: liquids and solutions.

The scattering of unpolarized light can be handled by considering the incident beam to consist of two components of equal intensity, one polarized along the z-axis and the other along the y-axis in the preceding figures. We then get Equation 12-27:

$$I' = I_0\alpha^2 \frac{8\pi^4}{r^2\lambda^4}(sin^2\theta_z + sin^2\theta_y)$$

EQUATION 12-27

Using simple trigonometry we get to Equation 12-28:

$$I' = I_0\alpha^2 \frac{8\pi^4}{r^2\lambda^4}(1 + cos^2\theta)$$

EQUATION 12-28

The angle θ is defined in Figure 12-17. Now a gas can be considered to be a collection of such small oscillators that are on average very far apart. The scattering from a unit volume of a gas is then simply the scattering from a single molecule times the number of molecules per unit volume (Equation 12-29):

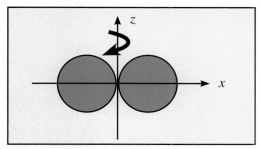

FIGURE 12-18 Schematic diagram of the scattering envelope.

$$I_\theta = \left[\frac{N}{V}\right]\left[\frac{I_0 8\pi^4}{\lambda^4}\right][\alpha^2]\left[\frac{(1 + cos^2\theta)}{r^2}\right]$$

EQUATION 12-29

The equation looks complicated, but you can see that it really consists of just four parts that are individually easily understood: first the number of molecules per unit volume; second, the characteristics of the incident light (its intensity and wavelength); third, the polarizability of the molecule (i.e., how easy it is for the incident beam to make the electrons slosh up and down); last, a term that depends upon the angle of observation and the distance from the scattering object. This understanding of light scattering was laid down by Lord Rayleigh and the equation given above can be rearranged in terms of a reduced intensity, R_θ, called the Rayleigh ratio (Equation 12-30):

$$R_\theta = \frac{I_\theta r^2}{I_0} = \left[\frac{N}{V}\right]\left[\frac{8\pi^4}{\lambda^4}\right][\alpha^2][1 + cos^2\theta]$$

EQUATION 12-30

This allowed Rayleigh to explain why the sky appears blue when the sun is overhead (Figure 12-19). The wavelength of blue light is about half that of red light. Because scattering depends on $1/\lambda^4$, the intensity of scattered blue light is then roughly 16 times that of red light. This also makes sunsets and sunrises have a reddish color. At these times of day you're observing the sun at an angle where there is a larger contribution from

FIGURE 12-19 Hilton Head Island, South Carolina, on a bad day! Here your authors sacrificed themselves on a sabbatical leave, laboring tirelessly and unceasingly on this masterpiece (do we have a great job or what?)

$$R_\theta = \left[\frac{2\pi^2}{\lambda^4}\right]\left[\frac{dn}{dc}\right]^2 [1 + cos^2\theta]\frac{N}{V}\left[\frac{M}{A}\right]^2$$

EQUATION 12-32

Liquids and Solutions

We hope you now have a basic feel for the nature of light scattering and an understanding of the origin of the equation that we will use as our basic result, modified appropriately to account for the nature of liquids and solutions. To do this rigorously we would have to delve into the mathematics of fluctuations, however. Like learning scientific German, life is too short for this! We will instead just try and give you a feel for the physics and introduce the modifications without proof.

Let's first look at the problem in an oversimplified manner. We can introduce additional molecules into a container holding a gas and change the number of molecules per unit volume (N/V) significantly. A liquid is not very "compressible," however, by which we mean that adding molecules to a liquid (at constant pressure) simply changes the total volume, but not N/V, or the concentration of molecules. The change in refractive index with concentration dn/dc, is therefore significant for a gas, but if liquids were "continuous" dn/dc would be zero.

Looking at the equation for scattering from a gas (Equation 12-32), you can see that $R_\theta = 0$ for $dn/dc = 0$. Physically what occurs is that in a liquid with a perfectly uniform density you can always pair molecules together in such a way that there is destructive interference from the light scattered from each. Liquids are not perfectly uniform in density, however. As we considered in our discussion of the T_g, there is something called free volume. This free volume is not evenly distributed throughout the volume of the liquid, so there are fluctuations in density. It is these fluctuations, the fact that at some instant of time there are a few more molecules per unit volume over here than over there, that give rise to light scattering in liquids.

transmitted light. Because more blue light is now "lost" to scattering, you see more of the red end of the visible spectrum.

But enough of the pretty pictures, let's get back to the nitty-gritty. The only unknown quantity in our equation for the Rayleigh ratio is the polarizability, α. However, it can be shown that[48] for a system like a gas, where the change in refractive index with concentration is small, the polarizability is related to the refractive index by Equation 12-31:

$$\alpha = \frac{1}{2\pi}\frac{dn}{dc}\frac{M}{A}$$

EQUATION 12-31

where n is the refractive index and M/A is the molecular weight divided by Avogadro's number. Thus the polarizability depends on the weight or size of an individual molecule. More on this shortly. Substituting for α we then get for the Rayleigh ratio (Equation 12-32):

[48] We hate using expressions like "it can be shown that," but we don't have room to explain everything. If you keep in mind that the refractive index is related to the phase shift as a result of combining the light scattered in the forward direction with the incident beam, then it makes physical sense that the refractive index depends on the polarizability.

Our interest is in dilute solutions, which unlike pure liquids are in some ways much more like a gas. There are solute molecules scattered randomly around and as long as their polarizability is different to that of the solvent you could imagine treating the system in the same way as scattering from a gas. Of course, you would have to subtract the "background" scattering from the solvent, but this is easily done experimentally. Unfortunately, the equations we obtained for a gas don't apply. The problem is that in liquids and solutions the molecules are relatively close-packed and the electric field created by other dipoles, as well as the incident light, must be taken into account. To get around this problem liquids and solutions are not considered in terms of their molecules, but as a dielectric medium that has local fluctuations in concentration and density. There can also be fluctuations in molecular orientation that can give rise to scattering effects, but we won't go into this as we can treat the random coils of flexible polymers as isotropic coils (if you are dealing with rigid-rod or liquid crystalline polymers, however, you do have to account for this).

So, the trick is to consider a solution to be a medium that can be subdivided into randomly fluctuating domains, each much smaller than the wavelength of the light. This allows us to consider the domains as scattering centers and apply the equation for scattering from a gas. We won't do this rigorously, but simply note that this results in replacing the α^2 term with a $\langle\Delta\alpha^2\rangle$ term, where $\Delta\alpha$ is the excess polarizability of one of these elements due to its deviation in concentration from the average. Then, just as α can be expressed in terms of the change in refractive index with concentration, so can $\langle\Delta\alpha^2\rangle$. The dn/dc term is now the change in refractive index of the solution with concentration. Also, $\langle\Delta\alpha^2\rangle$ now depends on the mean square concentration fluctuation, $\langle\Delta c^2\rangle$.

This is all very well, in that n_0 and dn/dc are known or can be measured, but what about $\langle\Delta c^2\rangle$? This can be obtained from the local fluctuations in the Gibbs free energy that are a result of the concentration fluctuations, resulting in a term in the second derivative of the free energy with respect to composition (Equations 12-33):

$$\langle\Delta\alpha^2\rangle = \frac{n_0^2}{4\pi^2}\left[\frac{dn}{dc}\right]\langle\Delta c^2\rangle$$

$$\langle\Delta c^2\rangle = \left[\frac{kT}{\partial^2 G/\partial c^2}\right]$$

EQUATIONS 12-33

or, equivalently, the first derivative of the chemical potential or osmotic pressure (Equations 12-34):

$$R_\theta^0 = K(1 + cos^2\theta)\left[\frac{M^2}{A}\frac{N}{V}\right]\left[\frac{RTV_s}{\partial\mu_s/\partial c}\right]$$

$$where:\ K = \frac{2\pi^2 n_0^2}{A\lambda^4}\left[\frac{dn}{dc}\right]^2$$

EQUATIONS 12-34

Hence, the last term in the expression for the Rayleigh scattering, given above. Note that we have used a superscript zero on the Rayleigh ratio, R_θ^0 to indicate that the domains we are considering are smaller than the wavelength of the light. Also, to simplify things, the parameters that are known or can be readily measured are lumped together in a constant, K.

The term in the second derivative of the free energy with respect to composition (or first derivative of the chemical potential or osmotic pressure), brings in a virial equation, usually expressed in a slightly different form than we used in our discussion of osmotic pressure (Equation 12-35):

$$\frac{\partial\mu_s}{\partial c} = V_s\frac{\partial\pi}{\partial c}$$

$$= \frac{RT}{M} + 2Bc + 3Cc^2 + \ldots$$

$$= \frac{RT}{M}[1 + \Gamma_2 c + \Gamma_3 c^2 + \ldots]$$

EQUATION 12-35

But, we're still considering small molecules that are all the same size. What if they have different chain lengths?

The scattering produced by a bunch of molecules having different chain lengths will simply be the sum of the scattering from each molecule. So we replace NM^2 with $\sum_i N_i M_i^2$. We can then use the definition of weight average molecular weight and also substitute for c, the concentration in weight/unit volume (keep in mind that here N is the number of molecules, so we have to divide by Avogadro's number, A, to get moles). Starting with Equation 12-34, rearranging and substituting we get the equation at the bottom of Equations 12-36:

$$\sum_i N_i M_i^2 = \overline{M_w} \sum_i N_i M_i$$

$$c = \frac{\sum N_i M_i}{VA}$$

$$R_\theta^0 = K(1 + cos^2\theta)\left[\frac{\overline{M_w}c}{[1 + \Gamma_2 c + \Gamma_3 c^2 + \ldots]}\right]$$

or:

$$\frac{K(1 + cos^2\theta)c}{R_\theta^0} = \frac{1}{\overline{M_w}}[1 + \Gamma_2 c + \Gamma_3 c^2 + \ldots]$$

EQUATIONS 12-36

Because of the dependence of the polarizability of a molecule on the size of that molecule and, in turn, the dependence of light scattering on α^2 or $<\Delta\alpha^2>$, we end up with a dependence on NM^2 for scattering from N molecules. This results in a dependence of light scattering on weight average molecular weight, in contrast to the number average determined by colligative properties measurements.

Light Scattering from Polymers

Now there's one last thing. A polymer that is larger than $\sim\lambda/20$ can no longer be considered a point source of radiation. There will be path differences, hence interference, from light scattered from different parts (i.e., different induced dipoles) of the same molecule, as illustrated in Figure 12-20. This results in an asymmetry in the scattering envelope, with more light being scattered in the forward direction than back towards the source (Figure 12-21). In order to account for this, a particle scattering factor, $P(\theta)$, is introduced into Equations 12-36 leading to Equation 12-37:

$$\frac{K(1 + cos^2\theta)c}{R_\theta^0} = \frac{1}{\overline{M_w}P(\theta)}[1 + \Gamma_2 c + \Gamma_3 c^2 + \ldots]$$

EQUATION 12-37

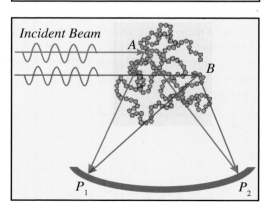

FIGURE 12-20 Schematic diagram of scattering from a polymer chain.

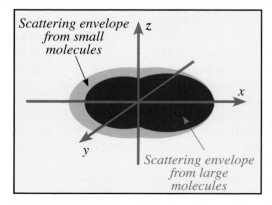

FIGURE 12-21 Schematic diagram of scattering envelopes.

$P(\theta)$ depends upon the shape of the molecule, which could be globular, a random coil, like most polymers in solution, and so on.

Light Scattering from Random Coil Polymers

So, Equation 12-37 is the general equation for light scattering from polymers. If you've jumped from page 367, you should know that the factor $P(\theta)$ depends upon the shape of the molecule. Our interest is in the most common case of random coils in dilute solution, where $P(\theta)$ can be simply expressed as a series in θ. It is usual to truncate this series after the second term, giving our final result, shown in Equation 12-38 in all its complex glory.

$$\frac{K(1 + cos^2\theta)c}{R_\theta^0} =$$
$$\frac{1}{M_w}[1 + \Gamma_2 c + \Gamma_3 c^2 + \ldots]$$
$$\times [1 + S\,sin^2(\theta/2)]$$

EQUATION 12-38

The term S depends on various constants and also the radius of gyration of the polymer chain. Note that light scattering experiments can therefore not only give you the weight average molecular weight, but also the second virial coefficient and the radius of gyration of the polymer chain (but not very accurately unless the molecular weight is > ~80,000 g/mole).

Equation 12-38 can be used in a couple of ways to measure molecular weight. One obvious approach is to measure scattering at $\theta = 0$, where $P(\theta) = 1$ ($sin^2\theta/2 = 0$) and $cos^2\theta = 1$, so Equation 12-38 reduces to Equation 12-39:

$$\frac{2Kc}{R_\theta^0} = \frac{1}{M_w}[1 + \Gamma_2 c + \Gamma_3 c^2 + \ldots]$$

EQUATION 12-39

Unfortunately, the intensity of the trans-

mitted incident beam is so strong compared to the light that is scattered, that this method is impracticable. However, at low angles $cos\,\theta$ and $P(\theta)$ are both ~1 (e.g., at 6°, $cos\,\theta = 0.994$) and the approximation that scattering at low angles is the same as at $\theta = 0$ has often been used.

Zimm Plots

Accurate measurements at low angles are not that easy, however, and more often a method described by Bruno Zimm is employed. Examine the scattering equation (Equation 12-38) carefully. A plot of:

$$\frac{K(1 + cos^2\theta)c}{R_\theta^0} \quad versus \quad sin^2(\theta/2)$$

for a solution of a given concentration would give an intercept of one over the weight average molecular weight at $\theta = 0$. Similarly, if we pick a given scattering angle (e.g., $\theta = 90°$) and measure the intensity of scattered light as a function of concentration, then a plot of:

$$\frac{K(1 + cos^2\theta)c}{R_\theta^0} \quad versus \quad c$$

would also give us an intercept of one over the weight average molecular weight, this time at $c = 0$.

Doing both would obviously give greater precision and Zimm suggested a way to do this on the same plot (Figure 12-22). The quantity $K(1 + cos^2\theta)/R_\theta$ is plotted versus $100c + sin^2(\theta/2)$, where the factor of 100 (or sometimes 1000) is introduced to spread out the data. This doesn't affect anything, because we are extrapolating to $c = 0$. Some data points for measurements made at various values of θ and c are shown as circles on the plot. There are two sets of lines, one corresponding to data obtained as a function of θ at constant c (the points that are the same color), while the other lines (more vertical) correspond to measurements as a function of c at constant θ. These trace out a charac ter-

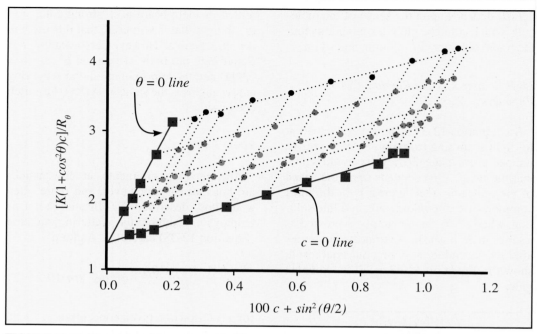

FIGURE 12-22 Zimm plot of data reported in Margerson and East (see suggestions for further reading).

istic grid pattern. The $c = 0$ and $\theta = 0$ points (squares) are then obtained (how?) and a double extrapolation to obtain the weight average molecular weight is performed. You probably won't see how this is done unless you do it yourself, however, so we sadistically explain no more, leaving you to construct your own Zimm plot as a homework problem.

FIGURE 12-23 "Blood's thicker than water", as the old saying has it. But so is pancake syrup!

VISCOSITY

Viscosity is another of those fundamental concepts that students seem to pick up imperfectly in their passage through the maw of the educational system. Many seem to confuse it with density, perhaps because motor oils of different viscosity have long been characterized according to a parameter called "weight." Also, in everyday life it is common to refer to a viscous fluid as being "thick," while a low viscosity fluid is described as being "thin" (Figure 12-23). Viscosity is actually a simple property to define, it is just a measure of the resistance to flow of a fluid. But within this simple statement lies a world of complexity, because matter, ranging in form from the lowest density gases to amorphous solids to the rock crusts that lie beneath the continents, flows. The study of the viscous properties of these materials is the science of *rheology*. We will discuss the fundamental rheological properties of polymers in Chapter 13. Here our concern is more limited and focused. Can we use viscosity to measure molecular weight?

It would seem intuitively obvious that we can. If viscosity is a measure of the resistance to flow of a fluid, then it must be related to frictional forces between the molecules. One would expect such frictional forces to increase with size or molecular weight (Figure 12-24). This is the way we will approach this topic to begin with, in a sort of semi-empirical manner to see what relationships can be established from experimental observation and the simplest of theories.

Poiseuille's Law

We will start with the French physicist and physician, Jean Poiseuille, who was interested in the circulation of blood in the body. This led him to study the flow of liquids through capillaries—pipes with a very fine diameter. He summarized his experimental results in an equation we call Poiseuille's law, where the rate of flow of fluid (volume/sec) is directly proportional to the pressure and inversely proportional to the viscosity (Equation 12-40):

$$v = \frac{\pi P r^4}{8L\eta}$$

EQUATION 12-40

where v is the volume of fluid passing through the pipe per second, P is the pressure, r is the radius of the pipe, L the length of the pipe and η the viscosity. The units of viscosity are *poise, P (dyne sec/cm²)*, in the cgs system. (To be more precise, this is the dynamic or absolute viscosity. There is also something called the kinematic viscosity, which is the dynamic viscosity divided by the density. Also, we have to specify that the flow is laminar or non-turbulent. You know, all the stuff you should learn in fluid mechanics.)

The Viscosity of Polymer Solutions

So, if we measure the rate of flow of polymer solutions through capillaries we can get a measure of their viscosity. Two simple

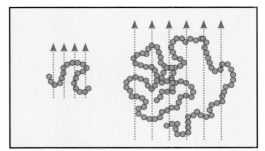

FIGURE 12-24 Which of these two molecules would display the highest viscous drag (drag—the ability to hold back the flow of the fluid)?

laboratory viscometers that allow one to do this are shown in Figure 12-25. Essentially, they are used to measure how long it takes the solution to pass between two marks (A and B) under the pressure imposed by gravity on the head of fluid. What we are actually trying to measure, however, is the frictional forces between the polymer and the solvent, not just the viscosity of the solution as a whole. We therefore determine the *relative viscosity*, η_{rel}, the viscosity of the solution divided by the viscosity of the pure solvent.

Furthermore, if what we want to determine is the frictional forces between a single polymer molecule and the solvent, then we need to make the measurements in dilute solution, so that there is no contribution from polymer/polymer interactions. In fact, just as in osmometry and light scattering, we measure the relative viscosity over a range of dilute solution concentrations and extrapolate to zero concentration.

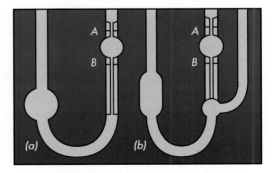

FIGURE 12-25 Schematic diagrams of (a) an Oswald viscometer and (b) an Ubbelholde or suspended-level viscometer.

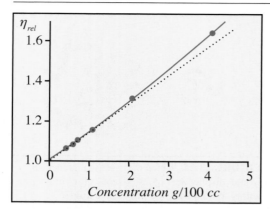

FIGURE 12-26 Plot of η_{rel} versus c for poly-(methyl methacrylate) in chloroform [replotted from the data of G. V. Schulz and F. Blashke, *J. fur Prakt. Chemie*, **158**, 130 (1941)].

The relative viscosity is easily obtained by rearranging Poiseuille's equation to express the viscosity in terms of the rate of flow, realizing that the pressure, P, maintaining the flow is simply proportional to the density of the solutions or pure solvent, while the rate of flow is simply proportional to the time taken to go between the two marks, as shown in Equation 12-41:

$$\eta_{rel} = \frac{\eta}{\eta_0} = \frac{t\rho}{t_0\rho_0} \approx \frac{t}{t_0}$$

EQUATION 12-41

(The subscript 0 refers to the pure solvent.) Because we are dealing with dilute solutions, we can make the additional assumption that the density of the solutions and solvent are (almost) the same and end up with the relative viscosity being equal to the ratio of the time it takes the solution to pass between the marks relative to the time taken by the pure solvent. If the relative viscosity depends in a simple way on the frictional forces between the polymer and solvent, then one might expect that a plot of η_{rel} versus c, the concentration, would be linear, as η_{rel} should increase with how much polymer is present. Its slope, however, should be proportional to the size of the polymer molecule, in that you would expect η_{rel} to increase at a faster rate with concentration for larger molecules

(Figure 12-26). At low concentrations, plots of relative viscosity versus concentration are indeed usually linear, but, as might be expected, show deviations from linearity at higher concentrations.

Just as in the treatment of osmotic pressure and light scattering, a power series can be used to fit the relative viscosity data, which at low concentrations can be truncated after a couple of terms (Equation 12-42):

$$\eta_{rel} = \frac{\eta}{\eta_0} = 1 + [\eta]c + kc^2 + \ldots$$

EQUATION 12-42

Both $[\eta]$ and k are constants, the square brackets around the former being part of the symbol. $[\eta]$ is called the *intrinsic viscosity* and is the quantity that we will relate to the molecular weight of the polymer.

There is actually a more fundamental reason for using an equation of the form shown in Equation 12-42 to fit the data. Einstein showed that for dilute solutions of solid particles the relative viscosity is given by (Equation 12-43):

$$\eta_{rel} = 1 + \gamma\phi$$

EQUATION 12-43

where ϕ is the volume fraction and γ is a constant equal to 2.5 for spheres. So it would seem entirely reasonable to simply change concentration scales and introduce higher order terms to account for interactions or other factors.

If it is the intrinsic viscosity that we want to determine, why don't we just plot the relative viscosity versus concentration and determine the initial slope? Because that would be too easy and we don't see any reason to pass up an opportunity to make your life difficult and miserable by introducing a few more viscosities and a more complex way of plotting the data. Actually, the plot itself is not too bad, and the reason we do all this is to try and minimize errors and obtain more accurate values of $[\eta]$. In the good old days, when we were just young whippersnappers, viscosity measurements were made using a stop-watch to time the flow between marks (today there are automated systems with lasers). If you had a slow-reaction-time klutz for a lab partner, your errors could indeed be large. Anyway, we digress, as is our wont, and the next definition we need to introduce is something called the *specific viscosity*, η_{sp}, defined in Equations 12-44:

$$\frac{\eta_{rel} - 1}{c} = \frac{1}{c}\left(\frac{\eta - \eta_0}{\eta_0}\right) = [\eta] + kc$$

$$\eta_{sp} = \frac{\eta - \eta_0}{\eta_0} = \eta_{rel} - 1$$

EQUATIONS 12-44

All we've done here is truncate the power series to neglect terms in c^3 or higher, and rearrange the equation so that the right-hand side is linear in c. The specific viscosity is then simply $\eta_{rel} - 1$. The intrinsic viscosity is just the intercept, the value of η_{sp}/c as c goes to zero. OK, you may think, we're done; get real! First, Huggins found that the slope of plots of η_{sp}/c versus c (i.e., k) are apparently proportional to $[\eta]^2$. Incorporating this assumption results in the Huggins equation (Equation 12-45):

$$\frac{\eta_{sp}}{c} = [\eta] + k'[\eta]^2 c$$

EQUATION 12-45

Now, if all these different viscosity definitions are making you depressed, then this will make you suicidal. Along comes Kraemer and defines the intrinsic viscosity a different way, as shown in Equation 12-46.

$$\frac{\ln\eta_{rel}}{c} = [\eta] + k''[\eta]^2 c$$

EQUATION 12-46

The quantity ($\ln \eta_{rel}/c$) is called the *inherent viscosity*. It too can be expressed as a power law in c in a similar fashion to the Huggins equation. In addition, $\ln \eta_{rel}$ can be

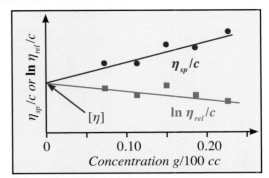

FIGURE 12-27 Schematic diagram of a "good" intrinsic viscosity plot.

expanded in powers of η_{sp}. Comparing these equations it can be shown that $k'' = k' - 1/2$. (This is an interesting homework question. To get an answer remember to neglect terms in c^2 and higher).

Both the Huggins and Kraemer equations lead to the same extrapolated value of $[\eta]$, so it has become common practice to use both equations on the same plot, as shown schematically in Figure 12-27. This, combined with the imposition of the requirement that $k'' = k' - 1/2$ increases confidence in the extrapolation, but you should be warned that not all data looks as good as that shown here.

And this may not just be due to the fact that you've got a ham-fisted lab partner. In systems with strong intermolecular interactions, such as solutions of ionic polymers or polyelectrolytes, severe deviations from linear behavior can be observed, even at low concentrations (Figure 12-28). This should

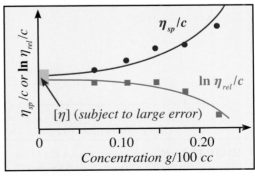

FIGURE 12-28 Schematic diagram of a "bad" intrinsic viscosity plot that one might observe from a solution of a polymer containing ions.

TABLE 12-3 VISCOSITY DEFINITIONS

NAME	SYMBOL & DEFINITION
Relative Viscosity (Viscosity Ratio)	$\eta_{rel} = \dfrac{\eta}{\eta_0} \approx \dfrac{t}{t_0}$
Specific Viscosity	$\eta_{sp} = \eta_{rel} - 1$ $= \dfrac{\eta - \eta_0}{\eta_0}$
Reduced Viscosity (Viscosity Number)	$\eta_{red} = \dfrac{\eta_{sp}}{c}$
Inherent Viscosity (Logarithmic Viscosity Number)	$\eta_{inh} = \dfrac{\ln \eta_{rel}}{c}$
Intrinsic Viscosity (Limiting Viscosity Number)	$[\eta] = \left(\dfrac{\eta_{sp}}{c}\right)_{c \to 0}$ $= \left(\dfrac{\ln \eta_{rel}}{c}\right)_{c \to 0}$

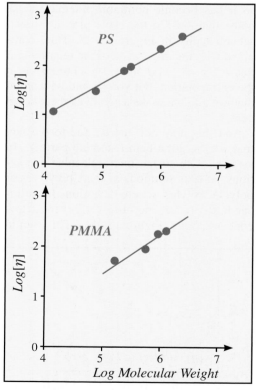

FIGURE 12-29 Plots of log $[\eta]$ versus log M for polystyrene (PS) and poly(methyl methacrylate) (PMMA) [replotted from the data of Z. Grubisic, P. Rempp and H. Benoit, *J. Polym. Sci., Polym. Letters*, **5**, 753 (1967)].

reinforce a point that is general to all extrapolation to zero concentration methods: you would like to perform your experiments at the lowest possible concentrations, but this is where the largest errors in measurement are bound to occur.

Now, if you've managed to keep track of all those viscosity definitions, you're a better man than I am Gunga Din! (You have to be familiar with the romantic imperialism of the books and poems of Rudyard Kipling to get this allusion.) So, we have summarized them all in Table 12-3. There are two sets of names, the ones we have used, which almost everybody else also uses, and a set devised by IUPAC, which they hoped everybody would use, but almost nobody does. Just remember that what you measure is the *relative viscosity* and what you want to know is the *intrinsic viscosity*.

Intrinsic Viscosity and Molecular Weight

This brings us to the relationship between the intrinsic viscosity and molecular weight. This was first established empirically from plots of *log*$[\eta]$ vs *log Mol Wt.*, which were found to be linear (Figure 12-29), meaning that the relationship between the two quantities has the form (Equation 12-47):

$$[\eta] = KM^a$$

EQUATION 12-47

This is called the Mark-Houwink-Sakurada equation and is the basis for determining molecular weight using solution viscosity measurements. The method is not absolute, but relative, requiring the initial determination of the intrinsic viscosity of a set of monodisperse samples of known molecular weight (determined by osmotic pressure or light scattering measurements). A plot of *log* $[\eta]$ versus *log M* then allows the determination of the constants K and a, *characteristic of that polymer in that solvent at that temperature* (Figure 12-30). If the intrinsic viscosity of a sample of unknown molecular weight is then measured in the same solvent at the same temperature, its molecular weight can be calculated. But, given that the unknown sample is *polydisperse*, what average are we calculating?

Viscosity Average Molecular Weight

You are actually measuring a new average, called the viscosity average (Equation 12-48):

$$[\eta] = K\overline{M_v}^a$$

EQUATION 12-48

This equation can be obtained in a straightforward manner by assuming that the specific viscosity of the (dilute) solution is simply the sum of the contributions from all the chains present, then you just work through the algebra shown in Equations 12-49. (We know that you won't do this unless we make you, so get ready for a nasty homework.)

$$\eta_{sp} = \sum_i (\eta_{sp})_i \qquad \frac{(\eta_{sp})_i}{c_i} = KM_i^a$$

$$\eta_{sp} = K\sum_i M_i^a c_i$$

$$[\eta] = \frac{\eta_{sp}}{c} = \frac{K\sum_i M_i^a c_i}{c}$$

$$c = \sum_i c_i \qquad w_i = \frac{c_i}{c} = \frac{N_i M_i}{\sum_i N_i M_i}$$

$$\overline{M_v} = \left[\frac{\sum_i N_i M_i^{a+1}}{\sum_i N_i M_i}\right]^{1/a}$$

EQUATIONS 12-49

We will show shortly that the constant *a* should have values between 0.5 (for a polymer in a theta solvent) and 0.8 (for a polymer in a "good" solvent), so that the viscosity average molecular weight is generally in-between the number average and the weight average. In fact, it can be shown that for a polymer with the most probable molecular weight distribution in a theta solvent the var-

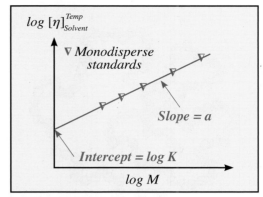

FIGURE 12-30 Schematic plot of *log* $[\eta]$ versus *log M* for a series of monodisperse polymers.

ious averages have the following proportions (Equation 12-50):

$$\overline{M_n} : \overline{M_v} : \overline{M_w} = 1 : 1.67 : 2$$

EQUATION 12-50

(Can you smell another one of those sneaky homework questions coming, one that asks you to remember bits and pieces of different parts of the subject?)

Well, this is all very nice to have this useful empirical relationship between intrinsic viscosity and molecular weight, you might say, but is there any underlying theory that would explain the form of the relationship and the nature of the constants *a* and *K*? Also, we have been very careful to specify that these constants depend upon the polymer, the solvent and the temperature: why is this?

Frictional Properties of Polymers in Solution

Theories of the frictional properties of polymer solutions, from which the relationship between viscosity and molecular weight must ultimately be derived, can proceed in various ways, but we will mention just two models. In the first, it is assumed that the velocity of the solvent is barely affected by the presence of the polymer, so that it streams or "freely drains" through the coil in a largely unperturbed fashion (Figure 12-31).

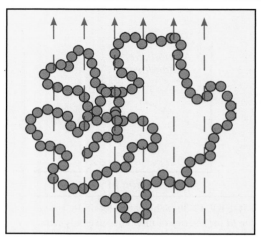

FIGURE 12-31 Schematic diagram of the freely draining model.

The polymer is then treated as a string of beads where the frictional force on each bead is described by Stokes law. However, this results in a relationship between $[\eta]$ and molecular weight, M, of the form $[\eta] \sim M^{1+\Delta}$, where Δ is a small fraction. Experimentally, we have seen that $[\eta] \sim M^a$, where a usually has values between 0.5 and 0.8. Obviously, the assumption that the polymer beads do not perturb the solvent is not a good one.

Going to the other extreme, it can be assumed that the polymer perturbs the medium so much that near the center of the coil the solvent molecules actually move with the same velocity as the beads. This means that the polymer coil can be treated as if it were a sphere that is, in effect, impenetrable to the solvent molecules that happen to lie outside its domain. Of course, this cannot be quite right, because (among other things)

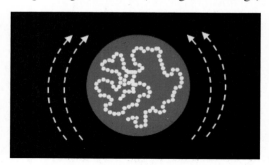

FIGURE 12-32 Schematic diagram of the equivalent hydrodynamic sphere.

the density of polymer segments decreases with increasing distance from the center of gravity of the coil (go back and check Conformations if you've forgotten all this stuff). Nevertheless, this is a useful and important start, because it leads to the concept of an *equivalent hydrodynamic sphere*, a hypothetical impenetrable sphere that would display the same properties as the polymer coil (Figure 12-32).

The nice thing about the assumption of an equivalent hydrodynamic sphere is that it allows you to do two things: first, use the Einstein relationship for the viscosity of a dilute solution of solid particles, given previously in Equation 12-43. Then we can use the definition of a volume fraction, ϕ (Equation 12-51):

$$c = \frac{N}{V}\left(\frac{M}{A}\right) \qquad \phi = \frac{NV_h}{V}$$

$$\phi = V_h\left(\frac{A}{M}\right)c$$

EQUATION 12-51

where A is Avogadro's number so that we arrive at Equation 12-52:

$$\eta_{rel} - 1 = \eta_{sp} = 2.5\phi = 2.5V_h\left(\frac{A}{M}\right)c$$

EQUATION 12-52

Hence, as $c \to 0$ this leads to Equation 12-53:

$$\left(\frac{\eta_{sp}}{c}\right)_{c \to 0} = [\eta] = 2.5V_h\left(\frac{A}{M}\right)$$

EQUATION 12-53

We now make the assumption that the volume occupied by a polymer coil is equal to that of a sphere whose diameter is defined by the root mean square end-to-end distance, $<R^2>^{0.5}$ (Equation 12-54):

$$V = \frac{4}{3}\pi\left(\langle R^2 \rangle^{1/2}\right)^3$$

EQUATION 12-54

For a polymer in dilute solution we have seen that $<R^2>^{0.5}$ is proportional to $M^{0.5}\alpha$, where α is the chain expansion factor (see Chapter 11; note that previously we related $<R^2>^{0.5}$ to the number of segments, but this is obviously equal to the molecular weight of the chain, M, divided by the molecular weight of a segment, M_0).

Accordingly, if we assume that V_h is proportional to V', then the intrinsic viscosity should be related to the molecular weight by Equation 12-55:

$$[\eta] = K' M^{0.5} \alpha^3$$

EQUATION 12-55

where the various constants of proportionality have been lumped into K'.

We know that for a good solvent α is proportional to $M^{0.1}$, while in a θ solvent $\alpha = 1$, so this gives us our final result (Equation 12-56) where α varies between 0.5 and 0.8.

$$[\eta] = KM^a \quad or: \quad [\eta] = K\overline{M}_v^{\,a}$$

EQUATION 12-56

This model is obviously flawed, we would not expect the solvent molecules within the volume defined by a polymer coil to move with exactly the same velocity as this coil, for example. But, in spite of its simplicity it gives a surprisingly good representation of the data. Also, it makes plain why the constants K and a vary with the polymer/solvent system and also the temperature. Finally, it is also a concept that is important in understanding SEC or GPC measurements, the final topic in our discussion of molecular weight determination.

SIZE EXCLUSION CHROMATOGRAPHY

When we talk about chromatography we are referring to a wide range of techniques that are used to separate the components of a mixture. The field has its origin in the work

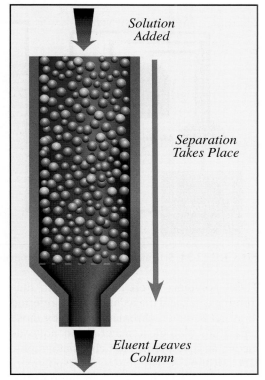

FIGURE 12-33 Schematic diagram of chromatographic technique.

of the Italian born Russian botanist, Mikhail Tswett, who (in 1906) separated plant pigments (chlorophylls) by pouring a petroleum ether extract of green leaves over a column of powdered $CaCO_3$ held in a glass column. The components of the mixture migrated down at different speeds, so that the column became marked with horizontal bands of different colors—a *chromatogram* (from the Greek, meaning *colored record*). The chromatographic technique (Figure 12-33) that concerns us here has its origin in the work published in 1959 by Flodin and Porath,[49] who developed cellulose based materials that act as "molecular sieves" and are capable of separating large molecules such as proteins and synthetic polymers according to size.

Initial work used various types of cross-linked, insoluble, polymer materials packed in a column in order to separate by size (we'll tell you how this happens in a moment).

[49] J. Porath and P. Flodin, *Nature* **183**, 657 (1959).

In the figure: *Solution Added* / *Separation Takes Place* / *Eluent Leaves Column*

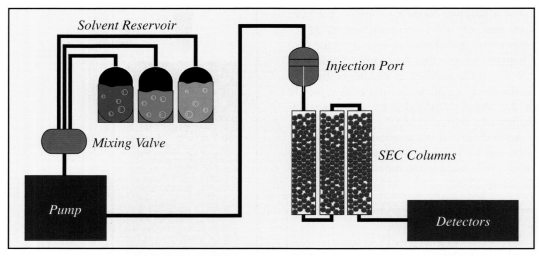

FIGURE 12-34 Schematic diagram of SEC chromatography.

These polymers swell in the solvent, thus forming a gel, and the technique was referred to as gel filtration chromatography by those separating biological materials in aqueous media, and gel permeation chromatography, GPC, by those separating synthetic polymers in organic solvents. However, the polymer gels initially used for separations were soft and could not stand up to the high pressures that are needed to reduce what were long separation times. Many modern instruments therefore use rigid silica-based materials in the form of porous beads packed into a column, or, more often, a set of columns, and the technique is more correctly referred to as size exclusion chromatography (SEC). Nevertheless, cross-linked polystyrene beads are still used and polymer scientists, often creatures of habit, more often than not still use the acronym GPC to refer to the technique, regardless of the nature of the beads packed in the column.

The experimental protocol is straightforward (Figure 12-34). Solvent or a mixture of solvents is pumped through the packed columns. There is an injection port at some point in the solvent stream before it reaches the columns. A polymer solution is injected at this point. The solution passes through the columns and through some sort of detector that measures the amount of polymer in what is called the eluant stream (e.g., something

that measures the refractive index of the solution, which varies systematically with polymer concentration).

So, how does separation according to size occur? It is only necessary to give you a simple qualitative picture, but you should keep in mind that there are a lot of subtleties associated with the details of the separation process. Essentially what happens is this: small particles or molecules not only flow around the beads as a solution is pumped through the columns, but can also diffuse in and out of the pores of the beads (Figure 12-35). Their average path length through the column is therefore much longer than that of a very large molecule that may be completely excluded from the pores. Molecules that are intermediate in size may diffuse in and out of large pores, but be excluded from smaller ones. Thus molecules have different path lengths through the column according to their size and are eluted at different times.

Although we will doubtless be castigated as hopeless sexists for saying this, it is rather like being dragged through the local mall by our wives, who for some reason seem inordinately fond of shoes (Figure 12-36). So their path from one end of the mall to the other is often lengthy, as women's shoe shops appear to be located at ten yard intervals (there must be a Law of Malls with Universal Constants that specifies the minimum number of these

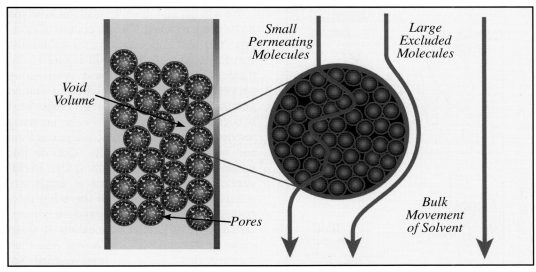

FIGURE 12-35 Schematic diagram of SEC chromatographic separation.

things per mall and their maximum distance apart). For us, on the other hand, there is a Sears at one end of the mall, which has some guys stuff like rototillers, but after that we could walk without deviation clear to the other end, pausing only for a moment as we pass Victoria's Secret to check out the window.

Accordingly, if you were to take a solution of a typical polydisperse polymer sample and inject it into a stream of pure solvent passing through a set of SEC columns, the molecules would start to separate according to size, with the larger ones eluting first. If you plot the concentration of polymer in a given volume of solvent as it comes off the columns (as measured by whatever detector your instrument utilizes), then you might get a chromatogram that appears like the one illustrated schematically in Figure 12-37. The amount of polymer eluted from the columns is plotted against what is called the *elution volume* (the amount of liquid that has

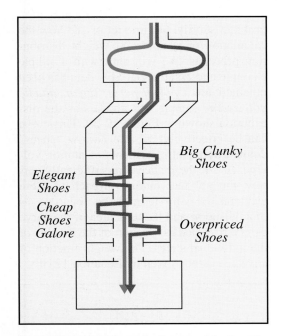

FIGURE 12-36 Schematic diagram of the "mall".

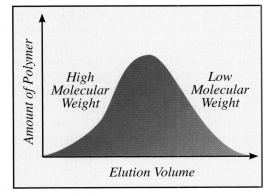

FIGURE 12-37 Schematic diagram of a size exclusion chromatogram.

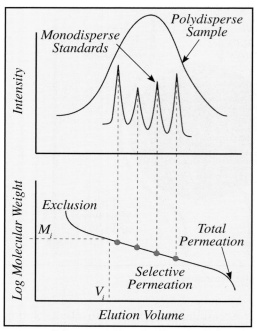

FIGURE 12-38 Schematic diagram illustrating the calibration of SEC.

come through the column from a given starting time, say, when the polymer was injected into the solvent stream). There may be just a small amount of very high molecular weight material that comes off first, and small amounts of low molecular weight material that come off last, and in-between a curve that reflects the molecular weight distribution of the sample.

Of course, SEC or GPC is not an absolute method of measuring molecular weight. Nevertheless, there must be a relationship between the molecular weight (or more accurately, size) of the molecules and the time it takes them to pass through the column, as measured by the elution volume, you just don't know what this is. A particular column or columns must therefore be calibrated by first passing through monodisperse samples of known molecular weight. A calibration curve can then be constructed by plotting the log of the molecular weight against the elution volume for these samples (Figure 12-38). This can then be used to determine the molecular weight distribution of an unknown sample.

We will get into the nitty-gritty of how you do this in a moment—first a couple of caveats. You will notice that on the calibration curve, shown at the bottom of Figure 12-38, there is an apparently linear relationship between log molecular weight and elution volume, but only over a certain range. This is because a particular set of columns has a restricted range where selective permeation and hence separation occurs. Outside this range, at small elution volumes, total exclusion of molecules larger than a certain size occurs. In other words, all these big molecules pass around the beads and are not separated. Similarly, total permeation of small molecules occurs—all molecules less than a certain size elute at the same volume. So, you have to use columns that are designed for specific molecular weight ranges. The second caveat is that this separation occurs by size and, just as we have seen in our discussion of viscosity, the size of a given polymer coil in solution varies with the nature of the polymer and solvent (hence their interactions) and the temperature. Accordingly, the type of calibration shown above is only good for a given polymer/solvent system at a specified temperature.

We will come back to this calibration problem shortly, but first let us see how the calculation of a molecular weight distribution proceeds. We will start with a simple example and assume that SEC data has been obtained for a *polydisperse, linear, atactic polystyrene* sample and appears as the distribution shown in Figure 12-39. This curve can be effectively cut into narrow "slices" defined by equal increments of elution volume, ΔV. If these slices are sufficiently narrow, the polymer eluting in each of these elements of volume may be regarded as monodisperse and the total area of the curve can be regarded as the sum of the heights of the individual slices. The weight fraction of any slice is then given by Equation 12-56:

$$w_i = \frac{h_i}{\sum h_i}$$

EQUATION 12-57

The molecular weight of the ith species is then obtained from the calibration curve at point V_i, and the molecular weight averages can be calculated from Equations 12-58:

$$\overline{M}_n = \frac{1}{\sum_i \dfrac{w_i}{M_i}}$$

$$\overline{M}_w = \sum_i w_i M_i$$

EQUATIONS 12-58

The weight average expression follows directly from its definition, but obtaining the number average in terms of w_i and M_i takes some manipulation, which we always leave as a homework problem. Also note that the intrinsic viscosity can be calculated from Equation 12-59:

$$[\eta] = \sum_i w_i [\eta]_i = K \sum_i w_i M_i^a$$

EQUATION 12-59

As you will see, this is more than an idle aside.

Hydrodynamic Volume

To summarize, the methodology we have described so far is fine for calculating the molecular weight distribution and molecular weight averages as long as monodisperse or, more accurately, very narrow molecular weight standards of the same polymer are available. To reiterate, this is because the size of a molecule in solution, its hydrodynamic volume, depends upon the nature of the polymer/solvent pair and the temperature. When we say the nature of the polymer, this also includes chain architecture: a linear polystyrene with the same number of segments as a star-shaped sample occupies a bigger hydrodynamic volume. In the star-shaped polymer, more of the segments are bunched together in the middle of the molecule where the arms of the star meet—Figure 12-40. If

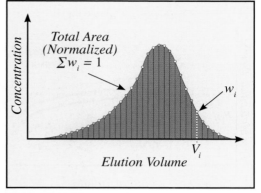

FIGURE 12-39 Schematic diagram illustrating the "slicing" of the SEC chromatogram.

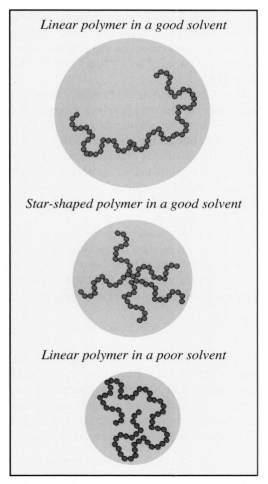

FIGURE 12-40 Schematic diagrams illustrating different hydrodynamic volumes of polymers.

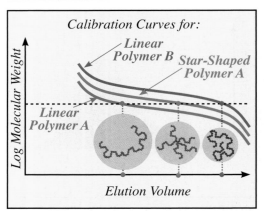

FIGURE 12-41 Schematic diagram illustrating different calibration curves at the same temperature in the same solvent.

these polymers are in a "good" solvent both may occupy a bigger hydrodynamic volume than a chemically different linear polymer in the same solvent at the same temperature, as this solvent may be a "poor" one for this polymer.

In effect, different calibration curves are necessary for different polymer solvent pairs, or even the same polymer/solvent pair if the microstructure of the polymer is different to that of the standards that are available (Figure 12-41). Sometimes, molecular weight averages based upon linear polystyrene standards are reported, but this can obviously lead to large errors.

These factors would seem to make SEC or GPC a tedious method that is only of limited use. Not so—there have been a couple of developments that make SEC the preferred method of obtaining molecular weight data.

The Universal Calibration Curve

First, in perhaps the most important discovery in this field, Benoit and coworkers[50] recognized that separations occur on the basis of hydrodynamic volume rather than molecular weight and made the connection to intrinsic viscosity, which also depends

[50] H. Benoit, Z. Grubisic, P. Rempp, D. Decker, J. G. Zilliox, *J. Chim Phys.*, **63**, 1057 (1966); Z. Grubisic, P. Rempp, H. Benoit, *J. Polym Sci.*, *Part B*, **5**, 753 (1967).

upon the size of a molecule in solution and, as we have seen, can be related to the volume of an equivalent hydrodynamic sphere, V_h. You should recall that the intrinsic viscosity of a dilute solution is related to V_h and the molecular weight by the Equation 12-60:

$$[\eta] = 2.5V_h \left[\frac{A}{M} \right]$$

EQUATION 12-60

where A is Avogadro's number and M is the molecular weight of the polymer.

Rearranging Equation 12-60 tells you that the product of the molecular weight and the intrinsic viscosity of a polymer is proportional to the hydrodynamic volume (Equation 12-61):

$$[\eta]M = 2.5V_h A$$

EQUATION 12-61

Benoit et al. then reasoned that all polymers, regardless of chemical structure and chain architecture, should fit on the same plot of $[\eta]M$ versus elution volume. And most of them do, as shown on the plot in Figure 12-42, which is called the *universal calibration curve*. (It is actually not quite universal, as data from things like liquid crystalline polymers that have extended chain rather than coil conformations in solution do not fall on this curve.)

The utility of the universal calibration curve is that it allows us to calculate the molecular weight of a polydisperse polymer sample for which we do not have monodisperse standards. First, a universal calibration curve is prepared for a given set of columns using, say, monodisperse polystyrene samples. This is plotted as *log J* versus elution volume (Figure 12-43) where for the *i*th sample (Equation 12-62):

$$J_i = [\eta]_i M_i$$

EQUATION 12-62

The intrinsic viscosity of the samples can

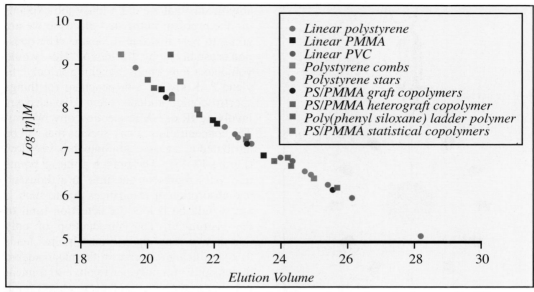

FIGURE 12-42 Universal plot of $log[\eta]M$ versus elution volume for a variety of polymers with different chemistries and architectures.

be measured directly or calculated if the Mark-Houwink-Sakaruda constants, K and a are known (for the same solvent and temperature used in the SEC experiments)—see Equations 12-63:

$$[\eta]_i = KM_i^a$$

$$J_i = [\eta]_i M_i = KM_i^{1+a}$$

$$M_i = \left[\frac{J_i}{K}\right]^{1/(1+a)}$$

EQUATIONS 12-63

SEC data for the unknown polydisperse sample can then be obtained on the same instrument using the same solvent and temperature; in other words we measure V_i and determine J_i. Then, as long as we know the values of the intrinsic viscosity parameters K and a for the unknown sample, the molecular weight distribution (the M_i's) and the number and weight average can be calculated.

This is wonderful stuff, but it would be even nicer if we did not have to calibrate the columns at all. This is now possible with new instruments that couple SEC with a light scattering apparatus (Figure 12-44). These have an array of detectors that are arranged around a flow cell and this allows the determination of the absolute molecular weight of the polymer in the eluant stream as it elutes from the column or columns.

LONG CHAIN BRANCHING

One thing you experience as a teacher is the ability of succeeding generations of students to give you the same wrong answer to the same question year after year. This prob-

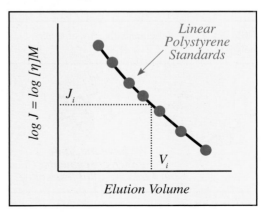

FIGURE 12-43 Schematic diagram of a universal calibration curve.

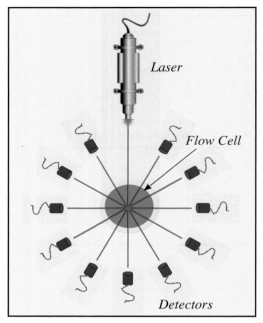

FIGURE 12-44 Schematic diagram of an instrument that couples SEC with a light scattering apparatus.

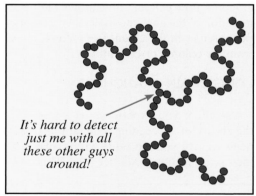

It's hard to detect just me with all these other guys around!

FIGURE 12-45 Schematic diagram of a star polymer.

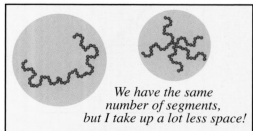

We have the same number of segments, but I take up a lot less space!

FIGURE 12-46 Schematic diagram of the size difference of a linear versus star polymer having the same number of segments.

ably means that we did a lousy job explaining the problem in the first place, so we are going to beat this one to death. The question is usually along the lines of "How would you detect long chain branching in polyethylene"? Knowing our penchant for things spectroscopic, students often give answers involving IR or NMR spectroscopy. Wrong! The concentration of any species that can be attributed to a long chain branch is very low (Figure 12-45). The branch point or points may only represent one unit in a thousand and absorption or resonances due to such a unit would be below the detection limit of the instrument. The introduction of only one or two long chain branch points leads to a significant decrease in the mean square dimensions of a polymer relative to a linear polymer of the same molecular weight (total number of segments), however, which means that such branching can be probed by scattering techniques or rheological/SEC methods; it is the latter that concerns us here.

The difference between the mean-square dimension of a polymer with long chain branching and an equivalent linear polymer with the same number of segments (Figure 12-46) can be expressed in terms of a parameter, g, equal to the ratio of the respective radii of gyration[51] (Equation 12-64):

$$g = \frac{\left\langle \overline{S}^2 \right\rangle_b}{\left\langle \overline{S}^2 \right\rangle_l}$$

EQUATION 12-64

where the subscripts b and l refer to branched and linear chains, respectively. (We didn't mention the radius of gyration in our discussion of conformations; it's simply a weight average measure of size of a polymer coil, $S = \sum m_i r_i / \sum m_i$, where m_i is the mass of segment i and r_i is its distance from the coil's center of gravity. It can be shown that $<S^2> = <R^2>/6$.

Statistical methods have been used to calculate the factor g for various branching

[51] W.H. Stockmayer and B. H. Zimm, *J. Chem. Phys.* **17**, 301 (1949)

architectures and some results are tabulated in the box below:

STAR-SHAPED POLYMERS WITH ARMS OF EQUAL LENGTH:

$$g = \frac{3}{f} - \frac{5}{f^2}$$

f = functionality of branch point

RANDOMLY BRANCHED MONODISPERSED POLYMERS:

$$g_3 = \left[\left(1 + \frac{\overline{m}_b}{7}\right)^{0.5} + \frac{4\overline{m}_b}{9\pi}\right]^{-0.5}$$

TRIFUNCTIONAL

$$g_4 = \left[\left(1 + \frac{\overline{m}_b}{6}\right)^{0.5} + \frac{4\overline{m}_b}{3\pi}\right]^{-0.5}$$

TETRAFUNCTIONAL

$\overline{m}_b$ = *number average number of branch points per molecule*

SEC has a major advantage over other techniques in the determination of long chain branching in that it essentially separates a sample into what can be considered monodisperse fractions. This allows these branching functions to be used in conjunction with the universal calibration concept. A lot of assumptions and approximations go into the methodology, however, so the results only give a relative number of long chain branches, which are nevertheless still useful. The procedure starts by introducing another branching function, g', this one related to the ratio of the intrinsic viscosities of the linear and branched polymers (Equation 12-65):

$$g' = \frac{[\eta]_b}{[\eta]_l}$$

EQUATION 12-65

The intrinsic viscosity of a branched sam-

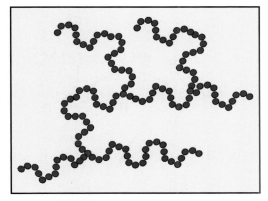

FIGURE 12-47 Schematic diagram of trifunctional branches.

ple is, of course always less than that of a linear chain containing the same number of segments, so that $g' < 1$. Also, g' is a function of the type and number of branches present (Figure 12-47). However, simple theoretical relationships between g and g' have not been developed and experimentalists have relied on empirical equations of the form (Equation 12-66):

$$g' = g^x$$

EQUATION 12-66

where x is typically in the range of 0.5 to 1.5. For randomly branched polymers it has been found that $x = 0.6$. The relative extent of long chain branching can now be determined using an ingenious method developed by Kurata et al.[52] This relies on the fact that for a branched sample the value of the experimentally determined intrinsic viscosity $[\eta]_{obs}$, will always be less than that determined from the universal calibration curve $[\eta]_{cal}$ using Equation 12-67:

$$[\eta] = \sum_i w_i [\eta]_i = \sum_i w_i \frac{J_i}{M_i}$$

EQUATION 12-67

[52] M. Kurata, H. Okamoto, M. Iwama, M. Abe, T. Homma, *Polym. J.* **3**, 729 (1972) and **3**, 739 (1972).

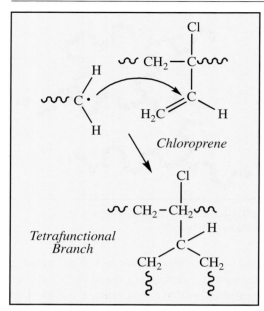

FIGURE 12-48 The formation of long chain branches in polychloroprene.

because this equation implicitly assumes the chains are linear.

A branching function $g'(\lambda, M)$ is now introduced (Equation 12-68):

$$[\eta]_b = g'(\lambda, M)[\eta]_l = g'(\lambda, M)KM^a$$

EQUATION 12-68

where λ is a branching parameter and M is the molecular weight of the chain (remember, we are going to consider SEC fractions that are assumed to be monodisperse). The branching parameter is simply equal to the (number) average number of branches per molecule per unit molecular weight (Equation 12-69):

$$\lambda = \frac{\overline{m}_b}{M}$$

EQUATION 12-69

The function g can now be expressed in terms of λ and then for a randomly branched sample g' can be obtained from $g' = g^{0.6}$. But by now you're probably lost, so we will back

off and show you how the whole procedure works by considering a concrete example— long chain branching in polychloroprene.[53]

Long-Chain Branching in Polychloroprene

When polychloroprene is synthesized free radically, branching occurs as a result of addition of a growing chain to the double bond contained in the minor amounts of 1,2- and 3,4- units present in chains formed up to this point in the synthesis. This branching process is favored at high degrees of conversion, when the monomer concentration is depleted (Figure 12-48). (In commercial processes the the polymer is actually isolated before 70% of the monomer is converted, otherwise gelation occurs.) Because the extent of branching increases as a function of conversion, it is possible to illustrate the SEC/$[\eta]$ method by isolating samples obtained from a reaction vessel at various stages of the polymerization. One of your authors actually did these experiments, so let's consider the procedure a step at a time.

The first thing you have to do is get yourself some linear polychloroprene standards. This was achieved by polymerizing chloroprene at low temperatures ($\leq$–20°C) and isolating the polymer after <5% conversion. Standards of narrow polydispersity were then obtained by fractionation of these poly(chloroprenes), using something called an automatic belt fractionator. These fractions were then characterized by light scattering to determine their molecular weight and their intrinsic viscosities (in THF at 30°C) were measured. The Mark-Houwink-Sakaruda viscosity parameters K and a were then determined from a plot of $log[\eta]$ versus molecular weight ($K = 4.18 \times 10^{-5}$ dl/g and $a = 0.83$)—see Figure 12-49.

The next step is to make sure the data fits a universal calibration curve prepared using polystyrene standards (also in THF at 30°C). The plot of $J = M[\eta]$ versus elution volume shown in Figure 12-50 shows that the fit is

[53] M. M. Coleman and R. E. Fuller, *J. Macromol. Sci.-Phys.*, **B11**, 419, (1975).

satisfactory. Each point on the normalized SEC curve then yields a weight fraction, w_i, also as a function of elution volume—see Figure 12-39. Now we get to the mathematical bit, so concentrate!

The idea is to use the SEC curve to calculate what would be the intrinsic viscosity of the whole sample by summing the intrinsic viscosities of all the individual fractions. This is done by taking each point on the SEC distribution and for each value of V_i obtain $J_i = M_i[\eta]_i$ from the universal calibration curve. We have assumed that the intrinsic viscosity of a branched polymer is equal to a branching function g', expressed in terms of a parameter λ equal to the (number) average number of branch points per unit molecular weight (see Equation 12-68).

We have presented equations for $g' = g^{0.6}$ and g previously and substituting we get the following two equations, the first relating $J_i = M_i[\eta]_i$ for each fraction (Equation 12-69):

$$[\eta]_i M_i = \\ KM_i^{(1+a)}\left[\left(1 + \frac{\lambda M_i}{6}\right)^{0.5} + \frac{4\lambda M_i}{3\pi}\right]^{-0.3}$$

EQUATION 12-69

The second sums the weighted values of $[\eta]_i$ to give the intrinsic viscosity of the whole sample (Equations 12-70):

$$[\eta]_b = \sum_i w_i[\eta]_i = \\ K\sum_i w_i M_i^a\left[\left(1 + \frac{\lambda M_i}{6}\right)^{0.5} + \frac{4\lambda M_i}{3\pi}\right]^{-0.3}$$

EQUATION 12-70

The value of the intrinsic viscosity of the sample calculated in this fashion, $[\eta]_{calc}$, can then be compared to that obtained experimentally for the whole sample, $[\eta]$. But you need a value of λ to do these calculations. So what you actually do is first calculate the intrinsic viscosity for the sample as if it were *linear*, i.e., $\lambda = 0$. Then, if the cal-

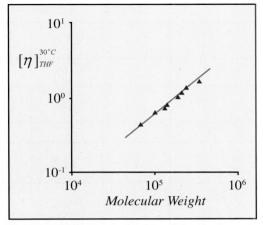

FIGURE 12-49 Determination of K and a for polychloroprene in THF at 30°C.

culated and observed values are identical, it is safe to say the samples are indeed linear. That is what was observed for polychloroprene samples obtained at conversions of 11.7% and 33.6%, as shown in Table 12-4. At higher values of conversion the calculated value of $[\eta]_{\lambda = 0}$ no longer agrees with the experimental value, however. Now what you do is apply an iterative procedure to the equations on the preceding page to obtain a value of λ that brings the calculated and observed values into agreement. Values of the branching parameter are given in Table

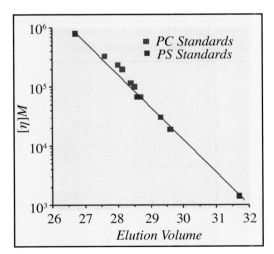

FIGURE 12-50 Universal plot of $log[\eta]M$ versus elution volume for linear polychloroprene (PC) and polystyrene (PS) standards.

TABLE 12-4 BRANCHING DATA FOR POLYCHLOROPRENE

CONVERSION (%)	11.7	33.6	55.9	62.8	71.9	82.3
$[\eta]_{THF}^{30°C}$	1.50_0	1.49_2	1.47_8	1.49_0	1.52_8	1.54_1
Calculated $[\eta]_{\lambda=0}$	1.49	1.48	1.60	1.67	1.82	1.92
$\lambda \times 10^5$	-	-	0.15	0.23	0.36	0.52
$\overline{M}_w$ ($\times 10^{-5}$)	3.25	3.25	4.05	4.44	5.49	6.15
$\overline{M}_n$ ($\times 10^{-5}$)	1.44	1.44	1.19	1.07	1.05	1.26
Poly-dispersity	2.3	2.3	3.4	4.2	5.2	4.9
γ	0	0	0.38	0.50	0.66	0.76

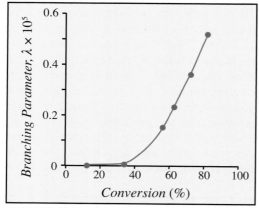

FIGURE 12-51 Plot of the branching parameter, λ, versus conversion for polychloroprene.

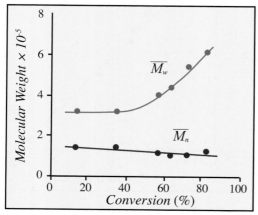

FIGURE 12-52 Plot of molecular weight averages versus conversion for polychloroprene.

12-4, as are other quantities that we will get to shortly.

Values of λ as a function of conversion are shown in Figure 12-51, while calculated values of the number and weight average molecular weights of the sample are displayed in Figure 12-52. At conversions greater than about 40% the branching parameter increases rapidly towards the point of incipient gelation. Note that the weight average molecular weight also increases rapidly for conversions >40%, while the number average remains almost constant, indicating that the polydispersity also increases with conversion as branching becomes significant.

Finally there is another parameter, γ, introduced independently by Stockmayer and Kilb, that is particularly useful for diene-type polymerizations, where the onset of gelation in a reactor can have important if not disastrous ramifications. This parameter, listed in Table 12-4, is defined by (Equation 12-71):

$$\lambda \overline{M}_w = \frac{\gamma}{1-\gamma}$$

EQUATION 12-71

If you play around with this equation you will see that γ is zero for a completely linear polymer and approaches values of unity at

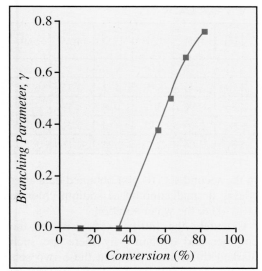

FIGURE 12-53 Plot of the branching parameter, λ, versus conversion for polychloroprene.

the incipient gel point. It is used as an indicator of how close you are to the gel point and it can be seen in Figure 12-53 that above 70% conversion polychloroprene is pretty damn close!

RECOMMENDED READING

H. G. Barth and J. W. Mays, *Modern Methods of Polymer Characterization*, Wiley-Interscience, New York, 1991.

P. J. Flory, *Principles of Polymer Chemistry*, Cornell University Press, Ithaca, New York, 1953.

D. Margerison and G. C. East, *Introduction to Polymer Chemistry*, Pergamon Press, Oxford, 1967.

L. H. Sperling, *Physical Polymer Science*, 3rd Edition, Wiley, New York, 2001.

STUDY QUESTIONS

1. The osmotic pressures at a series of concentrations from 2–10g of polymer per kilogram of solution of a fractionated vinyl chloride polymer in methyl amyl ketone at 27°C

have been determined. The experimental data are as follows:

π (cm of solvent)	0.4	1.1	2	3	4
c (g solute/kg solution)	2	4	6	8	10

Plot π/c versus c and determine the molecular weight of the polymer. (Pay careful attention to units!). Is this a number average or weight average? (Note ϱ of methyl amyl ketone = 0.8 g/cc)

2. The following data were obtained in terms of degree of polymerization versus the intrinsic viscosity of poly(methyl methacrylate) samples (with low monodispersities) in benzene at 25°C:

Degree of Polymerization	$[\eta]$ (dl/g)
700	0.334
1450	0.678
2240	0.929
3940	1.153
4080	1.305
9800	2.510

The molecular weight of the monomer is 100. Determine the constants K and a.

3. You are now given a sample of PMMA of unknown molecular weight. You determine that it has an intrinsic viscosity of 1.04 dl/g. What is its molecular weight? What average have you determined? If your polymer has the most probable molecular weight distribution, what would be its number and weight average molecular weight?

4. Consider the following values of $K(1 + \cos 2\theta)c/R_\theta$ (mole/g × 10^6 – i.e., the values in Table 12-5 on the following page are 3.18 × 10^{-6}, etc.) obtained in a light scattering experiment. The data were taken from the wonderful little book by Margerison and East

TABLE 12-5 LIGHT SCATTERING DATA

$\frac{c}{(g/cm^3)}$ θ	30°	37.5°	45°	60°	75°	90°	105°	120°	135°	142.5°	150
2.00×10^{-3}	3.18	3.26	3.25	3.45	3.56	3.72	3.78	4.01	4.16	4.21	4.22
1.50×10^{-3}	2.73	2.76	2.81	2.94	3.08	3.27	3.40	3.57	3.72	3.75	3.78
1.00×10^{-3}	2.29	2.33	2.37	2.53	2.66	2.85	2.96	3.12	3.29	3.38	3.37
0.75×10^{-3}	2.10	2.14	2.17	2.32	2.47	2.64	2.79	2.93	3.10	3.21	3.20
0.50×10^{-3}	1.92	1.95	1.98	2.16	2.33	2.51	2.66	2.79	2.96	3.11	3.12

(see suggestions for further reading) and are tabulated as a function of c and θ. Construct a Zimm plot and calculate $\overline{M}_w$.

This question kills our students every year, even though we go over the methodology in class. The data points in the table actually give the dots plotted in Figure 12-22. You need to get the squares. First take each row and plot the values of $K(1 + \cos2\theta)c/R_\theta$ versus $\sin^2(\theta/2)$. Extrapolate to $\theta = 0$. This gives you the data points shown as the squares that define the $\theta = 0$ line. You figure out how to get the $c = 0$ line and hence $\overline{M}_w$!

5. Because your polymer science professor wants to make you suffer as much as he or she did going through graduate school, you are forced to do old-fashioned viscosity experiments using an Ubbelholde viscometer.

Concentration (g/dl)	Time (secs)	
	(I)	(II)
0.40	743.3	515.9
0.35	711.3	475.1
0.30	673.3	440.2
0.25	620.0	402.5
0.20	583.0	373.1
0.15	542.2	346.7
0.10	466.4	323.6
0.05	367.3	306.1
Deionized water	291.5	291.5

In the first set of experiments (denoted I in the table above), you obtain data from a solution of a polyelectrolyte in deionized water.

In the second set (II) you obtained data after adding a small amount of sodium chloride (10^{-3} M) to the water solution.

Calculate the intrinsic viscosity of the polymer and explain why there are such marked differences between these two sets of data. Because this is "real" data the results may be equivocal. You will be expected to comment on your conclusions and be critical in your answer!

6. Below is intensity versus elution volume data that was recorded from a GPC curve of a sample of Polymer A in THF at 30°C.

Elution Volume	Intensity or Height	Elution Volume	Intensity or Height
32.5	0.0	26.0	147.1
32.0	0.8	25.5	120.0
31.5	4.5	25.0	88.0
31.0	15.0	24.5	67.8
30.5	30.2	24.0	50.1
30.0	60.0	23.5	36.2
29.5	105.5	23.0	22.1
29.0	165.3	22.5	12.3
28.5	225.1	22.0	6.9
28.0	256.0	21.5	2.0
27.5	250.3	21.0	0.1
27.0	220.6	20.5	0.0
26.5	181.5		

Under identical experimental conditions six monodisperse samples of polystyrene having molecular weights of 2,000,000, 850,000, 470,000, 105,000, 52,000 and 11,000 g/mole eluted at elution volumes of 23.3, 25.4, 26.7, 29.3, 30.3 and 32.2, respectively. The Mark-Houwink-Sakurada constants for polystyrene in THF at 30°C are $K = 3.71 \times 10^{-4}$ dl/g and $a = 0.64$.

A. Assume that the GPC curve of Polymer A is that of a polydisperse linear polystyrene. Calculate the weight and number molecular weight averages, the polydispersity and the intrinsic viscosity of the polystyrene.

B. Assume that the GPC curve of Polymer A is that of a polydisperse linear polychloroprene with Mark-Houwink-Sakurada constants in THF at 30°C of $K = 4.18 \times 10^{-5}$ dl/g and $a = 0.83$. Calculate the weight and number molecular weight averages, the polydispersity and the intrinsic viscosity of this polychloroprene.

C. Assume that the GPC curve of Polymer A is that of a polydisperse, tetrafunctionally star-branched polychloroprene (i.e. a polydisperse polymer made up of different molecular weight star-shaped polymers having four arms of equal length). Calculate the weight and number molecular weight averages, the polydispersity and the intrinsic viscosity of this star-branched polychloroprene.

D. To calculate the molecular weight averages and intrinsic viscosity you will have to extrapolate the calibration curves outside the range of the standards in order to obtain appropriate values of M_i or J_i at elution volumes that are less than 23.3. Would you expect this to introduce significant errors? Test your answer by calculations using different assumptions.

7. You want to make a star-like polymer with four branches. First, you polymerize styrene using a free radical polymerization process in the presence of the chain transfer agent bromotrichloromethane, CCl_3Br. This gives you a relatively low molecular weight product where essentially all of the chains have one Br end group. After washing and so on, you dissolve the polymer in a suitable solvent and add a stoichiometric amount of a tetrafunctional amine (based upon the concentration of Br groups). How would you attempt to determine whether or not you had indeed made a four-branched star polymer?

8. Briefly describe how SEC (size exclusion chromatography) separates polymers according to size. Explain how the universal calibration curve can be used to determine molecular weight.

If you are attempting to separate a sample consisting of two polymers each having the same molecular weight, but one is linear and the other highly branched, which would you expect to be eluted from the column first?

9. We told you this one was coming. A GPC curve such as that shown in Figure 12-39 can be cut into slices. As mentioned in the text, the molecular weight of the ith species, M_i, is obtained from the calibration curve at point V_i. Show that the number average molecular weight can be calculated from:

$$\overline{M}_n = \frac{1}{\sum_i \frac{w_i}{M_i}}$$

13

Mechanical and Rheological Properties

INTRODUCTION

For some reason, many students with backgrounds in chemistry or physics, who start studying polymers because they are interested in synthesis or structure, seem to be uninterested or even allergic to mechanical and rheological properties, perhaps regarding the subject as some branch of engineering that doesn't concern them. If you are one of these, dear reader, we hope to persuade you that this is not only an interesting and intellectually demanding subject in its own right, but that you need to know something about this field, because getting things wrong can have catastrophic consequences, as in the failure of Firestone tires (Figure 13-1) a few years ago.

An even more spectacular example of catastrophic failure involved the space shuttle *Challenger* (Figure 13-2). The disaster was due to problems with the "O-rings." Each segment of the booster rocket is sealed with two of these. Upon launch, the metal casing of the booster bows outwards. The O-rings should have responded to this by expanding and sealing the joint. But that day it was much colder than usual. The rubber was much less flexible (remember T_g?) and the joint didn't seal. Hot gases escaped, these were ignited, and the explosion followed.

The retarded elastic response displayed by the O-rings is an example of viscoelastic behavior, something everybody has come across in everyday life, usually without real-

FIGURE 13-1 A car tire that has seen better times!

FIGURE 13-2 An unmitigated disaster: the horrible explosion of the space shuttle *Challenger* (Source: NASA).

izing it. This ignorance can persist throughout college, as many students in the physical sciences study the properties of solids and liquids in entirely separate courses, usually in mechanics on the one hand and fluid mechanics on the other. This is OK if all you really want to know is the basics: Hooke's law for elastic solids, Newton's law for simple fluids in laminar flow, and so on.

HOOKE'S LAW

FOR ELASTIC SOLIDS

$$\sigma = E\varepsilon$$

NEWTON'S LAW

FOR SIMPLE LIQUIDS

$$\tau_{xy} = \eta\dot{\gamma}$$

But it's interesting to note that the Greek philosopher Heraclitus described rheology as *panta rei* (everything flows), you simply have to wait long enough (think of the earth's crust). When dealing with steel, for example, under low loads and in ordinary time frames, you don't have to worry about this a lot, but for polymers you always have to keep viscoelastic properties in mind.

Of course, we haven't explained precisely what we mean by viscoelasticity yet and we won't for a while. We are going to approach the subject in the conventional way, first by looking at the "elastic" properties of polymer "solids," then the rheological properties of polymer melts. This will remind you of some basic stuff you should know, but may have forgot, or, if you've been really sneaky, managed to avoid altogether.

MECHANICAL PROPERTIES— DEFINITIONS

Background

First let's consider the mechanical properties of polymers treated as solids. By this we mean that we will discuss the usual properties, strength, stiffness and toughness, although not necessarily in that order. We'll

start by looking at the subject in a historical context, as it always helps to have a feel for how ideas developed, rather than have them plonked on you as a matter of received wisdom.

If you ever wander around Europe, as did your authors collecting stories for some of our tales on polymer materials (it would kill us to have real jobs), you will immediately be struck by the size and magnificence of the cathedrals, most constructed in medieval times. You may be astonished to learn that these were not constructed using elaborate calculations of allowed stresses and strains, but on the basis of empirical "rules of thumb" that relied on experience and a traditional knowledge of appropriate proportions. Of course, they occasionally got things wrong, but the buildings that fell down aren't there to see. People would either start over or use the rubble to build something else.

Medieval stonemasons had a good excuse for ignoring theory—there wasn't any! But even when there was, practical engineers held it in contempt well into the 19th century, sticking to their rules of thumb (perhaps this is why at one point it was estimated that railway bridges in the United States were falling down at the rate of about 25 a year[54]). But we're jumping ahead of ourselves here, let's start with the first serious studies of the strength of materials, which were made by Galileo (Figure 13-3).

Galileo is, of course, most famous for his work on astronomy, which brought him to the attention of the Inquisition. This relentless body made it clear that they had some serious objections to his views and were quite prepared to do very nasty things unless he recanted. Discretion being the better part of valor, Galileo took up the intellectually safer pursuit of the strength of materials. He observed that a rod placed under a load has a strength that is proportional to its cross-sectional area. In other words, he almost came

---

[54] We got this statistic and a lot of the good historical stuff that follows from two marvelous books by J. E. Gordon, *The New Science of Strong Materials* and *Structures*, both published by Penguin Books in 1991.

up with the concept of stress. This may seem a bit trivial now, but Galileo was addressing difficult questions (in the context of the times) involving how an inanimate object supports a load and how you separate the mechanical properties of an object or structure that are a consequence of its shape from those that are due to the material from which it is constructed.

The next great scientist to address these problems was Hooke (Figure 13-4), a cantankerous and brilliant contemporary of Newton's (there is a long list of inventions and discoveries that can be attached to Hooke's name). So, the two laws we will be concerned with, Hooke's law for elastic solids and Newton's law for fluids, both had their genesis in the same time period, the latter part of the 17th century. But let's stick to solids, for now.

Hooke realized that a material or structure resists a load by pushing back with an equal and opposite force (Newton's third law). He also noted that this force is generated by deflections—an object under a load changes its shape. Crucially, he also realized that not only does a structure change shape when it is stretched or compressed, *so does the material from which it is constructed.* It is such *material* properties that concern us here.

We now know that the mechanical response of a material is a consequence of its atomic structure and the nature of its chemical bonds. Atoms strongly repel when pushed too close together as their orbitals overlap. Similarly, the attractive forces between them act against their being stretched apart. Of course, Hooke knew nothing about chemical bonds, his approach was a systematic experimental exploration of the relationship between forces and deflections in solids.

He studied various things: beams, wires springs, and out of these observations came his famous law "*ut tensio sic vis,*" as the extension so the force.

Like many a great man, Hooke was not overly modest and actually first published this law in 1676 as a Latin anagram, "*ceiiinossssttuv,*" in a book titled *A decimate of the centesme of the inventions I intend to*

FIGURE 13-3 Galileo (reproduced with the kind permission of the Edgar Fahs Smitt Collection of the University of Pennsylvania Library).

FIGURE 13-4 There is no surviving portrait of Hooke. This commemorative window in Saint Helen's, Bishopsgate, a formulaic representation of his likeness, also no longer exists; it was destroyed by an IRA bomb.

FIGURE 13-5 Newton (reproduced with the kind permission of the Edgar Fahs Smitt Collection of the University of Pennsylvania Library).

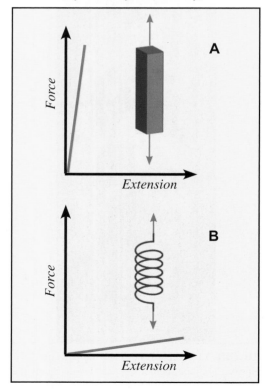

FIGURE 13-6 Schematic plots of force versus extension for (A) rectangular block and (B) spring.

publish, where Hooke sought to lay claim to priority in a number of fields. The world had to make of this what it could until he published a paper on the subject in 1679. But, among other things (like the nature of light and color), it was claims of priority (in the inverse square law of gravitation) that bought Hooke into conflict with Newton (Figure 13-5). It probably didn't help that the two men had very different personalities. Hooke was involved in various scandalous liaisons, including one with his niece, while Newton seemed to spend his spare time in strange and curious pursuits, like speculating on the Number of the Beast. It is safe to say that Newton despised Hooke and he could be implacable in his enmity.

Newton lived longer than Hooke and continued to denigrate Hooke's work after his death. Newton's vast influence and shadow thus lay over the field, which lay fallow for many years. This didn't help, but the main reason for the lack of progress probably had more to do with the fact that Hooke's law in its force/extension form doesn't distinguish between the properties of a structure and those of the material from which it is made. You can take a piece of steel, for example, and in the form of a rectangular block it is a very stiff object—it doesn't stretch very much under a load. The same piece of steel shaped into a helical spring would extend significantly under the same load, however (Figure 13-6).

This brings us to the concepts of stress and strain and their ratio, which we call Young's modulus. Like Hooke, Young (Figure 13-7) was a brilliant scientist who made major contributions to many fields (e.g., the wavelike nature of light, color perception), but his communication skills were somewhere between abysmal and incomprehensible. His definition of the modulus (~1800) given his name reads:

The modulus of the elasticity of any substance is a column of the same substance, capable of producing a pressure on its base which is to the weight causing a certain degree of compression as the length of the substance is to the diminution of its length.

Amazingly enough, later in life he made major contributions to deciphering Egyptian hieroglyphics. After reading his own work, hieroglyphics probably seemed easy. But, to be fair to Young, he was wrestling with a concept that needed the definition of stress and strain to be expressed in a simple form. These definitions were not to come until some years later, when Cauchy perceived that the load on an object divided by its cross-sectional area is a measure of the force at any point inside a material, rather like how we describe the pressure exerted by a gas on the walls of its container. Strain was defined as the normalized extension, the change in length divided by the original length of the object.

Building on this, Navier finally came up with the modern definition of Young's modulus, stress divided by strain, in 1826, and we now write Hooke's law in the form of Equation 13-1:

FIGURE 13-7 Young (reproduced with the kind permission of the Edgar Fahs Smitt Collection of the University of Pennsylvania Library).

$$\sigma = E\varepsilon$$
Stress = Young's Modulus × Strain
EQUATION 13-1

There are a couple of things about this relationship. First of all it is only an approximation. We'll get back to that in a while. Second, we have only considered simple elongation so far. There is a modulus associated with shear and also a bulk modulus. The most important point, however, is that the modulus determined this way, dividing stress by strain, is a material property and independent of the shape of an object. It is what we mean when we talk about the stiffness of a material. Stiffness is crucial in many engineering applications. If a strain of just 1.6% were allowed in an aircraft's wing spar booms, for example, it would look something like Figure 13-8.

Having reviewed the preliminary essentials, we can now go on and discuss some interesting stuff: the strength, stiffness and toughness of polymers and, for comparison, some other common materials like steel and glass. But there are some other definitions you will need, like shear stress, shear strain and compliance. In case you forgot (or never knew), compliance is simply strain divided by stress, so for a time-independent experiment, like the extension of a perfectly elas-

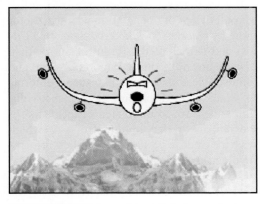

FIGURE 13-8 This would not help those who have a fear of flying!

tic solid, the compliance D is equal to $1/E$ (Equation 13-2):

$$D = \frac{\varepsilon}{\sigma} = \frac{1}{E}$$

EQUATION 13-2

This might seem trivial at this point, but just you wait until we get to viscoelastic behavior, where you will find that D is not equal to $1/E$! Bet you can't wait. In any event, if you've forgotten all this stuff or never knew it to begin with, you'd better not skip the rest of the definitions! First, we'll make some brief remarks about units.

An Aside on Units

English speaking peoples can be a perverse and idiosyncratic lot, preferring good old units like pounds per square inch to describe things like pressure and stress, rather than modern abominations like Pascals (Newtons/m²). We are as guilty as most, no doubt a lingering prejudice of our early education. Back then, in the dark ages, we had to take irrelevant academic subjects like History and Geography, rather than the far more useful things taught today's youth, like Drivers Ed. and Self Esteem. In any event, at some point we had to learn about the French Revolution. We also had to learn about the American Revolution, but we seem to recall that being taught as some sort of strategic withdrawal.

But we digress. The view of most English schoolboys of our generation was shaped far more by Charles Dickens (*A Tale of Two Cities*) and stories of the Reign of Terror than discussions of the more important social and philosophical changes that ensued. In terms of what we're discussing here, the revolution gave birth to three significant plans for the reform of measurement, one of which was the metric system (the other reforms involved the calendar and units of time and never caught on). The logical extension of the metric system is the modern SI (*Système International des Unités*), a meter-kilogram-second system of units.

The metric system was initially spread to other European countries in the wake of Napoleon's conquests, but did not start to became "common coin" for scientific and technological work until after the meeting of *The General Conference on Weights and Measures* in Paris in 1960 (boy, does that sound like a fun conference).

Accordingly, we will not be entirely consistent in our use of units, because some of the original data we will use was reported in pounds and inches or cgs units. So we we'd better discuss these units and how to switch between them.

Stress is force per unit area and this can create confusion right off the bat, because we have to distinguish between units of mass and weight. Originally the kilogram was a unit of weight. It was defined in terms of the force of gravity acting on a prototype platinum-iridium alloy cylinder presently stored under triple bell jars in Sèvres, France. (And in spite of all the precautions, traces of gunk are building up on its surface.) In the SI system, based on superior Gallic logic, the kilogram is now a unit of mass, of course. The unit of force is thus a kg m/sec² and is called a Newton (N). The unit of stress is then just a N/m² and this is called a Pascal (Pa). The antiquated British and US systems use lbs as a unit of weight so stress is in terms of lbs/square inch, or psi. We're not going to tell you what stress is in cgs units, because we just don't feel like it!

People are loathe to give up old and familiar units, because they assimilate an intuitive

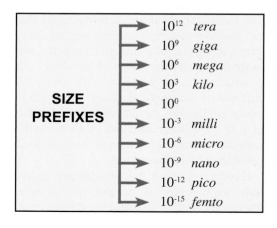

	10^{12} *tera*
	10^{9} *giga*
	10^{6} *mega*
	10^{3} *kilo*
SIZE	10^{0}
PREFIXES	10^{-3} *milli*
	10^{-6} *micro*
	10^{-9} *nano*
	10^{-12} *pico*
	10^{-15} *femto*

feel for the size of something as a result of accumulated experience. (How much effort it takes to walk a mile to the pub and how much beer is in the pint that you need to consume once you get there in order to restore your strength.)

And so it is with stress; we have a feel for the type of pressure or stress exerted by a 1 lb weight on a square inch. But a Pascal—give us a break! In fact a Pascal is not a very big stress at all, mainly because the units of area are meters squared—much bigger than a square inch. So in order to avoid dealing with embarrassingly large numbers we usually talk about Megapascals, MPa (10^6 Pascals) or Gigapascals, GPa (10^9 Pascals). The bodies that decide these things use steps of 1000 (10^3) in counting multiples (adopting prefixes from Greek roots) and submultiples (using prefixes from Latin roots)—see box at the bottom of the previous column.

Having meandered about a bit for our own and hopefully your amusement, we can finally get to the point, the relationship between various units that we will use in our discussion of mechanical and rheological properties that are given in the box below:

UNITS

Stress   1 Pa = 10 $dynes/cm^2$
                1 MPa = 145 psi

Strain
We don't have to worry about units of strain, because there aren't any (being a length divided by a length).

Modulus
It follows that units of modulus are the same as units of stress ($\sigma = E\varepsilon$).

Viscosity
We haven't discussed viscosity yet, but the units are named after the French physician and scientist Poiseuille.

1 P (*poise—cgs unit*) = 1 $dyne\ sec/cm^2$
1 Pl (*Poiseuille—SI unit*) = 10 P

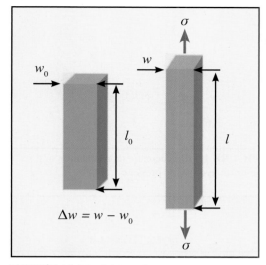

FIGURE 13-9 Definitions for Poisson's ratio.

Poisson's Ratio

In our introductory material, we talked about stresses and strains in simple stretching or tensile experiments. We neglected the fact that if you stretch a material lengthways, it will also contract in the direction perpendicular to the applied stress. The amount of this contraction is proportional to the extension (Figure 13-9) and we can write Equations 13-3:

$$\frac{\Delta w}{w_0} \propto \frac{\Delta l}{l_0} \quad or \quad \frac{\Delta w}{w_0} = \upsilon \frac{\Delta l}{l_0}$$

EQUATION 13-3

where υ is Poisson's ratio. It turns out that if we are considering isotropic, homogeneous materials, then the tensile modulus, E, and Poisson's ratio, υ, are the only constants we need.

Bulk Modulus

For example, the bulk modulus, B, which is what you measure if you subject a material to a uniform hydrostatic pressure (Figure 13-10), is defined by Equation 13-4:

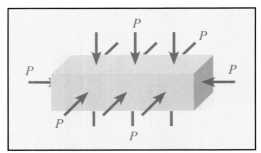

FIGURE 13-10 Schematic representation of a uniform hydrostatic pressure.

$$P = -B \frac{\Delta V}{\Delta V_0}$$

EQUATION 13-4

Where ΔV is the change in volume. The relationship between B, E, and v is then (Equation 13-5):

$$B = \frac{E}{3(1 - 2v)}$$

EQUATION 13-5

Shear Modulus

The jump from uniaxial elongation to uniform compression is a simple one in terms of defining all the stresses and strains. The final modulus we wish to define, the *shear modulus*, τ, is a little different and you have to pay

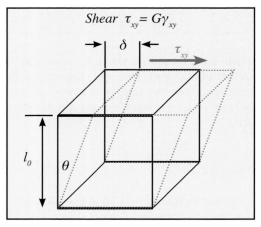

FIGURE 13-11 Schematic diagram depicting a material under shear.

attention to the direction of the forces and displacements. When you shear a material you exert a force parallel to one of its faces (Figure 13-11). It's like putting a book down on a table and pushing with your hand horizontally on the book, parallel to the cover and the table. The shear stress is defined as the component of the force parallel to the face on which you're pushing *divided by the area of that face* (if you press on your book you also have to exert a little bit of downward pressure in order to obtain a horizontal force). Note that in simple extension the force is applied perpendicular to a surface, here it is parallel and the stress is defined correspondingly.

Accordingly, our shear stress in the *xy*-plane is given by Equation 13-6:

$$\tau_{xy} = \frac{F}{A_{xy}}$$

EQUATION 13-6

You also have to think about how you define the strain, because now both the "vertical" and "horizontal" dimensions of the object are changing. The most convenient way is to define the strain in terms of the angle through which the material is twisted (Equation 13-7):

$$tan\,\theta = \frac{\delta}{l_0} = \gamma_{xy}$$

EQUATION 13-7

Hooke's law for shear stress and shear strain is then (Equation 13-8):

$$\tau_{xy} = G\gamma_{xy}$$

EQUATION 13-8

And the shear compliance, *J*, is given by Equation 13-9:

$$J = \frac{\gamma_{xy}}{\tau_{xy}}$$

EQUATION 13-9

The relationship between the shear modulus, Poisson's ratio and the tensile modulus is given by Equation 13-10:

$$G = \frac{E}{2(1 + v)}$$
EQUATION 13-10

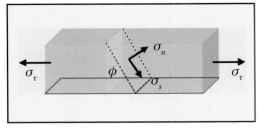

FIGURE 13-12 Schematic diagram depicting shear forces for a material under tension.

Keep in mind that this is for a uniform isotropic material. A lot of materials are not isotropic. Crystalline materials can be weaker or stronger along one crystallographic plane than another depending on the arrangement of atoms, relative strength of the bonds and the presence of defects. The modulus in each of these directions can also be different and we would have to write Hooke's law as (Equation 13-11):

$$\tau_{xy} = G_{xy}\gamma_{xy}$$
EQUATION 13-11

As you might guess, when describing the properties of anisotropic materials you now get into tensors and all sorts of horrible looking equations. Don't worry, we're not going there! We'll just finish this little review by considering a subtle feature of shear forces.

Shear Forces in a Material under Tension

It is important to realize that there are shear forces present in a material even when it is subjected to a simple tensile elongation. Consider an arbitrary plane in a rectangular specimen subjected to a tensile stress. Clearly, there are non-zero components of the load or force in directions parallel and perpendicular to this plane that result in a shear stress, σ_s and a normal stress σ_n (Figure 13-12 and Equations 13-12):

$$\sigma_s = \frac{F}{A} cos\,\phi\,sin\,\phi = \frac{F}{2A} sin^2\phi$$

$$\sigma_n = \frac{F cos\,\phi}{l^2 / cos\,\phi} = \frac{F}{A} cos^2\phi$$

EQUATIONS 13-12

The shear stress will be a maximum when $\phi = 45°$ and (Equation 13-13):

$$\sigma_s = \sigma_n = \frac{F}{2A}$$
EQUATION 13-13

That is why you see so-called shear bands in some polymers when they are stretched beyond the yield point (Figure 13-13). Don't worry if you don't know what the yield point is—we'll get to it! But if you're awake you should immediately grasp that these have something to do with shear forces being a maximum at angles of ± 45°.

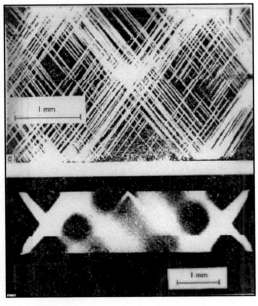

FIGURE 13-13 Shear bands [reproduced with permission from P. B. Bowdon, *Philos. Mag.*, **22**, 463 (1970)].

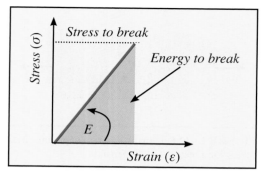

FIGURE 13-14 Schematic diagram of a stress/strain plot for an elastic material.

ELASTIC PROPERTIES OF MATERIALS

If materials were perfectly elastic, obeying Hooke's law up until they broke, then simple plots of experimentally obtained stress/strain data (Figure 13-14) would tell us most of what we would need to know: the strength (stress to break), stiffness (modulus, the slope of the line) and "toughness" of a material (given by the area under the curve, the strain energy or energy to break). Real materials are far more complex and interesting in their behavior, however, and we also have to consider things like yielding, creep, stress relaxation, fatigue and so on. We will get to some of this later, but first we will focus on the strength, stiffness and toughness of materials in general, as most students, particularly

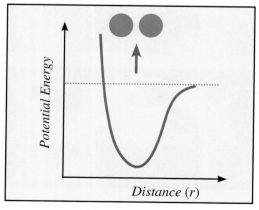

FIGURE 13-15 Schematic diagram of potential energy versus distance between two atoms.

those with a chemical bias, have a poor feel for these properties and their relationship to structure. This will provide a foundation for discussing the specifics of polymer mechanical properties. Let's start by reexamining Hooke's law.

Hooke's Law

It is important to realize that even if we had a perfectly ordered material with no defects whatsoever, then we would still get deviations from ideal linear behavior. To see why this is so, let's consider a simple model, a 1-dimensional array of atoms linked together by ordinary chemical bonds.

Now let's look at the forces between just two of these atoms taken in isolation (Figure 13-15 — we discussed this in Chapter 8, page 208). When the atoms are too close together they strongly repel. When they are more than a certain distance apart the force of attraction is small. There is some optimum distance apart where the potential energy (PE) is a minimum, which is to say it has its largest negative value and where the balance between repulsive and attractive forces is most heavily tilted in favor of the latter.

When arranged in a lattice, however, the atoms have neighbors on both sides and if you sum the contributions from each of these then locally the potential energy curve has a "U" shape (Figure 13-16) that at least near the bottom part of the curve can be approximated by a simple quadratic function of the form (Equation 13-14):

$$PE = k'' x^2 = \frac{k}{2} x^2$$

EQUATION 13-14

where x is the displacement of the bond from its minimum energy position and the factors k'' and k are constants (we use $k/2$ to get our final answer in a simple form). However, this is only an approximation and the PE can be represented more accurately by a power series (Equation 13-15):

$$PE = \frac{1}{2}kx^2 + \frac{1}{3}k'x^3 + \ldots$$

EQUATION 13-15

When our simple lattice is subjected to a (non-destructive) stretching force the atoms are pulled from their equilibrium positions and an equal and opposite restoring force is generated (Figure 13-16). This force is equal to the first derivative of the *PE* with respect to x, the displacement (think about it: energy is force times distance, so differentiating with respect to distance gives you a force). Differentiating the power series for the *PE* with respect to *x* gives us Equation 13-16:

$$f = kx + k'x^2 + \ldots$$

EQUATION 13-16

Or, normalizing the displacement with respect to the original length of the bond we can write this equation in terms of the strain, Δl (Equation 13-17):

$$f = k\Delta l + k'\Delta l^2 + \ldots$$

EQUATION 13-17

Clearly, for small strains we can neglect higher order terms (if the strain is 2%, then $\Delta l = 0.02$ and $\Delta l^2 = 0.0004$). So we see that at low strains we essentially have Hooke's law, but as the strain increases we would expect to see deviations, even in theoretically perfect materials.

And indeed we do. Of course, we cannot make perfect materials, there are always defects of some kind. But we can get close. What we would like to do for a polymer is grow true single crystals, where the chains are all fully extended. Remember, most crystalline solids are not single crystals, but collections of smaller crystals that grow from separate nucleation events until their boundaries impinge on one another. Furthermore,

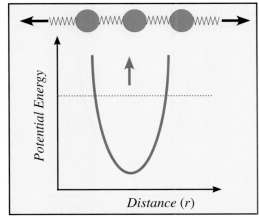

FIGURE 13-16 Schematic diagram of potential energy versus distance between three atoms.

polymers crystallized from both the melt and solution chain fold (see Crystallization, Chapter 12). But macroscopic single crystals of an unusual class of monomers, the diacetylenes, can be grown from solution and these can be polymerized in the solid state to give extended chain single crystals fibers. The stress/strain data obtained from one of these polymers shows that Hooke's law is indeed obeyed up to strains of about 2%, but beyond that a deviation occurs (Figure 13-17). Imperfections in these crystals no doubt contribute significantly to this deviation, but what you should take from this discussion is

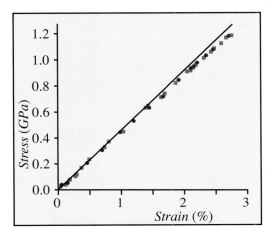

FIGURE 13-17 Experimental stress/strain plot for single crystals of polydiacetylene [redrawn from the data of C. Galiotis and R. J. Young, *Polymer*, **24**, 1023 (1983)].

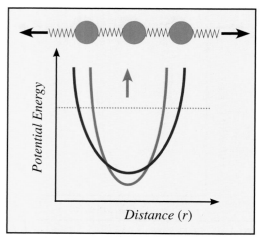

FIGURE 13-18 Schematic diagram of two potential energy versus distance curves for two different materials.

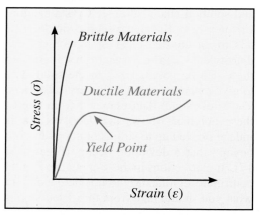

FIGURE 13-19 Schematic plots of stress versus strain for brittle and ductile materials.

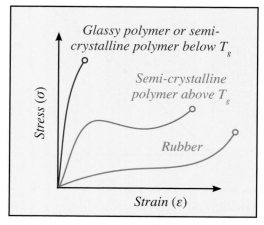

FIGURE 13-20 Schematic plots of stress versus strain for different polymers.

that Hooke's law is an approximation, good for small strains. Actually, for many engineering applications that is good enough.

In addition to using a model of a perfectly ordered solid to get an understanding of Hooke's law and its limitations from an atomic viewpoint, we can also start to look at properties like *stiffness* and *strength*. The stiffness of such a material should obviously be related to the stiffness of the chemical bonds within the system. The stiffness of each bond is given by the slope of the potential energy curve. Similarly, the strength of our "ideal" material should be related to the cohesive strength of the bonds, in turn proportional to the depth of the potential energy well (Figure 13-18). So, a strong stiff material should have a deep, narrow potential energy curve.

So, can the strength and stiffness of real materials be explained on the basis of such considerations? To a degree they can. In fact, some brittle[55] materials, like ceramics, at first glance appear to be almost ideal in their behavior. They are "elastic" up to stresses close to fracture (Figure 13-19), meaning that they will return to their original dimensions after stretching, not that they necessarily obey Hooke's law over the entire range of strain (which is seldom more than a couple of percent before fracture occurs). However, these materials are usually nowhere near as strong as they should be. Even more drastic deviations from ideal behavior are observed in ductile materials (e.g., many metals), where a *yield point* occurs well before fracture. At this point permanent or plastic deformation ensues (think of pulling on a piece of silly putty).

Stress/Strain Characteristics of Polymers

If we now look at the stress/strain characteristics of polymers (Figure 13-20), shown purely schematically just to give you a feel

[55] The word "brittle" to most people implies poor impact resistance. As we will see, it's a bit more complicated than this and in general brittle materials have poor resistance to the propagation of cracks.

for the shape of the curves (the plots should all be on different scales), we see similar types of behavior for glassy and semi-crystalline polymers. For elastomers, however, elastic but non-linear behavior is often observed up to strains of 500% or more. Furthermore, even in the initial, apparently Hookean regions of the stress/strain behavior of polymers, anelastic responses are often the norm. By this we mean the response is time dependent, whereas in other materials the elastic response is almost instantaneous. But, let's take one thing at a time and first make some general observations about stiffness, strength and toughness.

Stiffness

Materials vary enormously in their stiffness, the values of the modulus of a rigid material like diamond being about 2×10^5 times that of rubber (the values in Table 13-1 are rough approximations, just to give you a feel for how the modulus varies from one class of materials to another). For structural applications, you often want a very stiff material—recall Figure 13-8 showing what would happen to the wings of an aircraft having what you might consider to be very small amounts of strain in its wing spar booms.[56] (This also illustrates why a knowledge of the modulus of a material is so important—in engineering design you need to calculate deflections.)

In materials that are often used in construction—wood, concrete and steel—there is still an enormous variation in modulus, by a factor of 15 or so. You can also see that most polymer materials, at least in their usual melt processed form, are not very stiff at all, polyethylene having a modulus of about 150 MPa, while even a glassy polymer like atactic polystyrene has a modulus of only about 3000 MPa (about 1/20th that of window glass).

If trees were made out of polyethylene they would be positively lethal in a high

---

[56] We got the idea for this from J. E. Gordon, "*The New Science of Strong Materials*" Penguin Books (1991).

TABLE 13-1 APPROXIMATE MODULI OF VARIOUS MATERIALS

MATERIAL	E (psi $\times 10^{-6}$)	E (MPa)
Rubber	0.001	7
Polyethylene	0.2	150
Wood	2.0	14,000
Concrete	2.5	17,000
Glass	10	70,000
Steel	30	210,000
Diamond	170	1,200,000

wind, whipping about and lashing passers-by and dawdling dogs (Figure 13-21). You may also guess that if you reinforce polymers with glass fibers you should increase the modulus of the resulting composite significantly—that's one of the reasons you see so much glass fiber reinforced material about (there are other reasons that we'll get back to later).

Can these large variations in stiffness be explained in terms of the stiffness of the chemical bonds within a material? Qualitatively, they can, although it is not just the stiffness of the individual bonds that is important, but how they are arranged. Take a diamond, for example. In the crystal structure of diamond each carbon atom is covalently bonded to four other carbon atoms, as illustrated for the central carbon atom in Figure 13-22 (some bonds on the other atoms are omitted for clarity of presentation). The

FIGURE 13-21 If it's made of polyethylene, don't go near it!

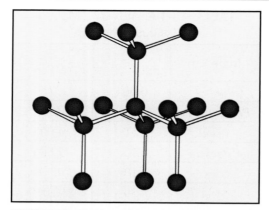

FIGURE 13-22 Diamond crystal structure.

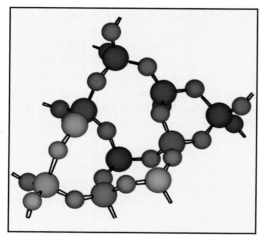

FIGURE 13-23 Structure of silica.

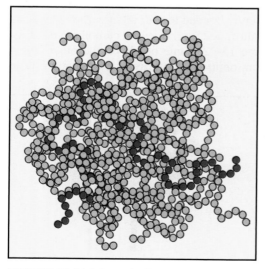

FIGURE 13-24 Schematic diagram of an amorphous polymer glass.

covalent bonds are strong and stiff, and the structure itself is a densely interconnected network of carbon atoms. Thus diamonds are rigid, strong materials. They are also brittle and can be readily cleaved and shaped by skilled craftsmen, but the property of toughness is something we will get back to later.

What about inorganic glassy materials? Silica, SiO_2, can exist in several crystalline forms (e.g., quartz), but also as a noncrystalline glass. Here each silicon atom is tetrahedrally bonded to four oxygen atoms and each bond is intermediate in character between ionic and covalent bonds. Like purely covalent bonds, these are rigid and strong, but now each tetravalent silicon atom is separated by divalent oxygen atoms and there is no regularity in the arrangement of the atoms (Figure 13-23). We would expect that there would be more "give" in an amorphous solid such as this (in terms of things like the cumulative effect of bond angle bending) and it should have a lower modulus than diamond. And it does. Furthermore, the modulus and other properties of glassy silica can be modified by breaking up the network. If sodium oxide is added, Si–O–Si linkages are broken and the resulting ends sealed by forming ionic bonds with sodium, $Si–O^- Na^+$, as in the manufacture of ordinary window glass. This breaking up of the network further decreases the modulus.

What about amorphous glassy polymers? If the polymer is thermoplastic, in other words there are no cross-links, then we have a bunch of entangled randomly coiled chains (Figure 13-24). The chains themselves are made up of covalently linked units, but the interactions between the chains are much weaker, just dispersion forces in materials like atactic polystyrene. If a piece of polystyrene or Plexiglas, poly(methyl methacrylate), is stressed, the load is borne by some complex combination of these bonds. Furthermore, even in the glassy state there can be local conformational rearrangements in these polymers, so overall we would expect that the modulus of glassy polymers should be much less than window glass. Is it? Can you bend sections of Plexiglas, such as those found in light fixtures, more easily than window glass? What is the effect of cross-linking? If we make a densely cross-linked network, as in Bakelite and some epoxies, would their modulus be higher than atactic polystyrene?

The answers to these questions can be gleaned from Table 13-2, which compares approximate values of the tensile modulus for various polymers. Rubbers or elastomer are also amorphous, of course, but they respond to a stress in an entirely different manner to all other types of materials. Because they have low T_gs, at ordinary temperatures, they respond to a load by changing their distribution of chain conformations, the chains becoming more extended as the material is stretched. A rubber has to be extended many times its original dimensions before the covalent bonds take the load. We will consider rubber elasticity as a separate topic later.

What about semi-crystalline polymers? On the basis of the arguments we have made so far we would expect that highly crystalline materials should be very stiff. But polymers are semi-crystalline. We therefore have to consider two situations, one where the polymer has a low T_g and one where it has a high T_g. In the latter case we would expect the material to be relatively stiff (as polymers go), as it would be some combination of the covalent bonds and intermolecular forces that would respond to a stress, as in amorphous glassy polymers. On the other hand, we would anticipate that polymers where the amorphous domains are well above the T_g at the temperature of use would have a lower modulus and be more flexible. And indeed they are. However, conformational freedom in the amorphous domains is restricted by the crystalline domains, so that the modulus depends significantly on the degree of crystallinity (Figure 13-25).

What if we could make perfectly extended chain fibers, so that the covalent bonds were being directly stressed? These should be very stiff and strong in the draw direction, but much weaker and with a much lower modulus perpendicular to this direction.

TABLE 13-2 Approximate Moduli of Polymers

Polymer	E (MPa)
Rubber	7
Polyethylene High Density Low Density	830 170
Polystyrene	3100
Poly(methyl methacrylate) (Plexiglas)	4650
Phenolic Resins (Bakelite)	6900

Highly oriented fibers have indeed been produced. However, these are not single crystals, with perfectly aligned chains extending from one end of the material to the other. The chains are of finite length and the drawn fibers have a microfibrillar and polycrystalline morphology. Accordingly, they are not as stiff and strong as an "ideal" material. Nevertheless, astonishing increases in both both the modulus and strength of polymer materials has been achieved using various processing techniques, Figure 13-26 shows a plot of the modulus of polyethylene as a function of draw ratio. We'll briefly look at the modulus of other fibers later; first let's discuss strength and toughness.

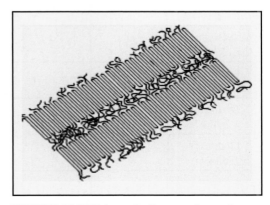

FIGURE 13-25 Schematic diagram of a semi-crystalline polymer illustrating restricted amorphous domains.

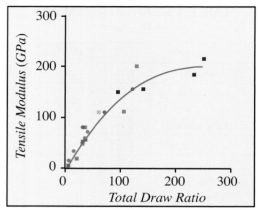

FIGURE 13-26 Plot of modulus versus draw ratio for polyethylene [redrawn from a plot by Roger Porter in *Materials Characterization for Systems Performance and Reliability*, J. W. McCauley and V. Weiss, Eds., Plenum Publishing (1986)].

TABLE 13-3 APPROXIMATE TENSILE STRENGTH OF POLYMERS

MATERIAL	TENSILE STRENGTH (PSI)	TENSILE STRENGTH (MPA)
Steel piano wire	450,000	3000
High-tensile steel	225,000	1500
Aluminum alloys	20,000 - 80,000	140 -550
Titanium alloys	100,000 - 200,000	700 - 1400
Wood—spruce (along grain)	15,000	100
Wood—spruce (across grain)	500	7
Ordinary glass	5000 - 25,000	30 - 170
Ordinary brick	800	5
Ordinary cement	600	4
Nylon fiber	140,000	950
Kevlar® 29 fiber	400,000	2800

Tensile Strength

When we talk about strength[56] we usually mean tensile strength, the stress required to pull a material apart and create two new surfaces. Materials can break in a brittle or ductile manner, but we will discuss that shortly. First let's look at the numbers in Table 13-3. They follow what you would intuitively expect from everyday experience, with a couple of exceptions. Glass (by which we will mean ordinary window glass) has a higher tensile strength than you would anticipate, given its relative fragility. Also, if you were not familiar with polymers you may not have expected the strength of organic fibers to be so great. (This, as you might expect, depends on how well you align the chains.) What makes certain materials stronger than others?

Most people don't think about the strength of materials, regarding, say, the high strength of steel compared to glass as a matter of divine imposition or one of those mysteries of nature that we'll never fathom. Those who got beyond that staple of the current high school science curriculum, Saving the Rain Forest, may guess that just like the modulus, it has something to do with the strength of the chemical bonds within a material. In fact, it's much more complicated than that and we have to look at the problem in a different way. The first question we should ask is "how strong should a material be?", rather than "why is steel stronger than polystyrene?"

The first person to start thinking about the problem in this way was A. A. Griffith in the 1920s. Griffith's brilliant idea was to relate the energy required to create two new surfaces when a material is fractured to the strain energy stored in the material before it broke. If you're having trouble with the concept of strain energy, think of compressing a spring or drawing a bow; you are storing potential energy that can be made to do work. Hopefully you will also recall that all surfaces have an energy (hence the phenomenon of surface tension in liquids).

The area under the stress/strain plot, illustrated above in Figure 13-14 for a hypothetical ideal elastic material, is actually the strain energy per unit volume. Now let's say we take a plane within our sample that separates two adjacent layers of atoms a distance d apart. To keep things simple let's assume that this plane has unit area, so the volume occupied by our two atomic layers is simply d. The strain energy stored in the bonds separating the layers is then (Equation 13-18):

$$\text{Strain Energy} = \frac{1}{2}\sigma\varepsilon d$$

EQUATION 13-18

Recalling Hooke's law we can substitute $\varepsilon = \sigma/E$ and obtain Equation 13-19:

$$\text{Strain Energy} = \frac{\sigma^2 d}{2E}$$

EQUATION 13-19

Now we let G be the surface energy of the material and let's also assume that the material will fail in an ideal brittle fashion by cleanly breaking the bonds across our selected plane (Figure 13-27). If we keep in mind that two surfaces are created when we break our sample and assume that all the strain energy is converted to surface energy we can write Equation 13-20:

$$2G = \frac{\sigma^2 d}{2E}$$

EQUATION 13-20

Hence the tensile strength is given by Equation 13-21:

$$\sigma = 2\sqrt{\frac{GE}{d}}$$

EQUATION 13-21

Of course, even for materials that fail in a nice brittle fashion Hooke's law is not obeyed right up to failure and the strain

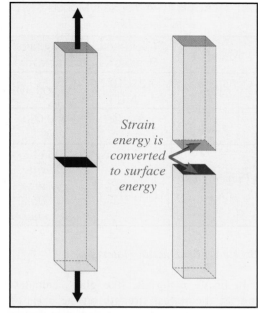

Strain energy is converted to surface energy

FIGURE 13-27 Schematic diagram illustrating the conversion of strain energy to surface energy.

energy is actually about half this, or (Equation 13-22):

$$\sigma \approx \sqrt{\frac{GE}{d}}$$

EQUATION 13-22

Young's modulus for a material can be measured easily, the distance apart of the atoms in most materials is about 2–3 Å, so that all that remains for a rough calculation of the theoretical strength is to obtain a measure of the surface energy. Griffith chose to study glass, first because it fails in a nice brittle fashion; second, because there is no change in structure on heating glass above its T_g. The surface tension of the melt, hence its surface energy, can then be measured and assumed equal to that of the amorphous solid. Griffith calculated that glass should have a theoretical strength of just less than 2×10^6 psi (~14 GPa), far greater than the measured tensile strength. There are similar large differences between the theoretical and actual strength of other materials (Table 13-4).

TABLE 13-4 YOUNG'S MODULI
FOR THREE MATERIALS

MATERIAL	THEORETICAL STRENGTH	MEASURED STRENGTH
Glass	$\sim 2 \times 10^6$ psi	$\sim 25,000$ psi
Steel	$\sim 5 \times 10^6$ psi	$\sim 400,000$ psi
Polyethylene Fibers	~ 25 GPa	~ 0.35 GPa (Tensile drawing) ~ 4 GPa (Gel spun)

Brittle versus Ductile Materials

In brittle materials, like glass, calculating the theoretical strength on the basis of breaking the bonds separating two layers in a material and forming two new surfaces makes sense in terms of the types of specific bonds in these systems. The metallic bond is non-specific, however, and failure of a theoretically perfect structure involves slip mechanisms along certain planes in the crystal (and therefore depends on crystal structure)—see Figure 13-28. This would lead to considerable permanent (plastic) deformation before failure. Even so, the theoretical strength of metals should be of the order of $E/10$, much higher than found in practice. As you might then guess, the reason that materials are weaker than they should be varies with structure and bonding. We'll consider brittle materials first.

Let's get back to glass. When Griffith went on to measure the actual strength of glass fibers, a funny thing happened. The tensile strength, measured as a stress to break, increased dramatically as he made the fibers thinner and thinner (Figure 13-29). The force required to break thinner fibers is, of course less, but the stress to break, force/unit area, should be the same, unless there is something different about thinner fibers. By plotting reciprocals Griffith found that the data could be extrapolated to a value close to the theoretically estimated strength.

Similar results have been obtained with other types of brittle materials, such as the poly(diacetylenes) mentioned earlier (Figure 13-30). So, what's going on? Griffith realized that the problem was not understanding why thin fibers are strong, but why thick ones are weak. He proposed that this weakness is due to defects, which in his

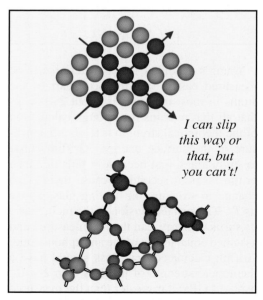

I can slip this way or that, but you can't!

FIGURE 13-28 Schematic diagram illustrating slip planes in a crystalline material versus an amorphous glass.

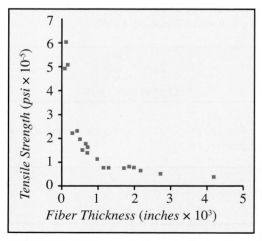

Fiber Thickness (inches × 10³)

Tensile Strength (psi × 10⁻⁵)

FIGURE 13-29 The tensile strength of glass fibers versus fiber thickness [redrawn from a figure in J. E. Gordon, *The New Science of Strong Materials*, Penguin Books (1991)].

glass fibers were probably surface cracks. In the poly(diacetylene) samples the flaws are largely within the crystal. (The thinner fibers that Griffith made had fewer surface cracks, perhaps because they bent more easily or were handled with greater care.)

If you've ever taken any mechanics classes you probably recall that a crack acts as a stress concentrator. In a hypothetical flaw-less material the lines of stress are uniformly spaced out and a load is evenly borne by all the atoms or molecules in the object. But the presence of a hole or a crack requires the stress to go around the opening (Figure 13-31). The stress concentration depends upon the size and shape of the defect. Inglis calculated the stress concentration factor for an elliptical hole to be given by Equation 13-23:

Stress Concentration Factor

$$= 1 + 2\sqrt{\frac{L}{R}}$$

EQUATION 13-23

---

where L is half the length of a crack and R is the radius of curvature at the tip. So for a long thin crack, not exactly an ellipse but close enough, the stress concentration can be very large.

This means that the bonds near the tip of the crack are carrying more than their fair share of the stress and can fail (Figure 13-32). The crack becomes longer and thinner, the stress concentration even greater, and the next bonds in line also fail. Thus a crack can propagate catastrophically through a material, traveling faster than the speed of sound (which is why you hear a bang when some brittle materials fail in a tensile test). Accordingly, the presence of cracks can concentrate an applied stress to the extent that at a given point the "actual" (as opposed to nominal or measured) strength of the material is exceeded and failure ensues. But surely such arguments apply to steel and Kevlar® as well as glass; what is different about these materials and why does glass fail more readily?

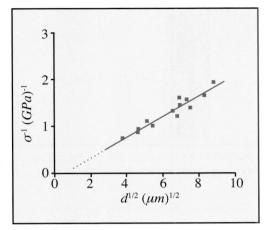

FIGURE 13-30 The reciprocal tensile strength of polydiacetylene single crystals [redrawn from the data of C. Galiotis and R. J. Young, *Polymer*, **24**, 1023 (1983)].

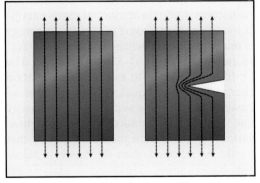

FIGURE 13-31 Schematic diagram illustrating stress concentration caused by a crack.

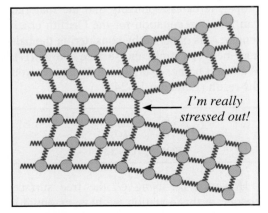

I'm really stressed out!

FIGURE 13-32 Schematic diagram illustrating the greater stress that the bonds carry at the tip of a crack.

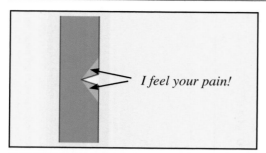

FIGURE 13-33 Schematic diagram illustrating relaxation of the area (shaded) around a crack.

To answer this we have to get back to Griffith and his seminal approach of using energy, rather than force or stress, to understand the problem of fracture. Let us do this by considering a constant strain (not stress) experiment, where a material is stretched a given amount and held there. The material then stores a certain amount of strain energy. If a crack appears, some of this strain energy is released and the material behind the crack, shaded in Figure 13-33, "relaxes." The Griffith approach is to assume that the strain energy released is equal to the work done on creating the two new surfaces. If the energy that has to be expended in creating these surfaces is more than the strain energy that would be released, then the crack is stable —it doesn't grow. This type of analysis led Griffith to propose that at any given stress there is critical crack length (the Griffith crack length). Cracks larger than this can propagate catastrophically and the material can fail. An equation for the Griffith crack length, l_g, is obtained by considering the balance between strain and surface energy (G). We won't bother to do this, but just give you the result (Equation 13-24):

$$l_g = \frac{2GE}{\pi\sigma^2}$$

EQUATION 13-24

However, by using G, the free surface energy, in this equation we have essentially assumed that the surface of the cracks is produced "cleanly," by breaking the bonds between just two layers of atoms. This is exactly the quantity we need in calculating the theoretical strength of a material, but not what we need for determining its resistance to crack propagation. Even in very brittle materials, the molecular structure of a material can be distorted and bonds broken to a considerable depth below the surface. If we let the energy that has to be used to produce a "real" fracture surface in a material be W, the work of fracture, we then get Equation 13-25:

$$l_g = \frac{2WE}{\pi\sigma^2}$$

EQUATION 13-25

Note the inverse square dependence on the stress (squared). As the load on a material increases, the critical crack length gets shorter.

Although there are problems in using this equation for engineering design, it does what we want: provides an understanding of why brittle materials are much weaker than they should be. In window glass the molecular structure is perturbed to a relatively shallow depth during fracture and W is about $6G$. This means that under any significant load the critical crack length is very short. The tensile strength of glass might be appreciable under a static load, but the glass is in a fragile state, the critical crack length is small and slight perturbations can lead to crack propagation and failure. You therefore don't want to use glass as a structural material—there's no margin of safety. Brittle materials in general have a low work of fracture or energy to break, as reflected in the area under the stress/strain curve (Figure 13-34).

Mild steel and other ductile metals, on the other hand, have an enormous work of fracture: W is of the order of 10^4–10^6 G and the critical crack length is very long. This high work of fracture makes them safe and dependable materials in load-bearing applications. As you might guess, the structure of these materials is perturbed to a significant depth below the fracture surface, hence the high work of fracture (Figure 13-35). The stress/strain plots of such a material display

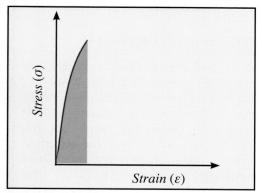

FIGURE 13-34 Schematic plot of stress versus strain for a brittle material.

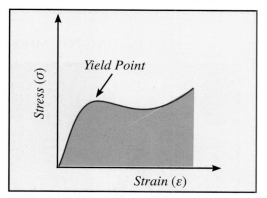

FIGURE 13-35 Schematic plot of stress versus strain for a ductile material.

a yield point, where a neck forms and the material deforms plastically.

As we mentioned previously, ductile failure in metals involves slip between layers of atoms in the crystal structure (yielding phenomena are also observed in polymers, but we'll treat that in detail in a separate section). The elastic limit is reached at much lower stresses than you would calculate for a hypothetically perfect material and again the problem is defects. Here the principal "weakening" mechanism is not microscopic cracks, however, but defects within the crystal structure known as dislocations (Figure 13-36). There are tricks that metal scientists use to strengthen or "harden" their materials, but, as you might guess, this then makes them more brittle and subject to the same type of failure mechanisms found in glass.

Summary: Strength and Toughness

It's time to pause a moment and see where we are. First, all materials are weaker than they could be because of cracks, defects in the crystalline lattice, and so on. This is something that some materials scientists like to tackle by making more perfectly ordered materials. Even so, it is not necessarily the measured tensile strength that is the most critical factor in choosing a material for a structural application, but its work of fracture, its resistance to the growth of cracks. This property also plays into impact strength. When you hit a ceramic vase with a hammer

you are really just applying a load at a point. (Perhaps it was a gift from your mother-in-law and you can't stand the sight of it.) The energy is actually conducted away very quickly in the form of stress waves traveling at about the speed of sound ($\sim E/\varrho$; about 11,000 mph in glass!). These reflect back and forth and accumulate at some spot, perhaps where there is a structural flaw or crack. Then if the material has a low work of fracture, it's history! Polymers, in general, are relatively tough, but at least in their melt-processed form, they are not as strong or as stiff as metals and ceramics—but they don't have to be if you're making film wrap or a toothbrush.

In the form of oriented fibers, however, polymers can be made very strong and stiff. Furthermore, their relatively low density results in high values of specific strength

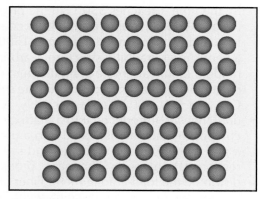

FIGURE 13-36 Schematic diagram of a dislocation.

FASCINATING POLYMERS—CARBON FIBER COMPOSITES

A 236-ft-long craft, one of the Visby Class built in Sweden, is a showpiece for composites. The laminate is vinyl-ester reinforced with fabrics of carbon fiber (Courtesy: Kockums AB).

Most modern carbon fibers are produced from either polyacrylonitrile (PAN) or mesophase (liquid crystalline) pitch. Originally, carbon fibers were derived from viscose rayon, but rayon became too expensive to produce compared to the many other organic polymer fibers that emerged after the end of WWII. The pitch-based carbon-fiber process is fascinating and analogous to the preparation of Kevlar® fibers. A high degree of orientation in melt-spun pitch is inherent and a natural consequence of the liquid crystalline (mesophase) nature of the material. However, here we are going to focus our attention on the fascinating chemistry that takes place when carbon fibers are produced from PAN fibers. It was Shindo and his group in Japan who first demonstrated that carbon fibers could be obtained by the carbonization of PAN. But the most important discovery was made by Watt and his colleagues at the Royal Aircraft Establishment in Farnborough, England. Watt showed that a markedly superior high-strength, high-modulus carbon fiber could be produced if PAN fiber was first stretched (to align the chains) and simultaneously oxidized at temperatures of 200–400°C, prior to the final carbonization step at 1500–3000°C. The chemical changes that occur during the process of making carbon fibers from PAN are very complex. Essentially, as the PAN fiber is stretched and heated above 200°C, cyclization reactions between adjacent nitrile groups occur. This stabilizes the structure and the material transforms into a dark, red-colored material with very little loss of weight. By the time it reaches 350°C in the presence of oxygen, the material is now black ("Black Orlon"), aromatization has occurred, and the polymer resembles a polypyridine ladder. Continued heating to 1500–3000°C yields a final carbon fiber having a very strong graphitic structure. Carbon fibers come in a variety of strengths and moduli. Some are more graphitic than others and they are often surface-treated or coated with a sizing. Typical ultimate strengths and moduli lie in a range of 1.5–3.5 GPa and 200–800 GPa, respectively. Of course, carbon fibers are pretty useless until they are introduced into a matrix, such as an epoxy or polyester resin, and the composite material fashioned into a useful article, such as the lightweight shaft of a golf club! There is much art and science that goes into designing carbon-fiber-reinforced composites. The matrix protects and binds the fibers together and acts to distribute the load among them. Fiber arrangement and orientation are dictated by the end use of the composite material.

FASCINATING POLYMERS—HIGH IMPACT ABS THERMOPLASTICS

The credit for discovering the high-impact ABS engineering thermoplastics that are produced in large quantities today goes to W. C. Calvert of the Borg Warner Company. In 1953, Calvert filed a patent for the preparation of ABS (acrylonitrile, butadiene, styrene) materials that were formed in the process of copolymerizing styrene (S) and acrylonitrile (A) in an emulsion of already formed, lightly cross-linked, polybutadiene (B). Calvert did not mention the presence of an in-situ graft copolymer in the initial patent application and it wasn't until some four years later in a second patent that the importance of grafting was acknowledged. The polybutadiene rubber was uniformly distributed as small (1–10 μ) size particles (the dispersed phase) in a matrix of SAN (the continuous phase) with the in-situ graft copolymer residing at the interface and acting as a "compatibilizer." In what was quaintly referred to as the "Whack" test, where a sample was whacked against the corner of a stone slab, thousands of formulations and variations in process conditions were crudely tested to optimize impact properties. The result was that in 1954,

An ABS pay phone circa 1960 (Source: www.retrocollection.com).

Borg Warner introduced Cycolac® resins to the market. They immediately found application in articles where impact strength was required, most notably in radio and appliance housings, telephones, luggage, refrigerator liners and pipes etc. In the good old days of the 1950s and 1960s (it helps if you are long in the tooth, think the Beatles invented music, and have a pair of rose-colored glasses!), companies, especially the large well-established ones, were stable and rather insular. Today, turn around, and you find another large traditional polymer company has disappeared (e.g., Celanese, Allied), divested its plastics interests (e.g., ICI, Monsanto) or formed joint ventures with another company to whom previously they would hardly speak (e.g., DuPont Dow Elastomers). General Electric, which has a major stake in engineering plastics, acquired Borg Warner's Cycolac® resins in the early 1980s and currently produces some 60 different grades of the material. Many other companies produce ABS materials and blends, including BASF (Terluran®), Bayer (Bayblend®, Lustran®, Novodur®), DSM (Electrafil®, Fiberfil®) and Dow (Magnum®, Prevail®, Pulse®, Retain®).

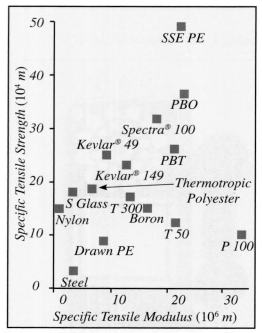

FIGURE 13-37 Plot of specific tensile strength versus specific tensile modulus [redrawn from an original plot by S. J. Krause et al., *Polymer*, **29**, 1354 (1988). T300, T50 and P100 are carbon fibers; Spectra 1000 is a gel-spun polyethylene and SSE PE is a solid state extruded polyethylene. Other polymers on the plot have stiff backbones and a liquid crystal character].

versus specific modulus, (the strength and modulus divided by the density), critical if you need a strong, stiff yet lightweight material (Figure 13-37).

We won't go into the details of how these

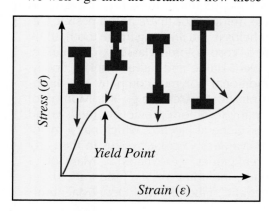

FIGURE 13-38 Schematic plot of stress versus strain for a semi-crystalline polymer.

materials are processed here (see our Chapter on "High Strength Fibers" in our *The Incredible World of Polymers* CD), but instead turn our attention to some of the details of the stress/strain behavior of polymers.

STRESS/STRAIN CHARACTERISTICS OF POLYMERS

We have discussed some general aspects of the mechanical properties of polymers in the context of a review of the stiffness, strength and toughness of materials. In polymeric materials as a class, these properties vary over an extraordinary range, from those typical of rubber to those typical of very high modulus, high strength fibers. In turn, this is a reflection of structure, which we have seen ranges from the random coil, purely amorphous states, to chain-folded semi-crystalline morphologies, to highly oriented fibers. By and large, however, polymers tend to deform irreversibly far more readily than the other materials we discussed in our general review and this deformation can increase dramatically even if the temperature is raised fairly modest amounts (often less than 100°C). Furthermore, there is a time dependence to their behavior. We'll neglect that for now and start by surveying general stress/strain behavior.

Let's start by looking at a simple polymer, polyethylene, that has a lot going on in its stress/strain plots (Figure 13-38). Flexible, semi-crystalline polymers such as this (where the T_g of the amorphous domains is below room temperature) usually display a considerable amount of *yielding* or "*cold-drawing*," as long as they are not stretched too quickly. For small deformations, Hookean elastic-type behavior (more or less) is observed, but beyond what is called the *yield point* irreversible deformation occurs.

Yielding in Polymers

At the yield point a neck forms and there is an actual decrease in stress while the sample stretches out considerably. (There are two things going on here. First, the testing

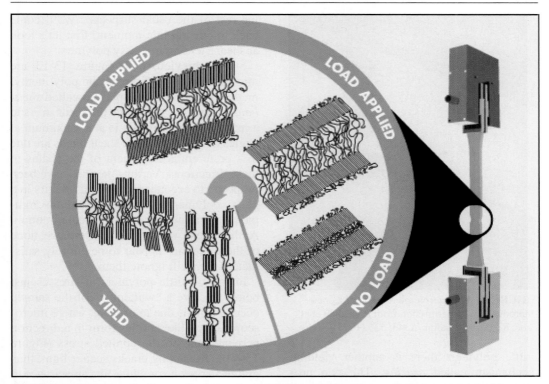

FIGURE 13-39 Schematic diagram depicting the yielding process in polymers [based on an original figure by J. Schultz, *Polymer Material Science*, Prentice-Hall, New Jersey (1974)].

machine usually applies a varying load in order to deform the sample at a constant rate; after the yield point less stress is required to stretch the sample a given amount. Second, the testing machine actually plots nominal stress, based on the original cross-sectional area of the sample; the actual stress changes once a neck is formed). After deformation has continued at an almost constant stress, the sample regains some resistance to strain and ultimately breaks at a higher load. So, what's the mechanism or molecular machinery behind all of this behavior?

Yielding in these polymers is a complex process (Figures 13-39), initially involving a stretching out and partial orientation of the chains in the amorphous domains, slip between lamellar planes of the crystal and a degree of chain unfolding. Eventually a fiber type of morphology is produced, albeit one with imperfect chain alignment, accounting for the upturn in the stress/strain curve. An expanded schematic view of the chains was

shown previously in Chapter 8, Figure 8-65. Note that these types of structural rearrangements, or relaxations as they are called, take time and if we pull on a sample of polyethylene quickly it fails in a more brittle fashion, with little yielding.

Polymers nearly always yield and plastically deform to some degree during failure and this is what makes them relatively tough materials (ever tried to break your old credit card, a glassy polymer, by bending it in half, backwards and forwards?—you couldn't do that with an inorganic glass). In the first example we considered, polyethylene, the process is called shear yielding, which also occurs in glassy polymers and semi-crystalline polymers below their T_g. However, the mechanism by which this occurs is different. In rigid polymers you usually don't see a yield point; instead as the stress on the sample increases there is a deviation from Hookean behavior followed by permanent deformation of the sample. Furthermore, in

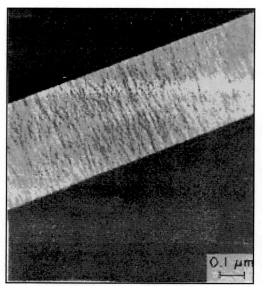

FIGURE 13-40 Electron micrograph of a craze [reproduced with permission from P. Beahan et al., *Proc. Roy. Soc.* London, **A343**, 525 (1975)].

brittle polymers there is another yielding mechanism called *crazing*. The two processes are not mutually exclusive and both can show up together or one can precede the other. The mechanism that dominates depends on temperature, morphology and

FIGURE 13-41 Craze formation in a rubber modified glassy polymer blend [reproduced with permission from R. P. Kambour and D. R. Russell, *Polymer*, **12**, 237 (1971)].

the rest of the usual suspects. We'll come back to crazing in a moment; first let's look at shear yielding in glassy polymers.

Shown previously in Figure 13-13 are micrographs of polystyrene and poly(methyl methacrylate) which display so-called *shear bands*. You may recall that the shear stress in a material under tension is at a maximum at angles of ±45° and these shear bands are due to a preferential alignment of the chains in these directions. Various theories have been proposed to account for how this occurs in a glassy polymer, involving free volume arguments or various types of molecular motion. At the time of writing this masterpiece, none of these theories appear to be entirely satisfactory, so we'll ignore them.

In many brittle polymers at stresses just below fracture a "whitening" of the sample occurs. This is due to *crazing*, where microscopic crack-like entities form in a direction perpendicular to the applied stress (Figure 13-40). These tiny cracks scatter light, thus giving an opaque or white-like appearance to the sample. Intriguingly, if examined under the microscope tiny fibrils can be seen spanning the cracks, helping to hold them together. Obviously, the formation of these tiny cracks and fibrils absorbs energy, helping to make "brittle" polymers tougher than other brittle materials, like window glass. But not tough enough for a lot of applications. Nevertheless, this suggests that one way of improving toughness and impact resistance in polymer materials is to "drive them crazy" (sorry!).

The formation of energy absorbing crazes in brittle polymers can be promoted by the addition of small amounts of rubber, either by blending, grafting or copolymerization. The rubber must be present as small, phase-separated domains, as this allows them to act as stress concentrators, a result of the large mismatch between the modulus of the rubber and the rigid matrix. This, in turn, promotes the formation of crazes throughout the body of the sample (Figure 13-41) rather than near a crack tip, where stress usually concentrates with catastrophic consequences in homogeneous rigid polymers and other materials like glass (we discussed this in our general

FIGURE 13-42 Schematic diagram depicting crack stopping.

review). This type of rubber reinforcement is now being used to produce all sorts of impact-resistant plastics.

There are other toughening mechanisms in polymers, the most interesting involving glass-fiber-reinforced polymers, often polyesters. These are intriguing because they are a composite of two brittle materials. Yet inebriated weekend boaters can often smack their fiberglass boat hulls into one another or the dock with minimal damage. Here crack stopping (Figure 13-42) plays a crucial role, as opposed to craze formation. The requirement for toughness in these composites is that the strength of the interface between the glass fibers and the polymer matrix be less than the strength of either bulk material. If this is not so, any propagating crack would just zip right across the interface without so much as a "by your leave."[57] However, if the interface is weaker than the bulk material, the stress concentration at the tip of the growing crack pulls the matrix away from the fiber, simultaneously absorbing energy, blunting the crack and thus reducing the stress concentration below that necessary for further crack propagation.

The processes of yielding we have briefly described above must obviously depend upon specific relaxation processes. By relaxation (not the type shown in Figure 13-43) we mean any sort of time-dependent molecular transition or rearrangement, such as a change in the conformation of a chain, crystalline slip, chain sliding, and so on. Accordingly, yielding phenomena will depend upon the rate at which a polymer is being strained relative to the time required for these relaxations to occur—the relax-

FIGURE 13-43 Not this type of relaxation!

ation rate. That is why polyethylene can be drawn slowly at room temperature to give a fiber-like morphology; the rate at which the sample is stretched is less than the relaxation times of the various molecular rearrangements involved. If the sample is stretched too quickly, however, these rearrangements do not have time to occur and the sample will fail in a brittle fashion.

This also results in a strong temperature dependence of stress/strain behavior. Consider poly(methyl methacrylate), for example (Figure 13-44). At temperatures around room temperature or less, it is a typical glassy polymer, failing in a brittle fashion.

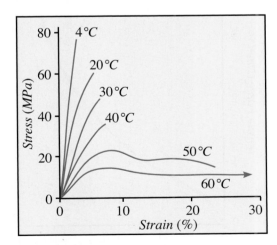

FIGURE 13-44 Stress/strain curves for poly(methyl methacrylate) as a function of temperature [redrawn from an original figure by T. S. Carswell and H. K. Nason, *Symposium on Plastics*, American Society for Testing Materials, Philadelphia (1944)].

---

[57] We nearly used a more vulgar and colloquial expression here, but we've pushed our publisher far enough!

Above about 45°C, however, considerable yielding can be observed. Note that the transition between brittle and ductile behavior occurs at a temperature that is significantly below the T_g. Various theories have been advanced to explain yielding phenomena in polymers, some involving free volume arguments while others involve various types of molecular motion. As far as we can make out, none of these are entirely satisfactory and we won't discuss them here. Instead, we will finish off our discussion of stress/strain behavior by considering rubber elasticity.

RUBBER ELASTICITY

Rubber is an extraordinary material with a rich history, shaped by a collection of intriguing characters and full of trials, tribulations and triumphs. There is so much good stuff that we devoted two of the chapters in our *The Incredible World of Polymers* CD to the subject. Here we will discuss the molecular basis of rubber elasticity.

First you should recall just how extraor-

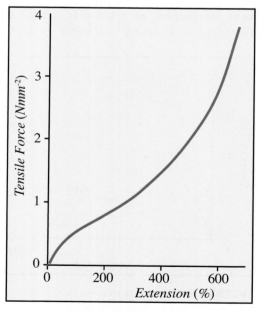

FIGURE 13-45 Stress/strain plot for natural rubber [redrawn from a figure in L. R. G. Treloar, *The Physics of Rubber Elasticity*, Third Edition, Clarendon Press, Oxford (1975)].

dinary elastomers are. "Ordinary" materials can stretch just a couple of percent elastically, by which we mean that they will then return to their original dimensions when a load is removed, not that they will necessarily obey Hooke's law over the entire range of elastic strain. We have also seen that this type of elastic behavior is (to a degree) related to the stiffness and strength of the chemical bonds in a material. Rubbers also have strong stiff (covalent) bonds, but they are capable of stretching more than 500% elastically (Figure 13-45). Furthermore, instead of just minor deviations from Hooke's law we see behavior that is markedly non-linear. Clearly the bonds in these materials cannot be stretching this much, so what's going on?

If you have been working your way through this subject in a more-or-less linear fashion, then you already know the answer to this question from our discussion of chain conformations. In amorphous polymers the chains are random coils with an average distance between the ends that goes as $N^{0.5}$, where N is the number of segments in the chain. You should also recall that we actually mean the root mean square end-to-end distance when we say this and if you don't know what we're talking about, go back and review this stuff! In any event, if we have a chain consisting of 10,000 segments, and let's say, for simplicity, that each segment is one unit long, then the average distance between the chain ends is 100 units (remember, above the T_g such chains are wiggling around and changing their shape all the time). This chain could be stretched many times its original length before the covalent bonds would have to bear a load, just by changing its conformation from a coiled to a more stretched-out shape.

It's therefore easy to see how a single chain in an amorphous polymer above its T_g can be stretched out, but what is the driving force for it to return to its original dimensions? Actually, in a real piece of rubber, consisting of many billions of chains, we have to specify that there is some cross-linking, so that the chains cannot slip past one another (Figure 13-46). Really what we are talking about is the (presumably relatively long) sections of chains between cross-link points, but the question remains the same, what makes a stretched rubber band contract when you let it go? We have an intuitive feeling for how a bond can be stretched and then go back to its original dimensions by imagining that it acts like a little spring, but this change of conformation business is not as intuitively obvious.

Thermodynamics Revisited

To understand rubber elasticity we have to revisit some simple thermodynamics (the horror! the horror!). Let's start with the Helmholtz free energy of our piece of rubber, by which we mean that we are considering the free energy at constant temperature and volume (go to the review at the start of Chapter 10 if you've also forgotten this stuff). If E is the internal energy (the sum of the potential and kinetic energies of all the particles in the system) and S the entropy, then (Equation 13-26):

$$F = E - TS$$

EQUATION 13-26

Stretching this sample, which we'll say has a length L, changes its free energy and the relationship of this change to the applied force, f, is given by Equation 13-27:

$$f = \left(\frac{\partial F}{\partial L}\right)_{V,T} = \left(\frac{\partial E}{\partial L}\right)_{V,T} - T\left(\frac{\partial S}{\partial L}\right)_{V,T}$$

EQUATION 13-27

If this relationship is not familiar or obvious to you, just keep in mind that energy has the dimensions of force times distance, so that differentiating with respect to distance gives you a force.

In an ideal crystalline material (Figure 13-47) it is the displacement of the atoms from their mean positions that generates an equal and opposite force when a load is applied. It is thus changes in the internal

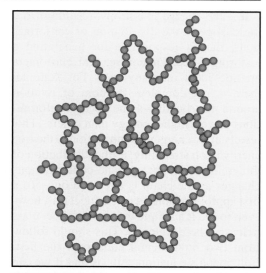

FIGURE 13-46 Schematic diagram of a rubber network.

energy (essentially the potential energy) that dominates the response of the material to a stress. We can then write Equation 13-28:

$$f \approx \left(\frac{\partial E}{\partial L}\right)_{V,T}$$

EQUATION 13-28

In an "ideal" rubber it is the other way around (Equation 13-29):

$$f \approx -T\left(\frac{\partial S}{\partial L}\right)_{V,T}$$

EQUATION 13-29

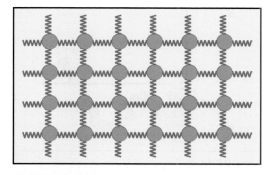

FIGURE 13-47 Schematic diagram of a crystalline lattice.

It is the change in entropy, resulting from the change in the distribution of conformations, that is responsible for generating a restoring force. Of course, real elastomers are not "ideal" in many ways. For example, there is not complete freedom of rotation around the bonds—some local conformations have a higher energy than others. This results in an energetic contribution to the free energy upon stretching, as the distribution of things like trans and gauche conformations changes as the chain is stretched out. To a first approximation we can neglect this, however, to see if a simple theory of rubber elasticity can be developed. This should follow from two fundamental equations: the first, Boltzmann's equation, tells us that if we can calculate the number of arrangements available to a system (Ω) then we can calculate its entropy from Equation 13-30:

$$S = k \ln \Omega$$

EQUATION 13-30

Then differentiating this we can obtain a force/extension relationship analogous to

Hooke's law (Equation 13-31):

$$f \approx -T\left(\frac{\partial S}{\partial L}\right)_{V,T}$$

EQUATION 13-31

The trick, of course, is calculating Ω. Fortunately, we don't have to know the absolute values of S and Ω, we're more interested in how these quantities change as a result of some process, like stretching a chain. This allows us to use probability distributions, which sounds rather awful, but for a lot of problems is fairly straightforward. For example, if we take a single chain and fix one of its ends at the origin of a Cartesian coordinate system and let the other end be at a position x_0, y_0, z_0 (Figure 13-49), then the distribution function, $P(x_0, y_0, z_0)$, is simply an expression describing the fraction of conformations that will fit between these end points. We can then write Equation 13-32:

$$\Omega_1 = P(x_0, y_0, z_0) \times constant$$

EQUATION 13-32

The fraction of conformations with this end-to-end distance is simply the number of such conformations divided by the total number of conformations available to the chain, which is a constant for a given chain. Similarly, the number of conformations that will fit between the origin and a point x,y,z (Figure 13-48) is the fraction of conformations that would have this end-to end distance multiplied by the same constant (which is related to the total number of conformations available to that chain, of course)—Equation 13-33:

$$\Omega_2 = P(x, y, z) \times constant$$

EQUATION 13-33

The change in entropy on stretching the chains from one position to the other is then just (Equation 13-34):

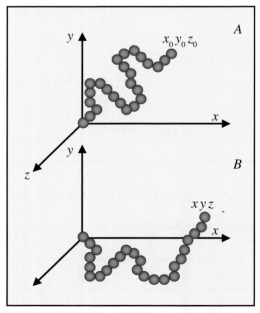

FIGURE 13-48 Schematic diagram showing a single chain fixed at the origin and the other end at positions (A) x_0, y_0, z_0 and (B) x, y, z.

$$S_2 - S_1 = \ln \Omega_2 - \ln \Omega_1$$
$$= \ln P(x,y,z) - \ln P(x_0, y_0, z_0)$$

EQUATION 13-34

The constant of proportionality cancels out so we can neglect it and just use the probability distributions. Naturally, this begs the question: "what is the probability distribution function?"

We discussed this when we considered conformations in Chapter 8, where we found that the distribution of chain end-to-end distances R is approximately Gaussian [meaning it goes as $\exp(-x^2)$] for a chain with a large number of segments, N. If we normalize everything correctly (so that $\sum P(R) = 1$ or, more accurately, $\int P(R) = 1$ for $N \sim \infty$) we have (Equation 13-35):

$$P(R) = \left[\frac{\beta^2}{\pi}\right] \exp(-\beta^2 R^2)$$

where:

$$\beta^2 = \frac{3}{2Nl^2} = \frac{3}{2\langle R_0^2 \rangle}$$

EQUATION 13-35

and $\langle R_0^2 \rangle$ is the root mean square end-to-end distance.

To keep the problem of stretching a single isolated chain simple, let's just consider uniaxial extension along a single coordinate axis, say the z-axis (Figure 13-49). If one end of the chain is at the origin $(0,0,0)$ and the other initially at a point $(0,0,z)$ then the number of arrangements available to the chain can be written (Equation 13-36):

$$\Omega_1 = P(R) = P(x,y,z) = P(0,0,z)$$

EQUATION 13-36

where the probability distribution function $P(x,y,z)$ is simply (Equation 13-37):

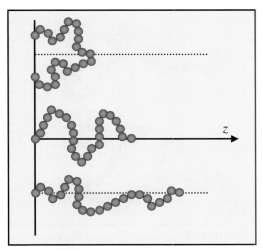

FIGURE 13-49 Schematic diagram showing a 1-dimensional stretch of a single chain.

$$P(x,y,z) =$$
$$\left[\frac{\beta^2}{\pi}\right]^{3/2} \exp(-\beta^2 [x^2 + y^2 + z^2])$$

EQUATION 13-37

Hence, for the one dimensional model we obtain Equation 13-38:

$$P(0,0,z) = \left[\frac{\beta^2}{\pi}\right]^{3/2} \exp(-\beta^2 z^2)$$

EQUATION 13-38

Using Boltzmann's equation, $S = k\ln\Omega$, we can then obtain an expression for the entropy of a chain (with respect to some arbitrary reference state)—Equation 13-39:

$$S = constant - k\beta^2 z^2$$

EQUATION 13-39

This can be differentiated to obtain the force necessary to hold the chain ends a distance z apart (Equations 13-40):

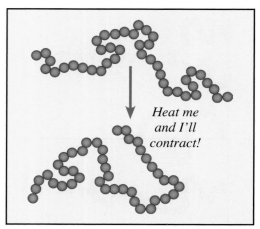

FIGURE 13-50 Schematic diagram depicting the contraction of a rubber chain upon heating.

$$f \approx -T\left(\frac{\partial S}{\partial z}\right)_{V,T}$$

Hence:

$$f = 2kT\beta^2 z$$

EQUATION 13-40

Note that the force required to maintain the chain ends a distance z apart is linear with z

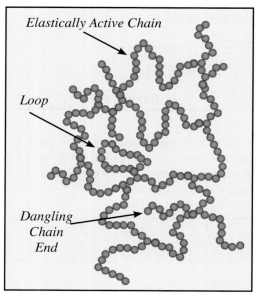

FIGURE 13-51 Schematic diagram of a rubber network.

and this force is zero when $z = 0$.[58] In other words we have force ~ extension, Hooke's law! Furthermore, unlike crystalline materials, this linear relationship should be good for deformations of more than a couple of percent; it should only fail at extensions where the assumption that the chain end-to-end distance can be described by a Gaussian distribution function breaks down (maybe extensions of 300% or more for long chains).

Also notice that by analogy with Hooke's law the modulus is proportional to $2kT\beta^2$ and *increases* with temperature. This would mean that if we stretched our chain a given amount then raised its temperature, it would contract, as illustrated in Figure 13-50 (as the modulus increases the amount of extension you get for a given force decreases). This is in contrast to crystalline and glassy materials, where the modulus *decreases* with temperature, because as a material expands the cohesive forces between the atoms and molecules is reduced. Elastomeric materials, as opposed to single chains, also expand thermally; how, then, do they behave? It depends on how much you stretch them. Both factors will contribute, but at high extensions they will contract; at low extensions, however, ordinary thermal expansion dominates and the material expands with temperature. There will also be some extension or strain where the two effects will cancel for a given temperature increase.

Networks

This brings us to the elastic behavior of rubber networks. There are a number of problems involved in developing a good theory, not the least of which is the fact that real networks are not perfect. If natural rubber is vulcanized there are all sorts of defects in the resulting network; dangling ends and

---

[58] This does not mean that at equilibrium there is zero displacement between the ends of the chain at some instant of time, but rather the average distance between the ends, $<z> = 0$. Remember that the shape of the chain fluctuates all the time and therefore so does the distance between its ends. Also, although $<z> = 0$, $<z^2>^{0.5} \neq 0$.

loops that are not part of the network, some chains that may not be cross-linked at all, and so on (Figure 13-51). There are also chain entanglements that can act to some degree as cross-links. However, by synthesizing (almost) monodisperse polymers and cross-linking them by their ends in solution (to minimize entanglements) it is possible to make good model networks.

But then there are theoretical difficulties, one of the most important being how you account for the displacement of the cross-link points as a function of how the sample as a whole is stretched or compressed. Remember, we are now considering a chain to be the set of segments between cross-link points, so the change in position of the cross-link points with deformation determines the change in the end-to-end distances of the chains. We will use the simplest assumption that the deformation is *affine*, meaning that the cross-link points are considered to be fixed in the body of the rubber and simply change their position in proportion to the change in dimensions of the sample as a whole (Figure 13-52).

Even with this assumption, you can tackle the problem in various ways. We will use the simple development described in Treloar's classic book on rubber elasticity,[59] but advanced students should start with Flory and go from there. Most simple models, including the one we will describe here, give the same force/deformation dependence, but the prefactors can be different. (Flory also gets an additional term that is important in describing swelling, but not simple deformation). We start with a block of lightly cross-linked rubber that is strained parallel to a set of Cartesian axes, as illustrated in Figure 13-53.

In rubber elasticity, it is usual to use extension ratios rather than strain. If a block of rubber is stretched parallel to a single axis, such that its length changes from l_0 to l (Figure 13-54), the extension ratio is defined to be (Equation 13-41):

[59] L. R. G. Treloar, *The Physics of Rubber Elasticity*, Third Edition, Clarendon Press, Oxford, 1975.

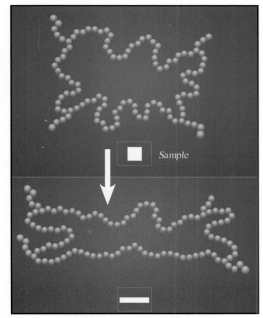

FIGURE 13-52 Schematic diagram illustrating affine deformation.

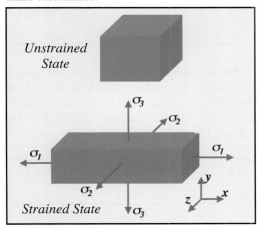

FIGURE 13-53 Schematic diagram illustrating the unstrained and strained states.

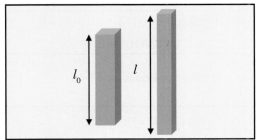

FIGURE 13-54 Schematic diagram illustrating the stretching of a block of rubber.

$$\lambda = \frac{l}{l_0}$$

EQUATION 13-41

The strain would be $(l - l_0)/l_0$. We also assume that there is no change in the volume of the sample so that we obtain Equation 13-42:

$$\lambda_1 \lambda_2 \lambda_3 = 1$$

EQUATION 13-42

and if the sample is stretched in one direction it will contract in the others.

Assuming that the chains within the network obey Gaussian statistics, the entropy of a single chain whose ends are at 0,0,0 and x_0, y_0, z_0 (see Figure 13-49) can be written (Equation 13-43):

$$s_0 = constant - k\beta^2 [x_0^2 + y_0^2 + z_0^2]$$

EQUATION 13-43

(We are using a lower case s to distinguish the entropy of a single chain from that of the network as a whole, which we will designate S.) If the chain is stretched to x, y, z then its entropy becomes (Equation 13-44):

$$s = constant - k\beta^2 [x^2 + y^2 + z^2]$$

EQUATION 13-44

Using the extension ratios (Equations 13-45):

$$x = \lambda_1 x_0 \quad y = \lambda_2 y_0 \quad z = \lambda_3 z_0$$

EQUATIONS 13-45

you can arrive at Equation 13-46:

$$s = constant$$
$$- k\beta^2 [x_0^2 \lambda_1^2 + y_0^2 \lambda_2^2 + z_0^2 \lambda_3^2]$$

EQUATION 13-46

and the change in entropy upon deformation is then (Equation 13-47):

$$\Delta s = - k\beta^2 [x_0^2 (\lambda_1^2 - 1) +$$
$$y_0^2 (\lambda_2^2 - 1) + z_0^2 (\lambda_3^2 - 1)]$$

EQUATION 13-47

If we now assume that all the chains between cross-link points have an identical number of segments (i.e., the parameter β is a constant), then the entropy of deformation of the network can be obtained by summing over all chains (Equation 13-48):

$$\Delta S = \sum \Delta s = - k\beta^2 [(\lambda_1^2 - 1) \sum x_0^2 +$$
$$(\lambda_2^2 - 1) \sum y_0^2 + (\lambda_3^2 - 1) \sum z_0^2]$$

EQUATION 13-48

where the terms $\sum x_0^2$, $\sum y_0^2$, $\sum z_0^2$, describe the positions of the chain ends relative to their other ends. This looks awful and to simplify the equation it would be nice to get rid of all the summation terms. This can be done by assuming that in the unstrained state (see top of Figure 13-53) there is a random distribution of chain end vectors. We can then write (Equation 13-49):

$$\sum R_0^2 = \sum x_0^2 + \sum y_0^2 + \sum z_0^2$$

EQUATION 13-49

Hence it follows (Equation 13-50):

$$\sum x_0^2 = \sum y_0^2 = \sum z_0^2 = \frac{1}{3} \sum R_0^2$$

EQUATION 13-50

This is because the chain end vectors are likely to be found with equal probability in any direction in space.

Upon substituting, we would still have a summation term, $\sum R_0^2$, but by definition the average of the square of the end-to-end distance, $<R_0^2>$, must be equal to the sum of the squares of the end-to-end distances of all the chains divided by the total number of chains, $\sum R_0^2/N$, hence (Equation 13-51):

$$\sum R_0^2 = N\langle R_0^2\rangle$$

EQUATION 13-51

We then get for the entropy (Equation 13-52):

$$\Delta S = -\frac{1}{3}Nk\beta^2\langle R_0^2\rangle(\lambda_1^2 + \lambda_2^2 + \lambda_3^2 - 3)$$

EQUATION 13-52

However we know that (Equation 13-53):

$$\beta^2 = \frac{3}{2\langle R_0^2\rangle}$$

EQUATION 13-53

Hence we obtain the important Equation 13-54:

$$\Delta S = -\frac{1}{2}Nk(\lambda_1^2 + \lambda_2^2 + \lambda_3^2 - 3)$$

EQUATION 13-54

Now let's see what this gives us for simple extension, say in the *x*-direction. We have mentioned that if there is no change in volume upon stretching, the rubber must contract in the *y* and *z* directions to account for its extension in the *x*-direction. Equations 13-55 follow directly from Equation 13-42:

$$\lambda_1 = \frac{1}{\lambda_2\lambda_3} \quad and \quad \lambda_2 = \lambda_3$$

EQUATIONS 13-55

The entropy change is then (Equation 13-56):

$$\Delta S = -\frac{1}{2}Nk\left(\lambda_1^2 + \frac{2}{\lambda_1} - 3\right)$$

EQUATION 13-56

and the force extension relationship is Equation 13-57 [using $\partial\Delta S/\partial l = (1/l_0)\,\partial\Delta S/\partial\lambda$]:

$$f = \frac{NkT}{l_0}\left(\lambda_1 - \frac{1}{\lambda_1^2}\right)$$

EQUATION 13-57

Dividing by the initial cross-sectional area of the sample, the stress/deformation relationship becomes (Equation 13-58):

$$\sigma = \frac{NkT}{V}\left(\lambda_1 - \frac{1}{\lambda_1^2}\right)$$

EQUATION 13-58

where $V = A_0 l_0$. Note that this expression is non-linear in the extension ratios (or strain) unlike the equation for a single chain. However, the temperature dependence of the modulus, E, defined in terms of this stress/extension ratio relationship (Equation 13-59):

$$\sigma = E\left(\lambda_1 - \frac{1}{\lambda_1^2}\right)$$

EQUATION 13-59

varies directly with T, like the single chain, so that a sufficiently stretched rubber band contracts when you heat it. A plot of σ (arbitrary units) versus λ is shown in Figure 13-55, demonstrating the large deviation from linearity predicted by the theory. But, is this particular model any good—how well does this plot compare to experiment?

A comparison is shown in Figure 13-56 for compression ($\lambda < 1$) and extension ($\lambda > 1$) data. It looks pretty good up to deformations of about 150%. At intermediate extension ratios (150% to about 300%—not shown here) the experimental curve lies above that predicted by theory, however. At higher extensions there are even bigger deviations, as shown in Figure 13-57. At higher extension ratios the experimental curve turns up significantly and very large deviations between theory and experiment are observed. For elastomers like natural rubber that have

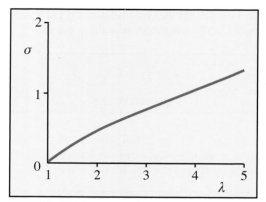

FIGURE 13-55 Theoretical plot of σ versus λ.

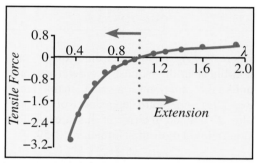

FIGURE 13-56 Experimental plot of tensile force versus λ [redrawn from a figure in L. R. G. Treloar, *The Physics of Rubber Elasticity*, Third Edition, Clarendon Press, Oxford (1975)].

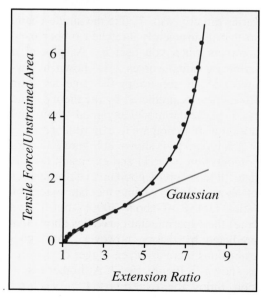

FIGURE 13-57 Experimental plot of tensile force/ unstrained area versus extension ratio (same source as in Figure 13-56).

a very regular microstructure, this deviation is at least in part due to crystallization, which can occur upon stretching and naturally increases the modulus of the sample. The same type of deviation is obtained for elastomers that are incapable of crystallizing, however, and here the deviation is due to the Gaussian approximation. If different distribution functions are used, reflecting the finite nature of the chain, then a good fit to the data can again be obtained.

The Mooney-Rivlin Equation

The non-linear response of elastomers to stress can also be handled by abandoning molecular theories and using continuum mechanics. In this approach, the restrictions imposed by Hooke's law are eliminated and the derivation proceeds through the strain energy using something called strain invariants (you don't want to know!). The result, called the Mooney-Rivlin equation, can be written (for uniaxial extension)—Equation 13-60:

$$\sigma = 2C_1\left(\lambda - \frac{1}{\lambda^2}\right) + 2C_2\left(1 - \frac{1}{\lambda^3}\right)$$

EQUATION 13-60

You should notice that the first term has the same form as that given by simple theories of rubber elasticity. The equation fits extension data for deformations up to about 300% very well, but cannot fit compression data using the same values of the constants C_1 and C_2. Attempts to obtain the second term (in this form) using a molecular theory have not, as yet, been very successful, so we'll say no more about it.

The most important fact that you should grasp from this discussion is the entropic nature of rubber elasticity. Although the agreement between the simple model described here and experiment is not that great you have to keep in mind that there are both theoretical assumptions (e.g., affine deformation) and mathematical approximations (Gaussian chain statistics) that have gone into the derivation. There are alternative theoretical approaches, such as the "phantom network" model of James and Guth that allows the junctions to fluctuate in position, as well as models that do not use the Gaussian approximation. These more complex approaches go a long way towards fixing up the agreement between theory and experiment, but the simple model is enough when you start out in this subject as it provides first and foremost what every good theory should: understanding.

POLYMER MELT RHEOLOGY

You may have heard the expression "as slow as molasses in February" (or was it January?), referring to the fact that this by-product of refining sugar (called treacle in Britain) flows really slowly when cold. Unfortunately for the people living in one area of Boston on January 15, 1919, not slowly enough (Figure 13-58). Actually, the temperature that eventful day had warmed up to about 40°F, from about 2°F the day before, when a tank holding more than 2 million gallons of molasses burst. The resulting 25 ft wall of gooey glop, moving at an estimated speed of 35 mph, inundated the area around the tank, trapping and destroying everything in its path. Twenty one people were killed and it took more than 6 months to clean up the mess. For the next 30 years there are reports that molasses would ooze out of cracks in the sidewalk on hot days.

The story of the Great Molasses Flood (it's true, you can't make up stuff like that) is a useful introduction to our discussion of the viscosity of polymer melts, because

FIGURE 13-58 The great Boston molasses disaster (Courtesy: AP/Wide World Photos).

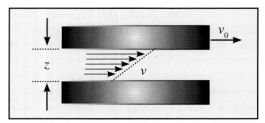

FIGURE 13-59 Schematic diagram of the velocity profile.

it should remind you of a couple of things that you probably know intuitively, but don't always think about when you approach the subject of rheology or fluid mechanics and start getting buried in all the equations. First, the obvious: some fluids, such as water, flow more easily than others, such as molasses or treacle. Second, the rate of flow increases with temperature. As you might guess with a moment's thought, viscosity must in some way be related to the frictional forces between molecules. In turn, this will be affected by free volume, which increases with temperature. If you increase the amount of free volume in the system, the molecules will find it easier to move to positions relative to one another.

Viscosity

The basic law of viscosity was formulated before an understanding or acceptance of the atomic and molecular structure of matter; although just like Hooke's law for the elastic properties of solids the basic equation can be derived from a simple model, where a fluid is assumed to consist of hypothetical spherical molecules. Also like Hooke's law, this theory predicts linear behavior at low rates of strain and deviations at high strain rates. But we digress. The concept of viscosity was first introduced by Newton, who considered what we now call laminar flow and the frictional forces exerted between layers within a fluid. If we have a fluid placed between a stationary wall and a moving wall and we assume there is no "slip" at the walls (believe it or not, a very good assumption), then the velocity profile illustrated in Figure

13-59 is established. Newton made the plausible assumption that the rate of deformation of the fluid, the strain rate, is proportional to the shear stress applied, but the motion is resisted by what he called "the lack of slipperiness of the parts of the liquid," what we now call the viscosity.

Let's look at this in a little more detail, to make sure that you understand what we mean by strain rate in a shear experiment. First, let's go back to Newton's good friend, Hooke. (If you've read the introduction to this chapter, you know we're being facetious!) We have seen above (Figure 13-11) that for shear the most convenient way to describe the deformation of a solid is in terms of the angle θ through which a block of the material is deformed (Equation 13-61):

$$tan\,\theta = \frac{\delta}{z} = \gamma_{xy}$$

EQUATION 13-61

A perfectly elastic solid subjected to a non-destructive shear force will deform almost instantaneously an amount proportional to its shear modulus and then deform no further, strain energy being stored in the bonds of the material. A fluid, on the other hand, continues to deform under the action of a shear stress, the energy imparted to the system being dissipated as flow.

If the top surface moves an amount δ in (say) the x-direction in t seconds and $v_0 = \delta/t$, the strain rate would be (Equation 13-62):

$$\dot{\gamma}_{xy} = \frac{\dot{\delta}}{z} = \frac{v_0}{z}$$

EQUATION 13-62

Or, more generally, if there is a velocity difference dv between two layers of a fluid dz apart (Equation 13-63):

$$\dot{\gamma}_{xy} = \frac{dv}{dz} = \frac{d}{dt}\left(\frac{dx}{dz}\right)$$

EQUATION 13-63

Just as stress is proportional to strain in Hooke's law for a solid, the shear stress is proportional to the rate of strain for a fluid; i.e., $\tau_{xy} \propto \dot{\gamma}_{xy}$. But, for a given shear stress, the rate of strain of treacle will be less than water, so there should be a constant of proportionality that is a material property and a measure of the frictional forces in a fluid; i.e. the viscosity, η. Hence, Newton's law is given by Equation 13-64:

$$\tau_{xy} = \eta\dot{\gamma}_{xy}$$

EQUATION 13-64

Newtonian and Non-Newtonian Fluids

In laminar flow, simple low molecular weight fluids, like water, are essentially Newtonian, but complex fluids like polymers and tomato ketchup show significant deviations from such linear behavior (Figure 13-60). Two types of deviations are observed, the most common being *shear thinning* behavior, where the viscosity decreases with increasing strain rate (which is also called shear rate in many texts, just to confuse you!). This is why you shake the living hell out of your bottle of ketchup in order to pour it over your fries, and also why it can suddenly release and splodge disastrously all over your plate. [If you were a civilized English person and used a nice Newtonian fluid like vinegar to enhance the flavor of your fries (chips), you wouldn't have this problem!] *Shear thickening*, where the viscosity increases with strain rate, occurs in some polymers that crystallize under stress, but is not common and we won't discuss this type of behavior any further.

For shear thinning fluids, it is usual to define an apparent viscosity η_a, or $\eta_a(\dot{\gamma})$, where the symbol in parentheses means that the viscosity is a function of the strain rate (Figure 13-61). The apparent viscosity at a particular strain rate is defined not as the tangent to the slope of the curve, as you might first guess, but as the slope of the secant drawn from the origin to that point. (We don't know why—its probably something to do with the

FIGURE 13-60 When the chips are down, what do you use?

inherent perversity of rheologists.)

If values of η_a are now plotted as a function of strain rate (Figure 13-62), it can be seen that although there is a dramatic decrease in apparent viscosity at high values of $\dot{\gamma}$, at low strain rates the apparent viscosity is essentially constant. Obviously, if you want to compare the rheological properties of different types of polymers, it is this strain rate-independent parameter that would be most useful, as it would presumably be a characteristic property of the polymer. This limiting value is called the *zero shear-rate viscosity*, η_m.

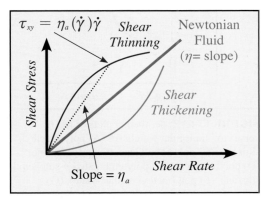

FIGURE 13-61 Schematic diagram of different shear stress versus shear rate curves.

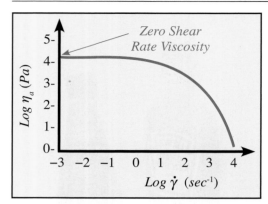

FIGURE 13-62 Schematic plot of log η_a versus log $\dot{\gamma}$.

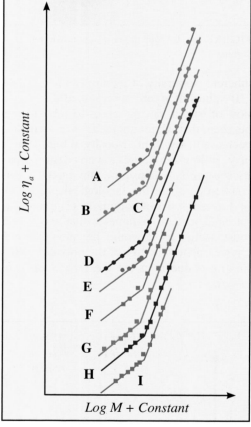

FIGURE 13-63 Experimental plots of log η_a versus log molecular weight for (A) polydimethyl siloxane, (B) polyisobutylene, (C) polyethylene, (D) polybutadiene, (E) polytetramethyl p-sulphenyl siloxane, (F) poly(methyl methacrylate), (G) poly(ethylene glycol), (H) poly(vinyl acetate) and (I) polystyrene [redrawn from G. C. Berry and T. G. Fox, *Adv. Polym. Sci.*, **5**, 261 (1968)].

The variation of η_m with molecular weight is particularly intriguing and important. Plots of log η_m against the log of the molecular weight display very similar curves for a wide range of polymers (Figure 13-63). For low molecular weight materials, there is a linear (first power) dependence of the (weight average) molecular weight or degree of polymerization (*DP*)—Equation 13-65:

$$\eta_m = K_L (DP)_w^{1.0}$$

EQUATION 13-65

(K_L is just a constant.) At high molecular weights the log/log plot is still linear, but with a different slope, meaning that η_m varies as a power of *DP* (Equation 13-66):

$$\eta_m = K_H (DP)_w^{3.4}$$

EQUATION 13-66

(The transition between the two regimes is not abrupt as the plots might imply, but occurs smoothly in a narrow molecular weight range.)

If you give some thought to the power law dependence of η_m with molecular weight, you might imagine that processing a high molecular weight polymer from the melt would be an imposing if not impossible task, because a doubling of the molecular weight would increase the melt viscosity by a factor of $2^{3.4}$, or about an order of magnitude! However, first keep in mind that η_m is the zero shear-rate melt viscosity. Because polymer melts are shear thinning, the melt viscosity encountered in a particular processing operation (see Chapter 14) can be considerably lower (the approximate range of shear rates encountered in some common operations is summarized in Figure 13-64). In addition, the melt viscosity decreases with increasing temperature. Unfortunately, polymers are susceptible to thermal degradation, so there is a limit to the processing temperature you can use that varies with the chemistry of the polymer. This means that you have to pay

FASCINATING POLYMERS—NON-DRIP PAINTS

Looking up at the majestic ceiling of the Sistine Chapel, an irreverent thought popped into your author's subconscious, "Wow! I bet Michelangelo would have given his right arm for non-drip paints." Anyone around who painted a ceiling in "the good old days," before the advent of non-drip paints, can attest that most of the paint actually ended up in the painter's hair. So what are non-drip paints? If you open a tin of a non-drip paint it appears jelly-like and if you dip your paint brush into it some of the jelly-like paint transfers to the brush. The relatively high viscosity of the paint prevents dripping, but high viscosity is not usually compatible with facile painting. However, when you brush the paint on the ceiling its viscosity magically decreases and the paint is easy to apply. What's going on? This is an example of a thixotropic paint. The viscosity of thixotropic paints depends not only upon temperature, but shear rate and time. In many non-drip emulsion paints associative thickeners are added to produce overall thixotropic properties (often an acrylic acid copolymer containing side groups with hydrophobic alkyl

One of your authors' brother-in-law doing something useful for a change and simultaneously painting both a cathedral ceiling and his hair!

ends—not unlike a macromolecular surfactant). In any event, under the influence of shear (brushing), the viscosity of the paint decreases. After you stop shearing (stop brushing), the paint reverts, after a period of time, to its original jelly-like state. Isn't science wonderful?

attention to the molecular weight of your polymer if you want it to be easily (hence cheaply) processed:; and extremely high molecular weight polymers cannot be melt-processed at all.

So, why are polymer melts shear thinning and what is the origin of this interesting variation of η_m with molecular weight? We will consider the latter question first.

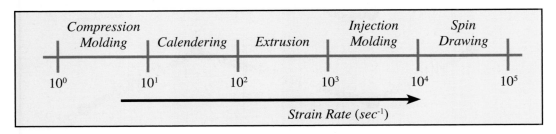

FIGURE 13-64 Strain rates encountered in polymer processing.

*Entanglements and the Dynamics of
Polymer Chains*

The viscosity of low molecular weight fluids depends largely on factors like free volume and this results in a linear dependence of η_m on *DP*. This makes intuitive sense, one would expect that doubling the size of the molecule would lead to proportional increases in frictional forces and hence viscosity. However, at chain lengths of the order of 300–500 main chain atoms things change and this is because of *entanglements*. Short chains don't entangle but long ones do— think of the difference between a nice linguini and spaghettios, the little round things you can get out of a tin. (We have some value judgements concerning the relative merits of these two forms of pasta, but on the advice of our lawyers we shall refrain from comment!)

A better analogy for a polymer melt would actually be a nest of very long, writhing, snakes (Figure 13-65), because unlike spaghetti the chains in the melt are constantly moving. It is the entanglements (the number of which increases significantly with molecular weight) that give rise to the 3.4 power law dependence and also shear thinning behavior. This latter phenomenon is easily explained. As the strain rate increases, the polymers become increasingly stretched out in the shear direction and this effectively decreases the number of entanglements relative to the unstrained state; thus the viscosity of the melt decreases.

FIGURE 13-65 Schematic diagram of a nest of snakes.

This is all well and good, but a deeper understanding of things like the 3.4 power law requires a study of the dynamics of polymer chains, an advanced topic that can involve mathematical techniques with which you may not be acquainted (and may never want to be!) Fortunately, the basic concepts are relatively straightforward and some key relationships can be obtained with beautiful simplicity using something called scaling arguments, applied to polymers by de Gennes. But before getting to this, it is useful to briefly consider an older model, the Rouse-Beuche theory, as this allows us to introduce the concept of relaxation modes, which we will employ later in our discussion of viscoelastic behavior.

In the Rouse-Beuche theory, a polymer chain is assumed to behave like a set of beads linked by springs (Figure 13-66). The beads vibrate with frequencies that depend upon the stiffness of the springs, but these vibrations are modified or damped by fric-

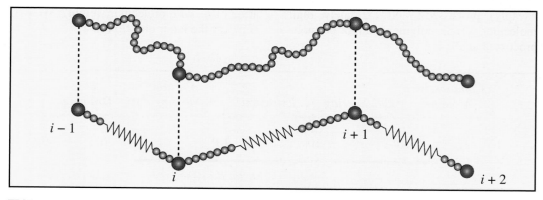

FIGURE 13-66 Schematic diagram illustrating the Rouse-Beuche model.

tional forces acting between the chain and the surrounding medium. There are various types of motion of a chain of this type, called the *normal modes of vibration*. You have probably calculated the frequency of vibration of a simple harmonic oscillator in your elementary physics class. This is the same type of thing, except you now have to consider coupled oscillations of the springs and the effect of damping. In any event, using straightforward classical mechanics a so-called relaxation time that is associated with each of these normal modes can be calculated. It was found that the flow behavior of polymer melts is dominated by the mode with the longest relaxation time, corresponding to a coordinated movement of the molecule as a whole.

This model, in its simplest form, predicts that viscosity should be directly proportional to molecular weight, which it is only below the entanglement limit. This is because chains are assumed to be moving independently, essentially allowing them to "pass through one another," thus neglecting entanglements. The model can be modified by allowing certain modes to be more strongly hindered or damped, but this introduces other problems and the theory then seems (to us) to get bogged down. Furthermore, de Gennes has questioned whether the concept of discrete modes is appropriate at all, as motions might be so coupled and mixed that there is, in effect, a continuum or spectrum of relaxation times. Instead, he proposed that the dynamics of a chain be treated as a snake-like motion among a set of obstacles (Figure 13-67) that represent the other chains in the melt. This he called *reptation* and proceeded to base his analysis on the tube model of Edwards.

Reptation

The tube model, as the name suggests, assumes that the motion of a polymer chain in a network of entanglements can be treated as if it were trapped inside a tube. There are then two types of motion: conformational changes within the confines of the tube and

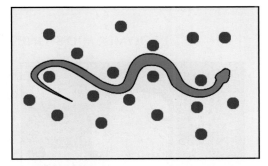

FIGURE 13-67 Schematic diagram illustrating a snake-like motion among a set of obstacles.

motion through the tube (Figure 13-68). It is the latter that interests us in discussing the rheology of polymer melts and this model provides a basis for detailed calculations. Even better from our perspective, de Gennes

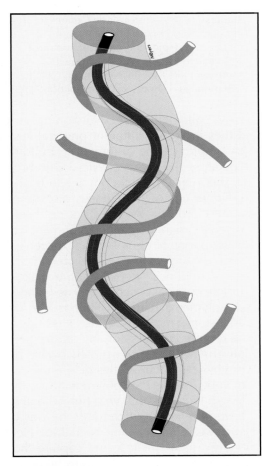

FIGURE 13-68 Schematic diagram illustrating the tube model.

POLYMER MILESTONES—PIERRE-GILLES DE GENNES

Pierre-Gilles de Gennes. [When your authors wrote to Professor de Gennes to ask permission to use his photograph, he replied with typical humor, "You are most welcome to use my photo—although somebody like Kim Basinger might be a better image of a polymer construction!"]

Pierre-Gilles de Gennes was born in Paris, France in 1932. He was educated at the Ecole Normale in Paris. After gaining experience primarily in the fields of neutron scattering and magnetism at the Atomic Energy Center (Saclay), he obtained his doctoral degree in 1957. He then went to Berkeley for a couple of years, served in the French Navy for 27 months and then accepted a position of assistant professor at Orsay. It was in Orsay that he initiated his studies on superconductors and liquid crystals. In 1971, de Gennes became a Professor of Physics in the prestigious College de France. His focus on polymer physics stemmed from his participation in STRASA-COL (a collaboration between Strasbourg, Saclay and the College de France). He continued to serve at the College de France and was the Director of ESPCI (Ecole Superiéure de Physique et de Chimie Industrielles de la Ville de Paris). Pierre-Gilles de Gennes has been called "the Isaac Newton of our time." He has a knack of seeing commonalities in wildly different physical systems and then reducing these observations to simple scaling arguments. Building on the seminal contributions of Sir S. F. Edwards, an English physicist who brought some new calculation techniques used in theoretical particle physics to bear on the problem of the spatial disposition of polymer chains, de Gennes discovered that there were analogies in the arrangement ("order in disorder") of polymer molecules and, amazingly, those observed when a system of magnetic moments moves from order to disorder. This led to new descriptions of complex phenomena in polymers. His so-called "blob" model, a simple but elegant concept, predicts the way in which polymers behave in concentrated solutions. Reptation, the snake-like motion of polymer chains, was another simple concept described by de Gennes that made an enormous impact on the field of polymer dynamics. He has also made seminal contributions in areas such as adhesion, gels, wetting, etc. His book, *Scaling Concepts in Polymer Physics*, should be required reading for any serious polymer scientist. In 1991, de Gennes' accomplishments were recognized by his peers and he was awarded the Nobel Prize in Physics. Pierre-Gilles de Gennes is a gem! Your authors, along with the other members of the PSU polymer faculty, had the opportunity to sit down and discuss at length their research with him when he visited our department in 1984. You would have immediately liked him, he was unpretentious, had a superb sense of humor and was eminently approachable. He was also an outstanding lecturer who could deftly "pitch" his talk and capture the interest of his audience regardless of their academic background. Sadly, he passed away at the time of writing this book.

adopted this approach in a simple way to determine what he called the main *scaling laws*, the power law relationships between things like viscosity and molecular weight. (Things like this always look simple after somebody a lot more clever than you points them out).

In order to follow the argument you need to recall a couple of things about Einstein's theory of diffusion. First, the diffusion coefficient depends upon a friction coefficient, ξ (Equation 13-67):

$$D = \frac{kT}{\xi}$$

EQUATION 13-67

Second, the distance traveled, l, by a particle diffusing in a medium in a time t goes as $t^{1/2}$. This is, of course, the random walk problem. When we applied this to a polymer chain we were concerned with the distance between the ends in a walk of N steps; here we are concerned with the distance traveled after a time t. Before we had $<R^2>^{1/2} \sim N$, here we have $<l^2>^{1/2} \sim t$, or, more formally (Equation 13-68):

$$\langle l^2 \rangle^{1/2} = 2Dt$$

EQUATION 13-68

Now we apply these equations to determine the time t_d required for a polymer chain to diffuse out of its tube. Naturally, the contour length of the tube is defined by the length of the chain, l_c, so based on the diffusion equations we can write (Equations 13-69):

$$D_{tube} = \frac{kT}{N\xi_s} \qquad \tau_d = \frac{l_c^2}{D_{tube}}$$

EQUATIONS 13-69

where ξ_s is the friction coefficient per segment of the chain (i.e., $\xi = N\xi_s$). The contour length of the chain depends on the number of segments that it contains; i.e., $l_c \sim N$, hence,

$\tau_d \sim N^3$.

Very nice, you might say, but where does the viscosity come in? Well the time required for a chain to diffuse out of its tube, τ_d, is something we will refer to as a *relaxation time*. In general, relaxation times in liquids measure the period necessary for a system to relieve an applied stress by the molecules slipping past one another. It is thus directly proportional to the viscosity and we can write Equation 13-70:

$$\eta_m \sim \tau_d \sim N^3$$

EQUATION 13-70

So scaling arguments predict that the melt viscosity goes as the third power of the molecular weight. Similar arguments predict that the overall diffusion coefficient (not the same as D_{tube} above) goes as N^2. This latter prediction is in excellent agreement with experiment, but the former is off by a bit. At this point, it is not clear if this is a minor problem that can be fixed by tinkering with the model, or a more fundamental difficulty. Regardless, the concept of reptation has played a seminal role in our understanding of polymer dynamics and the important thing at this point is that you get a feel for these ideas. Furthermore, this discussion has allowed us to introduce the concept of a relaxation time, something we need to explore in more detail. Before getting to this, however, we wish to consider some additional aspects of the behavior of polymer melts during processing and thus introduce the subject of *viscoelasticity*.

Melt Flow at High Shear Rates

We have already discussed one aspect of non-linear behavior in polymer melts, namely shear thinning. A second aspect manifests itself when we examine the flow of polymer melts through small diameter tubes or capillaries. This is the phenomenon of *jet* or *die swelling*, where a polymer forced into a narrow tube, diameter d_0, swells when

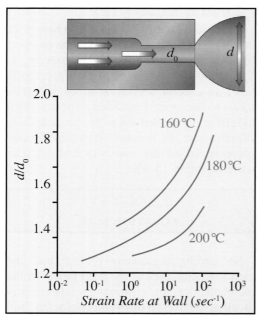

FIGURE 13-69 Jet swelling [drawn schematically from the data of Burke and Weiss, *Characterization of Materials in Research*, Syracuse University Press (1975)].

it emerges at the other end to a diameter d (this phenomenon is of obvious importance in things like extrusion). This is a complex phenomenon and depends on things like temperature, the strain rate, and the length of the capillary relative to its diameter (check the plots in Figure 13-69).

Clearly, this is an elastic rather than a simple viscous type of response and simple fluids like water do not behave in this fashion. A qualitative understanding of this is relatively straightforward: the chains in the melt are deformed as they are pushed into the capillary (or die, if we are talking about extrusion) and as they are sheared in the narrow channel. Normal stresses then develop in a direction perpendicular to the flow (i.e., axially in the capillary). These stresses are relieved and deformation is recovered when the polymer exits the die, so the polymer extrudate swells. (It's a bit more complicated than this, because some relaxations can also occur in the capillary, if it is long enough — but you get the idea.)

This elastic property of polymer melts also places a limit on the rate at which polymers can be extruded, because *melt fracture* can occur. Above a certain critical shear stress, polymers no longer extrude smoothly, as in the top picture of Figure 13-70, but start to display surface defects which, as the shear increases, results in a so-called "*sharkskin*" appearance (middle of Figure 13-70). The instability increases with strain rate and a banded structure appears. Ultimately, a loss of cohesion occurs and the melt fractures, distorting and breaking apart as it emerges

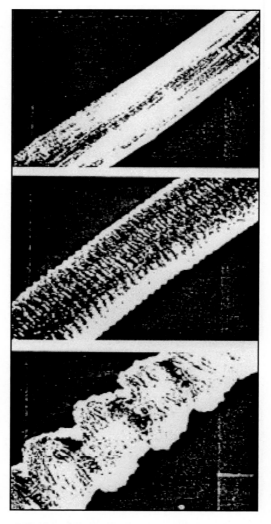

FIGURE 13-70 Melt fracture [reproduced with permission from J. J. Benbow, R. N. Browne and E. R. Howells, *Coll. Intern. Rheol.*, Paris, June-July 1960)].

from the die. Qualitatively, this phenomenon is also easily understood: above a certain strain rate the chains no longer have time to "relax" and the melt can behave like an elastic, almost glassy, material and fracture.

So, polymer melts display elastic as well as viscous behavior. In other words they are *viscoelastic*. Do polymer "solids" display some viscous behavior? Also, we've used the word "relaxation" when we talk about time-dependent behavior, but what do we mean by this? To find out we now need to explore the subject of viscoelasticity in more depth.

VISCOELASTICITY

An intuitive knowledge of viscoelasticity appears to have been with us throughout recorded history. The ancient Greeks, for example, liked to ride around in chariots that had light and flexible wheels made from thin wood (Figure 13-71). But Homer knew that the first thing to do on getting your chariot out in the morning was to put the wheels back on. Telemachus, in *The Odyssey*, would tip his chariot against a wall. The goddess Hebe had the job of putting the wheels on the chariot of gray-eyed Athene.[60] The Greeks were obviously familiar with a phenomenon called creep. If they left the weight of a chariot on the wheels for any length of time they would become permanently bent and eccentric. This would not be good if you were going out to battle Xerxes and the Persian hordes.

Similarly, Robin Hood, or the outlaw archers around whom the legends were constructed, knew never to leave their bows strung when not in use. This is because of another aspect of viscoelastic behavior,

---

[60] We found this stuff in *Structures*, J. E. Gordon, Penguin Books, 1991. Gordon also reported that the wheels of the coach of the Lord Mayor of London were at that time distinctly eccentric, presumably because the weight has been left on them for long periods. This was (and maybe still is) a general problem that apparently results in people becoming "seasick" when riding around in state coaches. No wonder the Queen often looks a bit under the weather!

FIGURE 13-71 It's a better ride if your wheels are round!

stress relaxation. Over time the stress in the bowstring would decrease (as the wood in the bow "relaxed"), seriously affecting the ability of an English archer to mow down noble French knights, who throughout most of the 100 years war apparently fought under the misapprehension that the English were equally chivalrous and would want to fight fair, knight against knight.

In this section we are going to examine such viscoelastic properties in some detail and we will start by examining in turn three important mechanical methods of measurement: creep, stress relaxation, and dynamic mechanical analysis. This will lead us to interesting things like *time-temperature equivalence* and a discussion of the molecular basis of what we have referred to as relaxation behavior.

Creep

Let's start with creep, which is the easiest to understand. In a creep experiment you simply subject a sample to a constant (non-destructive) load or stress and watch it deform (i.e., measure the strain) as a function of time (Figure 13-72). The first systematic studies of this type were conducted by the German physicist, Weber, in 1835, who noted that silk fibers exhibited an immediate deformation upon loading, behavior that we call elastic, followed by an extension that gradually increased with time. Similarly, upon removing the load both an immediate and delayed contraction were observed. In many

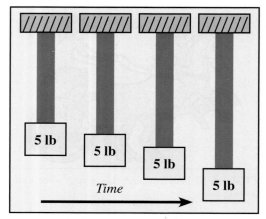

FIGURE 13-72 Creep—deformation under a constant load as a function of time.

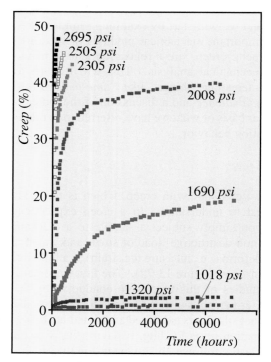

FIGURE 13-73 Creep curves for cellulose acetate [redrawn from the data of W. N. Findley, *Modern Plastics*, **19**, 71 (August 1942)].

which we now know are polymeric, the bulk of the deformation is time-dependent. Also, these materials are more susceptible to what is referred to as *secondary creep*, or irreversible deformation. In other words, after you remove the stress the material does not completely recover its original dimensions, no matter how long you wait.

As an example of creep, consider some classic experiments performed on cellulose acetate more than 50 years ago. The plots in Figure 13-73 show strain, measured as percent elongation, plotted as a function of time. Note how creep depends on stress. For loads of the order of 1000 psi (real units—we love it!) the strain/time plots are more or less flat, indicating that creep is not large (if you examine the initial parts of the plot, however, it is apparent that it took some time to reach maximum strain—the sample did not deform instantaneously, as an ideal elastic solid would). The extent of creep increases dramatically with stress and at loads in excess of 2000 psi the material fails in the time frame of the experiment, 7000 hours, or about a year. (It's hard to do experiments that last this long, because something usually goes wrong, like the power goes off and/or the temperature changes.)

It is important to realize that this type of behavior is not just a simple addition of linear elastic and viscous responses. An ideal elastic solid would display an instantaneous elastic response to an applied (non-destructive) stress (top of Figure 13-74). The strain would then stay constant until the stress was removed. On the other hand, if we place a Newtonian viscous fluid between two plates and apply a shear stress, then the strain increases continuously and linearly with time (bottom of Figure 13-74). After the stress is removed the plates stay where they are, there is no elastic force to restore them to their original position, as all the energy imparted to the liquid has been dissipated in flow.

Polymers can and do display both of these elements in their response to an applied stress (an elastic response and permanent deformation), but the defining feature of the creep curve of a viscoelastic solid is some sort of

of the materials studied by Weber, things like glass and silver, *primary creep*, the reversible component of the delayed response, was found to be just a fraction of the (almost) instantaneous elastic response (~3%). Conversely, in materials like silk and rubber, coupling of elastic and viscous responses to give a strain/time curve that is non-linear (Figure 13-75). A similar non-linear response is seen in recovery, when the load is removed. This type of behavior is called a *retarded elastic* or *anelastic* response. We will see that polymers show various combinations of these responses—elastic, viscous and retarded elastic—the extent of each depending on stress, the structure and morphology of the polymer (hence things like T_g) and temperature. We will also see that although the mechanical behavior is decidedly non-linear, for amorphous polymers it is possible to model behavior with a suitable combination of linear models, at least to a first approximation. But we're getting ahead of ourselves again—let's look at stress relaxation next.

Stress Relaxation

In contrast to creep, which is a constant stress experiment, stress relaxation is a constant strain experiment (and is usually somewhat easier to perform than a creep experiment)—Figure 13-76. A sample is deformed instantaneously (well, almost instantaneously) to a given value of the strain and the stress required to maintain that deformation is measured as a function of time. As the sample "relaxes" (i.e., as the chains change their conformations, disentangle and slide over one another) this stress decreases.

The data are not usually reported as a stress/time plot, but as a modulus/time plot. This time-dependent modulus, called the relaxation modulus, is simply the time-dependent stress divided by the (constant) strain (Equation 13-71):

$$E(t) = \frac{\sigma(t)}{\varepsilon_0}$$

EQUATION 13-71

Typical stress relaxation curves, plotted on a log/log basis, are shown in Figure 13-77 and vary considerably with temperature. We will see later how these various curves can

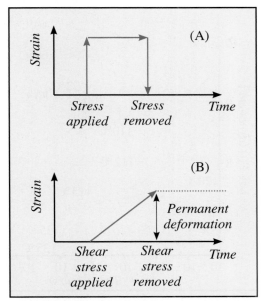

FIGURE 13-74 Schematic diagrams of (A) purely elastic response and (B) purely viscous response.

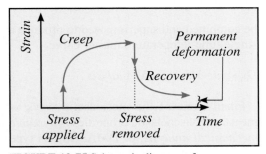

FIGURE 13-75 Schematic diagram of a viscoelastic response.

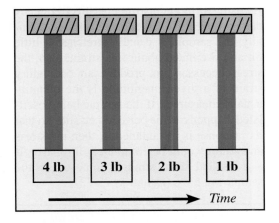

FIGURE 13-76 Stress relaxation—a constant strain experiment.

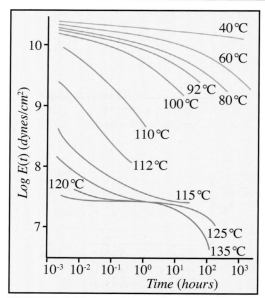

FIGURE 13-77 Stress relaxation of poly(methyl methacrylate) [redrawn from the data of J. R. McLoughlin and A. V. Tobolsky. *J. Colloid Sci.*, **7**, 555 (1952)].

be combined and superimposed to produce a single time-temperature master curve.

Dynamic Mechanical Analysis

Finally, one of the most useful ways of measuring viscoelastic properties is *dynamic mechanical analysis*, or DMA. In this type of experiment, an oscillating stress is applied to the sample and the response is measured as a function of the frequency of the oscillation. By using different instruments this frequency can be varied over an enormous range. Actually, the sample is usually stretched a little bit and oscillated about this strain; also, the stress necessary to produce an oscillatory strain of a given magnitude is the quantity usually measured. If the sample being oscillated happens to be perfectly elastic, so that its response is instantaneous, then the stress and strain would be completely in-phase. If a sinusoidal shear strain is imposed on the sample we have (Equation 13-72):

$$\gamma(t) = \gamma_0 \sin 2\pi f t = \gamma_0 \sin \omega t$$

EQUATION 13-72

where f is the frequency and ω the angular frequency of the oscillation. The time-dependent shear stress that would have to be applied to obtain this sinusoidal strain is given (for small strain amplitudes) by Hooke's law (Equation 13-73):

$$\tau(t) = G\gamma_0 \sin \omega t$$

EQUATION 13-73

In contrast, the shear required to produce sinusoidal strain in a Newtonian fluid would be 90° or $\pi/2$ out of phase with the strain, as Equations 13-74 would indicate.

$$\tau(t) = \eta\dot{\gamma} = \eta\frac{d}{dt}(\gamma_0 \sin \omega t)$$

$$\tau(t) = \eta\gamma_0 \cos \omega t$$

EQUATIONS 13-74

This should immediately suggest to you that in a viscoelastic solid the applied stress and resulting strain should be out of phase by some angle intermediate between 0 and $\pi/2$ radians. In other words, in order to obtain a sinusoidally varying strain (Equation 13-75):

$$\gamma(t) = \tau_0 \sin \omega t$$

EQUATION 13-75

A stress that is δ out of phase would have to be applied (Equation 13-76):

$$\gamma(t) = \tau_0 \sin(\omega t + \delta)$$

EQUATION 13-76

You could think about the experiment the other way around. An oscillating stress [$\tau(t) = \tau_0 \sin\omega t$] could be applied to the sample and the resulting strain would lag the stress by an amount δ [$\gamma = \gamma_0 \sin(\omega t - \delta)$]. But mathematically you can write it either way, the strain lagging the stress by δ or the stress leading the strain by δ. Everybody writes it in the latter form. If you remember your trigonometric relationships (we'll settle for you remembering that these relationships exist), then you will know that the expression for the applied stress (Equation 13-76) can be rewritten so as to separate the terms in δ from those in ω (Equation 13-77):

$$\tau(t) =$$
$$(\tau_0 \cos \delta) \sin \omega t + (\tau_0 \sin \delta) \cos \omega t$$

EQUATION 13-77

Look at this carefully; it has an in-phase component (the term in *sinωt*) and an out-of-phase component (the term in *cosωt*). This can be used to define the relationship between stress and strain in terms of two moduli. First writing Equation 13-78:

$$\tau(t) =$$
$$\gamma_0 \left[\left(\frac{\tau_0}{\gamma_0} \cos \delta \right) \sin \omega t + \left(\frac{\tau_0}{\gamma_0} \sin \delta \right) \cos \omega t \right]$$

EQUATION 13-78

Then defining an in-phase modulus, G', and out-of-phase modulus, G'' (Equations 13-79):

$$G'(\omega) = \frac{\tau_0}{\gamma_0} \cos \delta \qquad G''(\omega) = \frac{\tau_0}{\gamma_0} \sin \delta$$

EQUATIONS 13-79

we get Equation 13-80:

$$\tau(t) = \gamma_0 [G'(\omega) \sin \omega t + G''(\omega) \cos \omega t]$$

EQUATION 13-80

The in-phase component, G', is called the *storage modulus* while the out-of-phase component, G'', is called the *loss modulus*. It also follows that (Equation 13-81):

$$\tan \delta = \frac{G''(\omega)}{G'(\omega)}$$

EQUATION 13-81

The storage modulus is directly related to the energy stored in the sample (per cycle of oscillation) and can thus be thought of as the elastic component of the response. The loss modulus is related to the energy lost or dissipated per cycle and can therefore be thought of as the viscous component of the response. All of these terms vary with the frequency of oscillation of the sample, because δ depends upon ω. At some frequencies of oscillation there is little loss and the sample behaves like an elastic solid. At other frequencies the loss term is significant, it all depends on the frequency dependence of the relaxations in the material, which we will get to in a while. A schematic representation of the dependence of G', G'', and *tan δ* on frequency for an amorphous polymer is shown in Figure 13-78 (real data can look a bit more complex because the response of a polymer is governed by a spectrum of relaxation times, but don't worry about that for now).

You will notice that at low frequencies the storage modulus $G'(\omega)$ is characterized by low values, in fact those characteristic of a rubber. As the frequency increases, this modulus changes in value by several orders of magnitude and levels out at the value characteristic of the glassy state. At each of these extremes the loss modulus is low and the behavior is largely elastic. Clearly, there is a transition that is akin to a T_g and the loss modulus and tan δ are a maximum in this frequency range. Keep in mind that this is

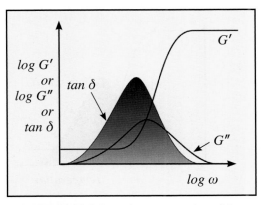

FIGURE 13-78 Schematic representation of the dependence of G', G'', and *tan δ* on frequency measured at constant temperature.

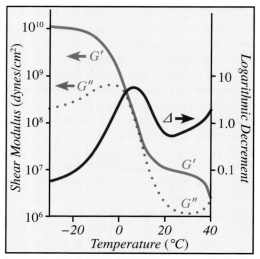

FIGURE 13-79 Experimental data of G', G'', and *tan δ* as a function of temperature [redrawn from the data of L.E. Nielsen, *Mechanical Properties of Polymers*, Reinhold, New York (1962)].

a constant temperature experiment and for the sake of argument let's assume it was conducted at room temperature. What may surprise you is that both glassy and rubbery polymers would give this type of curve, but they would be shifted along the frequency axis relative to one another.

Time-Temperature Equivalence

In fact, if you compare the results obtained at a constant temperature, where ω is varied, to those obtained by oscillating the sample

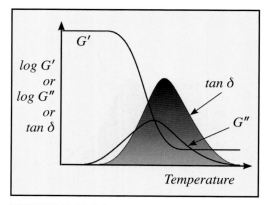

FIGURE 13-80 Schematic representation of the dependence of G', G'', and *tan δ* with temperature measured at constant frequency.

at a constant frequency and varying the temperature, there is a clear equivalence or correspondence. G', for example, increases several orders of magnitude as the sample is cooled. This occurs over the fairly narrow range of temperatures that define the T_g, of course. Some real data obtained as a function of temperature is shown in Figure 13-79. You cannot always find good data in the literature where a given polymer has been studied both as a function of time and temperature, so we'll just carry on with our idealized representation, shown in Figure 13-80.

So Figures 13-78 and 13-80 are our idealized representations. This correspondence is obviously interesting and important, but we will defer a discussion of the molecular origin of this until later. First we want to explore this time-temperature correspondence and the various regions of viscoelastic behavior in a little more detail.

Viscoelastic Behavior: Amorphous Polymers

In the next few paragraphs we will see that for cross-linked, high molecular weight amorphous polymers there are four regions of viscoelastic behavior: the *glassy region*, the *glass transition region*, a *rubbery plateau* and a *terminal flow region*. Before exploring these, we want to emphasize that properties do not change abruptly but gradually on going from one region to the next. A glassy polymer starts to display ductile behavior well below the T_g, for example. A rough summary of properties as a function of temperature is shown on the specific volume versus temperature plot seen in Figure 13-81.

Let's start our discussion of the range of viscoelastic properties of amorphous polymers by considering the modulus of an amorphous polymer measured as a function of temperature. We know that any measurement of stress, hence modulus, we make is going to vary with time (see Equation 13-71), so to compare values at different temperatures we make all our measurement

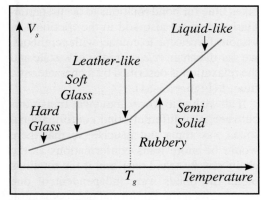

FIGURE 13-81 Summary of properties of amorphous polymers as a function of temperature.

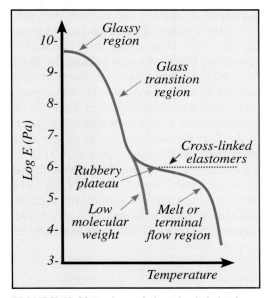

FIGURE 13-82 Regions of viscoelastic behavior.

over some arbitrary short period, say 10 seconds. In other words, the sample is stretched (almost) instantaneously to a chosen value of the strain and after 10 seconds the stress necessary to maintain this strain is measured and the modulus calculated.

At "low" temperatures (by which we mean well below the T_g of the sample) a value of the modulus characteristic of the glassy state is found (Figure 13-82). As the temperature is increased and the polymer approaches and passes through the glass transition region, the value of the modulus drops several orders of magnitude. In this region, the sample displays properties that are similar to leather, characterized by a marked anelastic or retarded elastic response. At temperatures above the T_g, three things can happen, depending on the molecular weight of the sample and whether or not it is cross-linked. Low molecular weight polymers (those whose chain length is less than the entanglement limit) become liquid-like in their behavior. High molecular weight polymers have a rubbery plateau, where the sample displays rubber-like elasticity in our short 10 second time span. The width of this more-or-less constant modulus region depends on the molecular weight of the sample and extends to the degradation temperature for lightly cross-linked materials. Finally, at higher temperatures a non-cross-linked polymer becomes liquid-like in its properties.

As you might expect from our discussion of dynamic mechanical properties, equiva-

lent results are obtained in a constant temperature experiment where time is the variable—in other words, the sample is stretched and the stress required to maintain a given strain is measured as a function of time. The modulus/time plot then appears as shown in Figure 13-83, the difference between one

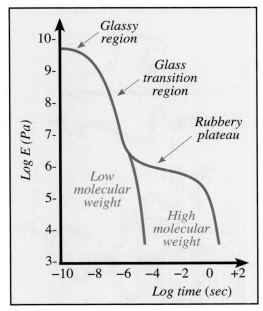

FIGURE 13-83 Schematic plot of modulus versus time.

polymer (say glassy) and another (say rubbery) being essentially a displacement along the time axis. Over "short" time periods the material is glassy. As time goes on the stress required to maintain a given strain decreases and the calculated modulus [$E(t) = \sigma(t)/\varepsilon_0$] drops orders of magnitude over a fairly well defined range, a transition that corresponds to the T_g in the modulus versus temperature experiment. Calculated values of $E(t)$ characteristic of a rubber are then obtained, and at longer times values characteristic of a melt. If our experiment started with a glassy polymer at room temperature it could take times of the order of 10^3 years to reach this level of stress relaxation.

Relaxation in Amorphous Polymers

So, what is the molecular basis for this range of behavior? To explore this let us first consider a hypothetical isolated chain in space, then imagine stretching this chain instantaneously so that there is a new end-to-end distance. The distribution of bond angles (trans, gauche, etc.) changes to accommodate the conformations that are allowed by the new constraints on the ends. Because it

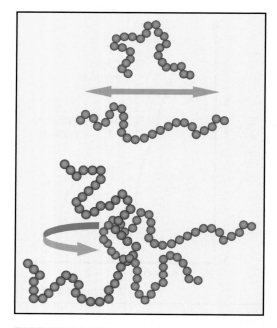

FIGURE 13-84 Relaxation processes.

takes time for bond rotations to occur, particularly when we also add in the frictional or viscous forces due to contact with neighbors, we say the chain *relaxes* to the new state and the relaxation is described by a characteristic time, τ (Figure 13-84).

If all we had to worry about was transitions between two different bond conformational states, say trans and gauche (so that there would be more trans conformations as the chain was stretched out), and if the rotations of all the bonds were independent of one another, then the conformational relaxation process we have described here could be described in terms of a single relaxation time τ_1 (top of Figure 13-84). But in real polymer materials there is a whole range or spectrum of time-dependent rearrangements.

Bond rotations are not independent of one another, but are affected by both short- and long-range steric interactions: a chain is not acted on by simple frictional forces between segments of different chains, but is also affected by entanglements (bottom of Figure 13-84) and so on! Accordingly there are all sorts of complex coupled motions or relaxation processes, often referred to as "modes," and each of these modes will have its own characteristic relaxation time (or times). Even this is a simplification and de Gennes has argued that the whole concept of modes may be invalid.

The various relaxation processes that occur in entangled chains may be so complex and coupled that only a smooth distribution or spectrum of relaxation times may actually exist. Nevertheless, the concept of modes and even models that are characterized by just one or two relaxation times is useful when you start out in this subject, as it gives you a good feel for what is going on. Just keep in mind that the dynamics of entangled polymer chains is a complex subject and one where we presently only have rough theoretical models.

Given all these caveats, how fast are relaxation processes in amorphous polymers? In low molecular weight liquids relaxations are very fast, occurring in time frames of the order of 10^{-10} secs. In a polymer melt ($T > T_g$)

FASCINATING POLYMERS—SILLY PUTTY

Unless you have just arrived from another planet, we bet you've all seen and played with Silly Putty®. It's fascinating stuff; you can shape, stretch or mold it into almost any form and it makes a great "stress reliever" for uptight individuals. It will bounce like a ball, but can shatter like glass. In 1989, at Alfred University in New York State, in what was called "The Great Silly Putty® Drop," an experiment was performed to determine what would happen to a 100 pound ball of silly putty if it was dropped from the roof of the McMahon Engineering Building. The burning question of the day was: "Will the ball bounce or shatter?" The crowd waited in restless anticipation and we wouldn't doubt that a little money exchanged hands. When the ball was dropped, it initially bounced about 8 feet into the air, but then shattered on the second impact. Well, what is this amazing material and how was it discovered? In 1943, James Wright, a Scottish engineer, was working in the labs of the General Electric Company in New Haven, Connecticut. For whatever reason, he happened to mix silicone oil with boric acid in a test tube, and, we presume, watched in amazement as the mixture transformed into a soft gooey solid. Moreover, he is purported to have tossed the gooey substance on the floor and watched it bounce like a vulcanized rubber ball. Although this bouncing putty had incredible properties, try as they might, General Electric scientists and engineers could find no practical applications for it. Academics, of course, loved the material, as it was a beautiful teaching tool with which to demonstrate the fundamentals of viscoelasticity. It wasn't until 1949 that it was marketed as a children's toy. And as they say, "the rest is history." Silly Putty® is still being produced in Pennsylvania and remains a favorite plaything for children and adults alike. But why has silly putty such unusual characteristics? It is thought that the relatively weak intermolecular bonds formed by the transfer of electron pairs from oxygen to boron are largely responsible. These bonds act like temporary or labile cross-links, which are continually making and breaking, and the physical and mechanical properties of the material then depend upon the time scale of the experiment.

Silly Putty® is a registered trademark of Binney & Smith (used with permission).

they are much slower. Maconnachie et al.[61] used neutron scattering to determine how quickly a stretched polystyrene sample ($M_w \sim 144,000$) would relax back to its unperturbed dimensions and found that it took about five minutes to approach completion. Relaxation processes are, of course, a strong function of temperature in polymers, so that at temperatures at and below T_g relaxation times are much longer. This brings us back to the four regions of viscoelastic behavior of amorphous polymers, which we will now briefly consider in terms of the types of relaxations that can occur.

In the *glassy state* (see Figure 13-82) conformational changes are severely inhibited, although they can occur over very long time periods, as we have seen. Some local conformational changes and motions involving side chains can and do occur, however. Nevertheless, for small loads and small strains it is some combination of the covalent bonds and secondary forces between the chains that are deformed and over short time spans the response can be regarded as essentially elastic.

As the temperature is raised to the glass transition region, thermal energy, hence free volume, increases and various cooperative motions involving longer chain sequences or segments can occur. Frictional forces are such that motions are sluggish, however, and retarded or anelastic responses are observed. In dynamic mechanical experiments, we are in a region where there is a severe mismatch between the imposed frequency of oscillation and the time scale of the relaxation processes. As a result, frictional losses are at a maximum, resulting in the observed peak in the *tan δ* curve that we described earlier.

If the temperature is increased above the T_g the time scale of conformational relaxation processes becomes shorter, so that the chains can much more readily adjust their shape or end-to-end distance to accommodate an imposed deformation in the material as a whole.

[61] A. Maconnachie, G. Allen, and R. W. Richards, *Polymer*, **21**, 1157 (1981).

At these temperatures, the time scale for chain disentanglement is longer than the time scale for conformational rearrangements, however, so that over short time periods the entanglements can act as cross-links. Accordingly, the chains can stretch out between these entanglement points and a rubbery plateau is observed in experiments conducted in the appropriate time frame. The behavior of Silly Putty® is a good way to illustrate this. Bounced on the floor the rate of loading is fast; the chains have time to change conformation, but don't have time to disentangle and the material behaves like a cross-linked rubber ball. Left standing, however, it will flow and change shape under its own weight as, over time, the chains disentangle.

Finally, as the temperature is increased still further the time necessary for disentanglement becomes shorter, chain diffusion becomes faster than the measurement time of the experiment and we enter the terminal flow region where a polymer melt is much more liquid-like in its properties. The various relaxation processes that are associated with chain conformational changes and disentanglement are still important, however, and polymer melts remain viscoelastic, accounting for phenomena like jet swelling and melt fracture that we discussed earlier.

Viscoelastic Behavior: Semi-Crystalline Polymers

What about semi-crystalline polymers? Their viscoelastic behavior is much more complex, because of the superposition of the behavior of the crystalline and amorphous domains. This superposition is not necessarily linear, and coupling of the responses can occur, particularly as the degree of crystallinity of a sample is increased and the amorphous domains become constrained by the crystalline domains. Because of this, the behavior of semi-crystalline polymers is much less uniform than their amorphous counterparts, displaying individual idiosyncrasies that often have to be described separately.

It is possible to make one generalization, however, and that concerns the large change in modulus that is observed at the crystalline melting point, T_m. Figure 13-85 displays schematically and roughly the type of (10 second) modulus vs. temperature behavior obtained from a sample of high density polyethylene. At low temperatures, the relatively small amorphous domains are constrained by the rigid, stiff crystalline regions and may also be below their T_g, so the sample displays a high modulus. The modulus decreases as the temperature is raised as smaller, less perfect, crystals melt and the chains in the now larger amorphous regions become less constrained. Depending on the degree of crystallinity of the sample, a T_g may also be observed at some temperature (T_g is always less than T_m). However, the largest change in modulus occurs over the range of temperature that defines the T_m.

Just as it is difficult to make broad generalizations about the viscoelastic properties of semi-crystalline polymers, so is it difficult to generalize about their underlying relaxation processes. We can say that at low temperatures (below the T_g) relaxations are usually associated with short chain segments or substituent groups, as in amorphous polymers. Above the T_g, however, relaxations are more complex, often involving coupled processes in the crystalline and amorphous domains and the behavior of individual polymers can be dominated by their own peculiarities. Compare, for example, *tan δ* data obtained from a linear (LPE) and a branched (low density) polyethylene (LDPE)—Figure 13-86. These are labeled, by convention, *α, β, γ*, starting from the highest temperature process.[62] The *β*-transition has the characteristics of a T_g, but is not observed in the linear sample, presumably because of the severe constraints to motion imposed by the crystalline domains in this highly crystalline material. The *γ*-process is observed in both samples and is a

[62] This can lead to some confusion, because in amorphous polymers the *α*-transition is the T_g, but because $T_g < T_m$, the T_g in semi-crystalline polymers, if a T_g is observed at all, is labeled the *β*-transition.

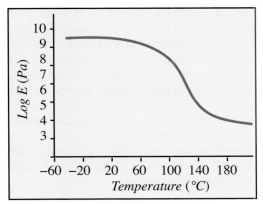

FIGURE 13-85 Schematic representation of the (10 second) modulus versus temperature behavior obtained from a sample of high density polyethylene.

local relaxation, but the *α*-transition is considerably more complex, involving coupled relaxations in both the crystalline and amorphous domains.

We have now examined some important aspects of the viscoelastic behavior of polymeric materials and discussed the relaxation processes responsible for these properties. There are two important points that you should take with you to the next section. First, time and temperature are inextricably intertwined when it comes to viscoelastic behavior. This will reemerge in the time-temperature superposition principle. Second, because all amorphous polymers have the

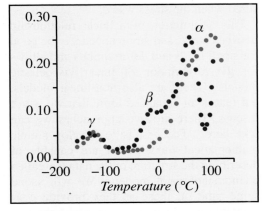

FIGURE 13-86 *Tan δ* data for a low density (blue) and linear (red) polyethylenes [redrawn from the data of H. A. Flocke, *Kolloid–Z. Z. Polym.*, **180**, 188 (1962)].

same structure, or, more accurately, because none of them have any, it is possible to discuss their properties in a general sense and expect our broad conclusions to apply to any polymer in this class. Semi-crystalline polymers can differ considerably in morphology and degree of crystallinity, however, and it is not possible to make broad generalizations of the same scope. This difference will carry through to our next discussion, where we consider mechanical and theoretical models of viscoelastic behavior.

Mechanical and Theoretical Models

If you step back and think about it, the mechanical and rheological properties of many solids and liquids can be modeled fairly well by just two simple laws, Hooke's law and Newton's law. Both of these are what we call linear models, the stress is proportional to the strain or rate of strain. If we examine viscoelastic properties like creep, the variation of strain with time appears decidedly non-linear (see Figure 13-75). Nevertheless, it is possible to model this non-linear time dependence by the assumption of a linear relationship between stress and strain. By this we mean that if, for example, we measure the strain as a function of time in a creep experiment, then for a given time period (say 1 hour) the strain measured when the applied stress is 2σ would be twice the strain measured when the stress was σ.

This assumption of a linear relationship between stress and strain appears to be good for small loads and deformations and allows for the formulation of linear viscoelastic models. There are also non-linear models, but that is an advanced topic that we won't discuss. There are two approaches we can take here. The first is to develop simple mathematical models that are capable of describing the structure of the data (so-called phenomenological models). We will spend some time on these as they provide considerable insight into viscoelastic behavior. Then there are physical theories that attempt to start with simple assumptions concerning the molecules and their interactions and

build a model from the ground up. This is also an advanced topic, so we won't get into the details, but will just try and give you a qualitative feel for some of the ideas and approaches that have been used. One such is the concept of reptation, so called because it assumes that the motion of a polymer is akin to that of a snake slithering through a nest of its neighbors. We discussed this when we considered polymer melts and we'll come back to ideas like this later.

We will first consider the parameters we are trying to model. Let us start with stress relaxation, where it is usual to describe properties in terms of a relaxation modulus, defined in Table 13-5 for tensile [$E(t)$] and shear [$G(t)$] experiments. The parameter used to describe the equivalent creep experiments are the tensile creep compliance [$D(t)$] and shear creep compliance [$J(t)$]. It is important to realize that the modulus and the compliance are inversely related to one another for linear, time-independent behavior, but this relationship no longer holds if the parameters depend on time.

This is an important point, so let's beat it to death with an example. Imagine that we perform a simple tensile creep experiment where a stress σ_0 is applied to a sample and after 10 hours the strain, $\varepsilon(10)$ is measured. Now let's take an identical sample and perform a stress relaxation experiment where the sample is stretched instantaneously to give

TABLE 13-5 MODEL PARAMETERS

TENSILE EXPERIMENT	SHEAR EXPERIMENT
Stress relaxation	
$E(t) = \sigma(t)/\varepsilon_0$	$G(t) = \tau(t)/\gamma_0$
Creep	
$D(t) = \varepsilon(t)/\sigma_0$	$J(t) = \gamma(t)/\tau_0$
Linear Time-Independent Behavior	
$E = 1/D$	$G = 1/J$
Time-Dependent Behavior	
$E(t) \neq 1/D(t)$	$G(t) \neq 1/J(t)$

this value of the strain [i.e., $\varepsilon_0 = \varepsilon(10)$]. The value of the stress required to maintain this strain after 10 hours, $\sigma(10)$, is then measured. We would find that (Equation 13-82):

$$E(10) = \frac{\sigma(10)}{\varepsilon_0} \neq \frac{\sigma_0}{\varepsilon(10)} = \frac{1}{D(10)}$$

EQUATION 13-82

This is because although $\varepsilon_0 = \varepsilon(10)$, in general, $\sigma(10) \neq \sigma_0$ (it will usually be less). In principle, the quantities we have defined, $E(t)$, $D(t)$, $G(t)$, and $J(t)$, provide a complete description of tensile and shear properties in creep and stress relaxation (and equivalent functions can be used to describe dynamic mechanical behavior). Obviously, we could fit individual sets of data to mathematical functions of various types, but what we would really like to do is develop a universal model that not only provides a good description of individual creep, stress relaxation and DMA experiments, but also allows us to relate modulus and compliance functions. It would also be nice to be able formulate this model in terms of parameters that could be related to molecular relaxation processes, to provide a link to molecular theories.

Modulus and compliance data can actually be related using *Boltzmann's Superposition Principle*. In addition to worrying about entropy, Boltzmann attempted to reduce the complex manifestations of primary creep to some simple scheme. Proposed in 1874, this principle essentially states that the deformation of a sample at any instant depends not only on the load acting at that instant, but on the entire previous loading history. Not only that; if the sample has been subjected to a number of loading steps, each of these steps makes an independent and additive contribution to the final deformation. Compliance, modulus and dynamic data can then be related to one another using integral and transform methods. You probably don't want to know, particularly as the superposition theory is actually a consequence of the assumption of linear viscoelasticity and we can derive these relationships using simple

mechanical models. Nevertheless, there are various important superposition principles that we can use in studying viscoelastic behavior and you should at least know what they are. Now let's talk about mechanical models.

Viscoelastic Behavior: Simple Models

Mechanical models are picture representations of various types of mechanical behavior and are rather like the combinations of resistors and capacitors used by electrical engineers to construct circuit diagrams. Many modern texts ignore these models or even advise their readers to avoid their use completely and just jump right into the equations. There is no doubt that these models are flawed and inconsistent, in that one such model might do a reasonable job of describing creep, but be totally inapplicable to stress relaxation. We think they are useful when you first start out in this subject, however, giving students a better feel for things, like retarded elastic behavior, than can be obtained from staring at an equation. Accordingly, we will start out with these simple models and then show the direction in which the mathematics needs to be developed in more advanced treatments.

The first thing we want to model is a simple linear elastic response, as described by Hooke's law. The equation for simple uniaxial extension is shown in Figure 13-87,

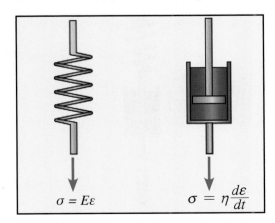

FIGURE 13-87 Schematic diagrams of a simple spring (left) and dashpot (right).

along with a drawing of a spring, which is the picture we will use to represent this type of behavior.

If we have a model for linear elastic behavior, we must surely have one for Newtonian viscous flow; and we do, the dashpot shown also in Figure 13-87. This is simply a piston in a cylinder that can be filled with various Newtonian fluids, each with a different value of the viscosity. Pulling (or pushing) on the piston causes it to move, as the fluid flows past the small gap between the piston and the cylinder walls, but the rate of deformation will depend on the viscosity of the fluid. (Some students who are a bit slow on the uptake or, more probably, trying to give us a hard time, ask what happens when the piston clunks to a stop at the bottom of the cylinder or pops out of the end; don't be too literal minded here, this is just a picture representing a type of behavior!)

In the same way as our dashpot represents the whole range of linear viscous flow, the spring represents the whole range of linear elastic behavior. If you are having a bad day and haven't wrapped your mind around the idea of a picture representation of a type of mechanical response, imagine that you have an unlimited supply of springs (or dashpots), each with a different value of the modulus (or viscosity), so that you can model any-

thing from the behavior of a rubber band (but remember, only for small deformations!) to a steel girder (or the flow of water compared to treacle).

We will first consider mechanical behavior in terms of plots of strain as a function of time, as this will allow us to see how various combinations of our basic linear elastic and viscous elements account for both primary and secondary creep. The behavior of an ideal spring is elementary. It should deform instantaneously when a load is applied and maintain a constant deformation until the load is removed, whereupon it immediately regains its original dimensions (see the top of Figure 13-74 above). The amount of deformation obtained for a given stress will, of course, depend upon the modulus of the spring. The dashpot displays perfectly viscous behavior, the strain increasing in a linear fashion with time. When the load is removed, the piston in the dashpot stays right where it finished up and does not return to its original position (see bottom of Figure 13-74). Thus there is some permanent deformation that depends upon the viscosity of the element and the time frame of the experiment.

The Maxwell Model

If you remember some of the stuff you were taught in basic electricity, you can probably guess what's coming next—we are going to consider placing our two basic linear elements first in series and then in parallel. With the two elements in series (Figure 13-88) we get the *Maxwell model*. Maxwell did not have this picture representation in mind when he formulated the model (reported in his classical paper *On the Dynamical Properties of Gases*), but rather considered the elastic and viscous forces acting on a general body. First he noted that if an applied stress varies with time, Hooke's law can be written (Equation 13-83):

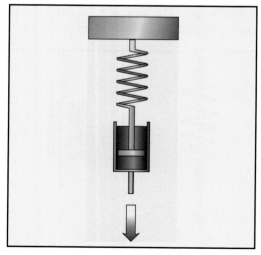

FIGURE 13-88 Schematic diagram depicting the Maxwell model.

$$\frac{d\sigma}{dt} = E\frac{d\varepsilon}{dt}$$

EQUATION 13-83

Maxwell then assumed viscous forces that were Newtonian, hence (Equation 13-84):

$$\frac{d\varepsilon}{dt} = \frac{\sigma}{\eta}$$

EQUATION 13-84

If the rate of strain is then the linear sum of these two components, which is what the mechanical model represents (Equation 13-85):

$$\frac{d\varepsilon}{dt} = \frac{\sigma}{\eta} + \frac{1}{E}\frac{d\sigma}{dt}$$

EQUATION 13-85

If we now perform a creep experiment, applying a constant stress, σ_0 at time $t = 0$ and removing it after a time t, then the strain/time plot shown at the top of Figure 13-89 is obtained. First, the elastic component of the model (spring) deforms instantaneously a certain amount, then the viscous component (dashpot) deforms linearly with time. When the stress is removed only the elastic part of the deformation is regained. Mathematically, we can take Maxwell's equation (Equation 13-85) and impose the creep experiment condition of constant stress $d\sigma/dt = 0$, which gives us Equation 13-84. In other words, the Maxwell model predicts that creep should be constant with time, which it isn't! Creep is characterized by a retarded elastic response.

The Maxwell model does a far more interesting job of modeling stress relaxation, however. If we again start with the basic equation (Equation 13-85) and impose the constant strain condition $d\varepsilon/dt = 0$ we get Equation 13-86:

$$\int_0^t \frac{E}{\eta} dt = -\int_{\sigma_0}^{\sigma} \frac{d\sigma}{\sigma}$$

EQUATION 13-86

and subsequently Equations 13-87:

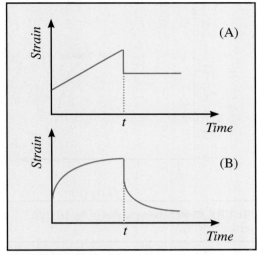

FIGURE 13-89 The Maxwell model gives the type of strain/time plot shown in (A) whereas real data looks like the plot shown in (B).

$$\ln\frac{\sigma}{\sigma_0} = -\frac{E}{\eta}t$$

or:

$$\sigma = \sigma_0\exp\left[-\frac{E}{\eta}t\right] = \sigma_0\exp\left[-\frac{t}{\tau_t}\right]$$

EQUATIONS 13-87

where τ_t is equal to η/E and has the units of time. A plot of the relaxation modulus $[\sigma(t)/\varepsilon_0]$, normalized to its initial value (i.e., divided by $E_0 = \sigma_0/\varepsilon_0$) is shown in Figure 13-90. At short time periods, by which we now mean $t << \tau_t$, the material is largely elastic; at long time periods, $t >> \tau_t$, it is largely viscous; when $t \approx \tau_t$, it is viscoelastic.

This is a decent bare-bones description of the gross features of stress relaxation and if we compare the Maxwell model to real data there appears to be a rough, qualitative fit in certain temperature ranges. In fact one could approximately fit the stress relaxation data shown in Figure 13-91 to two Maxwell curves, one describing the relaxation modulus from the glassy region through the T_g, and the second describing the behavior from the rubbery plateau through the terminal flow region. The Maxwell model only allows for a single value of the parameter τ_t, however.

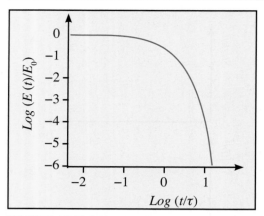

FIGURE 13-90 Schematic plot of $log\ [E\ (t)/E_0]$ versus $log\ (t/\tau)$.

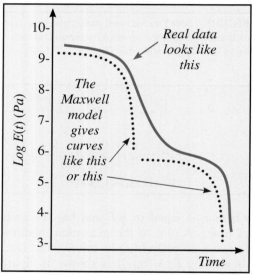

FIGURE 13-91 Schematic plot of $log\ E(t)$ versus time.

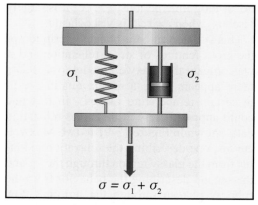

$$\sigma = \sigma_1 + \sigma_2$$

FIGURE 13-92 Schematic diagram of the Voigt model.

This parameter is called the *relaxation time* and what we hope you immediately grasp is that perhaps a model that incorporates a spectrum of relaxation times would provide a better representation of the data. We will discuss such a model later and show how it has the simple *form* of the Maxwell equation.

The Voigt Model

Of course, once you start playing with springs and dashpots, you can come up with all sorts of arrangements of the basic elements. The Maxwell model essentially assumes that a uniform stress acts on all parts of the system (i.e., both elements of the model). An alternative assumption is that the strain is identical in all parts of the system, which in the simplest model can be pictured as placing a spring in parallel with a dashpot (Figure 13-92). This is usually called the *Voigt model* (some call it the Kelvin model). It should be intuitively obvious that this arrangement of linear elements represents *retarded elastic behavior* (Figure 13-93). The spring wants to deform instantaneously when a stress is applied, but the dashpot won't let it. Similarly, when the applied load is removed the spring wants to return to its original dimensions instantaneously, but is prevented from doing so again by its connection to the dashpot, which slows it down to an extent that is determined by its characteristic viscosity.

If we apply the condition that the strain in each of the elements is the same (i.e., they both deform exactly the same amount) and

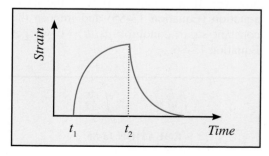

FIGURE 13-93 Schematic diagram of a retarded elastic response.

that the total stress applied to the model must equal the sum of the stress in each of the two elements, we can write (Equation 13-88):

$$\sigma(t) = E\varepsilon(t) + \eta\frac{d\varepsilon(t)}{dt}$$

EQUATION 13-88

If we now want to model a creep experiment we apply a constant stress, σ_0, hence we obtain Equation 13-89:

$$\frac{d\varepsilon(t)}{dt} + \frac{\varepsilon(t)}{\tau_t'} = \frac{\sigma_0}{\eta}$$

EQUATION 13-89

where we have again defined a characteristic time, $\tau_t' = \eta/E$, in this case called the retardation time. This linear differential equation has the solution (Equation 13-90):

$$\varepsilon(t) = \frac{\sigma_0}{E}[1 - exp(-t/\tau_t')]$$

EQUATION 13-90

or expressing this equation in terms of a compliance $D(t)$ we get Equation 13-91:

$$\frac{\varepsilon(t)}{\sigma_0} = D(t) = D_0[1 - exp(-t/\tau_t')]$$

EQUATION 13-91

But, just like the Maxwell model, the Voigt model is seriously flawed. It is also a single relaxation (or retardation) time model, and we know that real materials are characterized by a spectrum of relaxation times. Furthermore, just as the Maxwell model cannot describe the retarded elastic response characteristic of creep, the Voigt model cannot model stress relaxation—under a constant load the Voigt element doesn't relax (look at the model and think about it!) However, just as we will show that the form of the equation we obtained for the relaxation modulus from

the Maxwell model is of general applicability, so will the form of the equation for creep compliance correspond to that obtained from the Voigt model.

The Four-Parameter Model

The simplest flaws of the Maxwell and Voigt models, the fact that one cannot model creep while the other cannot model stress relaxation, can easily be fixed by combining our basic linear elements in different ways. One such is the so-called four-parameter model (Figure 13-94), which combines a Maxwell model in series with a Voigt model. The four parameters are the Maxwell modulus and viscosity, E_M and η_M, and the Voigt modulus and viscosity E_V and η_V.

Plots of strain versus time for a creep experiment were shown previously in Figure 13-75 and in this constant load experiment the strain obtained from each of the components can be simply summed to give Equation 13-92:

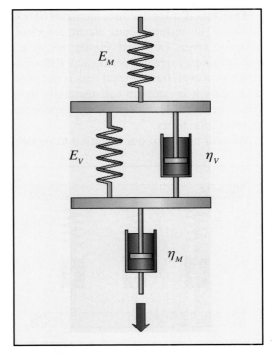

FIGURE 13-94 Schematic diagram of the four-parameter model.

STRESS RELAXATION

$$E(t) = E_0 exp(-t/\tau_t)$$

$$E(t) = \int_0^\infty E(\tau_t) exp(-t/\tau_t) d\tau_t$$

CREEP

$$D(t) = D_0 [1 - exp(-t/\tau_t')]$$

$$D(t) = \int_0^\infty D(\tau_t')[1 - exp(-t/\tau_t')]d\tau_t'$$

FIGURE 13-95 Comparison of the Maxwell and Voigt equations with ones that account for a continuous range of relaxation times.

$$\varepsilon(t) =$$

$$\frac{\sigma_0}{E_M} + \frac{\sigma_0}{\eta}t + \frac{\sigma_0}{E_V}[1 - exp(-t/\tau_t')]$$

EQUATION 13-92

This does display the three elements of real behavior, an instantaneous elastic response, primary creep (retarded elastic response) and secondary creep (permanent deformation). However, the fit to real data is not good and again it is because real materials have behavior that is characterized by a spectrum of relaxation times.

We will now turn our attention to models

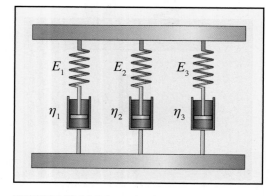

FIGURE 13-96 Schematic diagram of the Maxwell-Wiechert model.

that can accommodate this, but this discussion of simple mechanical models will hopefully give you a feel for creep and relaxation properties and the equations that describe them.

Distributions of Relaxation and Retardation Times

We have mentioned that although the Maxwell and Voigt models are seriously flawed, not least because they do not account for the spectrum of relaxation times that characterizes the behavior of real materials, the equations have the right form. What we mean by that is shown in the box above (Figure 13-95), where equations describing the Maxwell model for stress relaxation and the Voigt model for creep are compared to equations that account for a continuous range of relaxation times. These equations can be obtained by simply assuming that relaxation occurs at a rate that is linearly proportional to the distance from equilibrium and use of the Boltzmann superposition principle. Although this is a more rigorous approach, and although the use of mechanical models is anathema to many physicists (God bless their mathematical souls!), we will show how the same equations are obtained from the latter, as we believe this aids visualization.

One obvious way of introducing a range of relaxation and retardation times into the problem is to construct mathematical models that are equivalent to a number of Maxwell and/or Voigt models connected in parallel (and/or series). The Maxwell-Wiechert model (Figure 13-96), for example, consists of an arbitrary number of Maxwell elements connected in parallel. For simplicity let's see what you get with, say, three Maxwell elements and then extrapolate later to an arbitrary number, *n*.

The first thing you should remember is that the picture representation on the left assumes that the strain in each of the three Maxwell elements is identical, ($\varepsilon = \varepsilon_1 = \varepsilon_2 = \varepsilon_3$) while the stress, σ, applied to the system as a whole must equal the sum of the stress experienced by each of the individual elements ($\sigma = \sigma_1 + \sigma_2 + \sigma_3$). Using the first condition we can write Equation 13-93:

$$\frac{d\varepsilon}{dt} = \frac{\sigma_1}{\eta_1} + \frac{1}{E_1}\frac{d\sigma_1}{dt}$$
$$= \frac{\sigma_2}{\eta_2} + \frac{1}{E_2}\frac{d\sigma_2}{dt}$$
$$= \frac{\sigma_3}{\eta_3} + \frac{1}{E_3}\frac{d\sigma_3}{dt}$$

EQUATION 13-93

(Remember, the strain in each individual Maxwell element is just the sum of the strain in each of its two components, the spring and the dashpot.)

In a stress relaxation we can put $d\varepsilon/dt = 0$ and obtain for each of the Maxwell elements the equations shown in Equations 13-94.

$$\sigma_1 = \sigma_{01}exp(-t/\tau_{t1})$$
$$\sigma_2 = \sigma_{02}exp(-t/\tau_{t2})$$
$$\sigma_3 = \sigma_{03}exp(-t/\tau_{t3})$$

EQUATIONS 13-94

Then all we have to do is recall the definition of the stress relaxation modulus, $E(t) = \tau(t)/\varepsilon_0$, and the condition $\sigma_t = \sigma_1 + \sigma_2 + \sigma_3$ to obtain Equation 13-95:

$$E(t) = E_1exp(-t/\tau_{t1})$$
$$+ E_2exp(-t/\tau_{t2}) + E_3exp(-t/\tau_{t3})$$

EQUATION 13-95

Or, generalizing to an n-component system, where $E_n = \sigma_{0n}/\varepsilon_0$ (Equation 13-96):

$$E(t) = \sum_n E_nexp(-t/\tau_{tn})$$

EQUATION 13-96

The result we have just obtained, that the relaxation modulus is the sum of the responses of the individual elements, is essentially a statement of the superposition principle, in the sense that the dynamics of the system as a whole can be described as a superposition of a large number of independent modes of relaxation, each represented by a characteristic relaxation time, τ_t. What is intriguing is that even a simple model, consisting of just two Maxwell elements in parallel, reproduce some of the main features of stress relaxation in amorphous polymers (Figure 13-97). Amorphous polymers show a less abrupt and more gradual decline in modulus over each of the two steps, however, as we will show later, indicating that a range of relaxation times is necessary to more accurately describe behavior.

As you might expect, you can also create a model using a bunch of Voigt models (in this case, in series) to obtain an equation for creep compliance (Equation 13-97):

$$D(t) = \sum_n D_n[1 - exp(-t/\tau_{tn}')]$$

EQUATION 13-97

These models can be extended to the situation where there is a continuous distribu-

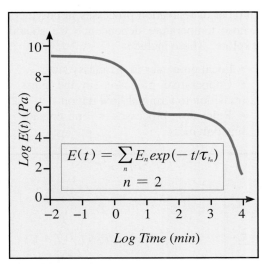

FIGURE 13-97 Plot of log E(t) versus log time (calculated using $E_1 = 3 \times 10^9$ Pa, $E_2 = 5 \times 10^5$ Pa, $t_1 = 1$ min, $t_2 = 10^3$ min [following J. Aklonis and W. Macknight, *Introduction to Polymer Viscoelasticity*, John Wiley and Sons, New York, (1983)].

tion of relaxation and retardation times by replacing the summations with integrals (see Figure 13-95). This allows relationships to be established between variables using transform techniques. Don't worry, we won't go there. Also, as we mentioned at the beginning of this section, these equations can also be obtained without recourse to mechanical models, and that is something you should start to look at if this part of the subject interests you. Here we will turn our attention to the time-temperature superposition principle.

Time-Temperature Superposition Principle

In the preceding sections, we have looked at the various types of relaxation processes that occur in polymers, focusing predominantly on properties like stress relaxation and creep compliance in amorphous polymers. We have also seen that there is an equivalence between time (or frequency) and temperature behavior. In fact this relationship can be expressed formally in terms of a superposition principle. In the next few paragraphs we will consider this in more detail. First, keep in mind that there are a number of relaxation processes in polymers whose temperature dependence we should explore. These include:

• Local processes in the glassy state
• Cooperative processes in the glass transition to terminal flow region
• Relaxation processes in semi-crystalline polymers

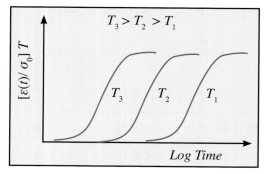

FIGURE 13-98 Schematic creep plots of $[\varepsilon(t)/\sigma_0]T$ versus log time at three different temperatures.

We will focus nearly all of our discussion on the second of these, as this is where most work has been done and some interesting relationships established.

Because you are probably brain dead after working your way through all of those mechanical models, we will first remind you of the type of data you observe in, for example, a stress relaxation experiment as a function of time (shown schematically in Figure 13-83) compared to an equivalent experiment conducted as a function of temperature (see Figure 13-82); there is an obvious equivalence.

The correspondence in time-temperature behavior has considerable significance and, as we have mentioned, can be expressed formally in terms of a superposition principle. To begin with, we will confine our attention to amorphous polymers and consider some creep experiments.

If the strain measured as a function of time in a constant stress experiment is plotted against time, curves of (almost) exactly the same shape are obtained at different temperatures—they are just shifted along the time axis (as you might expect, creep is faster or more pronounced at higher temperatures). Note that in these plots, the strain, $\varepsilon(t)$, is normalized by dividing by the stress (σ_0), and multiplying by the temperature (to account for the small temperature dependence of the modulus). Note also that in order to obtain complete creep data at low temperatures the experiment may have to be conducted over very long time periods, perhaps even years, so that only parts of the curves shown in Figure 13-98 may be obtained experimentally. However, because these curves have essentially the same shape, they can be superimposed by picking data obtained at some arbitrary reference temperature and shifting the other curves or parts of curves along the time axis so that they all superimpose to form a master curve. The amount that the data set (curve) corresponding to a specific temperature is translated is called the *shift factor*, a_T. Thus, in principle, we could use data obtained at higher temperatures and shorter times to determine behavior at lower temperatures and much longer time periods, providing that we know a_T. Of course, the shift factor can be measured experimentally and a master curve plus a typical plot of log a_T versus temperature is shown schematically in Figure 13-99. To reiterate it is not essential to obtain the full creep curve for data obtained (say) at low temperature—all that is necessary is that a sufficient part of the creep curve be obtained to allow the shift factor to be determined unambiguously.

To make this more apparent, let's consider some stress relaxation data. Previously we have shown (almost) complete curves, like those shown previously in Figure 13-83, going from the glassy region (actually, just above the T_g) to the terminal flow region. For a glassy polymer at room temperature the complete curve might take years to obtain. This would not only entail selfless application, but also luck, as things always go wrong (the electricity goes off in a storm and the temperature of the sample changes).

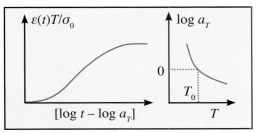

FIGURE 13-99 Schematic master curve and plot of log a_T versus temperature.

So, how were these curves obtained?

Figure 13-100 shows how a stress relaxation curve is constructed. If you examine the stress relaxation data shown on the left of this figure, you can see that the curve obtained at −80.8°C can be shifted along the time (x) axis to partially overlap and extend the curve obtained at −76.7°C, which, in turn, can be shifted to overlap and extend the curve obtained at −74.1°C, and so on. Each of these shifts gives a value of the shift factor, a_T, for a particular temperature.

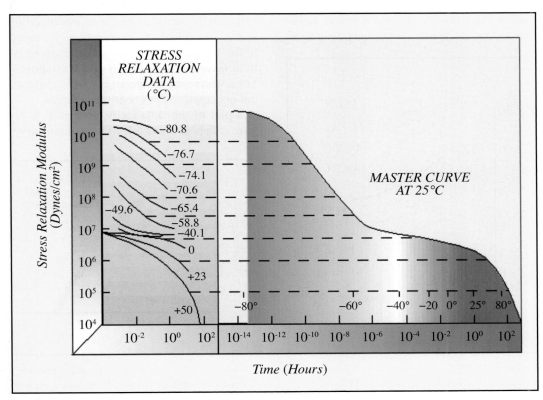

FIGURE 13-100 Summary of the preparation of a typical master curve.

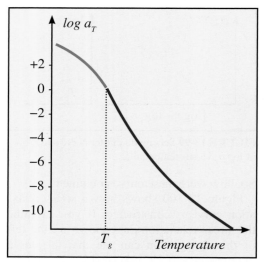

FIGURE 13-101 Schematic plot of the shift factor a_T versus temperature.

A plot of the log of the shift factor against temperature is shown schematically in Figure 13-101. The shift factor is obviously important, in that it can now be applied to the entire master curve, allowing a prediction of the behavior of, say, a glassy polymer at low temperature over very long time

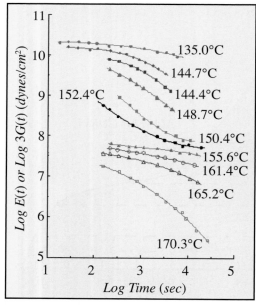

FIGURE 13-102 Stress relaxation and creep data for polycarbonate [redrawn from the data of J. P. Mercier, J. J. Aklonis, A. V. Tobolsky, *J. Appl. Polym. Sci.*, **9**, 447 (1965)].

periods. There are a number of additional things about the shift factor and its temperature dependence that are interesting, however. First, note that the shape of the curve changes near the T_g. There is a difference in the temperature dependence of a_T above and below this temperature. We'll focus on the temperature range above the T_g (blue line) to begin with.

Let's consider a concrete example. Stress relaxation and creep data obtained for polycarbonate (weight average molecular weight = 90,000 g/mole; $T_g = 150°C$) in the temperature range 135°–173°C, are shown in Figure 13-102. (Check the original paper to see how the authors converted the measurements of a shear creep modulus to a relaxation modulus.) Data were also obtained for a sample with a (weight average) molecular weight of 40,000 ($T_g = 145°C$). Master curves were then constructed by shifting the data along the time axis until they superimposed (Figure 13-103). Of course, this means that you have to pick a reference temperature and shift data obtained at other temperatures onto this curve. A reference temperature, T_0, equal to the temperature at which the characteristic relaxation time was 10 sec, was chosen. These reference temperatures were close to, but not equal to the T_gs of the samples.

A plot of the shift factor against $T - T_0$ gave data points that fell on a common curve,

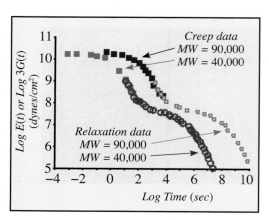

FIGURE 13-103 Master curves for polycarbonate [redrawn from the data of J. P. Mercier, J. J. Aklonis, A. V. Tobolsky, *J. Appl. Polym. Sci.*, **9**, 447 (1965)].

as shown in Figure 13-104. The equations below (Equations 13-98) both gave an excellent fit to this data. (Remember T_0 was close to but not equal to the T_gs of the samples.)

$$log\ a_T = \frac{-16.14(T - T_0)}{56 + (T - T_0)}$$

$$log\ a_T = \frac{-17.44(T - T_g)}{51.6 + (T - T_g)}$$

EQUATIONS 13-98

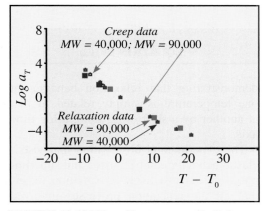

FIGURE 13-104 Plot of log a_T versus T - T_0 for polycarbonate [redrawn from the data of J. P. Mercier, J. J. Aklonis, A. V. Tobolsky, J. *Appl. Polym. Sci.*, **9**, 447 (1965)].

The WLF Equation

This bottom equation of Equations 13-98 is called the WLF equation, after Williams, Landel and Ferry, who found that for amorphous polymers the curve describing the temperature dependence of the the shift factor a_T has the general form (Equation 13-99):

$$log\ a_T = \frac{-C_1(T - T_s)}{C_2 + (T - T_s)}$$

EQUATION 13-99

The WLF equation applies to amorphous polymers in the temperature range of T_g to about T_g + 100°C. In this equation T_s is the reference temperature, these days taken to be the T_g, while C_1 and C_2 are constants, initially thought to be universal (with C_1 = 17.44 and C_2 = 51.6), but now known to vary somewhat from polymer to polymer. These experimental observations bring up a number of interesting questions. What is the molecular basis of the time-temperature superposition principle? What is the significance of the log scale and what does the superposition principle tell us about the temperature dependence of relaxation behavior? And what about the temperature dependence of a_T at temperatures well below T_g?

To answer these questions, it is convenient to examine a particular type of behavior, say stress relaxation, and note that an equiva-

lent treatment can be presented for creep and dynamic mechanical properties. Also, to simplify, let's consider the expression for $E(t)$ written in terms of a sum of relaxation processes, rather than the integral form necessary to describe a continuous distribution (Equation 13-100):

$$E(t) = \sum_n E_n exp(-t/\tau_{tn})$$

EQUATION 13-100

We now consider a particular mode of relaxation (therefore leaving out the subscript n) and let the characteristic time describing the molecular rearrangements associated with this mode be τ_{t0} at temperature T_0. At temperature T_1 this relaxation time will, of course, be different, τ_{t1} (remember, the molecules move faster at higher temperatures). Now let's simply define the ratio of the two relaxation times to be (Equations 13-101):

$$a_T = \frac{\tau_{t1}}{\tau_{t0}} \quad Hence: \quad \frac{t}{\tau_{t1}} = \frac{t}{a_T \tau_{t0}}$$

EQUATIONS 13-101

Taking logs we get Equation 13-102:

$$log\left(\frac{t}{\tau_{t_1}}\right) = log\left(\frac{t}{\tau_{t_0}}\right) - log\, a_T$$

EQUATION 13-102

demonstrating that relaxation behavior at one temperature is simply related to that at another by a shift of a_T along a log time axis.

The experimental observation that viscoelastic behavior can be shifted and superimposed, at least to a first approximation, means that all the relaxation processes involved have the same (or nearly the same) temperature dependence. Because the relaxation processes that occur in polymers involve a whole range of length scales, by which we mean they involve coupled motions that range from local rearrangements of just a few chain segments to motions of the chain as a whole, this implies that frictional forces encountered by a chain act in the same or a very similar manner on all the segments. This is important if you intend to be in the business of developing theories of polymer dynamics, but concerns us no further here.

The question we now wish to address concerns the WLF equation. Why does the temperature dependence of the shift factor have this form for temperatures ranging from the T_g into the terminal flow region?

We discussed the nature of glass transition only qualitatively in the section on thermal properties (Chapter 10). We did, however, mention a couple of essentially empirical equations that describe the viscosity of a fluid. One such is the Doolittle equation, which we rewrite here in a somewhat different form (Equation 13-103):

$$\eta = A\, exp\, B\left[\frac{V - V_f}{V_f}\right]$$

EQUATION 13-103

V is the total volume of the system and V_f is the free volume. At the T_g, $V_f \rightarrow 0$ and $\eta \rightarrow \infty$, so that a glass is formed. Recall that molecular relaxation times are related to viscosity by the definition $\tau_t = \eta/E$; and because

the modulus changes only slightly with temperature relative to the viscosity the shift factor can be written (Equation 13-104):

$$a_T = \frac{\tau_{t_1}}{\tau_{t_0}} \approx \frac{\eta_1}{\eta_0}$$

EQUATION 13-104

where η_0 and η_1 are the viscosities at T_0 and T_1, respectively. We can now substitute from the Doolittle equation for the viscosity, but before doing so we first need an expression for how V_f varies with temperature. Williams, Landel and Ferry assumed that the free volume is a linear function of temperature and could be given by Equation 13-105:

$$f = f_0 + \Delta\alpha(T - T_g)$$

EQUATION 13-105

where f is the fractional free volume (V/V), f_0 is the fractional free volume at T_g and $\Delta\alpha$ is the difference in the coefficients of thermal expansion of the liquid and the glass, i.e., $\Delta\alpha = \alpha_l - \alpha_g$. Substituting all the relevant terms into the expression for a_T and taking logs, the following equation is obtained (Equation 13-106):

$$log\, a_T = \frac{(-B/2.303\, f_0)(T - T_g)}{(f_0/\Delta\alpha) + (T - T_g)}$$

EQUATION 13-106

It can now be seen that the constants C_1 and C_2 in the WLF equation are related to free volume. Clearly, this equation can only apply to amorphous polymers and to temperatures close to the T_g and above (remember, T_g is a kinetic phenomenon and manifests itself over a range of temperatures). But what about transitions below the T_g and semi-crystalline polymers?

Relaxation Processes in the Glassy State

Relaxation processes in amorphous polymers below the glass transition involve local motions, which can occur not only in the main chain of the polymer (Figure 13-105), but also in the side chain. These also obey a time-temperature superposition principle, but the temperature dependence of the shift factor is not governed by a WLF-type equation but has an Arrhenius form (Equation 13-107):

$$\tau_t^{-1} \sim exp(-\Delta G_d/RT)$$

EQUATION 13-107

where different conformational states are separated by a (free) energy barrier ΔG_a, and τ_t^{-1}, the reciprocal of the relaxation rate is a measure of the frequency of the transitions between these states.

Relaxation Processes in Semi-Crystalline Polymers

As you might expect, the temperature dependence of transitions in the amorphous regions of polymers with a low degree of crystallinity are still governed by WLF or Arrhenius-type expressions, depending on the nature of the transition. Certain transitions in more crystalline polymers, such as the α-transition in polyethylene (a coupled crystalline/amorphous relaxation) also have an Arrhenius temperature dependence. In general, however, time-temperature superposition often involves "vertical" as well as "horizontal" shift factors, usually because of a change in the degree of crystallinity with temperature. This obviously complicates time-temperature behavior considerably and does not allow broad generalizations, so we will go no further into this subject.

RECOMMENDED READING

J. J. Aklonis and W. J. MacKnight , *Introduction to Polymer Viscoelasticity*, Second Edition, Wiley-Interscience, New York, 1983

J. E. Gordon, *The New Science of Strong Materials*, Penguin Books, 1991.

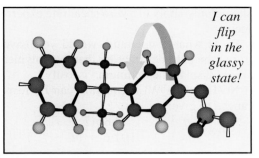

I can flip in the glassy state!

FIGURE 13-105 Chemical structure of polycarbonate illustrating a local motion that occurs in the glassy state.

L. H. Sperling, *Physical Polymer Science*, Third Edition, Wiley, New York, 2001.

G. Strobl, *The Physics of Polymers*, Springer-Verlag, Berlin, 1996.

L. R. B. Treloar, *Physics of Rubber Elasticity*, Second Edition, Clarendon Press, Oxford, 1975.

R. J. Young, "Strength and Toughness" in *Comprehensive Polymer Science, Vol. 2, Polymer Properties*, C. Booth and C. Price, Editors, Pergamon Press, 1989.

STUDY QUESTIONS

1. Sketch a plot of the modulus of an amorphous polymer as a function of temperature, labeling the different regions of viscoelastic behavior. Briefly describe the types of relaxation behavior that occur in each region.

2. A particular extruder is found to be most effective when the melt viscosity of the polymer being processed is 2×10^4 Poise. Polycrud ($T_g = 75°C$) was found to have this viscosity at 145°C when its $(\overline{DP})_w$ (weight average molecular weight) is 700.

A. What would be the viscosity of a sample of polycrud whose $(\overline{DP})_w$ is 500. [Still above the entanglement limit; assume you can use the equation $\eta_m = k(\overline{DP})^?$, (figure out what the exponent is), even though this should only apply to a zero shear rate experiment!]

B. At what temperature would you have to run the extruder in order that the polymer would display the optimum viscosity?

NOTE: The WLF equation can be rearranged to give:

$$log_{10}\left[\frac{\eta}{\eta_{ref}}\right] = -\frac{8.86(T - T_{ref})}{101.6 + (T - T_{ref})}$$

3. Sketch the stress/strain curves you would expect for the following polymers:

    A. Atactic polystyrene
    B. A butadiene/styrene copolymer (90% butadiene)
    C. High density polyethylene

Assume slow strain rates at room temperature. Put all three curves on the same plot and label the curves.

4. Briefly comment on the difference between the elastic properties of a metal and a rubber band. Explain what gives rise to these differences.

5. A. Briefly explain the time-temperature superposition principle and how it can be used to predict creep properties.

B. Obtain an equation for the shift factor, a_T, (Equation 13-106 in the text) from the Doolittle equation.

6. If a strip of lightly cross-linked natural rubber initially at room temperature is heated in an oven, it can be observed to *expand*. If the same piece of rubber is stretched at room temperature to twice its original longitudinal dimensions then placed in the same oven, it can be observed to *contract*. How do you explain these phenomena?

7. Consider a piece of rubber subjected to a shear stress such that $\lambda_x = 1, \lambda_y = 1$, and $\lambda_z = 1/\lambda$. Derive an equation relating the shear stress, τ, to the shear strain, γ. Assume $\gamma = 1 - 1/\lambda$. Comment on the difference between this stress/strain relationship and the one for simple elongation.

8. As a graduation present, your parents give you the cheap, old clunker of a car that has been sitting in the garage for the last five years. The car is so old that it has flexible vinyl seat coverings (i.e., plasticized PVC). You notice some yellow gunk has built up on the inside of the windshield during this time. Because nobody in your family smokes, you guess that the plasticizer has evaporated out of the PVC and condensed on your windshield. Given that the T_g of pure PVC is about 80°C, what changes in the stress/strain behavior of your seat coverings would you anticipate over time? What do you think will happen to your cheap seats and why do you wish you could afford real Corinthian leather?

9. How would you expect the viscosity of a polystyrene melt to vary with molecular weight? How does it vary with strain rate?

10. Out of idle speculation, you drop a metal ball from a height of 2 yards on to a sample cut from a plastic sheet. (You are an old-fashioned engineer and can have no truck with units such as meters.) You observe that the plastic sheet punctures in a ductile fashion, leaving a hole roughly the size of the ball. You find that if the same metal ball is dropped onto another sample of the same plastic sheet from a height of 4 yards, the sample shatters in a brittle manner. How would you explain these observations?

11. Consider the following mechanical models

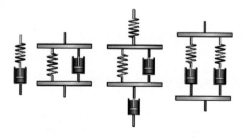

Which of these models does not do a good job of describing the retarded elastic response that is characteristic of creep? Identify the model (Voigt? Maxwell?) and explain why.

Which of these models would not do a good job of describing stress relaxation? Again, identify the model and explain why.

Which of these models would best give a description of viscoelastic behavior in terms of a spectrum of relaxation times (well, say 2). Again, explain why.

12. What is melt fracture and under what circumstances does it occur?

13. In the text, we defined the shear stress and shear strain in a dynamic-mechanical experiment in terms of sinusoidal functions.

A. Believe it or not, using the real and imaginary parts of an exponential function:

$$e^{i\omega t} = \cos \omega t + i \sin \omega t$$

can be simpler for certain problems, because of the ease of differentiating exponential functions. Really! Look it up! Anyway, you are about to find out! Using:

$$\dot{\gamma}(t) = \gamma_0 e^{i\omega t}$$
$$\dot{\tau}(t) = \tau_0 e^{i\omega t + \delta}$$

derive expressions for the complex shear modulus ($G^* = G' + iG'$) and compliance ($J^* = ?$) and show how the real and imaginary parts of these equations (G', G') correspond to expressions given in the text (Equations 13-79 for the modulus; components of the compliance are not given!).

B. Although $G(t) \neq 1/J(t)$, show that $G^* = 1/J^*$ (at the same temperature and frequency).

C. Consider the Maxwell model (Equation 13-85) subjected to a sinusoidal tensile stress:

$$\dot{\sigma}(t) = \sigma_0 e^{i\omega t}$$

Differentiate and substitute this equation into Equation 13-85 and obtain the following expression for the complex tensile compliance D^*, defined as the difference in strain at times t_2 and t_1 divided by the difference in stress at these two times:

$$D^* = \frac{\varepsilon(t_2) - \varepsilon(t_1)}{\sigma(t_2) - \sigma(t_1)} = D - \frac{iD}{\tau\omega}$$

D. Show that the complex tensile modulus at the same temperature and frequency is given by:

$$E^* = \frac{E\tau^2\omega^2}{1 + \tau^2\omega^2} + i\frac{E\tau\omega}{1 + \tau^2\omega^2} + E' + iE''$$

Also show that:

$$\tan \delta = \frac{1}{\omega\tau}$$

14

Processing[63]

INTRODUCTION

In this chapter we are going to describe the methods used to shape polymers into the various forms in which they are used in everyday life, such as automotive bumpers, medical tubing or the hodge-podge of stuff shown in Figure 14-1. These methods are generally referred to as processing, because the polymer undergoes a process of operations while being transformed from raw material to final product. A solid chip (pellet) of a polyamide might proceed through drying, melting, extrusion, stretching, and cooling on its way to becoming fibers for a nylon carpet, for example.

The field of polymer processing is large in its own right and just parts of the subject have books devoted to them. Here our goal is to simply give you an overview of the subject. Our opinion is that even a hard-core polymer chemist or physicist should know what an extruder is and how it operates. You never know, one day you may have to talk to a polymer engineer!

Many of you may be familiar with simple forms of processing equipment, such as the *"Play-Doh Fun Factory: An Extruder Toy,"* which you may have played with when younger. (We were not so lucky, being given bits of broken bottles and rusty razor blades and told to go play in the middle of the road.) This is an example of what is called pri-

FIGURE 14-1 Plastic stuff.

[63] Written in collaboration with Professor Kirk Cantor, Penn College.

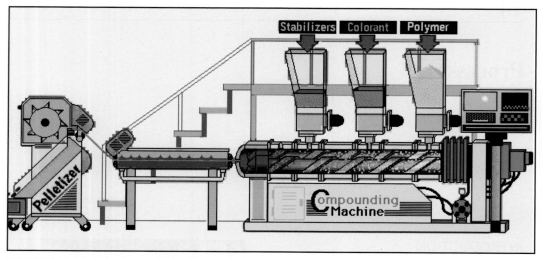

FIGURE 14-2 Schematic diagram of a machine that produces polymer pellets (Source: www.pct.edu/prep/).

mary processing equipment, which includes things like extruders and injection molders, which you've probably heard of. There are also various types of auxiliary equipment that is used to perform supplementary tasks on the material, either upstream or downstream from the primary equipment. Most of these will just be mentioned in passing. Also, for the most part, polymer is delivered to processors in pellet form, so we will focus on what is called pellet processing. Figure 14-2 shows an example of some of the things we've just mentioned. Polymer is mixed with various additives in a twin screw extruder. More on this shortly. Commercial polymers are almost invariably mixed with things like antioxidants, plasticizers and so

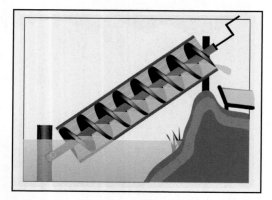

FIGURE 14-3 Schematic diagram of Archimedes' screw.

on. Note how the extruded polymer is then solidified by passing through a water bath. It is then chopped into pellets for sale to plastics manufacturers who may be making things like buckets. We will start our discussion of processing by considering the operation shown here—extrusion.

EXTRUSION

Extrude means to "push out." When we put toothpaste on our toothbrush, we are extruding it through the "nozzle" of the tube. Archimedes (probably) invented a form of the extruder which consisted of a screw in a barrel and was used to raise water from a river to irrigate surrounding fields (Figure 14-3). For polymers, an extruder is used to continually push molten material through a shaping die located on the discharge end of the machine. Depending on the shape of the die orifice through which melted polymer passes, various products can be produced.

Many billions of pounds of polymer are extruded annually into products ranging from household plumbing pipe to thin film grocery sacks, textile fibers to weather stripping. In fact, almost all polymer passes through an extruder at some point, since many polymerization reactors feed extruders in order to include additives and to pelletize. Furthermore, most primary processing

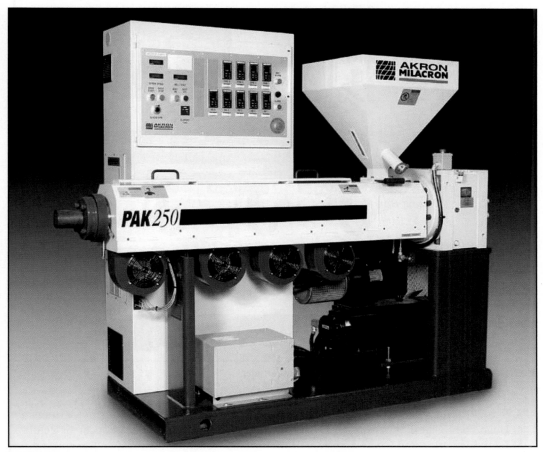

FIGURE 14-4 A commercial PAK 250 single screw extruder (Courtesy: Cincinnati Milacron).

equipment, like injection and blow molders, uses extruders on the front-end to melt the polymer pellets and push them into the mold or through the die.

Single Screw Extruders: Hardware

As the name implies, a single screw extruder has one screw that rotates continuously inside of a barrel, melting, mixing, and metering the polymer as it conveys the material toward the die (Figure 14-4). The purpose of single screw extrusion is to feed a die with a homogeneous material at constant temperature and pressure. Accordingly, these machines are used primarily for plasticating pellets and forming products to final (or near final) shape. Single screw extruders are rated by their screw diameter (inches,

millimeters). Typical production machines range from about 2–8 inches.

The most important system characteristic determined by screw diameter is machine throughput, the amount of material that can be processed per unit time (pounds per hour, kilograms per hour). Because the extruder holds an amount of polymer equal to the volume of free space between the screw and the barrel, the throughput of an extruder is proportional to the screw diameter cubed. So, while a 2-inch extruder may process 100 lb/hr of polyethylene, an 8-inch extruder may process 6400 lb/hr of the same material.

Single screw extruders essentially consist of five components;

1. Drive System
2. Feed System

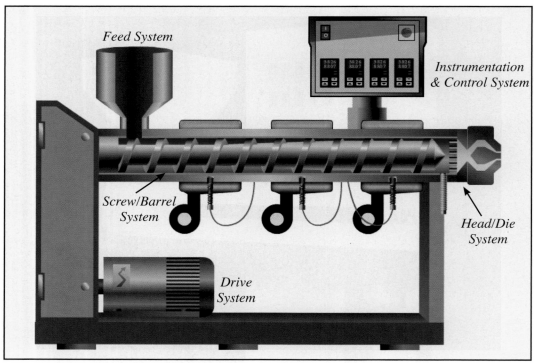

FIGURE 14-5 Schematic diagram of a single screw extruder.

3. Screw/Barrel System
4. Head/Die System
5. Instrumentation and Control System

The drive system consists of a motor, a speed reducer, and thrust bearing (Figure 14-5). Obviously, the motor turns the screw and is often fairly large, as it takes a lot of power to push high viscosity molten poly-

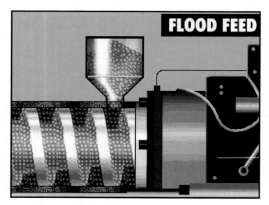

FIGURE 14-6 Schematic diagram of the feed system for a single screw extruder.

mer through the die. The motor transmits power through a speed reducer or gearbox, which generally gears the motor speed down by a factor of about 12:1. High screw speeds would lead to lots of shear, generally beating the living hell out of the polymer and degrading it. The gear reduction also aids in torque generation, important for turning the screw against very high viscosity polymer melts.

The feed system is much less complex than the drive system. It consists of a hopper and a feed throat. Pellets can be fed manually or through vacuum lines into the hopper, which are sometimes integrated with dryers to remove unwanted moisture from hygroscopic materials. Pellets flow or tumble into the feed throat, the part just above the barrel shown in Figure 14-6. This unit is generally cooled to prevent the polymer from softening too early, which might clog the system and inhibit the free flow of the pellets through the throat area into the barrel of the extruder. The hopper is usually maintained at a near-full level since most single screw extruders are "flood fed" keeping the throat area full and allowing the screw channels to completely fill with each screw rotation.

The screw/barrel system is the heart of the operation. Not only does it pump the polymer through the die, but it must also ensure that the melt is homogeneous and has a uniform temperature and pressure. Any compositional inhomogeneities or fluctuations in temperature or pressure lead to variations in the final product.

The screw, illustrated in Figure 14-7, is a long shaft with a thread wrapped helically around it. The thread is called a flight. Between adjacent sections of the flight is the channel. Most screws have three primary sections: feed, compression (or transition), and metering and each can have a different channel depth. The channel depth is generally largest (and constant) throughout the feed section. It is the smallest and also constant in the metering section, but decreases along the compression section. The amount of screw compression is quantified by the compression ratio:

$$\text{Compression Ratio} = \frac{\text{Feed Channel Depth}}{\text{Metering Channel Depth}}$$

Typical values of the compression ratio range from 2:1 to 4:1, depending on the type of polymer and the bulk density of the feed material.

Another important characteristic of screw geometry is the *L/D* ratio:

$$L/D \text{ Ratio} = \frac{\text{Screw Length}}{\text{Screw Diameter}}$$

Typical values of the *L/D* ratio range from 18:1 to 32:1, with 24:1 being the most common. The *L/D* ratio is based on the number of functions that a screw performs. For example, 24:1 is normal for the three standard functions: conveying pellets, melting, and pumping (generating pressure to feed the die). Fewer functions would require a shorter screw and more functions would require a longer screw. The screw fits closely inside the barrel. Clearance, the distance between the flight tip and the barrel wall, is generally 0.1% of screw diameter. Accordingly, a 4-inch screw will have a clearance of 0.004 inches. Clearances smaller than this will result in excessive metal-to-metal wear, while higher clearances reduce the melting and pumping capacities of the screw.

The barrel is a hollow cylinder extending from the end of the feed throat to the tip of the screw (Figure 14-8). The exit end of the barrel is referred to as the head. Coating the entire inside surface of the barrel is a very

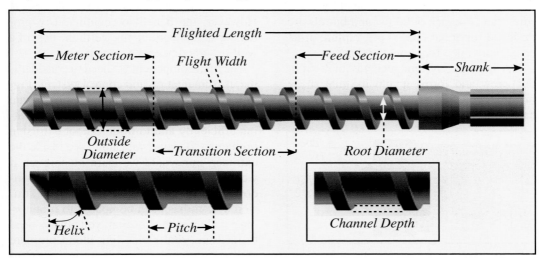

FIGURE 14-7 Schematic diagram with labels of the screw of a typical single screw extruder.

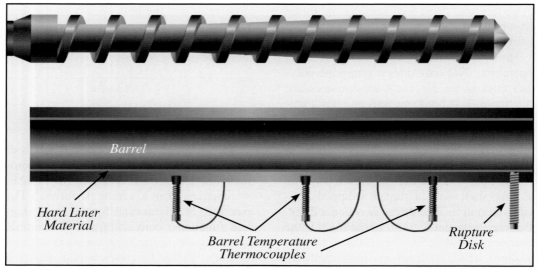

FIGURE 14-8 Schematic diagram of the screw and barrel of a typical single screw extruder.

hard liner material, tungsten-carbide alloy, for example. This lining reduces wear on the barrel, extending its life. Barrel replacement is very costly and time-consuming. Just before the head end of the barrel, located at the six o'clock position (toward the floor) is a hole in the barrel fitted with a device called a rupture disk. This device is an important safety mechanism. In the event that excessive pressure builds up at the head, a weld in the rupture disk will fail allowing molten polymer to escape onto the floor, relieving the pressure. Since normal operating pressures can approach 5000 psi and barrels are designed for about 10,000 psi, rupture disks are usually rated for about 9000 psi.

The head/die system receives melt from the screw and forms it into its final shape just prior to cooling (Figure 14-9). The die has been called the "brains" of the operation,

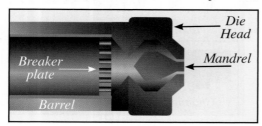

FIGURE 14-9 Schematic diagram of the die head and mandrel for a single screw extruder.

since the product's final shape is determined by die forming of the melt. Usually a nozzle and/or an adaptor are employed to guide the melt stream from the barrel exit into the die entrance. The internal flow geometry and exit orifice geometry are critical in determining the shape and solid-state properties of the product. Much analysis and experience goes into good die design. Product geometry and properties, material properties, throughput rates, heating/cooling requirements, and die assembly and maintenance are some of the considerations that are part of die design. However, the benefit of continuously producing sellable product is worth the cost of a well-designed head/die system.

The final component we need to consider is the instrumentation and control system (Figure 14-10). Good instrumentation is vital to ensure that an extrusion line is operating efficiently and to troubleshoot extrusion problems. Because we cannot see inside the extruder, nor would we learn much even if we could, instrumentation serves as a "window" onto the process. Monitoring an extrusion system has been compared to monitoring a patient in the hospital: you always need to measure vital signs, like blood pressure and body temperature. Likewise, an extrusion line that is not properly monitored can

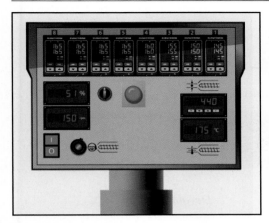

FIGURE 14-10 Schematic diagram of the instrument panel for a single screw extruder.

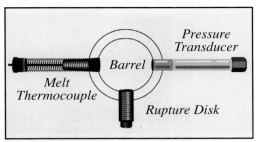

FIGURE 14-11 Schematic diagram illustrating the important measurements required for single screw extrusion.

become unstable, producing lots of expensive scrap.

The three most important parameters to measure are temperature, pressure, and current. Using thermocouples, hardware temperature is usually measured in multiple zones along the barrel, in the head, and in the die (Figure 14-11). Melt temperature and pressure are measured just prior to exiting the barrel, using a thermocouple and transducer, respectively. Electric current on the motor is measured with an ammeter.

Process control is essential for producing high-quality finished product. The temperature should be maintained within a range of +/– 10°F (plastic processing guys like traditional units), which is accomplished with a closed-loop system comprised of a microprocessor temperature controller, a thermocouple, a heating unit, and a cooling unit (see Figure 14-12). Pressure is not usually controlled directly, but is a function of three process variables: screw rotational speed, polymer melt viscosity, and die flow geometry. Motor current, also, is not directly controlled, but depends on the sources of power consumption mentioned earlier. Other important process variables that are controlled include product geometry, cooling rate, and raw material moisture content. In summary, the instrumentation and control system is crucial in minimizing product variation.

Single Screw Extruders: Functionality

Now we turn to a discussion of how the hardware interacts with the material on its way to becoming an extruded product, such as vinyl siding or a heart catheter. This is somewhat inelegantly called the functionality of the extruder, which we will categorize into six zones shown in Figure 14-13 and listed below.

1. Solids Conveying
2. Melting
3. Melt Pumping
4. Mixing
5. Devolatilization
6. Die Forming

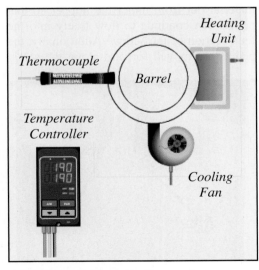

FIGURE 14-12 Schematic diagram of process control for single screw extrusion.

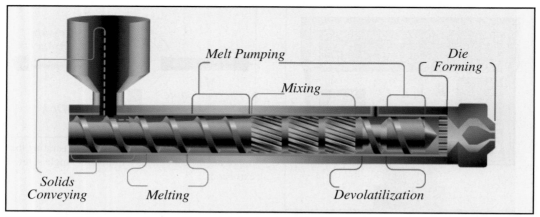

FIGURE 14-13 Schematic diagram illustrating the functionality of a typical single screw extruder.

Solids conveying is the process of moving raw material, such as pellets, prior to melting. It can be divided into gravity-induced and drag-induced solids conveying. Gravity-induced solids conveying occurs down through the hopper as the pellets flow toward the screw (Figure 14-14). Low friction between the pellets as well as between the pellets and the hopper aids in free flow. Any flow restrictions due to excessive friction, poor hopper design, or irregular particle shape can lead to a stoppage of flow in the hopper, called bridging.

Drag-induced solids conveying occurs along the screw. Its purpose is to move material out from under the hopper so that pellets can continue to flow freely into the feed section of the screw. Additionally, the pellets are moved forward along the screw toward the compression section, where melting primarily takes place. You might therefore anticipate that high friction between the pellets and the screw would facilitate this. Not so; if there is high friction, then the pellets simply travel circularly around the rotating screw without being conveyed toward the outlet. Instead, there must be low friction between the pellets and the screw and high friction between the pellets and the barrel. With this ideal condition, the pellets slip along the screw root while the barrel retards their circular motion, so that the flight pushes them longitudinally toward the compression section.

Melting begins about three to five diameters down the screw (Figure 14-15). (The term melting is used here regardless of whether the initial polymer is amorphous or semi-crystalline.) The mechanism by which pellets melt is quite interesting, since we have few other examples of this process in everyday life. Soon after the pellets move out from under the hopper, they become

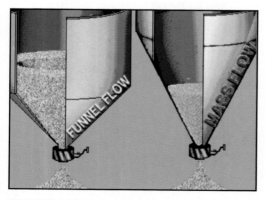

FIGURE 14-14 Schematic diagram illustrating funnel and mass flow.

FIGURE 14-15 Schematic diagram illustrating melting of the polymer as it proceeds down the screw.

compacted together into a solid bed (right hand side of Figure 14-15). As the solid bed moves down the barrel, the shear (friction) between the solid bed and the barrel wall creates a thin layer of melt at the barrel surface called the melt film. Because of the rotation of the screw and the helix angle of the flight, polymer in the melt film is carried toward the back of the channel where it is deposited into a melt pool. As melt is continually created by shear at the interface of the solid bed/melt film, the melt pool grows in volume, pushing the solid bed toward the front of the channel. As this process moves down the screw, the channel depth in the compression section is reducing, acting to drive the solid bed toward the barrel, which further aids melting. In this manner, all solids are completely melted near the end of the compression section. This is known as a contiguous melting mechanism. Of course, it is dependent on proper screw design and processing conditions (e.g., temperatures and screw speed).

Melt pumping is the process of generating enough pressure in the melt to overcome the resistance created as the polymer flows into and through the head/die tooling (Figure 14-16). As mentioned above, the combination of flow rate, melt viscosity, and die restriction can create substantial resistance, usually called head pressure (values between 1000 and 5000 psi are common). It is the job of the screw metering section to overcome that resistance by pumping the melt into the head. In most cases, the pressure is highest at the tip of the screw and lower back stream. This leads to the situation where the motion of the screw rotating in the barrel acts to drag the polymer toward the die, but the differential pressure acts to push the polymer back toward the hopper. Obviously, drag flow forward dominates and material exits the die (unless the end of the extruder is closed off!). However, this does lead to a circulation of flow through the metering section, where melt flows primarily downstream when it is at the top of the channel and somewhat back stream when it shifts to the bottom of the channel.

The equation in the box on the right can

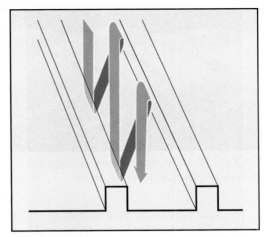

FIGURE 14-16 Schematic diagram illustrating melt pumping.

be used to calculate the flow rate from the extruder. It is known as the combined flow equation, because it accounts for the combination of drag flow forward (the first term, which is positive) and pressure flow backward (the second term, which is negative). The equation implicitly assumes that the melt is Newtonian, with the viscosity independent of shear rate. Most polymers are non-Newtonian, of course, but the equation is still valuable as an approximation and in understanding the dependence of flow rate on each of the critical parameters.

EXTRUDER FLOW RATE EQUATION

$$Q = \alpha N - \frac{\beta \Delta P}{\mu L}$$

Q = *Volumetric flow rate* (in^3/sec)
Screw geometry constants
  $\alpha = (\pi^2/2)D^2H \sin\phi \cos\phi$ (in^3)
  $\beta = (\pi/2)DH^3 (\sin\phi)^2$ (in^4)
D = *Screw diameter* (*in*)
H = *Metering channel length* (*in*)
ϕ = *Flight helix angle*
N = *Screw speed* (*revs/sec*)
ΔP = *Pressure increase over the metering section* (*psi*)
μ = *Melt viscosity* (*psi·sec*)
L = *Length of metering section* (*in*)

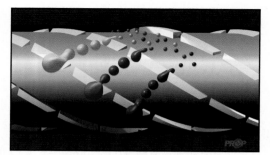

FIGURE 14-17 Schematic diagram illustrating dispersion through an on-screw devise.

The next functional zone is mixing. This is a very well developed discipline and one could devote an entire course to this topic. Here we will just focus on understanding two types of fluid mixing important to polymer processing: distributive mixing and dispersive mixing. Distributive mixing is primarily used for fluid/fluid mixtures, such as when blending two polymer melts. Here the fluid is spatially rearranged or "distributed" to produce a uniform composition. This is best accomplished by subjecting the mixture to many flow direction changes, such as when the flow travels through a device that splits and recombines the flow field many times. It is also used to incorporate color into a melt and to reduce temperature gradients. Dispersive mixing, on the other hand, is used to reduce the size of particulate additives in a polymer matrix. For example, calcium carbonate is often used as a filler or reinforcement in polymers. Dispersion is used to breakup this solid additive for subsequent distribution, which results in even smaller-scale composition uniformity. This occurs best when the fluid is subjected to high shear or stretching (elongational) flow, which transfers energy to the particle. Many on-screw devices have been designed to accomplish this mechanism (Figure 14-17). High deformation rates, coupled with high matrix viscosity, produce high dispersion force.

Devolatilization is the process of removing unwanted gases (volatiles) from the melt. If these gases were to exit through the die with the melt, undesired foaming and/or surface imperfections would occur in the final product (Figure 14-18). (Of course, polymers are often foamed by design, but here we are discussing conditions when foaming is to be avoided.) Conditions leading to gas in the melt include moisture that turns to steam and a chemical reaction that liberates a by-product. A specially designed screw and barrel system is required for devolatilization (Figure 14-19). This system, known as a two-stage extruder, includes the typical first three screw sections (the first stage) followed by an extraction section, recompression, and a pumping section (the second stage). A vent port is located in the barrel at the extraction section. When the melt enters the deep-channeled extraction section, the pressure in the polymer drops to zero, allowing the volatiles to exit through the vent without the melt being pushed out. Generally, a vacuum pump is attached to the vent port to aid the process. Vent flow of the melt can occur, but under proper design and operating conditions, only gases exit through the vent port.

The final functional zone is die forming. Shaping of the melt just prior to solidification takes place here. There are hundreds of different types of dies for producing extruded products, but this section will focus on three phenomena that are general to polymer flow through all dies: die swell, melt fracture,

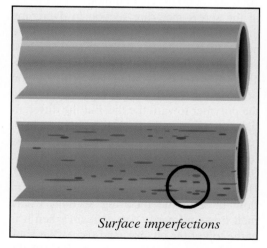

Surface imperfections

FIGURE 14-18 Schematic diagram illustrating (top) a "clean" extrudate and (bottom) surface imperfections in an extrudate.

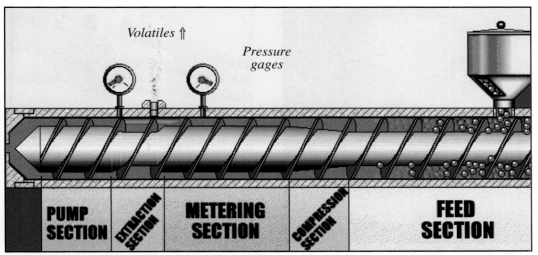

FIGURE 14-19 Schematic diagram of a two-stage extruder screw.

and die drool. (We also discussed aspects of die swelling and melt fracture in Chapter 13.) Die or jet swelling is when polymer exiting the die (the extrudate) expands to a size larger than the orifice through which it came (Figure 14-20). Although commonly thought to be the result of decompression, die swell actually occurs due to the relaxation of shear-oriented molecules. Inside the die, shear causes the long chain molecules to orient in the direction of flow. However, when shear is removed, the molecules tend to relax toward a random coil. Die swell thus occurs to some degree in all polymers. Design of the die must consider not only the amount of swell, but also the different amount of swell in areas of different shear.

As we also mentioned in Chapter 13, melt fracture is a product defect that ranges from surface roughness on the extrudate to large-scale inconsistencies in shape. Unlike die swell, melt fracture is avoidable. This defect is related to high shear stress in the die. When the melt exceeds a critical value of shear stress (on the order of 20 psi), fracture begins to occur. Since shear stress is equal to the product of shear rate and viscosity, conditions leading to either high shear rate (e.g., high screw speed, narrow flow geometry) or high viscosity (e.g., high molecular weight, low melt temperature) can lead to fracture.

Die drool is the name of the condition when, over time, a build-up of material occurs on the die lips and die face (Figure 14-21). This is problematic, since the drool can either scratch the extrudate or stick to it, in either case causing blemishes on the final product. The cause of die drool is thought to be the migration of low molecular additives or polymeric species to the surface of the extrudate. In order to solve this problem, internal lubricants are added to the raw materials to be extruded, or a low-friction coating is used on the die lips.

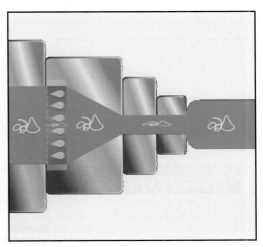

FIGURE 14-20 Schematic diagram illustrating die swell in a polymer extrudate.

FIGURE 14-21 An example of die drool.

FIGURE 14-23 PVC pipes.

Single Screw Extruders: Applications

In this section we will cover some of the products made by extrusion. While there are hundreds of different products, most fall into one of six general areas, categorized by die types:

1. Pipe and Tubing
2. Sheet and Cast Film
3. Coating
4. Fibers
5. Blown Film
6. Profile

Pipes and tubes find application in household plumbing, medical devices, and irriga-tion. These products are made using annular dies (Figure 14-22). Material exiting the extruder enters the die either from the back (in-line die) or the side (crosshead die). In either case, the melt flows around a mandrel (or tip) located in the center of the die, which hollows out the extrudate. In-line dies result in smoother flow streamlines, important for temperature and shear-sensitive materials such as poly(vinyl chloride) (PVC), which is prone to degradation (Figure 14-23). Cross-head dies allow access to the back of the die. This provides an entry point for air, for example, which can be used to inflate a mol-ten tube to size.

Once the polymer exits the die, it is gen-erally solidified rather quickly by pulling it through a chilled-water trough (called free extrusion) or through a closed water

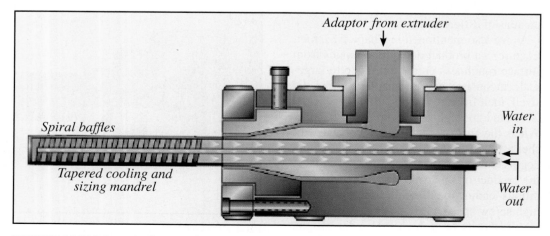

FIGURE 14-22 Schematic diagram of a crosshead die.

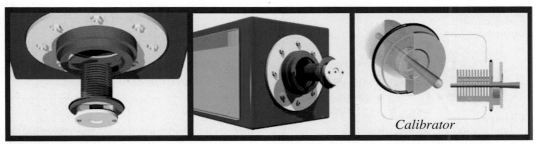

FIGURE 14-24 The calibrator used in the extrusion of plastic pipes.

tank under vacuum (called calibration) Upon entering a vacuum tank, the extrudate encounters a fixture known as a calibrator (Figure 14-24) that assists in controlling the outside diameter of the pipe while it expands under vacuum. Final product dimensions depend on process conditions and tooling geometry, where the relationship with the latter is often expressed as the *draw down ratio*, shown in the box below.

DRAW DOWN RATIO

$$= \frac{D_D^2 - D_T^2}{D_O^2 - D_I^2}$$

D_D = Die exit diameter
D_T = Tip diameter
D_O = Product outside diameter
D_I = Product inside diameter

The proper draw down ratio depends on the polymer being extruded.

Sheet and cast film are widely used in applications ranging from adhesive tape to baby diapers. An industrial example of extruded film production is shown in Figure 14-25. The difference between sheet and film is defined, somewhat arbitrarily, as sheet being thicker than 25 mil (1 mil = 0.001 inch) and film being thinner than 25 mil. The two are extruded similarly through *slot or sheet dies* (Figure 14-26). An important aspect of a slot die is the shape of the manifold, the flow region in which the melt is distributed across the die lips after entering from the extruder. A well-designed manifold will ensure a uniform flow velocity across the die width, leading to a uniform thickness distribution

FIGURE 14-25 Polymer extruded rolled film production (Courtesy: ExxonMobil).

in the product. Slot dies are often identified by their manifold design, such as a *T-die* or a *coat hanger die*. Flat extrusions are often solidified over rollers, such as a *chill roll stack* (Figure 14-27). Chill rolls are cored so that a heat transfer fluid can be circulated internally.

Extrusion coating can be divided into two main categories: *substrate coating* and *wire coating*. Substrates are generally flat, such as paper or a polymer sheet. Therefore, slot dies are used to coat substrates in the manu-

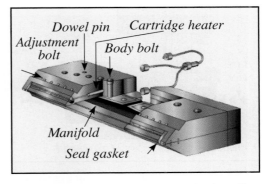

FIGURE 14-26 Schematic diagram of a sheet die.

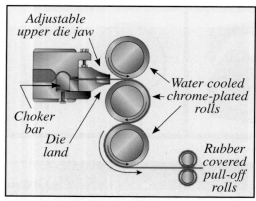

FIGURE 14-27 Schematic diagram of cast film rollers.

facture of products such as paperback book covers. Wire coating is an important process in the production of extension cords, electrical service cable, and telephone wire. This process employs a crosshead die with the wire entering the die through the back (Figure 14-28). Coating takes place in what is called the gum space between the wire guide (tip) and the die. The thickness of the coating is determined by both the die geometry and the pulling speed on the wire, also called the line speed.

An important process variable in all types of extrusion is the take-up ratio, defined in the box at the top of the next column. The take-up ratio not only controls product geometry, but it also has a significant effect on molecular orientation in the final product. Whenever polymer melt is stretched, such as an extrudate under a high take-up ratio,

TAKE-UP RATIO

$$= V_L/V_M$$

V_L = Velocity of the line (pulling speed)
V_M = Velocity of melt exiting die

molecules tend to orient in the direction of stretching. This is very important for producing high tensile strength in fibers.

Fiber (filament) extrusion is used to manufacture products such as textiles for clothing and carpeting, and fishing line. Many filaments may be extruded at one time through a die called a *spinneret*, containing many holes (Figure 4-29). The spinneret does not actually rotate, but gets its name from the organ through which silkworms "spin" their threads. That is why filament extrusion is often called fiber spinning. To improve tensile strength, many fibers are post-treated after extrusion. A series of temperature-controlled rollers called *godets* are used to heat-treat and stretch the fibers as a final step in processing.

Blown film extrusion is perhaps the most widely used extrusion technique, by production volume. Billions of pounds of polyethylene are processed annually by this method to make products such as grocery sacks and trash can liners. In a blown film system (Figure 14-30), the melt is generally extruded vertically upward through an annular die. The thin tube is filled with air as it travels up to a collapsing frame that flattens it before it enters the nip rollers, which pull the film away from the die. The flattened tube then travels over a series of idle rollers to a slitter,

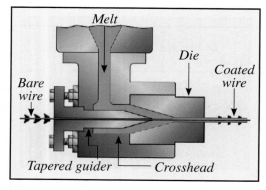

FIGURE 14-28 Schematic diagram of a crosshead die for wire coating.

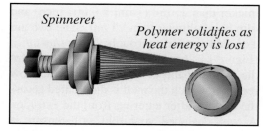

FIGURE 14-29 Schematic diagram of a spinneret used in melt spinning.

FIGURE 14-30 Commercial blown film production (source: ExxonMobil).

winder, or other conversion equipment, such as a bag-making machine.

Although the bubble of film may be several stories high, in order to transfer heat from the polymer, the important conditioning of the film takes place at the base of the bubble between the die face and the so-called frost line, where solidification begins (Figure 14-31). The bubble geometry in this region is crucial for establishing final product geometry, as well as solid-state properties. As the melt exits the die gap, the internal air

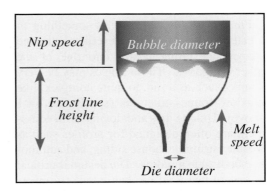

FIGURE 14-31 Schematic diagram of the bubble.

pressure stretches the molten tube circumferentially to a large diameter. At the same time, the high nip speed stretches the tube longitudinally. This biaxial stretching both thins the melt to final thickness and imparts a biaxial orientation in the product, leading to a "rip-stop" effect we term toughness. For example, when the bagger at the grocery store drops a can of soup in our bag, biaxial orientation prevents the bag from splitting open under the impact.

Biaxial orientation leads to isotropic properties in blown film, that is, properties that are equal in the two primary directions the film was stretched (i.e., parallel to the flat bit!). Orientation in the machine direction of the film is controlled primarily by take-up ratio, defined above. To control orientation in the transverse direction, we measure something called the *blow-up ratio*, while the *forming ratio* provides an indication of the degree of isotropy.

BLOW-UP RATIO

$$= D_B/D_D$$

D_B = *Bubble diameter above the frost line*

D_D = *Die diameter*

FORMING RATIO

$$= TUR/BUR$$

TUR = *Take-up ratio*
BUR = *Blow-up ratio*

Finally, *profile extrusion* is something of a catch-all category that describes extruded shapes not fitting neatly into one of the other categories. These shapes may be very simple, such as a rod, to quite complex, like window frame structures (see Figure 14-1 above). Unique dies and downstream equipment are often designed for profiles such as weather stripping, house siding, and automotive body side molding. Die design is critical to ensure that final product geometry meets specification. Die swell and draw down are two important considerations here. Special cooling techniques, calibrators, and pulling units are often used with profiles.

Twin Screw Extruders: Hardware

Twin screw extruders are primarily mixing and compounding devices, as schematically illustrated previously in Figure 14-2. A commercial twin screw extruder is shown in Figure 14-32 and a pair of typical twin screws in Figure 14-33. The melt flow mechanisms through these machines can provide for various types of mixing, down to near-molecular level. The majority of twin screws extrude strand that is subsequently pelletized for delivery to another process, such as injection molding or single screw extrusion. However, an increasing number of forming operations are taking place on twin screw lines, generally the job of single screw lines. This process is called direct extrusion or in-line compounding.

Like single screw extruders, their twin screw counterparts are rated by their screw diameter, in this case, one of the screws. However, unlike single screws, twin screws are highly modular machines. Screws are comprised of individual screw elements, such as conveying elements, kneaders, and pin mixers, that are fitted on a splined shaft and perform a multitude of functions. Barrels are built of temperature-controlled sections, each only a few diameters long. They contain various input or output ports, for functions such as liquid injection or devolatilization. This design modularity leads to a very versatile machine, capable of being reconfigured for a broad range of mixing applications. Adding to this versatility, it is typical to see *L/D* ratios on twin screws that range from 30–50, providing for a large number of functions. We will discuss twin screw hardware by dividing the topic into four major categories:

1. Auxiliary Equipment
2. Co-rotating twin screws
3. Counter-rotating twin screws
4. Conical twin screws

FIGURE 14-32 A commercial twin screw extruder (Source: Cincinnati Milacron).

Auxiliary equipment plays a crucial role in twin screw extrusion, because of the multiple input and output streams that can be built into the system. Upstream you can find feeders, liquid injectors, and side stuffer extruders. While most single screws are flood-fed, twin screws are generally starve-fed, as illustrated in Figure 14-34. Starve feeding is when the throat area is kept only partially filled with raw material, resulting in the screw channels being only partially filled. This serves two purposes. The various feeders can be metered, to provide for close control of composition consistency. Also, since the flight of one screw fits into the channel of the adjacent screw, filling the channels completely with polymer would result in dangerously high barrel pressures.

Downstream equipment generally involves a vacuum pump, for devolatilization, and pelletizing hardware. Strand pelletizers are located at the end of a water trough, while hot-face pelletizers cut the extrudate immediately upon die exit. When direct extrusion

FIGURE 14-33 A typical pair of twin screws (Source: Cincinnati Milacron).

FIGURE 14-34 Schematic diagram illustrating stave feeding.

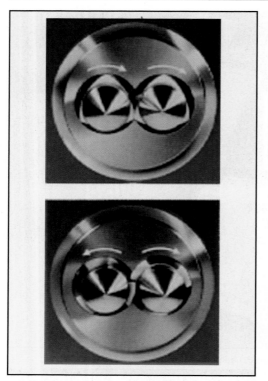

FIGURE 14-35 Co-rotating (top) and counter-rotating (bottom) screws.

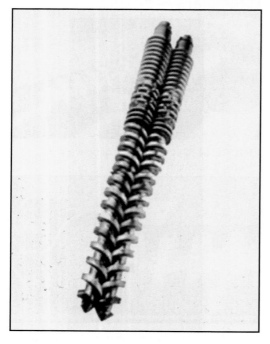

FIGURE 14-36 Conical twin screws.

is conducted, it is customary to have a short single screw extruder or a gear pump, located at the end of the twin screw, to generate pressure for the die. Remember that twin screws are starve-fed, so are usually maintained at low operating pressures.

There are two primary types of twin screw extruders: *co-rotating* and *counter-rotating*. In co-rotating twin screws, the screws rotate in the same direction, as illustrated in Figure 14-35. This requires that the "flight handedness" (e.g., right-handed flights) be the same for each screw. Because of the flow pattern created by co-rotation, the material travels in a figure eight path along the screws and very little material goes through the intermesh region between them. This allows the screws to be operated at very high speeds (over 1200 RPM). Co-rotators do not generate a lot of pressure, however, and so find their largest use in high speed compounding operations.

As you might guess, counter-rotating twin screws rotate in the opposite direction, usually up through the intermesh. This requires that the two screws have opposite flight handedness. The flow mechanism in counter-rotators yields a much higher fraction of material passing through the intermesh than in co-rotators. Also, the geometry creates a closed C-shaped volume of material in the channel. This closed volume results in much higher pumping capability, leading to a greater use of counter-rotators in direct extrusion applications.

Finally, conical twin screws are tapered counter-rotators, where the screw diameter decreases from the shank to the tip, as shown in Figure 14-36. These machines are almost exclusively used for direct extrusion of poly(vinyl chloride), in applications such as making PVC pipe or window frames. The geometry of this processing equipment provides a large surface area at the back of the screws, which is good for dry mixing the multiple powdered feed components found in most PVC systems (polymer, stabilizers, colorant, etc.). Furthermore, the shear rate in the melt decreases along the screw length as the diameter decreases, an advantage given the sensitivity of PVC to shear.

FASCINATING POLYMERS—PVC: "THE POISON PLASTIC"?

Rodney Dangerfield, the renowned American comedian who starred in that classic intellectual film *Caddy Shack* (single-handedly destroying all vestiges of etiquette while elevating golf to the level of a contact sport!), is known for his trademark, "I get no respect!" This could be said of the polymer poly(vinyl chloride) (PVC). It is the third largest commodity plastic produced. PVC is relatively cheap and has a set

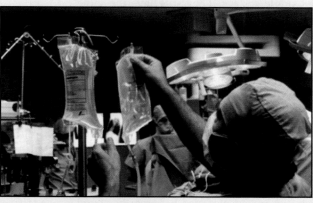

Plasticized PVC IV bags (Source: Solvay).

of chemical and physical properties that renders it particularly useful for numerous applications. So why no respect? It's actually worse than that, PVC is considered by some to be the pariah of the plastics world, a hazardous material, and has been called "The Poison Plastic" by Greenpeace, the environmental watchdog. A little radical, perhaps, but do they have a point? Or, is PVC perfectly safe, as many in the industry assert? If Greenpeace and similar organizations had their druthers, PVC (and ultimately, chlorine) would be banned! Why, you may ask? To quote a "fact sheet" entitled "The Poison Plastic" published by Greenpeace (www.greenpeace.org): "Most common plastics pose serious threats to human health and the environment. The problems of plastics include extreme pollution from production, toxic chemical exposure during use, hazards from fires, and their contribution to the world's growing waste crisis. But one plastic stands alone: PVC, throughout its lifetime, is the most environmentally damaging of all plastics." Of course, it is not difficult to find an authoritative opinion that takes the opposite view. For example, in a chapter on PVC in his 1994 book, *The Consumer's Good Chemical Guide*, John Emsley, Science Writer in Residence, Department of Chemistry at Cambridge University, states, "As far as I am aware, no member of the public has ever been harmed by PVC, and many people owe their lives to it. It is time we learned to live in peace with a rather wonderful plastic." The Commonwealth Scientific and Industrial Research Organization of Australia concluded in 1998, ". . . The balance of evidence suggests that there is no alternative material to PVC in its major product applications that has less overall effect on the environment." So who's right? There are no easy answers and the interested reader is encouraged to read the information from all sides (a good start is the chapter on PVC on our *Incredible World of Polymers* CD) and make up their own minds. For the record, your authors would not ban PVC, except in the specific case of phthalate plasticized infant teething toys. It is a very useful material that can be produced safely and economically. If it is used sensibly, PVC poses no extraordinary risks to humans, animals or the environment. Disposal, while more difficult than other common commodity plastics, is not an overwhelming problem, if common sense and dedication is the order of the day.

Twin Screw Extruders: Functions

Twin screw extruders operate differently than single screws. They still have the same six functional zones, described earlier for a single screw extruder, but the way things happen within each zone is often quite different. For example, while single screws are flood-fed, twin screws are starve-fed. This leads to partially filled screw channels, a condition that can both improve mixing and minimize barrel pressure generation, as mentioned above. In addition, contiguous melting occurs in single screw machines. Twin screws, on the other hand, produce a more efficient dispersed melting mechanism, where unmelted particles become dispersed in the melt matrix, leading to complete melting over a relatively short distance.

Perhaps the most important difference between single screw and twin screw operations deals with the decoupling of throughput and shear. On a flood-fed single screw, throughput is increased by increasing screw speed. However, this results in a proportional increase in the shear rate experienced by the material. This may result in a detrimental increase in heating. On a starve-fed twin screw, however, throughput is increased by increasing the metering rate of the feeder. This allows better control of shear, hence mixing, regardless of throughput. The ability to customize screw design using different elements provides yet another level of mixing (deformation) control.

One of the more interesting aspects of twin screw functionality is the longitudinal nature of the system. The equipment has been described as a series of unit operations taking place along the length of the machine. Operations include material transfer, both in and out, material deformation, and/or heat transfer (in/out) (a chemical engineer's dream!). Polymer pellets are usually introduced in an initial zone to accomplish melting. A little further along, while the temperature is still relatively low and the matrix viscosity relatively high, an additive to be dispersively mixed can be fed (see arrows on Figure 14-37). Following this, when the matrix viscosity has become lower, a liquid color concentrate needing distributive mixing might be added, and so on.

Twin Screw Extruders: Applications

Twin screw extrusion of polymers is most often used for:

1. Compounding
2. Blending
3. Reactive Extrusion
4. Direct Extrusion

Compounding is the highest volume use for twin screw extruders. By compounding, we mean the incorporation of one or more additives into a polymer matrix for the purpose of tailoring material properties to final product use. There are numerous additives used for a wide range of property enhancements, including stabilizers, lubricants, color concentrates, fillers, and reinforcements.

Blending is the process of bringing together two or more different polymer melts to create a single product that has some property or properties that are superior to the "parent" polymers. For example, in polycarbonate/acrylonitrile-butadiene-styrene (PC/ABS) blends, PC provides toughness and environmental stability, while ABS is a lower cost polymer. In blending, there is no chemical bonding between the individual components, only a physical mixing. Alloys are like blends, but include a compatibilizer, an additive that creates some domain interaction. Important measurements of the quality of a blend are the size and arrangement of domains (coarse or fine, as illustrated schematically in Figure 14-38). Twin screw extruders have the ability to provide a great deal of control over these parameters.

Reactive extrusion takes place when the twin screw system is designed to incorporate one or more chemical reactions during the process, such as grafting or even polymerization. Grafting is the process of attaching pendant molecules onto the backbone of a polymer chain. For example, a small concentration of maleic anhydride can be incorporated into polypropylene to make the

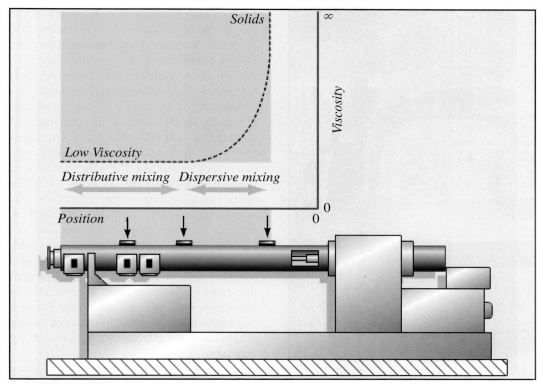

FIGURE 14-37 Schematic diagram illustrating the changes in viscosity and type of mixing in a twin screw extruder.

normally non-polar polymer readily accept a coating of paint. The continuous nature of reactive extrusion makes it attractive for certain polymerizations. However, the number of monomers that can be polymerized this way is restricted, due to heat transfer limitations.

Finally, *direct* extrusion combines two often separate functions: mixing and forming. In this processing method, a forming die (other than a strand die) is located at the end of the twin screw extruder for making products such as tubing and sheet. There are several advantages to this technique, including reduced shipping and storage, and exposure of the polymer to only one heat cycle. Disadvantages are that twin screws are significantly more expensive than single screws and are more complicated to operate.

INJECTION MOLDING

Injection molding is the most common method for processing polymers in the world. The vast majority of plastic products that we encounter in everyday life were injection molded. Look around, you might see a computer keyboard, a speaker housing, a calculator case, watch dial, mouse buttons, lamp base . . . well, you get the idea. Injection molding is not limited to typical consumer items, though. Large automotive body parts (see Figure 14-39), contact lenses, miniaturized electronic components, and millions of other products are injection molded.

The purpose of an injection molding machine is to plasticize the polymer (i.e.,

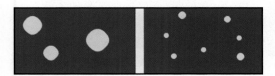

FIGURE 14-38 Schematic diagrams of different dispersions: course (leftt) and fine (right).

FIGURE 14-39 Many of the parts on this car were injection molded (Courtesy: Bayer).

make it deformable), inject the melt into a closed, cooled mold, and solidify the part prior to the mold opening and the part ejecting. An example of a rather large commercial injection molding machine is shown in Figure 14-40. The mold is designed to

produce the 3-dimensional geometry of the part shape, as opposed to the (essentially) 2-dimensional nature of extruded products, like pipe. In this section, we will cover various injection molding machines and their components, as well as injection molds.

An injection molding machine can be simplistically divided into an injection side and a clamping side (Figure 14-41). The injection side closely resembles an extruder, with one major exception: the screw can move forward and backward. In what are often called reciprocating screw machines, the screw sequentially plasticizes the pellets, then acts as a ram to quickly inject a shot of polymer into the mold. Early injection molders employed a simple plunger instead of a screw. This yielded much lower plastication rates (throughput) and a lower mixing efficiency than current machines. Plunger machines have all but disappeared today, except for some rare cases where no mixing is desired.

Traditionally, injection units have been controlled with hydraulics. Today, however, the market also includes electronic as well as hybrid control of the injection side. It is important that the injection unit be capable of delivering the entire shot of material to the mold in a short time (seconds) against very high pressure (on the order of 20,000 psi). Regardless of the control mechanism, the primary method for rating injection units is

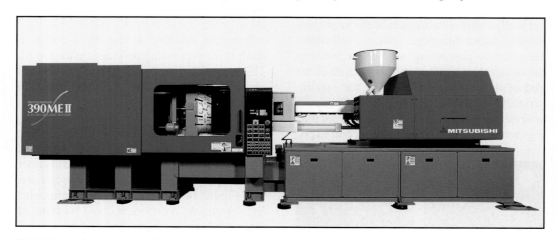

FIGURE 14-40 A commercial injection molding machine (Courtesy: Mitsubishi).

by the maximum shot size. This is the maximum mass of material that can be delivered to the mold in one forward stroke of the screw (e.g., ounces of polystyrene).

The clamping side is responsible for keeping the two halves of the mold closed together during injection. This requires a substantial amount of force, because the high-pressure injection is acting to drive the mold halves apart. In fact, the primary method for rating injection molding machines is by the amount of clamping force the unit is capable of developing. For this reason, an injection molding machine is sometimes called a press, for short. Commercial machines have clamping forces ranging from 20 tons to 10,000 tons.

Two large plates (or *platens*) are the distinguishing features on the clamping side (see Figure 14-42). Each half of the mold is bolted to one of the platens. The stationary platen has a hole through which the melt enters the mold from the injection unit. The moving platen pushes the mold closed prior to melt injection and pulls the mold open for part ejection. Housed within the moving platen is an ejector assembly for knocking the part off of the mold. In most cases, hydraulics are used on the clamping side to open and close the mold and to lock the mold closed during injection. On many machines a mechanical linkage, called a *toggle*, provides the primary locking force on the clamp. Electric clamping mechanisms are also becoming popular today. Whatever type of clamping system is used, it is important that it provide fast, accurate positioning of the moving platen and adequate clamping force.

The injection mold is probably the most important component in the machine. A different mold is required for every product and the product quality is most dependent on the design and machining of the mold from which it is produced. Machining tolerances for the building of molds are often on the order of ten-thousandths (0.0001) of an inch. Because of the machining accuracies required and the complexities of the assembly, molds often cost three to five times the price of the molding machine. A typical mold, one used to make a construction cone, for example, as

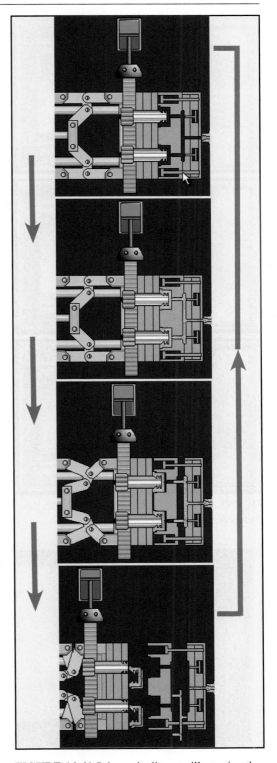

FIGURE 14-41 Schematic diagram illustrating the sequence of events in injection molding.

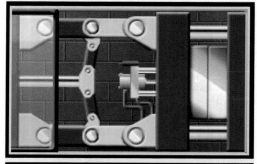

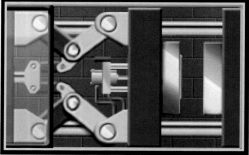

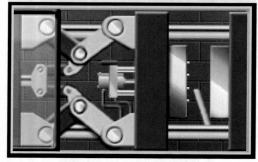

FIGURE 14-42 Schematic diagram illustrating the clamping side of an injection molder.

FIGURE 14-43 A commercial injection mold for construction cones (Courtesy: Product Design Inc.).

illustrated in Figure 14-43, includes a core and a cavity. The core forms the internal features of the part and is located on the moving half of the mold. This ensures that as the part shrinks during cooling, it tightens onto the core and remains with the ejector mechanism. The cavity on the stationary half of the mold forms the external features of the part.

To maximize the production rate, most molds are designed to produce multiple parts for each shot. To accomplish this, the melt is injected through the nozzle on the end of the barrel. It then enters the mold through a channel called the sprue bushing (or *sprue*). The melt then divides, flowing through a series of *runners* and entering each cavity through a small restriction called, naturally enough, a *gate*. This is illustrated in Figure 14-41. The runner system should be designed such that every cavity receives material that has experienced the same flow history. This will ensure the lowest variation in part quality. The gate is small enough that the material it contains generally solidifies quickly, keeping the cavity closed while the material in it solidifies.

The type of auxiliary equipment used for injection molding is typically the same as that used in most other polymer processing operations. Hoppers, dryers, and vacuum loaders are utilized upstream, while conveyors and grinders are located downstream. Most injection molded products incorporate some amount of in-plant regrind from runners, sprues, and rejected parts. An auxiliary device, growing in popularity these days, is the robot. High-speed robot arms are widely used for everything from picking parts and runners out of the press to packaging the final product.

Injection Molding: Features

Injection molding is a cyclic process. As an arbitrary starting point, assume that the mold is closed and the screw is in its retracted position with a shot of melt ready to be injected. This is the position at the top of Figure 14-41. The four steps that comprise the process are:

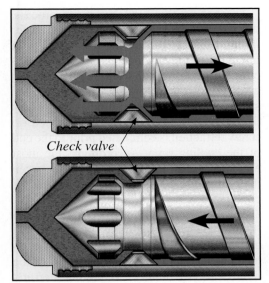

FIGURE 14-44 Schematic diagram of a check valve in injection molding machines: (top) during plastication and (bottom) during injection.

1. Filling
2. Packing
3. Plastication/Cooling
4. Ejection

The filling step begins with the screw moving forward (without rotating), pushing melt through the sprue, runners, gates, and into the cavities (second from the top of Figure 14-41). Filling ends when material reaches the last place to fill in the cavities (assuming they all fill equally). At the end of fill, the screw remains about one-quarter inch back from the nozzle, leaving a cushion of melt to prevent metal-to-metal contact between the screw tip and nozzle. Because the mold is cooled, it removes heat from the polymer as it fills. It is therefore important that the filling step be completed within a few seconds, or the polymer will solidify before it finishes filling the mold. If this happens, a defect known as a *short shot* is created.

Due to the low filling times and high melt viscosities, injection pressures are quite high, as mentioned previously. Generation of this pressure would not be possible without a *check valve* located at the tip of the screw, as illustrated in Figure 14-44. This

valve remains closed during injection (bottom), preventing the melt from back flowing over the screw. When the screw is moving backwards during plastication (top of Figure 14-44), the valve opens, allowing melt to flow over the screw tip.

An interesting effect occurs while the cavity is filling with polymer (Figure 14-45). One might expect that as material enters the gate, it would be continually pushed ahead by the material that enters afterward. However, the material that first enters the cavity immediately spreads out to contact the cool mold wall and solidifies. The material that

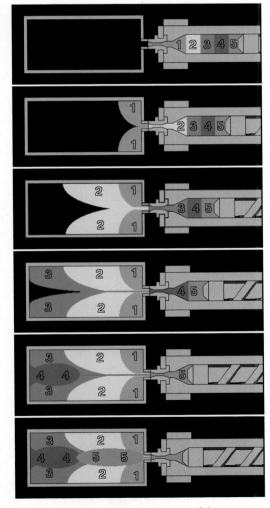

FIGURE 14-45 Schematic diagram of the sequence of events (from top to bottom) that occurs during the filling of the cavity.

follows then flows through the solid layer just created, eventually reaching mold wall on which to solidify. This process, called *fountain flow*, continues until the mold is filled, leaving a core of melt to solidify inside the frozen layer. The result is a laminar molecular structure across the thickness of the part, high orientation near the walls and low orientation in the core.

After the filling step comes a *packing step*. This is when pressure is kept on the material in the mold until the appropriate time to retract the screw. In the packing step, two things occur: compensation for shrinkage and gate freeze-off. As material in the mold cools, it tends to shrink in volume. In order to match the machined dimensions of the cavity, additional material is pushed into the mold during packing to compensate for volume lost to shrinkage. Additionally, packing holds the material in the mold until the gate solidifies. Otherwise, if pressure were released immediately after fill, the pressure in the cavity would push some of the melt out through the gate.

The next step is the plastication and cooling step. With the gate frozen, the screw can begin to retract for developing the next shot of polymer. The screw now begins to rotate, pulling pellets down from the hopper, plasticating them along the screw, and conveying the melt through the check valve, mentioned previously, and over the tip of the screw into a melt pool (Figure 14-44). As this melt pool fills, it begins to push the screw backwards until it reaches a predefined position, where it stops rotating and comes to rest. The next shot is ready to be delivered at this point.

FIGURE 14-46 Plastic bottle caps.

Throughout the screw retraction process, the part has been cooling by heat transfer through the mold. The mold is cored with internal cooling channels through which circulate a heat transfer fluid. Any blockages can lead to a "hot core." Depending on the size of the part, cooling times vary widely from seconds to minutes. In most cases, screw retraction is completed prior to the end of the cooling time. Accordingly, the most efficient cycles minimize the time between the end of screw retraction and the end of cooling.

The final step in the process is ejection (bottom diagram in Figure 14-41). At the end of cooling, the mold opens and ejector pins actuate to push the parts out of the mold. This is generally a very quick step and is followed by the mold closing and the cycle beginning again. For some mold designs, the parts ejected are still attached to the runner and the two must be separated. In other mold designs, the parts and runners are separated.

Injection Molding: Applications

There are far too many applications for injection molded products to thoroughly cover here. However, a few examples will help provide an idea of some of the important mold design and machining considerations.

Consider the cap from a soda bottle (Figure 14-46). Often it is molded at a different facility than the bottle, and filling and capping takes place at yet a third facility. If the cap molder produces a product that has a diameter slightly above specification, while the bottle molder produces a bottle with a diameter slightly under specification, the company filling the bottle with Coke®, or Pepsi®, or whatever, is going to be seriously displeased with the ensuing mess. This emphasizes the need to design and machine molds, and process parts, to the very close tolerances, as discussed above.

Let's consider the humble bottle cap just a little further. It has threads on the inside to attach it to the bottle. Recall that when this part is molded, a core forms the inside of the product. After cooling, ejection is actuated to take the part off of the core. However, you can't just push the part off the core with ejector pins, as this would damage the threads. Therefore, the core (in fact, all the cores) must be rotated while the cap is held in place, in order to unscrew the cap from the mold. This, of course, makes the mold more expensive.

Now consider a PVC pipe elbow, a commonly used product in residential plumbing (Figure 14-47 is what we saw, among some disgusting-looking stuff that we had forgotten was there, when we looked under our bathroom sink). This seems like a fairly reasonable part to mold, until one begins to think about the shape of the core and removing the part from the core. Fortunately, a relatively straightforward approach has been developed that uses a core that is split in two halves. Each half is moveable and the two halves can be pulled apart, hence out of each end of the elbow, for ejection.

These are just a few examples of the enormous range of products made by injection molding. The next section will cover variations to the injection molding process. Some of these variations are so innovative they make possible unique products that would otherwise be enormously expensive or practically impossible to produce.

INJECTION MOLDING VARIATIONS

Even though it is possible to produce a seemingly countless number of products using conventional injection molding techniques, variations to this process open up even more opportunities. This section will cover some of the injection molding variations that are used to improve the properties, increase the production efficiency, and expand the usefulness of injection molded products.

Hot Runners

First, we will consider hot runners. These are injection molds in which the runner material remains molten after each shot, as illustrated in Figure 14-48. The runner sys-

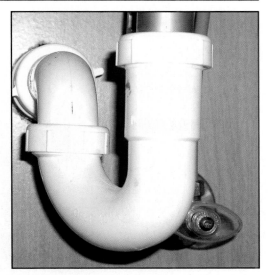

FIGURE 14-47 A PVC pipe elbow.

tem is heated and isolated from the cavity, which is cooled, as in conventional injection molding. When the mold opens for ejection, only the part is ejected, leaving hot material in the runner for the next shot.

The primary advantage of these systems is that all material goes into the part, that is, none is lost to scrap in runners and sprues. For molds designed to run millions of cycles, a tremendous cost savings is realized. The main disadvantage is that the molds are more costly to design and build, and they require a separate temperature-control system.

Insert Molding

Insert molding, also called overmolding,

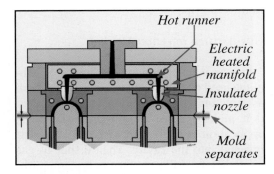

FIGURE 14-48 Schematic diagram of a hot runner.

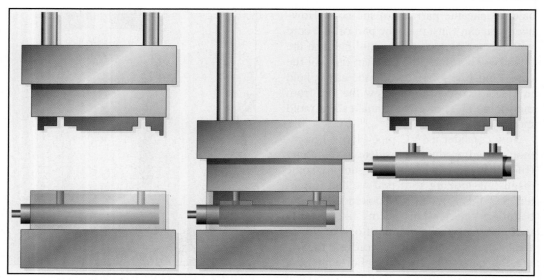

FIGURE 14-49 Schematic diagram of insert molding: part inserted (left), mold closes and plastic injected (middle), then mold opens and part is removed (right).

is a technique for encapsulating a component with plastic. For example, consider the plug on the end of an extension cord. The metal blades that fit into the outlet on the wall are insert molded into the plastic plug body. During insert molding, the insert, which is typically metal, is either manually or robotically placed in an open injection mold. The mold then closes and the part is molded around the insert, as shown in Figure 14-49. Many parts are produced this way, to capture the combined effects of the multiple components. The extension cord plug, mentioned above, provides a conductive product that is protected with an insulating layer. A metal screwdriver can be insert molded with a plastic handle for lighter weight and an improved grip.

Gas-Assisted Injection Molding

Gas-assisted injection molding is a technique used to hollow out thick sections, such as structural ribs, common in many injection molded parts. In a typical gas-assist application, a short shot of polymer is first injected into the cavity. Then a shot of gas immediately follows the polymer, pushing the melt into the unfilled regions of the

mold and leaving thick areas hollow (Figure 14-50). (Recall how fountain flow proceeds in conventional injection molding to understand how the hollow sections are formed.) Nitrogen gas is usually used, to prevent oxidation.

There are numerous advantages to using gas-assisted injection molding. Less material is used to produce each part; therefore, lower raw material costs are realized and products are lighter in weight. Less material also means less heat to remove from each part, so cycle times are reduced. Because lower pressures are required to push gas through the mold compared to a full shot of high viscosity polymer, lower clamping forces are necessary. This means that a smaller, less expensive press can be used. Finally, the surface quality of parts may be improved. The slow cooling that occurs in solid, thick parts can lead to shrinkage marks on the surface (called *sinks*) and warpage. However, hollow sections produced by gas-assist solidify quickly under gas pressure, producing defect-free surfaces (Figure 14-51). Although gas-assist has been around for some time, its use is only now beginning to grow. This comes at the end of about a decade of legal battles between multiple developers of gas-assist

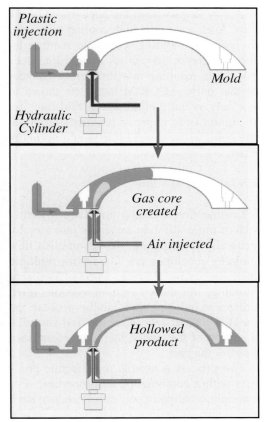

FIGURE 14-50 Schematic diagram of gas-assisted injection molding.

FIGURE 14-51 Examples of products made by gas-assisted injection molding.

processes. Look for more widespread use of gas assist in the years ahead.

Shear Controlled Orientation Injection Molding

Shear controlled orientation injection molding (SCORIM) is a novel technique for establishing the alignment of polymer molecules and fiber reinforcements in molded parts. Solid-state properties, which are highly dependent on molecular and additive alignment, can be improved through the use of controlled orientation.

The SCORIM process employs a unique hardware component that is retrofitted to a conventional injection molding machine. Located between the injection nozzle and the mold, this unit contains a pair of pistons that exert pressure on the melt after it is injected

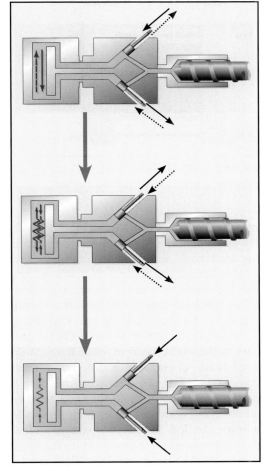

FIGURE 14-52 Schematic diagram of shear controlled orientation injection molding.

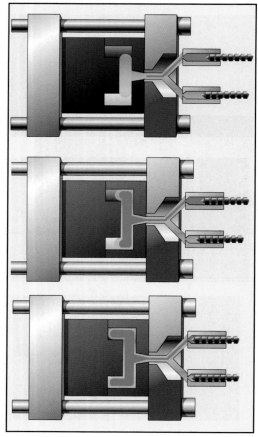

FIGURE 14-53 Schematic diagram illustrating the sequence of events in co-injection molding.

line may detract from the appearance of the part. Also, due to the high cooling rates used in injection molding, complete reentanglement of molecules across a weld line does not occur, resulting in a structural weakness at that point. SCORIM has been shown to not only visually eliminate a weld line, but to significantly improve product strength at the point where a weld line would normally exist.

Co-Injection Molding

Co-injection molding involves injecting two or more different materials into a mold, generally in sequence. To accomplish this, molding machines are fitted with multiple injection units (Figure 14-53). Two-shot molding, the most common version, often utilizes an indexing mold. In this process, the mold is rotated 180° after the first shot fills a portion of the cavity, then the second shot finishes the part.

This process is used to manufacture products with a combination of properties. For example, a ballpoint pen barrel needs to have the rigidity of a stiff material, but it is convenient to have attached a more easily gripped area consisting of an elastomeric material. Another example is the set of keys on a computer keyboard. To improve wear resistance, the characters are co-injected rather than printed onto the key top. Finally, co-injection can be used to encapsulate a low cost material that has poorer properties, such as regrind, inside a shell of prime material.

Reaction Injection Molding

Reaction injection molding (RIM) is a process where a controlled chemical reaction occurs in a mold after injection of the material components. Commercially, the process is used almost exclusively for production of polyurethane parts by injecting a two-component system of isocyanate and alcohol (Figure 14-54). Polyurethanes can be produced with a wide range of properties, from very low density foams to reinforced, structural components. RIM is widely used

into the mold, as illustrated in Figure 14-52. By operating out of phase, the pistons move the melt back and forth inside the cavity, establishing a shear field that orients the polymer molecules in the direction of flow. A final packing step, where the pistons act in phase, can be used while the part solidifies.

An additional benefit to the SCORIM process is the obliteration of weld lines. Weld lines occur wherever two flow fronts meet one another. For example, when molding a part that will contain a hole, the mold includes an obstruction around which melt flows. As a result, the flow front splits to travel around the obstruction and then rejoins on the other side, forming a weld line. Weld lines can be problematic for two reasons: aesthetics and structural integrity. The presence of a weld in various automotive, recreation, and industrial applications.

In the most common RIM process, two low viscosity materials are delivered to a mixing chamber through nozzles. The nozzles atomize the liquids into fine droplets, maximizing surface area for optimum mixing. The mixture is then injected into a closed mold, as in conventional injection molding. Besides the inherent property advantages gained from using polyurethanes for certain applications (see *The Incredible World of Polymers* CD), an important processing advantage is the low clamping pressures that are required to inject a low viscosity material. This provides a significant energy savings.

Compression Molding

Conventional injection molding and most of its variations are primarily used to process thermoplastic materials. Thermoplastics are much more adaptable to processing through a screw/barrel system than are most thermosets, because of the potential problems that can occur as a result of premature cross-linking. Nevertheless, there are many applications for molded thermoset products. For example, parts requiring high temperature resistance, such as under an automotive hood (Figure 14-55) or around an oven, are best made with cross-linked materials. Also, thermosets can contain very high percentages of reinforcing materials, such as glass fiber, useful in structural applications. Accordingly, a technique is necessary for molding these materials.

Compression molding is the method used to accomplish this. Just as in the clamping side of an injection molder, a compression molder is generally configured in a vertical arrangement. Since the molder does not have an injection unit, or even a single point for material entry into the mold, each cavity is loaded individually when the clamp is open. This can occur either manually or robotically.

The charge of material placed into each cavity is known as a *preform*. Unlike the lower viscosity of melt injected in conven-

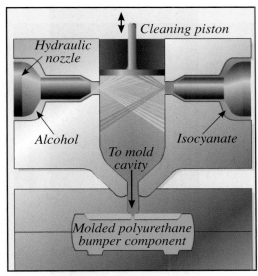

FIGURE 14-54 Schematic diagram illustrating the reaction injection molding process.

tional injection molding, preforms are closer to solids, having the consistency of a thick dough. These thermosetting preforms, which are designed to crosslink while in the mold, often contain many additives, such as colorant, reinforcement, catalyst, and reaction accelerator. These materials are, therefore, known as bulk molding compounds (BMC).

Once the cavities are loaded, the clamp is closed and molding begins. Heat and shear initiates cross-linking, which controls the cycle time. The parts do not need to be cooled

FIGURE 14-55 A peek under the hood.

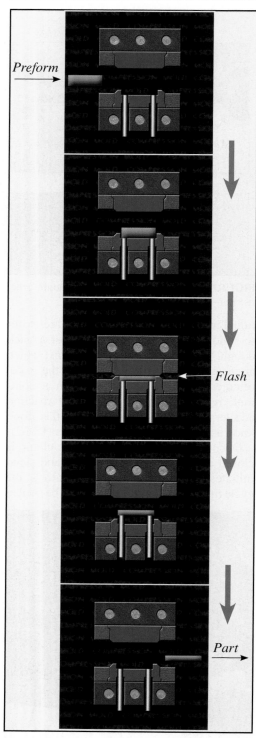

Preform

Flash

Part

FIGURE 14-56 Schematic diagram of the sequence of events that occurs in compression molding.

(nor does the mold), since cross-linking solidifies the polymer. At the end of molding, the clamp opens and the parts are removed from the cavities. In many cases, *flash* must then be removed from each part. Flash is the name for a thin edge of polymer that escaped from the cavity, was pressed between the mold halves, and remains attached to the part (see Figure 14-56).

Compression molding is used mainly for relatively simple parts requiring no inserts and no intricate geometry, because of the high viscosity of BMCs and the relatively low shear rate generated during compression. In order to produce more complex parts, a variation of compression molding that employs a higher shear rate, transfer molding, is used.

Transfer Molding

Transfer molding has elements of both compression molding and injection molding. (In fact, in the historical development of plastics molding, transfer bridged the gap between compression and injection.) Like compression molding, transfer molding is used for thermoset materials, often delivered as BMCs. However, like injection molding, transfer molding utilizes a single material entry point into the mold (sprue) and a material distribution system (runners) to deliver polymer to multiple cavities.

A transfer molding machine looks very much like a compression molder, except that the system contains a transfer chamber, for loading the polymer, and a piston to drive the material through the sprue and into the mold (see Figure 14-57). The primary advantage of this process over compression molding is the ability to produce more complex parts, including those with inserts. This is because the additional shear generated as the material travels through the sprue and runners reduces the viscosity of the polymer, allowing it to better flow through the mold and around inserts. An additional benefit is that the increased shear can help increase the rate of the cross-linking reaction, thus completing the process in a shorter time.

Blow Molding

Blow molding is used to produce hollow products and the nature of the process is somewhat like glass blowing. Although a range of items is produced by blow molding, such as automotive items, toys, and household appliances, the vast majority of blow molded products are containers. Just glance at the soda bottles, milk jugs, etc., that can be observed as you walk down the aisles of your local grocery store and you will be convinced of the importance of blow molding.

Blow molding is a two-step process. In the first step, a preform is made either by extrusion or injection molding (top two diagrams of Figure 14-58), as we will discuss below. The second step utilizes air pressure to inflate the preform inside a closed, hollow mold. The polymer expands to take the shape of the cooled mold and solidifies while air pressure remains in the part (third diagram from the top of Figure 14-58). After cooling, the mold halves open, the part is ejected (bottom diagram) and the next preform enters the mold. Several variations of blow molding are used, depending on the type of product being made and the polymer being processed. We will look at a few of the most important of these variations in this section.

Extrusion blow molding utilizes an extruder to produce the preform to be blown. In a typical arrangement for producing juice bottles, the tube is extruded vertically downward to some predefined length, as shown in Figure 14-59. This tube is commonly called the *parison*, from the Latin word for wall. When the parison is the proper length, the two open mold halves shuttle to a position surrounding it. With the parison in position, the mold closes and the tube is cut from the extruder. The mold is designed to pinch the tube and form a seal at the bottom, but leave the top open. The mold then shuttles back to the blow position. At this point, a blow pin is inserted into the hole at the top of the parison, pressurizing the parison against the mold walls with air. After sufficient cooling, the blow pin is removed and the part is ejected (Figure 14-60). While the part was cooling,

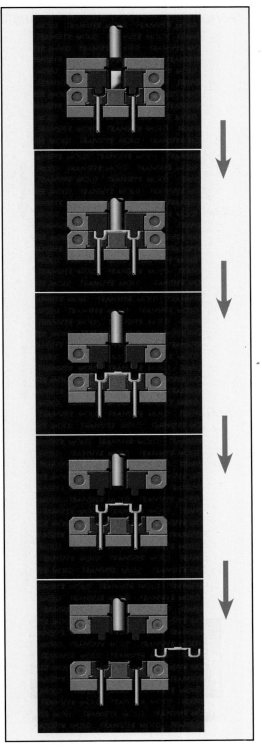

FIGURE 14-57 Schematic diagram of the sequence of events that occurs in transfer molding.

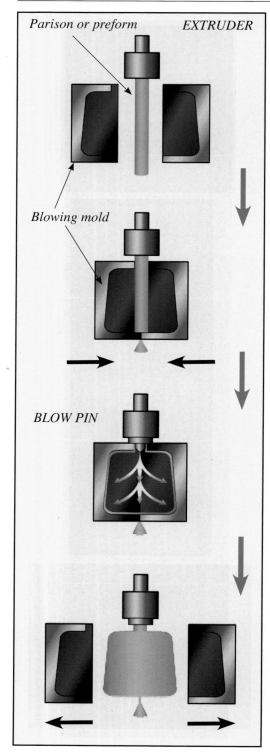

FIGURE 14-58 Schematic diagram of the sequence of events that occurs in blow molding.

FIGURE 14-59 Extrusion blow molding: formation of parisons.

the next parison was being extruded, so that it is ready to be taken by the mold upon its return.

This type of extrusion blow molding machine uses something called a shuttle machine to get all the parts in place. Alternatively, a rotary machine, illustrated in Figure 14-61, can be used. Rotary extrusion blow molders are used in high volume applications, as in the production of laundry detergent bottles.

One of the more important things to understand about extrusion blow molding involves the freely suspended, molten parison. In some operations, the parison may be several feet long and suspended for many seconds before the mold captures it (see Figure 14-62). In all cases, the parison is acted upon by gravity, so the polymer being used must

FIGURE 14-60 Extrusion blow molding after mold is closed and bottles have been ejected.

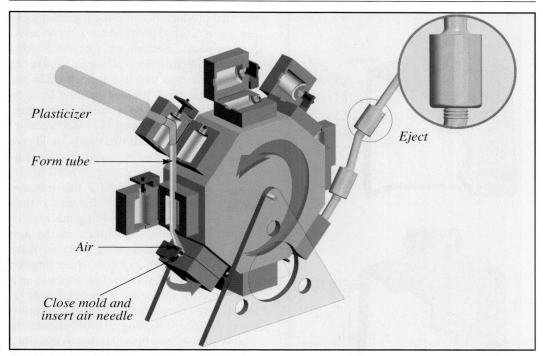

FIGURE 14-61 Continuous tube blow molding process.

therefore have a good *melt strength*. Melt strength can be defined as a high resistance to (stretching) flow in the absence of shear, and can be thought of in terms of an extrudate that has a consistency that is more like taffy than water. Certain polymers have been synthesized specifically for good parison melt strength. For example, some grades of high molecular weight, high density polyethylene have good melt strength.

Another effect of gravity acting on a parison is *parison sag*. This results in a decrease in the wall thickness along the parison length, due to its increasing weight. This, in turn, leads to a product with varying wall thickness, as illustrated in Figure 14-63, often resulting in poor properties in thin areas and wasted material in thick areas. The problem of parison sag can be solved using *parison programming*. A moveable mandrel is positioned by a programmable controller, allowing the system to produce a specified wall thickness at each of several locations along the length of the parison. As the mandrel is moved downward, the die gap increases,

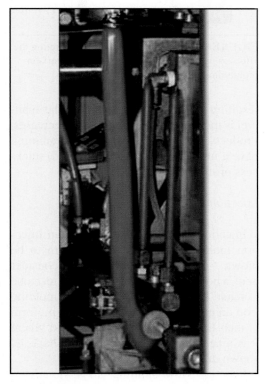

FIGURE 14-62 This bloody long parison better have good melt strength!

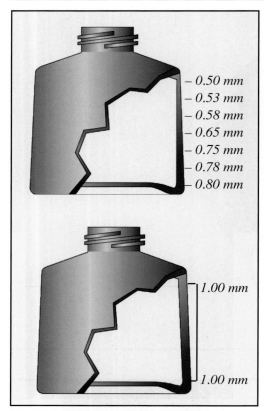

- 0.50 mm
- 0.53 mm
- 0.58 mm
- 0.65 mm
- 0.75 mm
- 0.78 mm
- 0.80 mm

1.00 mm

1.00 mm

FIGURE 14-63 Schematic diagram illustrating the effect of parison sag (top) and the use of parison programming (bottom).

resulting in thicker extrudate. As the mandrel is moved upward, the die gap decreases. Products made with parison programming have a uniform (or another specified) thickness profile.

Injection Blow Molding

Injection blow molding utilizes an injection molder to produce the preform to be blown. In the example we will consider here, a hollow preform resembling a test tube is made by conventional injection molding (top diagram of Figure 14-64). The preform is then transferred to a second mold where it is reheated and blown into final shape, as shown in Figure 14-64.

Injection blow molding has some advantages over extrusion blow molding. First, it is capable of greater dimensional accuracy in

the final product, because the injection mold can be machined to produce a very accurate preform. Second, an injection molded preform contains no weld lines, unlike an extruded parison that has a pinch-off at the bottom. If a blow molded product is to be pressurized, like a soda bottle, the weakness of a weld line may be unacceptable. Of course, the fact that two costly molds are required for injection blow molding is a significant disadvantage.

The two molds needed for this process, an injection mold and a blow mold, may be located in the same molding machine or in different molding machines. In the first case, known as a one-stage process, there are multiple stations in the machine through which the material travels. For example, in a four-station machine the polymer would be injected into the preform, shuttled to a conditioning (profiled heating) station, shuttled to a blow station, and then shuttled to ejection.

In the two-stage process, injection of the preform is performed on one machine and blowing is performed on a second machine. Although the obvious disadvantage to this process, compared to one-stage, is the high cost of two machines, there are some significant benefits. First, because of the small size of preforms compared to bottles, they are much more cost-effective to transport. This allows beverage bottlers to buy preforms from injection molders and blow them in-house, prior to filling. Additionally, when the two steps are decoupled, it is possible to injection mold a larger number of preforms for inventory or shipping, using the type of complicated-looking machine shown in Figure 14-65.

Stretch Blow Molding

Stretch blow molding is a variation of injection blow molding used primarily for processing polyethylene terephthalate (PET), the polymer of choice for soda bottles. In this process, a stretch rod located inside the blow mold elongates the heated preform, either before or during the blowing step (see Figure 14-66). Of course, as in all blow molding

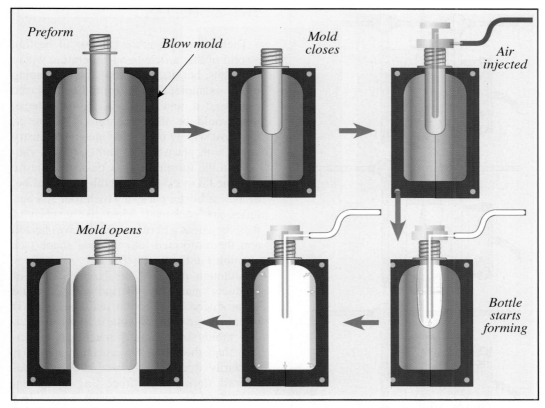

FIGURE 14-64 Schematic diagram of the sequence of events that occurs in injection blow molding.

processes, the preform is designed to provide the specified product thickness at the end of the blowing step.

Stretch-blow processing is used to develop biaxial molecular orientation. If blow pressure alone were used, the stresses in the preform wall would be predominantly circumferential, producing orientation only in the hoop direction of the bottle. However, the additional stresses generated longitudinally by the stretch rod result in biaxial orientation.

Biaxial orientation improves the barrier properties of PET. (Barrier strength is a term used to describe the ability of a layer to inhibit the diffusion of gases, such as the carbon dioxide in soda.) Improvements in processing PET over the last decade have significantly improved the shelf life of soda bottles and led to the large increase in the production of smaller bottles (single-serving sizes).

FIGURE 14-65 A complicated machine that produces a large number of preforms (Courtesy: Husky).

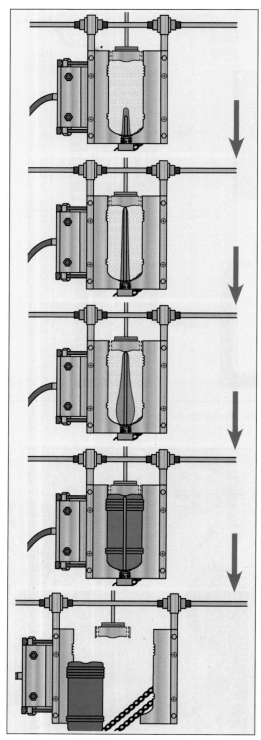

FIGURE 14-66 Schematic diagram of the sequence of events that occurs in stretch blow molding.

THERMOFORMING

Thermoforming is the process of heating a solid plastic article to a temperature where it softens, but does not flow, then reshaping it. For example, if you were to heat a plastic rod, bend it, and cool it in the new shape, that would be thermoforming. However, the most widely used techniques in industry begin with plastic sheet and employ a vacuum in the forming step. Consider a plastic cup, the lid to a one-pound tub of margarine, or a clear blister pack in which four AA batteries are packaged. Most likely, each of these began as a plastic sheet that was heated and thermoformed into its final shape (see examples in Figure 14-67).

Although many of the large number of products made by thermoforming could be produced by injection molding, there is often a large cost advantage to the former. The molds are generally much less expensive than those used in injection molding, primarily due to the lower stresses experienced during this processing operation. Also, these lower stresses result in much less energy expended during the molding operation, another large cost savings over time. However, thermoforming is somewhat more limited in the range of product shapes that can be produced.

There are primarily two types of hardware systems used in thermoforming. In the shuttle system, cut sheets of plastic are first loaded into a heating station that softens the poly-

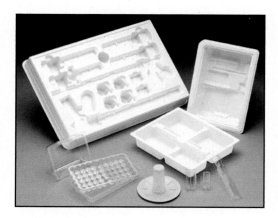

FIGURE 14-67 Thermoformed articles.

FASCINATING POLYMERS—PS & STYROFOAM®

In the years leading up to WWII, a limited amount of styrene was polymerized to produce amorphous polystyrene (PS) in both Germany and the United States. This brittle, glassy polymer found application mainly as an electrical insulator and as a dielectric material for condensers. But with the scarcity of natural rubber in WWII, styrene was to have a much more important role as a comonomer in the production of styrene/butadiene (SBR or Buna S) synthetic rubber. After the war, there was an abundance of styrene, as natural rubber again became available, but advances in the technologies of injection molding, foaming and impact modification helped polystyrene become the commodity plastic that it is today. Anyone who has spent too much time in low-life bars guzzling beer will be only too aware that polystyrene is a beautifully clear glass-like material. It finds application in many "throwaway" articles, in addition to other, more permanent, clear packaging items such as "blister packs" for pharmaceuticals and food. But we are also likely to run across PS in another of its ubiquitous guises, as foamed packaging and insulation materials. You have all opened a package and had those annoying little plastic insulating beads fly all over the place. Or you have bought, say, a new television set or computer and found it packed securely between pieces of incredibly light molded foamed plastic pieces. Or purchased eggs in a foamed plastic container; drunk hot coffee from a foamed plastic cup or eaten a hamburger from foamed clam-shell container. Chances are that this is foamed polystyrene, or as it is generically called, Styrofoam®. How is PS foam made? Actually it is quite elegant and simple. If styrene is polymerized in the presence of about 5% of a volatile solvent (like pentane) using suspension polymerization, spherical PS beads containing the solvent are produced. If these beads are now subjected to steam, they soften and expand like popcorn (some 40 times their original dimensions), as the volatile solvent tries desperately to get out. To form a molded sheet or article, these expanded beads are placed in a heated mold, pressure is applied, and the beads are fused together. A continuous process is used to make commercial Styrofoam® insulation materials.

A thermoformer machine that mass produces polystyrene cups (Courtesy: Brown Machine).

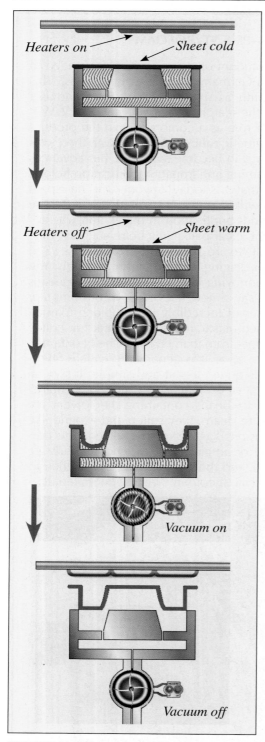

Heaters on —— →   Sheet cold

Heaters off ——— →   Sheet warm

Vacuum on

Vacuum off

FIGURE 14-68 Schematic diagram of the sequence of events that occurs in vacuum thermoforming.

mer. The softened sheet is then shuttled to a forming station where it is reshaped into the final product, often by pulling a vacuum in a mold cavity into which the sheet is drawn. Alternatively, an in-line system employs a thermoforming station that receives plastic directly from a continuous sheet extruder. As the line pauses to thermoform product, slack builds in the extruded sheet, but this is quickly taken up as the forming unit pulls in sheet for the next set of parts (see picture in the *Fascinating Polymers* panel).

Many variations of the forming process are used to accomplish specific shaping of the sheet. We will describe three basic techniques: *vacuum forming*, *drape forming*, and *plug-assist forming*. Vacuum forming utilizes a cavity mold shaped like the convex side of the product to be formed, as shown in Figure 14-68. The softened sheet is located over the top of the cavity and a seal is created where the sheet and mold meet. The seal prevents leakage during evacuation of the air in the mold. As vacuum is drawn in the cavity, the sheet pulls down into the mold, acquiring its shape. Since the mold is cooled, the part solidifies quickly after contacting the mold walls.

Two important considerations in the design of the mold are the *draw ratio* and *draft angle*. The draw ratio is the depth of the cavity divided by the width at the top of the cavity. Because the sheet is stretched as it is drawn into the cavity, excessive thinning in the part walls can occur if the draw ratio is too high. A reasonable draw ratio for vacuum forming is 1 or less. The draft angle is defined as the angle between the wall of the cavity and a perpendicular to the base. This angle does not need to be very large (angles of 5° are typical), but it is necessary to aid in removal of the part from the mold.

Notice that during thermoforming only one side of the original sheet touches the mold. Accordingly, this process is most often employed for products that will be used on the concave side, so that the mold does not require a highly polished, expensive surface finish. For example, whirlpool tubs are vacuum formed so that the side that the bather sits on never makes contact with the mold (Figure 14-69), retaining the glossy surface finish of the original sheet.

Of course, some products require that the convex side be visible, such as the display panel on the front of a vending machine (Figure 14-70). For such products, *drape forming* is used. In drape forming, a mold core is created to shape the underside (concave side) of the product and the heated sheet is drawn around the core via vacuum (Figure 14-71).

The third and final type of thermoforming process we wish to mention is *plug-assist forming*. In this process, which is similar to vacuum forming, a plug is used to stretch the sheet down into the cavity before vacuum is applied (Figure 14-72). Once the plug seals off the top of the cavity, vacuum accomplishes the final forming of the part. As the plug is pushing the sheet into the cavity, the section of sheet under the plug is not being stretched. This allows parts with deeper draw ratios to be formed.

ROTATIONAL MOLDING

Rotational molding is a process used to make large, hollow objects such as playground equipment and storage tanks. In this process, a clamshell-style mold is filled with powdered polymer (most often polyethylene) and then rotated on two axes simultaneously in order to coat the interior surface of the mold with polymer as shown in Fig-

FIGURE 14-70 Vending machine panels produced by drape forming.

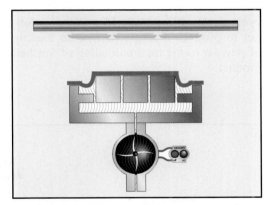

FIGURE 14-71 Schematic diagram of the stage where the vacuum is applied in drape forming.

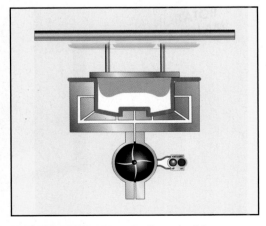

FIGURE 14-72 Schematic diagram of the stage where the vacuum is applied in plug-assist forming.

FIGURE 14-69 A thermoformed acrylic bath (Source: Alcove, www.alcove.ca).

ure 14-73. The rotational speed is relatively slow, so that the entire mold surface is coated evenly. Rotation of the mold takes place inside an oven to cause the powder particles to fuse together, creating a solid part. Once the part is formed, the rotating mold is transferred to a cooling station where the entire mold, while still rotating, is brought to room temperature. The part is then transferred to an unloading/loading station.

There are advantages and disadvantages to this process compared to other methods of polymer processing. Generally, large plastic parts are produced this way. Fully hollow objects can be made, without requiring holes in the part, a necessity in blow molding. Rotational molds are much less expensive to produce than injection molds. However, a major drawback to this process is the long cycle times. It may take 15 minutes or longer to produce a single part, from loading of powder to the final unloading of finished product.

RECOMMENDED READING

A. Brent-Strong, *Plastics: Materials and Processing*, Second Edition, Prentice Hall, New Jersey, 2000.

N. G. McCrum, C. P. Buckley and C. B. Bucknall, *Principles of Polymer Engineering*, Second Edition, Oxford Science Publications, Oxford, 1997.

D. H. Morton-Jones, *Polymer Processing*, Chapman and Hall, London, 1989.

STUDY QUESTIONS

1. Describe how plastic beverage bottles can be produced.

2. Discuss the process of blown film extrusion.

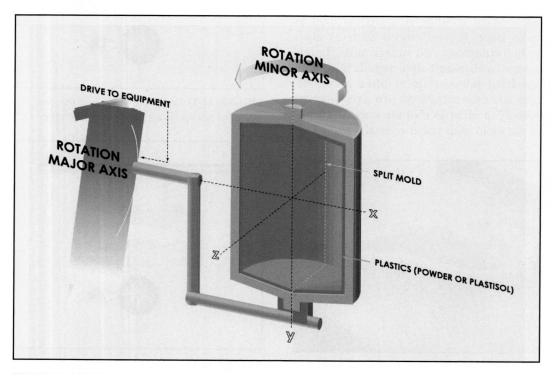

FIGURE 14-73 Schematic diagram of rotational molding.

3. Using other sources describe (a couple of typed pages) the use of reaction injection molding to produce polyurethane foams. Include a discussion of the chemistry of the reactions.

4. Describe and discuss the processing methods you would use to produce rigid PVC pipes and elbow joints. Include in your answer a description of the problems that can be encountered when processing PVC.

5. Discuss and describe the difference between injection blow molding and extrusion blow molding.

6. Discuss the differences between single screw and twin-screw extrusion.

7. Go to the literature and find out how rubber tires are made. Write a short essay on the process.

8. Describe how thermoforming techniques are used to make foamed polystyrene cups.

Index